T0348830

Microwave De-embedding

La Clairvoyance

The painting "La Clairvoyance" of the Belgian artist Rene Magritte can be viewed as an intriguing, captivating, and eye-catching picture symbolizing the book's purpose: to provide the reader with all the ingredients required for extracting the full potential of the microwave de-embedding concept from its theoretical background (i.e., the egg) to recent applications (i.e., the bird) like waveform engineering. The de-embedding concept allows accessing the behavior of the actual device and then predicting (i.e., La Clairvoyance) its performance to design amplifiers.

Microwave De-embedding

From Theory to Applications

Giovanni Crupi
University of Messina, Italy

Dominique M.M.-P. Schreurs
KU Leuven, Leuven, Belgium

AMSTERDAM • BOSTON • HEIDELBERG • LONDON
NEW YORK • OXFORD • PARIS • SAN DIEGO
SAN FRANCISCO • SINGAPORE • SYDNEY • TOKYO

Academic Press is an imprint of Elsevier

Academic Press is an imprint of Elsevier
The Boulevard, Langford Lane, Kidlington, Oxford OX5 1GB, UK
225 Wyman Street, Waltham, MA 02451, USA
Radarweg 29, PO Box 211, 1000 AE Amsterdam, the Netherlands
525 B Street, Suite 1800, San Diego, CA 92101-4495, USA

British Library Cataloguing-in-Publication Data
A catalogue record for this book is available from the British Library

Library of Congress Cataloging-in-Publication Data
A catalog record for this book is available from the Library of Congress

ISBN: 978-0-12-401700-9

For information on all Academic Press publications
visit our website at http://store.elsevier.com

14 15 16 17 18   10 9 8 7 6 5 4 3 2 1

Contents

Foreword

The very title of this book, *Microwave De-Embedding: From Theory to Applications* brings to me some memories from my early research career. When I started working at Politecnico di Torino on active microwave device physics-based modeling (that was, alas, more than 30 years ago), on-wafer measurements were hardly available and the measured scattering parameters were in package or with coaxial text fixtures. At that time, I was doing physics-based 2D simulations of gallium arsenide metal–semiconductor field effect transistors (GaAs MESFETs), and the simulated scattering parameters I was deriving from the frequency-domain solution of the linearized continuity and Poisson equations were no match at all with the measured ones. This was causing me considerable puzzlement and distress: Why did the simulated S_{11} look so regular, while the measured one had sudden wiggles when plotted on the Smith chart? Not to speak of the vagaries of (however small) S_{12}, one of the keys to transistor stability. For my senior, and now happily retired, colleague in charge of the measurements, the explanation was simple enough: What I was simulating was (more or less) the intrinsic device, the unobservable core of transistor operation; to obtain a fair comparison, I should shift the reference planes and de-embed the test fixtures.

I went on to apply this wise recipe by embedding the simulated *S* parameters into a network of intrinsic parasitics; then I added pad capacitances, and after this I added low-pass sections simulating the test fixtures and some transmission lines for plane shift.... Finally, I tried to optimize the resulting network in order to get back the measured *S* parameters. The FORTRAN code implementing gradient optimization was slow, there were too many parameters to optimize, and the program was running on a PDP-11 DEC computer with a hard disk as large as 500 Mbytes (or so, I recollect)—all those factors made the optimization painful, but, in the end, I got, after (too) many false starts, an embedding network that gave me back, starting from the simulated data, the measured ones. Was that the right one? Nobody knows, but it looked reasonable. A few years later, with the advent of on-wafer probes and cold-FET parasitic characterization techniques, the situation improved dramatically—finally, measurements matched simulations (and not, mind you, the other way round…).

This is to say that reading through the table of contents of *Microwave De-Embedding: From Theory to Applications* had a sort of Proustian madeleine effect on me—although I never had to do measurements myself; modeling and simulation was my field. To quote from Chapter 1, "To de-embed properly…is a work of art that requires knowledge and experience"—a sentence that I can fully endorse. The concept of embedding and de-embedding is indeed shown, through a series of interesting and informative chapters, to be far richer than the simple "mathematical shift of reference planes" (see Chapter 1, *A Clear-Cut Introduction to the De-embedding Concept: Less is More* by G. Crupi, D. Schreurs, and A. Caddemi), that is, transmission line embedding or de-embedding. De-embedding concerns the removal of linear parasitics that can be directly measured not only through cold-transistor techniques but also (and more conveniently, in technologies, with lossy or complex substrates) by characterizing dummy text structures developed ad hoc, not to mention the possibility of deriving their parameters through accurate EM solvers, ultimately obtaining integrated transistor models that can be effectively scaled (see Chapter 4, *High-Frequency and Microwave Electromagnetic Analysis Calibration and De-embedding* by J. C. Rautio and Chapter 8, *Electromagnetic-Analysis-Based Transistor De-embedding and Related Radio-Frequency Amplifier Design* by M. Yarlequé, D. Schreurs, B. Nauwelaers, D. Resca, and G. Vannini). However, de-embedding can be extended to the inclusion or exclusion of nonlinear parasitics within the framework of waveform-based measurements, see Chapter 5, *Large-Signal Time-Domain Waveform Based Transistor Modeling* by I. Angelov, G. Avolio, and D. Schreurs, and Chapter 6, *Measuring and Characterizing Nonlinear Radio-Frequency systems* by W. Van Moer, L. Lauwers, and K. Barbé), ultimately becoming a circuit design tool (see Chapter 9, *Nonlinear Embedding and De-embedding: Theory and Applications* by A. Raffo, V. Vadalà, and G. Vannini). Embedding and de-embedding also enter noise characterization, as discussed in Chapter 2, *Millimeter-Wave Characterization of Silicon Devices Under Small-Signal Regime: Instruments and Measurement Methodologies* by G. Dambrine (a chapter that is a really nice and self-contained introduction to microwave measurements) and Chapter 3, *Characterization and Modeling of High-Frequency Active Devices Oriented to High-Sensitivity Subsystems Design* by E. Limiti, W. Ciccognani, and S. Colangeli. Embedding and de-embedding pervasively enters not only gray-box or equivalent circuit modeling but also black-box or behavioral modeling, as discussed in Chapter 7, *Behavioral Models for Microwave Circuit Design* by J. C. Pedro and T. R. Cunha, again an excellent and self-contained introduction to black-box modeling at large.

Devoting a book to a very specific subject like microwave de-embedding is like visiting a historical city (Turin, for instance...) in search of a particular architectural item (say, baroque churches). There are lots of baroque churches in Turin, so the visitor is not to be disappointed, but going through the streets and the piazzas the visitor will see a lot of other artistic items and also will stop here and there to rest and to eat some local delicacy, or will go through the same quarters again and again seeing monuments he or she is already familiar with, maybe each time from a different angle. But let us go back to the real subject, microwaves; a book on de-embedding is ultimately bound to include a wealth of important topics in microwave modeling and design, but all of them see from a certain perspective and are endowed with a certain distinctive flavor. In this respect, the editors (Giovanni Crupi and Dominique Schreurs) and all the authors have done an excellent job in assembling this book, which has the quality of being, so to speak, both specific and inclusive (shall I say *broadband*?). I would recommend it to all researchers working in the evergreen and fascinating field of active microwave and millimeter wave device modeling.

Giovanni Ghione
Dipartimento di Elettronica e Telecomunicazioni
Politecnico di Torino
July 2013

Foreword

The problem of measurement calibration has always been with us. Before the days of computer-controlled instrumentation, it was inevitably a manual process: measure the input and output losses and correct the measurements manually. For a vector measurement, the best one could usually do was to insert a short circuit at some convenient point, adjust the reflectometer to indicate a short circuit, and correct the measurements to put the reference plane where it was really needed. This process simply was not good enough to obtain, for example, scattering parameters of transistors for use in circuit design. The accuracy fell short of what was required, and the laborious process made the cost of characterizing a device or component far too great.

With the advent of small computers in the late 1970s and early 1980s, and the standardization of the IEEE 488 instrument-control bus in 1975, computer-controlled measurements became a practical reality. The IEEE 488 bus was intended primarily for the control of instruments and data collection, but once those data were safely stored in the computer's memory, that computer could easily be used for the calibration and de-embedding of the test system.

De-embedding is the process of removing the effects of undesired circuit elements from measurements so that the measurement describes the device of interest alone. That device is invariably embedded in some type of test fixture, often with wire bonds, ribbons, or other types of connection. By replacing the device with a set of well-defined standards, measuring them, and applying post-measurement computation, it is possible to remove the parasitic effects virtually completely. An early technique, for example called *through-short-delay* (TSD), had three types of standards used in the process: a direct connection, a short, and a length of transmission line. Today, those early methods are largely obsolete, as subsequent methods are more versatile and accurate. It is no longer necessary, for example, to have highly precise test standards; it is possible to determine their imperfections by analysis and include them in the de-embedding process. As a result, modern microwave measurements, even at millimeter wavelengths, are impressive. They are well beyond the capabilities many of us would have thought possible 30 years ago.

Progress in this area has provided more than simply improved accuracy and versatility; it has given us a more sophisticated view of the problem. De-embedding has been extended in concept from use exclusively in measurements to being an integral part of the design process. Modern electromagnetic simulation, for example, would not be very useful without the ability to remove port parasitics and interconnections from the results of an analysis. Even in circuit simulation, the ability to remove an element mathematically from an analysis is often essential.

Most remarkably, the technology of de-embedding has moved from the exclusively linear domain into the nonlinear. As such, the idea has necessarily become broader, including techniques necessary for large-signal network analysis and new technologies, such as *X* parameter and waveform engineering approaches. These involve kinds of measurement that were inconceivable in the early days of automated measurement technology.

In view of the recent history of this technology, I am enthusiastic about its growth in the coming years. It seems inevitable that the more advanced methods described in this book will become everyday technology, and that new methods going well beyond these will be developed. It's an exciting time to be working in this field.

Stephen Maas
AWR Corporation
July, 2013

About the Editors

Giovanni Crupi is an assistant professor at the University of Messina, Italy, where he teaches microwave electronics and optoelectronics. Since 2005, he has been a visiting scientist with KU Leuven and IMEC, Leuven, Belgium. Giovanni's main research interests include small and large signal modeling of advanced microwave devices. He is the chair of the IEEE Microwave Theory and Techniques Society (MTT-S) Fellowship program and serves as an associate editor of *International Journal of Numerical Modelling: Electronic Networks, Devices and Fields*.

Dominique Schreurs is a full professor at KU Leuven, Leuven, Belgium. Previously, she had been a visiting scientist at Agilent Technologies (USA), Eidgenössische Technische Hochschule Zürich (Switzerland), and the National Institute of Standards and Technology (USA). Dominique's main research interests concern linear and nonlinear characterization and modeling of microwave devices and circuits, as well as linear and nonlinear hybrid and integrated circuit design for telecommunications and biomedical applications. She is the technical chair of ARFTG and serves as the editor of the *IEEE Transactions on Microwave Theory and Techniques*.

Authors' Biographies

Iltcho Angelov

Chapter 5. Large-Signal Time-Domain Waveform-Based Transistor Modeling

Iltcho Angelov was born in Bulgaria. He received his MSc in electronics (honors) and PhD in mathematics and physics from Moscow State University. From 1969 to 1991, he was with the Institute of Electronics, Bulgarian Academy of Sciences (IE BAS) as a researcher and research professor, and head of the Department of Microwave Solid State Devices from 1982. Since 1992 he has been with Chalmers University of Technology, Göteborg, Sweden as a research professor. As a researcher he worked on various microwave devices: Impatt, Gunn, BJT, FET, low-noise & power amplifiers, oscillators, synchronization & phase modulators, frequency dividers, multipliers, and low-noise receivers up to 220 GHz. In recent years his main activity has been related to FET and HBT modeling. Together with CAD companies FET GaAs, and later GaN, a HEMT model was implemented associated with various CAD tools.

Gustavo Avolio

Chapter 5. Large-Signal Time-Domain Waveform-Based Transistor Modeling

Gustavo Avolio was born in Cosenza, Italy, in 1982. In 2006 he received the MSc degree in electronic engineering from the University of Calabria (UniCAL), Italy. In 2008 he joined the TELEMIC Division of the KU Leuven, Leuven, Belgium, where he obtained the PhD degree in electronic engineering in 2012. His research work focuses on large-signal measurements and nonlinear characterization and modeling techniques for microwave active devices. His research is supported by Fonds Wetenschappelijk Onderzoek – Vlaanderen (FWO).

Kurt Barbé

Chapter 6. Measuring and Characterizing Nonlinear Radio-Frequency Systems

Kurt Barbé received her Masters in electrical engineering in 2005 from the Vrije Universiteit Brussel (VUB), Brussels, Belgium. In 2011, she obtained her PhD at the Fundamental Electricity and Instrumentation (ELEC) department of the VUB. Her PhD work was focused on nonlinear block-oriented modeling. Currently, she is working as a postdoctoral researcher of the Flemish Research Foundation (FWO-Vlaanderen). Her main research interests are in the field of nonlinear modeling and non-Gaussian signal analysis for biomedical applications. She holds strong expertise in parameter estimation of nonlinear models and the statistical postprocessing of fMRI signals.

Alina Caddemi

Chapter 1. A Clear-Cut Introduction to the De-embedding Concept: Less is More

Alina Caddemi has been an assistant professor in the field of Microwave Electronics from 1990 to 1998 at the University of Palermo, Italy. In 1998, she joined the University of Messina, as an associate professor of Electronics. Her current research interests are: temperature-dependent and noise characterization techniques for solid-state devices, cryogenic measurements and modelling of HEMT's for radio-astronomy applications, noisy circuit modelling of field-effect transistors, neural network modelling of devices, CAD and realization of hybrid low-noise circuits. Prof. Caddemi serves as an associate editor of the *International Journal of Numerical Modelling: Electronic Networks, Devices and Fields and Microwave Review*, as well as a reviewer of many international journals.

Walter Ciccognani

Chapter 3. Characterization and Modeling of High-Frequency Active Devices Oriented to High-Sensitivity Subsystems Design

Walter Ciccognani received the MS degree in electronic engineering from the University of Roma "Tor Vergata" in 2002 and a PhD in telecommunications and microelectronics therefrom in 2007. From 2007 to date, he has collaborated with the same university, where he has been a researcher since 2012. His research interests include linear microwave circuit-design methodologies, linear and noise analysis/measurement techniques, and small-signal and noise modeling of microwave active devices.

Sergio Colangeli

Chapter 3. Characterization and Modeling of High-Frequency Active Devices Oriented to High-Sensitivity Subsystems Design

Sergio Colangeli received the MS degree in electronic engineering in 2008 from the University of Roma "Tor Vergata" where he has recently received a PhD degree in telecommunications and microelectronics. His research interests include low-noise design methodologies for microwave applications, and small-signal and noise measurement and modeling of microwave active devices. He was a recipient of the EuMIC Young Engineer Prize in 2012.

Telmo Reis Cunha

Chapter 7. Behavioral Models for Microwave Circuit Design

Telmo Reis Cunha was born in Porto, Portugal, in 1973. He received the diploma and doctorate degrees in electronics and computer engineering from the University of Porto, Portugal, in 1996 and 2003, respectively. Before 2004 he was first involved with the Astronomical Observatory of the University of Porto and, afterward, he was a technical director and research engineer with Geonav Lda., a private company near Porto. Since 2004 he has been an assistant professor with the Department of Electronics, Telecommunications and Informatics, University of Aveiro, Portugal, and also a research engineer with the Institute of Telecommunications of Aveiro. He has been lecturing in the areas of control theory and electronics, and has been involved in several national and international research projects. His current research interests include behavioral modeling and linearization applied to radio frequency and microwave devices, and also integrated-circuit signal integrity analysis.

Gilles Dambrine

Chapter 2. Millimeter-Wave Characterization of Silicon Devices under Small-Signal Regime: Instruments and Measurement Methodologies

Gilles Dambrine is currently the deputy director of the Institute of Electronics, Microelectronics and Nanotechnology (IEMN-University of Lille-CNRS). He is a full professor of electronics at Lille 1 University and specialist in the characterization of active devices under high-frequency regimes. Over the years, his research interests have been oriented to the study of the microwave and millimeter-wave properties and applications of nanodevices such as CNTs, graphene, and semiconducting nanowires. He is also an IEEE senior member, the author and coauthor of about 120 papers and communications and four book chapters in the field of microwave devices, and also the scientific coordinator of Nanoscience-Characterization-Center ExCELSiOR (www.excelsior-ncc.eu).

Lieve Lauwers

Chapter 6. Measuring and Characterizing Nonlinear Radio-Frequency Systems

Lieve Lauwers received a PhD in electrical engineering and Masters in mathematics (option statistics) from the Vrije Universiteit Brussel (VUB), Brussels, Belgium in 2009 and 2005 respectively. Presently, he is an assistant professor at the VUB, department ELEC. Also, he is the VUB coordinator of Flanders Training Network for Methodology and Statistics (FLAMES) and fellow of the Flemish Research Foundation (FWO-Vlaanderen). His main interests are in the field of system identification, time series analysis, and statistical signal processing for biomedical engineering applications. He is currently an associate editor for the *IEEE Transactions on Instrumentation and Measurement* and the Hindawi *Journal of Stochastics*. He is the recipient of the 2011 Outstanding Young Engineer Award from the IEEE Instrumentation and Measurement Society.

Ernesto Limiti

Chapter 3. Characterization and Modeling of High-Frequency Active Devices Oriented to High-Sensitivity Subsystems Design

Ernesto Limiti has been a full professor of electronics in the EE Department of the University of Roma "Tor Vergata" since 2002. His research activities are focused on three lines, all in microwave and millimeter-wave electronics. The first concerns active devices oriented to small-signal, noise, and large-signal modeling. Novel methodologies are developed for noise characterization and modeling; equivalent-circuit modeling strategies are implemented both for small- and large-signal operating regimes for GaAs, GaN, SiC, Si, InP MESFET/HEMT devices. The second line is related to design methodologies and characterization methods for low-noise devices and circuits. His main focus is on extremely low-noise cryogenic amplifiers. His collaborations run with major radio-astronomy institutes in Europe (FP6 and FP7 projects). The third line is analysis and design methodologies for linear and nonlinear microwave circuits. He is a referee of several international microwave and millimeter-wave electronics journals and is a member of several international conference and workshop steering committees.

Bart Nauwelaers

Chapter 8. Electromagnetic-Analysis-Based Transistor De-embedding and Related Radio-Frequency Amplifier Design

Bart Nauwelaers is a full professor at the KU Leuven, where he currently heads the Department of Electrical Engineering (ESAT). Since 1981 he has been with the TELEMIC division of the same department, being involved in research on microwave antennas, passive components, interconnects, microwave integrated circuits and MMICs, linear and nonlinear device modeling, MEMS, ultrasonics, and phononics. He is a former chair of IEEE AP/MTT-Benelux and past chair of URSI-Benelux. Bart Nauwelaers teaches courses on microwave engineering, information theory and transmission, digital and analogue communications, and design in electronics and telecommunications. For the last 20 years he has served education in several functions, the last being the program director for the Bachelor and Masters programs in electrical engineering.

José Carlos Pedro

Chapter 7. Behavioral Models for Microwave Circuit Design

José Carlos Pedro received the diploma in 1985, doctorate in 1993, and habilitation degree in 2002 in electronics and telecommunications engineering, from University of Aveiro, Portugal, where he is now a full professor. His research interests include modeling, design, and testing of various nonlinear microwave circuits. He is the leading author of *Intermodulation Distortion in Microwave and Wireless Circuits* (Artech House, 2003) and has authored more than 250 technical papers and served as an associate editor for the *IEEE MTT Transactions* and reviewer for major international microwave journals and symposia. He served his university department as the coordinator of the scientific council and as the department head. He received the Marconi Young Scientist Award in 1993 and the 2000 IEE Measurement Prize. In 2007 he was elected as a fellow of the IEEE for his contributions to the nonlinear distortion analysis of microwave devices and circuits.

Antonio Raffo

Chapter 9. Nonlinear Embedding and De-embedding: Theory and Applications

Antonio Raffo was born in Taranto, Italy, in 1976. He received an MS degree (honors) in electronic engineering and a PhD in information engineering from the University of Ferrara, Ferrara, Italy, in 2002 and 2006, respectively. From 2006 to 2010 he was with the Engineering Department, University of Ferrara, as a post-doctoral researcher. He is currently a research associate at the Engineering Department, University of Ferrara, where he teaches the courses in semiconductor devices and electronic instrumentation and measurement. He has coauthored over 90 papers in international scientific journals and conferences and serves as reviewer for many international journals. His research activity is mainly oriented to nonlinear electron-device characterization and modeling and circuit-design techniques for nonlinear microwave and millimeter-wave applications. Antonio Raffo is a member of the IEEE MTT-11 Technical Committee (Microwave Measurements).

James C. Rautio

Chapter 4. High-Frequency and Microwave Electromagnetic Analysis Calibration and De-embedding

James C. Rautio is CEO, president, and founder of Sonnet Software, which has been developing, selling, and supporting the world's highest accuracy planar EM software for over 30 years. He received his BSEE from Cornell University, an MS from the University of Pennsylvania, and his PhD from Syracuse University, advised by Professor Roger Harrington. He has published numerous papers on the topics of EM analysis, error, and calibration. In a related area, he has published and presented many times on the life of James Clerk Maxwell as well as taking an active role in preserving Maxwell's home, Glenlair.

Davide Resca

Chapter 8. Electromagnetic-Analysis-Based Transistor De-embedding and Related Radio-Frequency Amplifier Design

Davide Resca is currently a senior microwave design engineer at MEC srl, Bologna (Italy). He joined MEC in January 2009, as an MMIC designer after his PhD and postdoctoral studies with the Department of Electronics, Computer Science and Systems (DEIS) of the University of Bologna. He is responsible for MMIC design, hybrid module design, passive and active device characterization, and modeling. His current research interests are advanced CAD techniques for microwave integrated-circuit design, EM subcircuit advanced modeling, and planar filter design synthesis and analysis.

Valeria Vadalà

Chapter 9. Nonlinear Embedding and De-embedding: Theory and Applications

Valeria Vadalà was born in Reggio Calabria, Italy, in 1982. She received an MS degree (honors) in electronic engineering from the Mediterranea University of Reggio Calabria, Reggio Calabria, Italy, in 2006 and a PhD in information engineering from the University of Ferrara, Ferrara, Italy, in 2010. Since 2010, she has been with the Engineering Department, University of Ferrara, as a post-doctoral researcher. Her research interests include nonlinear electron-device characterization and modeling and circuit-design techniques for nonlinear microwave and millimeter-wave applications. She has wide experience in microwave electronic instrumentation for linear and nonlinear characterization of electron devices. She serves as reviewer for many international journals.

Wendy Van Moer

Chapter 6. Measuring and Characterizing Nonlinear Radio-Frequency Systems

Wendy Van Moer was born in Belgium in 1974 and received Master of Engineering and PhD degrees in engineering from the Vrije Universiteit Brussel (VUB), Brussels, Belgium, in 1997 and 2001, respectively. She is currently an associate professor with the Department of Electrical Measurement (ELEC), VUB, and a visiting professor at the Department of Electronics, Mathematics and Natural Sciences, University of Gävle, Sweden. Her main research interests are nonlinear measurement and modeling techniques for medical and high-frequency applications. She has published over 100 related conference/peer reviewed journal articles. She was the recipient of the 2006 Outstanding Young Engineer Award from the IEEE Instrumentation and Measurement Society. Since 2007, she has been an associate editor for the *IEEE Transactions on Instrumentation and Measurement*. From 2010 till 2012, she was an associate editor for the *IEEE Transactions on Microwave Theory and Techniques*. In 2012, she was elected as a member of the administrative committee of the IEEE Instrumentation and Measurement Society for a 4-year term.

Giorgio Vannini

Chapter 8. Electromagnetic-Analysis-Based Transistor De-embedding and Related Radio-Frequency Amplifier Design

Giorgio Vannini is a full professor of electronics and currently the head of the Engineering Department of the University of Ferrara, Italy. He joined the University of Ferrara in 1998 after being a research associate at the University of Bologna and the National Research Council. He holds a PhD degree in electronics and computer science engineering. During his academic career he has been a teacher of applied electronics, electronics for communications, and industrial electronics as well as advanced courses on microwave circuit design. He has coauthored over 200 papers devoted to electron-device modeling, computer-aided design techniques for MMICs, and nonlinear circuit analysis and design. He is a co-founder of the academic spin-off MEC (Microwave Electronics for Communications).

Manuel Yarlequé

Chapter 8. Electromagnetic-Analysis-Based Transistor De-embedding and Related Radio-Frequency Amplifier Design

Manuel Yarlequé is an associate professor at Pontificia Universidad Católica del Perú. He teaches graduate and undergraduate courses on electromagnetics, microwaves, and antennas. He is a member of the IEEE MTT-S and ComSoc. In 2001, he joined KULeuven ESAT-Telemic research, focusing on the design of highly efficient and linear RF amplifiers, obtaining a doctorate degree in 2008. During that period, he performed research on RF transistor modeling and de-embedding. His current research interests are electromagnetics, surface plasmon resonance applications, RF/microwave circuit design, microwave imaging, and phase array antennas.

Authors

Iltcho Angelov
Chalmers University of Technology, Göteborg, Sweden

Gustavo Avolio
ESAT-TELEMIC, KU Leuven, Leuven, Belgium

Kurt Barbé
Dept. ELEC, Vrije Universiteit Brussel, Brussels, Belgium

Alina Caddemi
University of Messina, Messina, Italy

Walter Ciccognani
Dipartimento di Ingegneria Elettronica, University of Roma "Tor Vergata", Roma, Italy

Sergio Colangeli
Dipartimento di Ingegneria Elettronica, University of Roma "Tor Vergata", Roma, Italy

Giovanni Crupi
University of Messina, Messina, Italy

Telmo R. Cunha
Instituto de Telecomunicações, Universidade de Aveiro, Aveiro, Portugal

Gilles Dambrine
IEMN, CNRS, University of Lille, France

Lieve Lauwers
Dept. ELEC, Vrije Universiteit Brussel, Brussels, Belgium

Ernesto Limiti
Dipartimento di Ingegneria Elettronica, University of Roma "Tor Vergata", Roma, Italy

Bart Nauwelaers
ESAT-TELEMIC, KU Leuven, Leuven, Belgium

José C. Pedro
Instituto de Telecomunicações, Universidade de Aveiro, Aveiro, Portugal

Antonio Raffo
Dipartimento di Ingegneria, University of Ferrara, Ferrara, Italy

James C. Rautio
Sonnet Software, Inc., North Syracuse, NY, USA

Davide Resca
MEC – Microwave Electronics for Communications, Bologna, Italy

Dominique M.M.-P. Schreurs
ESAT-TELEMIC, KU Leuven, Leuven, Belgium

Valeria Vadalà
Dipartimento di Ingegneria, University of Ferrara, Ferrara, Italy

Wendy Van Moer
Dept. ELEC, Vrije Universiteit Brussel, Brussels, Belgium

Giorgio Vannini
Dipartimento di Ingegneria, Università di Ferrara, Ferrara, Italy

Manuel Yarlequé
Departamento de Ingeniería Pontificia Universidad Católica del Perú, Lima, Perú

CHAPTER

1 A Clear-Cut Introduction to the De-embedding Concept: Less is More

Giovanni Crupi[1], Dominique M.M.-P. Schreurs[2], Alina Caddemi[1]

[1] *University of Messina, Messina, Italy,* [2] *ESAT-TELEMIC, KU Leuven, Leuven, Belgium*

1.1 Introduction

To provide an adequate background for those readers who are making their first steps in this fascinating but also challenging field of knowledge, the present chapter is meant to be a general introduction to the concept of de-embedding, which will be extensively expanded and deepened in the subsequent chapters. Before starting to read this book, the first question that you should ask is: What is de-embedding? As a broad definition, the term *de-embedding* can be conceptually used to refer to any mathematical shift of the electrical reference planes. The possibility of moving reference planes is of great utility in the areas of the microwave measurements, modeling, and design. This is because the electrical characteristics are not always directly measurable at the reference planes of interest. Depending on whether the reference planes are shifted closer to or further away from the actual device under test (DUT), a distinction can be made between de-embedding and embedding: The former allows cutting down on unwanted contributions, whereas the latter allows the inclusion of additional contributions to investigate the tested device as if it is placed in a hypothetical network. Hence, de-embedding and embedding allow access to precious information by subtracting or adding contributions. De-embedding and embedding can be seen as cases where "less is more"—more information can be made readily available by subtracting the contributions or blending together two or more networks.

To target an introductory and comprehensive overview, all aspects concerning de-embedding will be covered, going from the theoretical background to experimental results illustrating its practical application. This goal is accomplished by structuring the present chapter into five main sections. The outline of the chapter is as follows.

Section 1.2 is dedicated to introducing the fundamentals in the field of microwave measurements, which can be broken down into two main types: linear (also noise) and nonlinear measurements. The latter play a fundamental role because these measurements allow one to obtain a complete characterization of the transistor behavior under realistic microwave operating conditions.

Microwave De-embedding. http://dx.doi.org/10.1016/B978-0-12-401700-9.00001-X

Section 1.3 is focused on addressing the basic principles of the microwave transistor modeling and its usefulness as a support for a quick development and improvement of device fabrication and circuit design. In particular, attention will be given to models based on the equivalent circuit representation, going from linear (also noise) to nonlinear modeling.

Section 1.4 is aimed at addressing the importance of the de-embedding concept. The differences in performing the de-embedding of linear and nonlinear data will be discussed, with emphasis on its use for waveform engineering.

Finally, Section 1.5 is devoted to present experimental results for FinFET, which is a cutting edge transistor based on advanced silicon technology. The analysis of the experimental results represents a valuable support to achieve a clear-cut understanding of the studied theoretical background.

1.2 Microwave measurements

Although the division of the electromagnetic (EM) spectrum into bands is manmade and rather arbitrary, the microwave frequency range can be commonly defined as the portion of the EM spectrum going from 300 MHz to 300 GHz. This is a broad definition because it is not limited only to the super-high-frequency band from 3 GHz to 30 GHz but embraces also the ultra-high-frequency band from 300 MHz to 3 GHz and the extremely high frequency band from 30 GHz to 300 GHz. The latter band is also known as millimeter waves, whereas the upper region from 300 GHz to 3 THz is called submillimeter waves. Depending on the adopted convention, microwaves can be considered either as a subset of radio frequency (RF) band ranging from 30 kHz to 300 GHz or as the region above the RF band, which ranges from 30 kHz to 300 MHz.

Extensive and growing interest is devoted to microwave frequencies, due to the electronics and telecommunications applications that require incessantly higher operating frequencies and wider bandwidths. Because the transistor is the key component in most high-frequency applications, this section focuses on the measurement techniques for transistors operating at microwave frequencies. Nevertheless, the reported study can be straightforwardly extended to other electronic devices and circuits.

Since its inception in 1948, the transistor has been, and still is, the workhorse of the electronics industry. Over the years, various types of transistors have been proposed and they can be classified into two main categories: bipolar junction transistors (BJTs) and field-effect transistors (FETs). The transistor is an active semiconductor device with four terminals. Nevertheless, only three terminals are typically considered because the substrate is usually connected to ground. These terminals are called base, emitter, and collector for BJTs and gate, source, and drain for FETs. To characterize the transistor as a two-port network, one of the three terminals should be kept common to the input and output ports. In this chapter, attention will be mostly paid to FET devices in common source configuration.

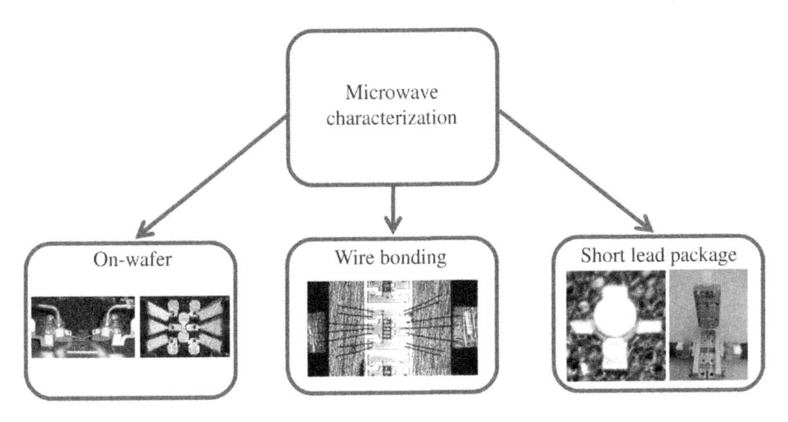

FIGURE 1.1

Illustration of on-wafer and coaxial characterizations: a GaAs HEMT with the input and output on-wafer pads contacted by coplanar probes, a GaN HEMT attached to the aluminum carrier of the text fixture and wire bonded to the input and output coaxial connectors, and a GaAs pseudomorphic HEMT in a short lead package placed in a transistor test fixture with input and output coaxial connectors.

To access the transistor behavior with experimental instrumentation, electrical and mechanical interconnections should be made between the transistor and the outside world. The present chapter will focus on on-wafer characterization, which is the most convenient technique to measure transistor behavior. As a matter of fact, this technique consists of measuring the transistor behavior straightforwardly on the wafer, right away after fabrication. The on-wafer characterization is cheaper and faster than the coaxial solution (see Figure 1.1). The latter is based on inserting the device into a test fixture or package with coaxial connectors. In case of transistors having the package with leads, the microwave measurements should be carried out by placing the packaged device in a coaxial transistor test fixture (TTF). A fundamental benefit of on-wafer measurements is given by the possibility of accessing the real performance of the transistor. This is because the extrinsic effects around it are minimized by avoiding encompassment of the extrinsic contributions arising from the test fixture or package. To illustrate the difference between on-wafer and coaxial characterizations, Figure 1.1 shows on-wafer, wire-bonded, and packaged devices based on high-electron-mobility transistor (HEMT) technology, which is the workhorse behind many microwave applications.

To carry out on-wafer measurements, the wafer should be placed on a dedicated probe station, which is based on mechanical arms and precision micropositioners to connect the probes having coplanar waveguide tips to the device pads and, in turn, the probes are connected to the instruments with coaxial cables.

Hence, the probes can be considered as adapters accomplishing the transition from coaxial to coplanar waveguide.

A fundamental prerequisite to accurately measure the microwave transistor behavior is represented by the calibration procedure. The calibration consists of a mathematical procedure aimed at retrieving the true behavior of the tested device from the raw measurements. To accomplish this goal, known precision standards are measured in place of the DUT. Based on the comparison between the measured and the known characteristics of the standards, the calibration process allows identifying the systematic errors that are assumed to be time invariant (e.g., imperfections in the measurement system). After applying the calibration corrections to remove the systematic errors, the accuracy of the microwave measurements can be still affected by residual systematic errors, which arise mainly from imperfections in the calibration standards, along with random and drift errors. The effects of random errors that are nonrepeatable (e.g., noise) can be minimized by making more measurements and taking the average, while the drift errors that are due to performance changes in the measurement system after the calibration (e.g., temperature variations) can be removed by repeating the calibration. It should be highlighted that the calibration process is strongly entangled with the de-embedding concept. By switching from uncalibrated to calibrated measured data, the electrical reference planes can be moved from the instrument ports to the calibration planes at which known precision standards have been connected. Typically, the calibration planes do not coincide with the instrument ports. To obtain the true behavior of the DUT, the calibration planes should be located as close as possible to the DUT ports. As an example, the calibration planes for on-wafer DUTs correspond to the probe tip planes. Hence, a shift of the reference planes is obtained when the raw data are properly corrected by using the calibration information.

Depending on how the device is characterized, microwave measurements can be subdivided into two main categories: small- and large-signal measurements, which are also known as linear and nonlinear measurements. Small-signal measurements allow complete capture of the behavior of a linear network or of a nonlinear network, if the amplitude of the excitation signal is sufficiently small such that the nonlinear network can be considered to respond linearly and, therefore, the superposition principle can be applied (i.e., the response of a linear circuit to a sum of excitations is given by the sum of the responses of the circuit to each excitation acting alone). Hence, small-signal measurements can be used to characterize the behavior of a nonlinear device, like a transistor, only under small-signal approximation of the time varying signal around the DC operating point. On the other hand, large-signal measurements allow capturing the transistor behavior under realistic microwave operating conditions, namely when harmonics are generated from the device nonlinearities (see Figure 1.2).

The remainder of this section is organized into two parts. The first subsection is dedicated to linear measurements, including noise, whereas the second subsection is focused on nonlinear measurements.

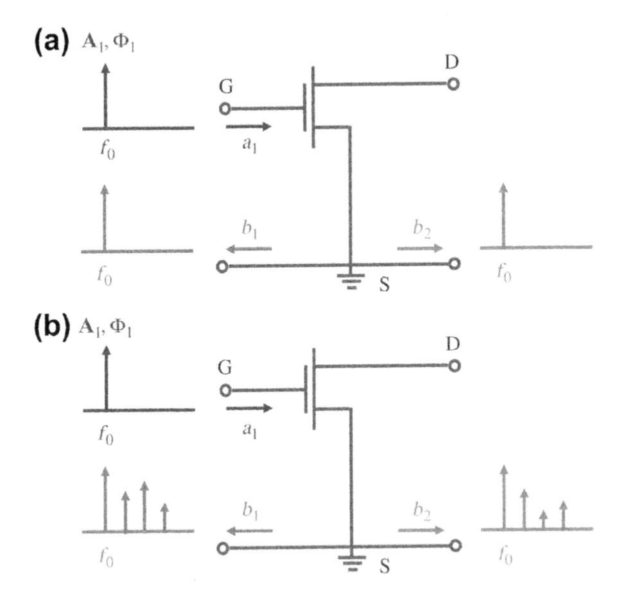

FIGURE 1.2

Illustration of an FET operating under (a) small-signal and (b) large-signal conditions: a_1 represents the excitation, whereas b_1 and b_2 represent the network response.

1.2.1 Linear measurements: from scattering to noise parameters

The small-signal behavior of a two-port network, like a transistor, can be described with various linear matrix representations consisting of four complex parameters. These parameters are used to describe the relationships between the four variables associated to any two-port network: two independent and two dependent variables representing, respectively, the network excitation and the corresponding network response. The choice of a particular set of parameters depends on the desired independent and dependent variables. The matrix representations based on using voltages and currents as the four variables—impedance (*Z*), admittance (*Y*), hybrid (*H*), and transmission (ABCD or *A*) parameters—require short and open circuit terminations, which can be difficult to implement at high frequency and over a broad frequency band. Furthermore, these highly reflective terminations might cause an active device, such as a transistor, to oscillate, thus rendering measurements impossible. This drawback is overcome with the scattering (*S*) parameters based on defining the four variables in terms of ratios of power waves. They consist of reflection and transmission coefficients measured by terminating the network in a reference impedance Z_0, which is typically 50 Ω. Hence, *S* parameters are the most suitable to be measured, and therefore they are the most appropriate way to represent a linear microwave network. Simple and well-known conversion formulas enable

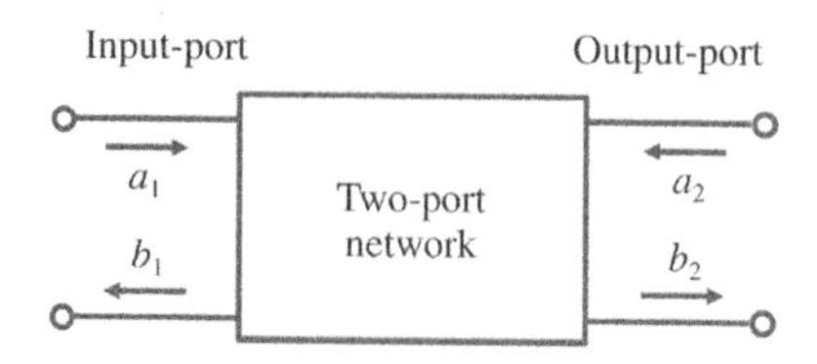

FIGURE 1.3

Illustration of the scattering parameters for a two-port network.

one to obtain the other equivalent parameter sets from the S parameter measured with the vector network analyzer (VNA). S parameters for a two-port network are defined by the following linear matrix equation:

$$\begin{bmatrix} b_1 \\ b_2 \end{bmatrix} = \begin{bmatrix} S_{11} & S_{12} \\ S_{21} & S_{22} \end{bmatrix} \begin{bmatrix} a_1 \\ a_2 \end{bmatrix} \tag{1.1}$$

where a_i and b_i (i.e., with i being 1 or 2) represent the incident and scattered travelling voltage waves at port i, as illustrated in Figure 1.3. The term *scattering* refers to the fact that each incident wave is scattered (i.e., reflected or transmitted). The squared magnitudes of the four travelling voltage waves represent the incoming and outgoing power at the two ports. The S parameters of an FET are typically measured in common source configuration and, conventionally, the gate-source and drain-source ports, known also as input and output ports, are assumed to be ports 1 and 2, respectively.

It is worth mentioning that the **Y**, **Z**, and **A** matrices are, respectively, useful to calculate the resulting matrix of two two-port networks connected in parallel, in series, or in cascade:

$$\mathbf{Y} = \mathbf{Y}_1 + \mathbf{Y}_2 \tag{1.2}$$

$$\mathbf{Z} = \mathbf{Z}_1 + \mathbf{Z}_2 \tag{1.3}$$

$$\mathbf{A} = \mathbf{A}_1\mathbf{A}_2 \tag{1.4}$$

where the subscripts 1 and 2 refer to the two networks to be connected.

Regarding noise characterization, any noisy linear two-port network can be represented as a noiseless two-port network with two additional partially correlated noise sources at the input and/or output ports (see Figure 1.4). The two noise sources and their complex correlation can be described with the noise correlation matrix, which consists of four real quantities: the diagonal terms are real values representing the power spectrum of each noise source, whereas the off-diagonal terms are complex conjugates of each other and represent the cross-power spectra of the noise sources [1,2]. Depending on the matrix representation used for the noiseless two-port network, different forms of the noise correlation matrix can be defined. The three most common forms are the admittance, impedance, and chain representations,

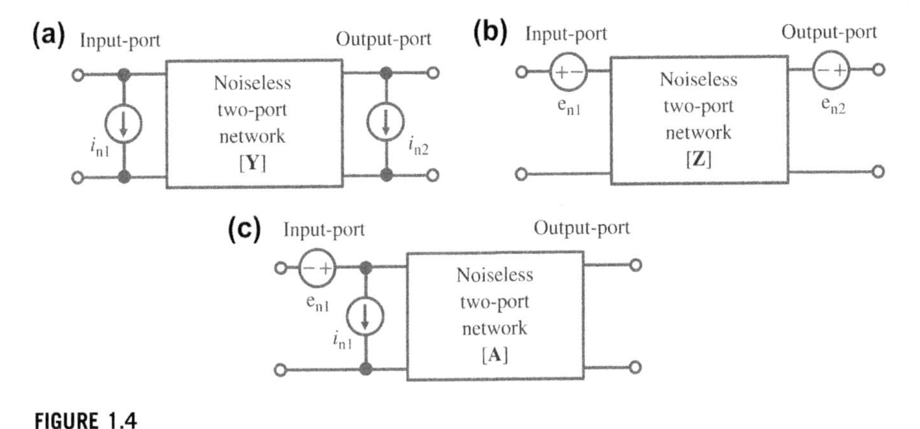

FIGURE 1.4

Illustration of the (a) admittance, (b) impedance, and (c) chain representations of a noisy linear two-port network.

which are based on using the conventional electrical matrices **Y**, **Z**, and **A** to describe the noiseless two-port network and the corresponding correlation matrices $\mathbf{C_Y}$, $\mathbf{C_Z}$, and $\mathbf{C_A}$ to describe the noise sources. As depicted in Figure 1.4, the admittance representation uses two current noise sources in parallel to the input and output ports of the noiseless network, the impedance representation uses two voltage noise sources in series at the input and output ports of the noiseless network, and the chain representation uses a voltage noise source in series and a current noise source in parallel at the input port of the noiseless network. These three representations $\mathbf{C_Y}$, $\mathbf{C_Z}$, and $\mathbf{C_A}$ are, respectively, useful to calculate the resulting correlation matrix of two two-port networks connected in parallel, in series, or in cascade:

$$\mathbf{C_Y} = \mathbf{C_{Y1}} + \mathbf{C_{Y2}} \tag{1.5}$$

$$\mathbf{C_Z} = \mathbf{C_{Z1}} + \mathbf{C_{Z2}} \tag{1.6}$$

$$\mathbf{C_A} = \mathbf{A_1 C_{A2} A_1^+} + \mathbf{C_{A1}} \tag{1.7}$$

where the plus sign is used to denote the Hermitian conjugation (i.e., transposed conjugate).

It should be mentioned that, similar to the case of *S* parameters, simple and well-known conversion formulas allow transforming a noise correlation matrix from one representation to another. In particular, the conversion formulas for the noise correlation matrix are based on the following equation [2–6]:

$$\mathbf{C'} = \mathbf{TCT^+} \tag{1.8}$$

where **C** and **C′** represent the noise correlation matrices before and after the conversion, whereas **T** represents the transformation matrix depending only on the conventional electrical matrix of the network.

An important figure of merit to evaluate the noise performance of a linear noisy two-port network is given by the noise factor F, which is defined as the available signal-to-noise power ratio at the input port divided by the available signal-to-noise power ratio at the output port [7]. The noise factor is a function of the source reflection coefficient Γ_S, and the analytical representation of this dependence requires four real quantities, known as noise parameters. Depending on the particular formulation used, different sets of the noise parameters can be adopted. Nevertheless, these noise parameter sets can be easily converted to each other because they represent equivalent ways of describing the same dependence. One typical representation of $F(\Gamma_S)$ is given in terms of the minimum noise factor F_{min}, magnitude and phase of the optimum source reflection coefficient Γ_{opt}, and the noise resistance R_n [7–10]:

$$F(\Gamma_S) = F_{min} + \frac{4\frac{R_n}{Z_0}\left|\Gamma_S - \Gamma_{opt}\right|^2}{\left|1+\Gamma_{opt}\right|^2\left(1-\left|\Gamma_S\right|^2\right)} \tag{1.9}$$

where the reference impedance Z_0 is usually 50 Ω. It can be seen from Eqn (1.9) that F_{min} represents the minimum value of F, which is achieved when Γ_S is equal to Γ_{opt}, and R_n represents how rapidly F increases as Γ_S moves away from its optimum Γ_{opt}. It should be pointed out that the noise factor and the minimum noise factor are, respectively, called noise figure NF and minimum noise figure NF_{min} when expressed in dB. Figure 1.5 shows a block diagram illustrating a typical noise measurement setup for on-wafer microwave transistors, where the noise figure is measured with a noise figure meter (NFM) along with a noise source (NS). To remove the contribution of the receiver of the NFM, the calibration process can be performed by connecting the NS directly to the receiver port. The determination of the noise figure of on-wafer transistors is a challenging task because the measurements should be properly de-embedded from the contributions of the input and

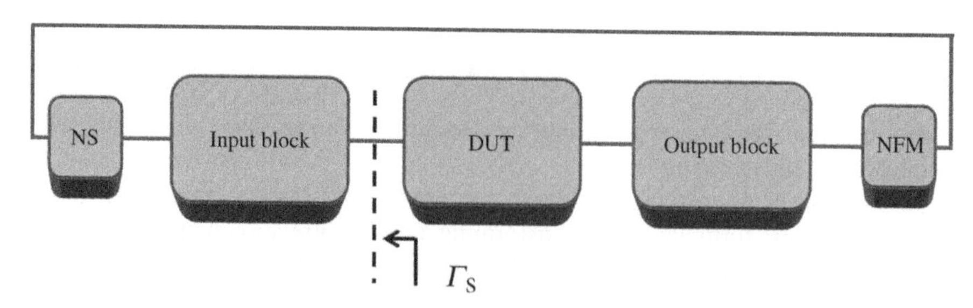

FIGURE 1.5

Block diagram of a typical noise measurement setup for on-wafer microwave transistors. The noise figure of the whole system (i.e., transistor, the input and the output stages) is measured by using a noise figure meter (NFM) with a noise source (NS). A source tuner can be used to change the source reflection coefficient Γ_S.

output stages, which are used to connect the device, respectively, to the noise source and to the receiver [8]. To determine the four transistor noise parameters, the classical approach consists of using appropriate data processing techniques applied to at least four noise figure measurements for different source impedances synthesized by a source tuner [11–14]. Nevertheless, this source-pull based method is time consuming and an expensive automatic tuner system with an associated sophisticated calibration is needed. Alternative strategies have been developed over the years to obtain the noise parameters from a single measurement of the noise factor with a 50 Ω source impedance, which is indicated as F_{50} [15–18]. A typical solution consists of simulating the noise parameters with a noise model based on expanding the small-signal equivalent circuit with appropriate noise sources, which are determined to reproduce the measured F_{50}.

Finally, it should be pointed out that the noise parameters and the noise correlation matrix are equivalent representations consisting of four real numbers and simple transformation formulas allow passing from one representation to the other [2–6]:

$$\mathbf{C}_A = 2kT\begin{bmatrix} R_n & \dfrac{F_{min}-1}{2} - R_n Y_{opt}^* \\ \dfrac{F_{min}-1}{2} - R_n Y_{opt} & R_n\left|Y_{opt}\right|^2 \end{bmatrix} \tag{1.10}$$

where k is Boltzmann's constant, T is the absolute temperature (i.e., Kelvin scale), Y_{opt} is the optimum source admittance associated to Γ_{opt}, and the asterisk sign denotes the complex conjugate.

Further details on the noise characterization will be provided in Section 1.3 from a modeling perspective. The reader is referred to Chapters 2 and 3 for a more in-depth analysis of this subject.

1.2.2 Nonlinear measurements

Because microwave transistors are inherently nonlinear devices, linear measurements, as discussed in the preceding section, are limitative. S-parameter measurements are able only to characterize the transistor behavior under small-signal conditions at the set DC operating point. In order to characterize the full behavior of the transistor, either such S-parameter measurements have to be repeated at a large range of operating points (hundreds of measurements are common) or nonlinear microwave measurements can be adopted. The latter is preferred because then the small-signal operating condition, meaning that the response is linear to the excitation, is no longer required. Instead, the device can be characterized under operation conditions (bias and power level) that are realistic for the final application. Nonlinear measurements allow characterizing the new spectral components (harmonics, intermodulation products) that are generated by the transistor under nonlinear operation.

To achieve complete characterization, these measurements should be vectorial (i.e., both amplitude and phase) [19,20]. Scalar nonlinear microwave measurement systems such as power meters and spectrum analyzers are adequate for evaluating the performance of fabricated circuits, but vectorial information is essential in the modeling and design phase, as will be elaborated on in Chapters 5 and 9, respectively. The most suitable instrumentation to obtain such kind of measurements is the sampler-based large-signal analyzer [21] or the mixer-based nonlinear vector network analyzer [22]. The architectures of both instruments as well as the calibration procedure are documented in Chapters 6 and 7. More illustrations on the uses of such vectorial nonlinear microwave measurements can be found elsewhere [23–25]. For completeness, also the vector signal analyzer is able to obtain vectorial information, but it is a narrowband instrument primarily aimed for characterizing finished circuits and systems under modulated excitations.

1.3 Microwave modeling

The transistor represents the active component at the heart of most high-frequency electronic circuits. In light of that, to release a successful microwave circuit, the transistor should be fabricated, characterized, and modeled before moving to the design, realization, and characterization of the circuit (see Figure 1.6). Hence, microwave transistor modeling plays a fundamental role because of its utility not

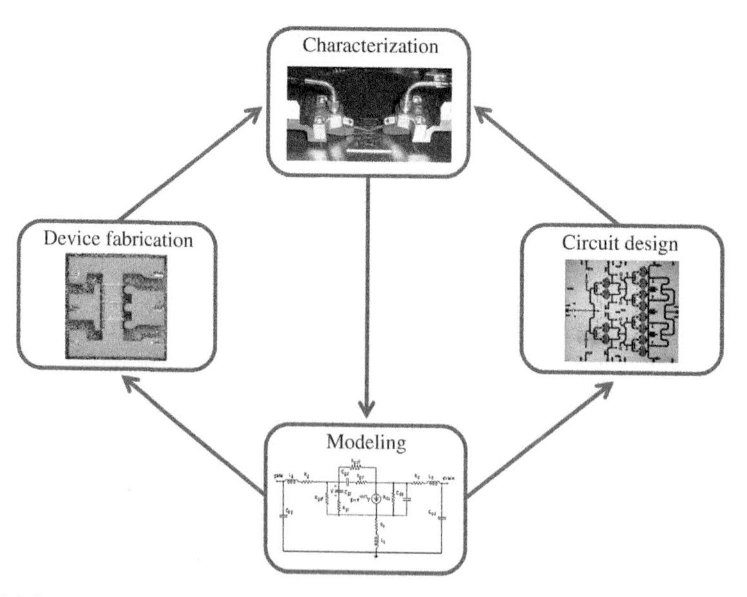

FIGURE 1.6

Illustration of the importance of microwave transistor modeling as helpful feedback for device fabrication and as a valuable tool for circuit design.

only in providing precious feedback to improve transistor fabrication but also in enabling a quick and reliable optimization of microwave circuit design. The latter allows one to minimize expensive and time-consuming cycles of design and realization of the microwave circuit that hopefully should be characterized only once at the end, to verify its real performance with respect to the predicted behavior. If the measured performance does not fulfill the required constraints, efforts should be done to find the weak link in the chain going from device fabrication to circuit design.

Microwave transistor modeling can be defined as the art of determining an accurate representation of the electrical characteristics for a transistor operating at high frequencies. Over the last decades, a plethora of modeling techniques have been proposed to describe successfully the electrical behavior of microwave transistors. However, existing procedures may often turn out to be inadequate to take into account the rapid and incessant evolution of transistor technologies. In light of that, intensive and endless research is required to develop innovative modeling strategies tailored to the latest transistor technologies. Basically, transistor models can be broken up into three main groups: physical models [26–30], behavioral models [31–35], and equivalent circuit models [36–40]. Physical models can be considered as transparent boxes because their extraction is based on the study of the physical phenomena occurring in the device. Opposite to physical models, behavioral models can be treated as black boxes. This is because their extraction is based on using behavioral input–output relations to mimic the experimental data of the device, regardless of its inner physics. The behavioral models will be further considered in Chapter 7. A tradeoff between physical and behavioral modeling approaches is given by the equivalent circuit models that are also known as grey box models. As a matter of fact, equivalent circuit models are extracted by using experimental data but maintaining the link with physical operating mechanisms. Depending on the specific use of the model, these three types of modeling approaches offer different advantages and disadvantages. Nevertheless, equivalent circuit models are widely used because they provide a valuable compromise solution. In comparison with physical models, equivalent circuit models allow achievement of a higher simulation speed, which is a crucial requirement to optimize circuit design. On the other hand, in comparison with black-box models, equivalent circuit models allow one to obtain better feedback for improving the device fabrication processes. This is because the circuit elements are physically meaningful because the equivalent circuit models can be clearly linked to the physical structure of the device.

As illustrated in Figure 1.7, the equivalent circuit modeling of microwave transistors is a complex research area that needs an interdisciplinary know-how. In particular, researchers who wish to work actively in this field should develop profound competencies in the following four main expertise areas. First, a strong background in semiconductor device physics is required to obtain physical meaningful models. Secondly, an intensive expertise in microwave measurement techniques and setups is necessary to gather the experimental measurements for model extraction and validation. Thirdly, a deep knowledge of the circuit network theory is

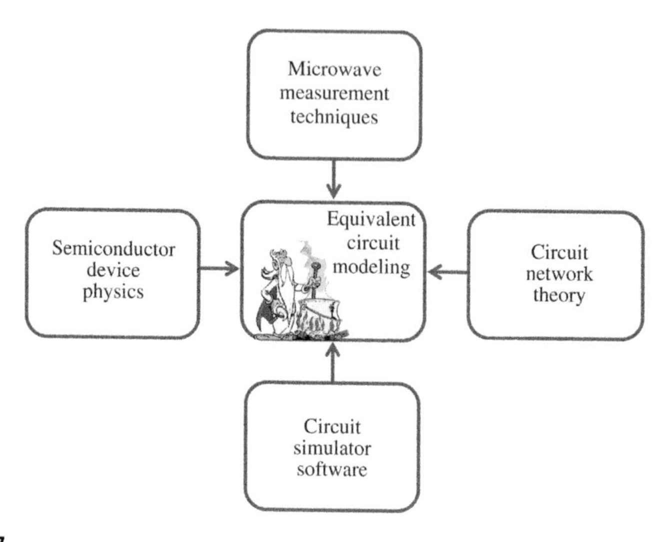

FIGURE 1.7

Illustration of the interdisciplinary know-how required for the equivalent circuit modeling of microwave transistors.

essential to identify the circuit topology that fits the specific case. Finally, good skill in the use of different software packages is indispensable. As a matter of fact, to save time and increase accuracy, the processes of measurement acquisition and model extraction should be automated. Furthermore, the obtained model should be implemented into commercial simulators to be easy to use for technologists and designers.

To determine an accurate and efficient circuit model, a crucial role is played by the choice of circuit topology that should be appropriate for the case under study because even the values of the extracted elements can depend on the selected topology. The appropriate topology should be determined by carefully analyzing the available information on the physical structure and layout of the investigated transistor and the available DC and microwave measurements. Nevertheless, this starting topology could be modified during the model extraction step to get better results. In general, the selection of the appropriate number of circuit elements is a challenging task. This is because a higher number of circuit elements can help to achieve a better agreement between measurements and simulations but, on the other hand, the model complexity gets increased. As a consequence, a good tradeoff between model accuracy and complexity should be targeted.

1.3.1 Small-signal equivalent circuit

Typically, microwave transistor modeling starts with extracting the small-signal equivalent circuit, which can be used as a cornerstone to build both noise and

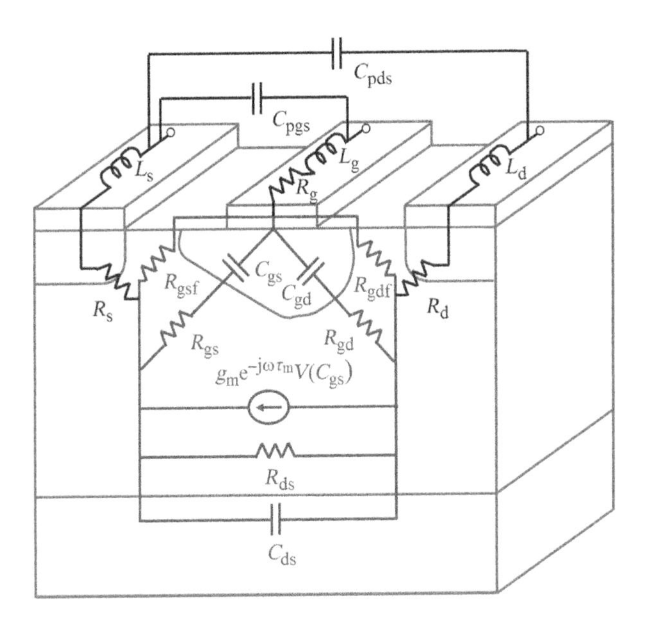

FIGURE 1.8

Illustrative sketch of the correlation between the equivalent circuit model and the physical structure for an on-wafer microwave FET.

large-signal models. The small-signal equivalent circuit is commonly obtained from *S* parameters, which can be straightforwardly measured with a VNA as already mentioned in Section 1.2.1. Although *S* parameters allow characterizing completely the transistor behavior under a small-signal excitation regimen, the determination of the corresponding small-signal equivalent circuit model is of great utility. This is because the circuit model brings numerous and valuable advantages. First, a deeper investigation of the microwave characteristics can be achieved because each equivalent circuit parameter (ECP) is intended to model the electrical characteristics of a particular region of the transistor, as sketched in Figure 1.8.

Secondly, the scaling of the microwave characteristics can be quickly estimated by using the conventional scaling rules of the ECPs. For example, enlarging the gate width can be roughly thought of as equivalent to adding more transistors in parallel and, consequently, all ECPs associated with admittances (i.e., capacitances and conductances) should be directly proportional to the gate width, whereas all ECPs associated with impedances (i.e., inductances and resistances) should be inversely proportional to it. Nevertheless, these oversimplified scaling rules need to be adapted, especially for extrinsic elements, to account for the particular transistor layout and physics. Thirdly, the ECPs are frequency independent, resulting in a more compact representation. Fourthly, the frequency independence of the ECPs implies also that the implementation of the equivalent circuit in circuit simulators allows extrapolating the small-signal behavior at frequencies beyond the limits of the available instrumentations. Although the validity of the model extrapolated far out of the frequency range of the used measurements cannot be guaranteed, the frequency independence of the ECPs allows one to achieve a trustworthy extrapolation in the frequency range up to which the model topology is appropriate. Fifthly, the equivalent circuit can be used as starting point to determine both noise and large-signal models.

However, the extraction of the small-signal equivalent circuit from *S*-parameter measurements is an ill-conditioned problem because there are more unknowns than equations, namely only eight equations representing the real and imaginary parts of the four *S* parameters expressed as a function of ECPs at each frequency. The number of circuit elements depends on the specific circuit topology but a standard model is usually composed by around 18 elements, as shown in Figure 1.8. To solve this ill-conditioned problem, many techniques have been proposed, which can be divided into two main categories: optimization and analytical methods. Although the best choice between these two approaches depends on the particular application, the optimization procedures have the drawback that the extracted element values could be physically meaningless and critically depending on the starting parameter values, local minima, and optimization technique itself. These drawbacks are overcome by using the analytical approach that allows extracting directly and straightforwardly the ECPs. In particular, the analytical extraction techniques are based on decomposing the ill-conditioned problem into two subproblems and then solving them subsequently. This is achieved by splitting the equivalent circuit elements into two main groups: the intrinsic elements, which are bias dependent, and the extrinsic or parasitic elements, which are assumed to be bias independent. The latter elements represent the contributions arising from the interconnections between the real device and the outside world. As an illustrative example, Figure 1.9 shows a typical topology of small-signal equivalent circuit for a microwave FET. The

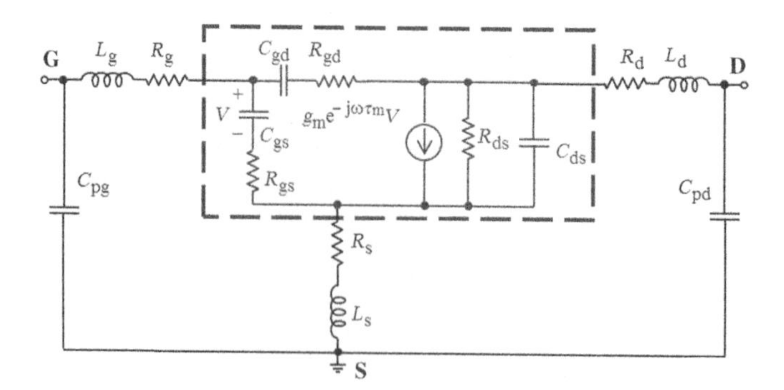

FIGURE 1.9

Typical topology of the small-signal equivalent circuit for a microwave FET.

reported circuit is composed by eight extrinsic elements (i.e., C_{pg}, C_{pd}, L_g, L_s, L_d, R_g, R_s, and R_d) and 10 intrinsic elements (R_{gsf}, R_{gdf}, C_{gs}, R_{gs}, C_{gd}, R_{gd}, C_{ds}, R_{ds}, g_m, and τ_m).

The input and output capacitances C_{pg} and C_{pd} account for the capacitive coupling respectively between gate-source and drain-source pads, whereas the feedback extrinsic capacitance between gate-drain pads is commonly omitted. This is because its value is typically rather small and also barely distinguishable from the intrinsic feedback capacitance C_{gd}. The inductances L_g, L_s, and L_d represent the inductive contributions of the metal contact pads and access transmission lines. Usually, the value of L_s is quite small, in the order of less than a few pH. In the case of wire-bonded or packaged transistors, the inductance values can be in the nanoHertz range of magnitude, thus making their effects much more pronounced. The resistances R_s and R_d are mostly due to the drain and source ohmic contacts, whereas the resistance R_g originates mainly from the gate metallization. As will be illustrated in Section 1.5.1, to reduce the gate resistance that can significantly affect the transistor's microwave performance, the gate current path width can be increased by using the multifinger layout based on connecting many gate fingers in parallel.

The intrinsic network is constituted by an input series resistor–capacitor (RC) branch C_{gs} and R_{gs}, a feedback series RC branch C_{gd} and R_{gd}, an output parallel RC branch C_{ds} and R_{ds}, and a voltage-controlled current source characterized by the transconductance g_m and its time delay τ_m. The transconductance quantifies the device capability to amplify by converting changes in the input voltage into changes in the output current with the output voltage kept constant. Because the transconductance cannot respond instantaneously to changes in the input voltage, the delay inherent to this process is described by τ_m. Furthermore, the input and feedback resistances R_{gsf} and R_{gdf} associated with conduction current contributions respectively at the gate-source and gate-drain sides should be included when the measured direct current (DC) gate current is not negligible. Contrary to DC insulated gate devices (e.g., metal-oxide-semiconductor field-effect transistor, MOSFET), the gate conduction current can increase significantly by applying a forward gate bias voltage to Schottky gate devices (e.g., metal-semiconductor field-effect transistor, MESFET and HEMT).

The first step of the analytical modeling techniques consists of determining the extrinsic elements. Typically, their contributions can be determined with three different approaches based on exploiting the "cold" transistor (i.e., $V_{DS} = 0$ V, passive device) [38–45], dedicated dummy test structures (e.g., open and short) [46–50], and electromagnetic simulations [51–54]. Despite their differences, all three approaches have one thing in common: they enable making use of the de-embedding concept to subtract the extrinsic effects from the data with simple and straightforward matrix manipulations, thus making it possible to obtain the intrinsic elements.

The first approach consists of extracting the extrinsic ECPs by exploiting *S* parameters measured on the cold transistor. The advantage of using the cold condition is that the equivalent circuit network can be significantly simplified. Such a bias condition is known as "cold" because the equivalent temperature, representing the average kinetic energy of the channel carriers, is cold at $V_{DS} = 0$ V with respect to the typical operating bias conditions. The absence of carriers drifting in the active channel from source to drain allows one to disregard the voltage-controlled current source. Hence, two unknowns (i.e., g_m and τ_m) vanish. Nevertheless, also the two equations associated to the real and imaginary parts of S_{21} vanish (i.e., $S_{12} = S_{21}$) because the cold transistor behaves as a reciprocal device with $[S] = [S]^T$ (i.e., it does not contain any nonreciprocal media, such as ferrites or plasmas, or active devices [55]). Additionally, the gate-source and gate-drain intrinsic circuits can be approximated to be equal to obtain a simplified circuit like the one reported in Figure 1.10, where the investigated device is a HEMT [41]. Although this approximation strongly depends on the specific transistor, the intrinsic device structure is commonly considered to be highly symmetric at $V_{DS} = 0$ V. A further simplification of the equivalent circuit can be achieved by using an appropriate V_{GS} and frequency range. Generally, the extrinsic capacitances are extracted from the imaginary parts of the Y parameters of the capacitive Π-network representing the device at low frequencies with V_{GS} below pinch-off, while the inductances and resistances are respectively obtained from the real and imaginary parts of the Z parameters of the resistive inductive T-network representing the transistor at high frequencies under high V_{GS} [41].

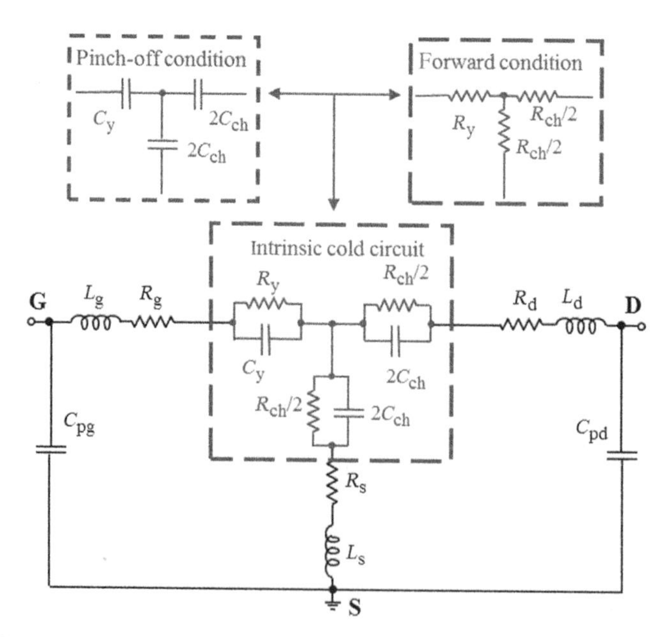

FIGURE 1.10

Typical topology of the small-signal equivalent circuit for a microwave FET under cold condition.

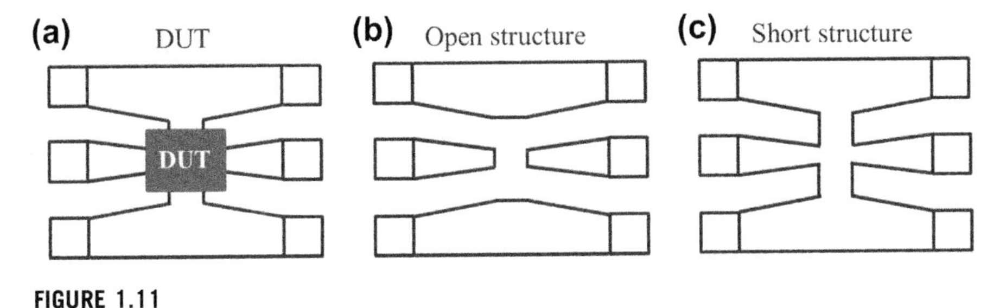

FIGURE 1.11

Illustration of two dummy test structures for (a) an on-wafer DUT: (b) open structure, and (c) short structure.

The second approach consists of using *S* parameters measured on dedicated dummy test structures to remove extrinsic contributions from the experimental data of the transistor. These parasitic effects for on-wafer DUTs are mainly due to the contact pads, the metal interconnections, and the substrate (see Figure 1.11). However, these test structure-based techniques allow access to the intrinsic transistor data only when it is possible to disregard the extrinsic effects originating from the interaction between the extrinsic and the intrinsic transistor sections (e.g., the drain and source contact resistances). The proposed de-embedding techniques can be classified depending on the number of the required test structures. For instance, one-step de-embedding is based on open structure, two-step de-embedding is based on open and short structures, and three-step de-embedding is based on open, short, and thru structures [47].

The last approach consists of characterizing the extrinsic passive part of the transistor through its *S* parameters obtained with electromagnetic simulations. As an example, the electromagnetic simulations can be exploited to obtain the *S* parameters associated to the test structures when they are not available. For an extensive discussion on electromagnetic analysis, the reader is referred to Chapters 4 and 8.

The method, or combination of methods, should be chosen to suit the needs of the particular case accounting for the fact that each extraction procedure has its own advantages and disadvantages. Regarding the cold approach, an accurate determination of the extrinsic circuit elements is required to remove their contributions. Although the extraction of the extrinsic ECPs can be very useful for investigating the transistor physics, this step can be extremely challenging when the circuit complexity is too high (e.g., silicon transistors with lossy substrate) or when the lumped element approximation is not applicable (e.g., large multifinger transistors working at millimeter-wave frequencies). This drawback is solved by adopting the methods based on test structures and electromagnetic simulations because the extrinsic contributions can be removed with matrix manipulations even without having to explicitly determine the corresponding circuit representation.

Nevertheless, after applying the procedures based on measured or simulated S parameters of the test structures, the cold methods can be still required for removing residual extrinsic contributions. Furthermore, the test structure based approach has the drawback that additional test structures, which are also area consuming in the case of on-wafer transistors, are required, whereas the electromagnetic simulation-based approach requires detailed information regarding the transistor layout and simulation expertise.

After determining the contributions associated to the extrinsic elements, their effects can be removed from the measurements, as will be discussed in detail in Section 1.4.1. Hence, the subsequent step of the analytical modeling techniques consists of calculating the intrinsic elements from the intrinsic Y parameters at each bias point. Commonly, the Y parameter representation is adopted because the intrinsic circuit section is based on a Π topology. The two conductances g_{gsf} and g_{gdf}, which are respectively defined as the inverse of the resistances R_{gsf} and R_{gdf}, can be evaluated from the real parts of the intrinsic Y_{11} and Y_{12} at low frequencies by considering the capacitances as open circuits [56]:

$$g_{gsf} = \text{Re}(Y_{11} + Y_{12}) \tag{1.11}$$

$$g_{gdf} = -\text{Re}(Y_{12}) \tag{1.12}$$

Next, the experimental data can be de-embedded from the contributions of these two conductances by subtracting the following **G** matrix from the intrinsic admittance matrix:

$$\mathbf{G} = \begin{bmatrix} g_{gsf} + g_{gdf} & -g_{gdf} \\ -g_{gdf} & g_{gdf} \end{bmatrix} \tag{1.13}$$

Finally, the remaining eight intrinsic circuit elements can be calculated at each frequency point by using the following relationships between them and the intrinsic Y parameters:

$$\begin{aligned} Y_{11} &= \frac{1}{R_{gs} + \frac{1}{j\omega C_{gs}}} + \frac{1}{R_{gd} + \frac{1}{j\omega C_{gd}}} \\ &= \frac{\omega^2 R_{gs} C_{gs}^2}{1 + \omega^2 R_{gs}^2 C_{gs}^2} + \frac{\omega^2 R_{gd} C_{gd}^2}{1 + \omega^2 R_{gd}^2 C_{gd}^2} + j\omega\left(\frac{C_{gs}}{1 + \omega^2 R_{gs}^2 C_{gs}^2} + \frac{C_{gd}}{1 + \omega^2 R_{gd}^2 C_{gd}^2}\right) \end{aligned} \tag{1.14}$$

$$\begin{aligned} Y_{12} &= -\frac{1}{R_{gd} + \frac{1}{j\omega C_{gd}}} \\ &= -\frac{\omega^2 R_{gd} C_{gd}^2}{1 + \omega^2 R_{gd}^2 C_{gd}^2} - \frac{j\omega C_{gd}}{1 + \omega^2 R_{gd}^2 C_{gd}^2} \end{aligned} \tag{1.15}$$

$$Y_{21} = \frac{\frac{g_m e^{-j\omega\tau_m}}{j\omega C_{gs}}}{R_{gs} + \frac{1}{j\omega C_{gs}}} - \frac{1}{R_{gd} + \frac{1}{j\omega C_{gd}}} = \frac{g_m\left[\cos(\omega\tau_m) - \sin(\omega\tau_m)\omega R_{gs}C_{gs}\right]}{1 + \omega^2 R_{gs}^2 C_{gs}^2} +$$

$$- \frac{\omega^2 R_{gd} C_{gd}^2}{1 + \omega^2 R_{gd}^2 C_{gd}^2} - j\left\{\frac{g_m\left[\cos(\omega\tau_m)\omega R_{gs}C_{gs} + \sin(\omega\tau_m)\right]}{1 + \omega^2 R_{gs}^2 C_{gs}^2} + \frac{\omega C_{gd}}{1 + \omega^2 R_{gd}^2 C_{gd}^2}\right\} \tag{1.16}$$

$$\begin{aligned} Y_{22} &= g_{ds} + j\omega C_{ds} + \frac{1}{R_{gd} + \frac{1}{j\omega C_{gd}}} \\ &= g_{ds} + \frac{\omega^2 R_{gd} C_{gd}^2}{1 + \omega^2 R_{gd}^2 C_{gd}^2} + j\omega\left(C_{ds} + \frac{C_{gd}}{1 + \omega^2 R_{gd}^2 C_{gd}^2}\right) \end{aligned} \tag{1.17}$$

To obtain the intrinsic ECPs in terms of the intrinsic Y parameters, Eqns (1.14)–(1.17) can be rewritten as follows:

$$R_{gd} = -\mathrm{Re}\left(\frac{1}{Y_{12}}\right) \tag{1.18}$$

$$C_{gd} = \frac{1}{\omega \mathrm{Im}\left(\frac{1}{Y_{12}}\right)} \tag{1.19}$$

$$R_{gs} = \mathrm{Re}\left(\frac{1}{Y_{11} + Y_{12}}\right) \tag{1.20}$$

$$C_{gs} = -\frac{1}{\omega \mathrm{Im}\left(\frac{1}{Y_{11}+Y_{12}}\right)} \tag{1.21}$$

$$R_{ds} = \frac{1}{\mathrm{Re}(Y_{22} + Y_{12})} \tag{1.22}$$

$$C_{ds} = \frac{\mathrm{Im}(Y_{22} + Y_{12})}{\omega} \tag{1.23}$$

$$g_m = \left|\frac{(Y_{11} + Y_{12})(Y_{21} - Y_{12})}{\mathrm{Im}(Y_{11} + Y_{12})}\right| \tag{1.24}$$

$$\tau_m = -\frac{1}{\omega}\mathrm{phase}\left\{(Y_{21} - Y_{12})\left[1 + j\frac{\mathrm{Re}(Y_{11} + Y_{12})}{\mathrm{Im}(Y_{11} + Y_{12})}\right]\right\} \tag{1.25}$$

The intrinsic section of the small-signal equivalent circuit can be significantly simplified, as shown in Figure 1.12, when the operational frequency is low enough to adopt the quasi-static (QS) approximation. This assumption means that the

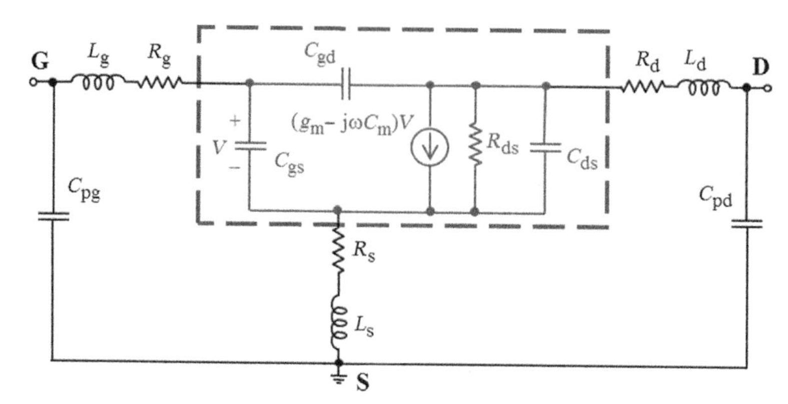

FIGURE 1.12

Simplified topology of the small-signal equivalent circuit for a microwave FET by adopting the quasi-static approximation.

intrinsic transistor can respond instantaneously to any change in the time-varying signal at the intrinsic terminals without depending on its history. Hence, the QS approximation allows neglecting the intrinsic time constants modeling the non-quasi-static (NQS) effects, which represent the inertia of the intrinsic transistor in responding to signal changes [57–61]. In the case of the circuit topology in Figure 1.9, the intrinsic QS model is obtained by omitting the resistances R_{gs} and R_{gd} and approximating the delay τ_m with the transcapacitance C_m, which is given by the product between g_m and τ_m [61–63]. By disregarding the NQS effects, the equations representing the intrinsic Y parameters in terms of the intrinsic ECPs can be strongly simplified. In particular, the QS assumption implies that the real parts of the intrinsic Y parameters should be frequency independent and the imaginary parts of the intrinsic Y parameters should be proportional to the frequency:

$$Y_{11} = g_{gsf} + g_{gdf} + j\omega(C_{gs} + C_{gd}) \tag{1.26}$$

$$Y_{12} = -g_{gdf} - j\omega C_{gd} \tag{1.27}$$

$$Y_{21} = g_m - g_{gdf} - j\omega(C_{gd} + C_m) \tag{1.28}$$

$$Y_{22} = g_{ds} + g_{gdf} + j\omega(C_{ds} + C_{gd}) \tag{1.29}$$

The upper limit of validity of Eqns (1.26)–(1.29) is determined by the onset frequency of NQS effects represented with the three time constants τ_{gs} (i.e., $R_{gs}C_{gs}$), τ_{gd} (i.e., $R_{gd}C_{gd}$), and τ_m (i.e., $g_m^{-1}C_m$). As a rule of thumb, the QS approximation can be in first approximation adopted at frequencies relatively low with respect to the intrinsic gain cutoff frequency f_T. As a matter of fact, the values of both the upper limit of validity of the QS approximation and the intrinsic f_T should be higher in shorter channel transistors and independent of the gate width. Nevertheless, a detailed analysis of the intrinsic Y parameters and time constants at each bias point are required to accurately determine the upper limit of validity of the QS approximation [61].

Many techniques have been proposed to expand the small-signal equivalent circuit to model the high-frequency noise properties of the transistors at the selected bias condition. These various methods reflect the differences among the studied cases, such as different device technologies and investigated frequency ranges. Generally, attention is mainly devoted to the modeling of the noise properties of the intrinsic section of the device. This is because the noisy behavior of the extrinsic section can be straightforwardly represented by adding a noise source for each resistance to account for the associated thermal noise. Hence, only the values of the extrinsic resistances and the ambient temperature should be known to obtain these uncorrelated noise sources from Nyquist's formula. On the other hand, noise modeling of the intrinsic FETs is based on using two noise current sources and the complex correlation coefficient between them.

Over the years, several modifications of this noise model have been proposed to improve its accuracy and/or to simplify its extraction. Typically, the correlation coefficient between the intrinsic noise sources has been assumed to be purely imaginary or even equal to zero by assuming that the correlation arises from the intrinsic circuit itself, which accounts for the capacitive coupling between the channel and the gate [64–69]. One of the most popular and widely used approaches consists of reducing the noise model to two uncorrelated noise sources, namely a gate noise voltage source at the input and a drain noise current source at the output. Such a model is often referred to as the two-parameter noise model because the unknowns are reduced from four to two parameters by assuming the complex correlation coefficient to be equal to zero [67]. These two uncorrelated noise sources are usually implemented by assigning two equivalent noise temperatures, known as T_{gs} and T_{ds}, to the resistances R_{gs} and R_{ds}, respectively. This type of description of the high-frequency noise performance for transistors is known as the noise temperature model because it is based on using equivalent noise temperatures. By considering T_{gs} to be equal to the ambient temperature, the extraction of a noise model starting from a small-signal equivalent circuit is reduced to the determination of only one parameter, namely T_{ds} [16] and [67,68]. Figure 1.13 shows a typical noise model, which is based on five uncorrelated noise sources obtained by assigning an equivalent temperature to each resistor of the small-signal equivalent circuit. Nevertheless, the number of noise sources has been often increased by using circuit topologies containing a higher number of resistances, such as R_{gd} and additional extrinsic resistances [70,71].

As mentioned in Section 1.2.1, the reader can find a more comprehensive treatment of this subject in Chapters 2 and 3.

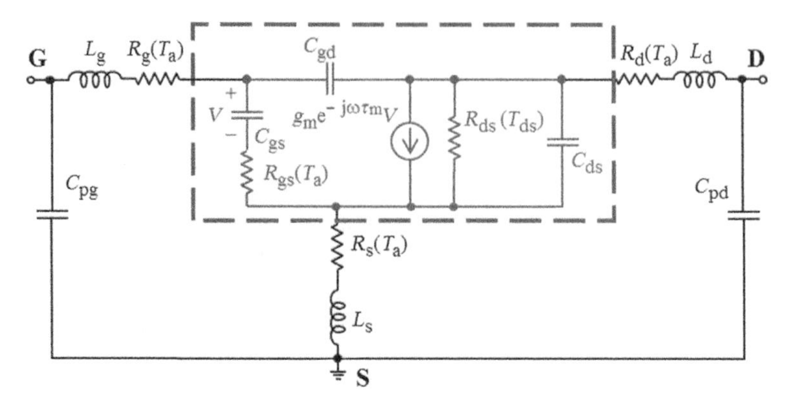

FIGURE 1.13

Noise equivalent circuit for a microwave FET.

1.3.2 Large-signal equivalent circuit

The small-signal equivalent circuit determined under a wide range of operating bias points can be used as cornerstone for building the large-signal model. One approach is based on simplifying the equivalent circuit by adopting the QS approximation. Under this condition, the intrinsic core of the large-signal equivalent circuit can be represented with four nonlinear sources: two charge sources and two current sources (i.e., Q_{gs}, Q_{ds}, I_{gs}, I_{ds}) modeling, respectively, the displacement and the conduction intrinsic current contributions (see Figure 1.14) [72–76]. However, the gate current source can be often omitted because of its negligible role, especially in the case of DC insulated gate transistors.

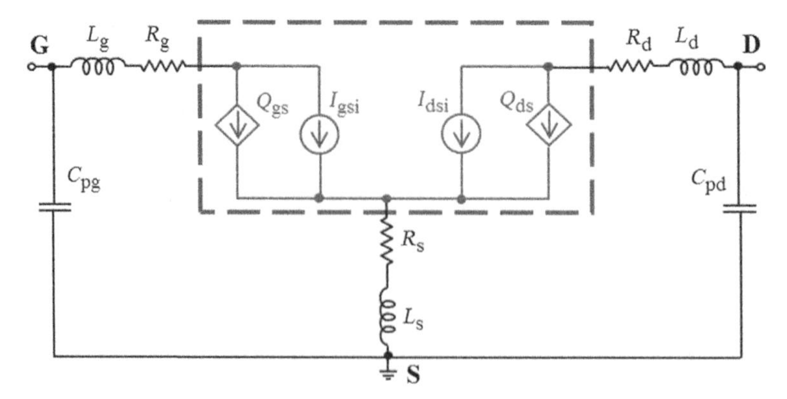

FIGURE 1.14

Quasi-static large-signal equivalent circuit for a microwave FET.

The current and charge sources can be obtained by integrating, respectively, the real and imaginary parts of the intrinsic Y parameters versus the intrinsic gate-source and drain-source voltages:

$$I_{gs}(V_{gsi}, V_{dsi}) = I_{gs}(V_{gsi0}, V_{dsi0}) + \int_{V_{gsi0}}^{V_{gsi}} \mathrm{Re}\{Y_{11}(V, V_{dsi0})\} \mathrm{d}V + \int_{V_{dsi0}}^{V_{dsi}} \mathrm{Re}\{Y_{12}(V_{gsi}, V)\} \mathrm{d}V \quad (1.30)$$

$$I_{ds}(V_{gsi}, V_{dsi}) = I_{ds}(V_{gsi0}, V_{dsi0}) + \int_{V_{gsi0}}^{V_{gsi}} \mathrm{Re}\{Y_{21}(V, V_{dsi0})\} \mathrm{d}V + \int_{V_{dsi0}}^{V_{dsi}} \mathrm{Re}\{Y_{22}(V_{gsi}, V)\} \mathrm{d}V \quad (1.31)$$

$$Q_{gs}(V_{gsi}, V_{dsi}) = \int_{V_{gsi0}}^{V_{gsi}} \frac{\mathrm{Im}\{Y_{11}(V, V_{dsi0})\}}{\omega} \mathrm{d}V + \int_{V_{dsi0}}^{V_{dsi}} \frac{\mathrm{Im}\{Y_{12}(V_{gsi}, V)\}}{\omega} \mathrm{d}V \quad (1.32)$$

$$Q_{ds}(V_{gsi}, V_{dsi}) = \int_{V_{gsi0}}^{V_{gsi}} \frac{\mathrm{Im}\{Y_{21}(V, V_{dsi0})\}}{\omega} \mathrm{d}V + \int_{V_{dsi0}}^{V_{dsi}} \frac{\mathrm{Im}\{Y_{22}(V_{gsi}, V)\}}{\omega} \mathrm{d}V \quad (1.33)$$

By using Eqns (1.26)–(1.29) to express the intrinsic Y parameters in terms of the intrinsic ECPs, Eqns (1.30)–(1.33) can be rewritten as follows:

$$I_{gs}(V_{gsi}, V_{dsi}) = I_{gs}(V_{gsi0}, V_{dsi0}) + \int_{V_{gsi0}}^{V_{gsi}} \left[g_{gsf}(V, V_{dsi0}) + g_{gdf}(V, V_{dsi0})\right] \mathrm{d}V + \int_{V_{dsi0}}^{V_{dsi}} g_{gdf}(V_{gsi}, V) \mathrm{d}V \quad (1.34)$$

$$I_{ds}(V_{gsi}, V_{dsi}) = I_{ds}(V_{gsi0}, V_{dsi0}) + \int_{V_{gsi0}}^{V_{gsi}} \left[g_{m}(V, V_{dsi0}) - g_{gdf}(V, V_{dsi0})\right] \mathrm{d}V + \int_{V_{dsi0}}^{V_{dsi}} \left[g_{ds}(V_{gsi}, V) + g_{gdf}(V_{gsi}, V)\right] \mathrm{d}V \quad (1.35)$$

$$Q_{gs}(V_{gsi},V_{dsi}) = \int_{V_{gsi0}}^{V_{gsi}} [C_{gs}(V,V_{dsi0}) + C_{gd}(V,V_{dsi0})]dV - \int_{V_{dsi0}}^{V_{dsi}} C_{gd}(V_{gsi},V)dV \tag{1.36}$$

$$Q_{ds}(V_{gsi},V_{dsi}) = -\int_{V_{gsi0}}^{V_{gsi}} [C_{gd}(V,V_{dsi0}) + C_{m}(V,V_{dsi0})]dV + \int_{V_{dsi0}}^{V_{dsi}} [C_{gd}(V_{gsi},V) + C_{ds}(V_{gsi},V)]dV \tag{1.37}$$

As the operating frequency increases, the large-signal model should be extended to include the NQS effects. A solution consists of using directly the bias dependence of the intrinsic RC elements, as will be discussed in Section 1.5.3 [77]. A different strategy to account for the NQS effects consists of extending the number of the charge and current nonlinear sources, namely a combination of zero order sources and infinite number of higher order sources. Nevertheless, the model can be implemented only by truncating the number of nonlinear sources (see Figure 1.15). The higher order sources are determined by accounting for the NQS effects when integrating the intrinsic small-signal equivalent circuit elements versus the intrinsic voltages [78]. By including the high-frequency NQS effects, the model accuracy is improved at higher frequencies. The price to be paid is a higher model complexity that negatively affects the simulation convergence and speed. In light of that, the QS model has a use in the lower RF frequency range.

For the sake of completeness, it should be mentioned that the complexity of the large-signal model could be further increased by accounting for the low-frequency dispersive effects, which are due to traps and thermal phenomena. Different

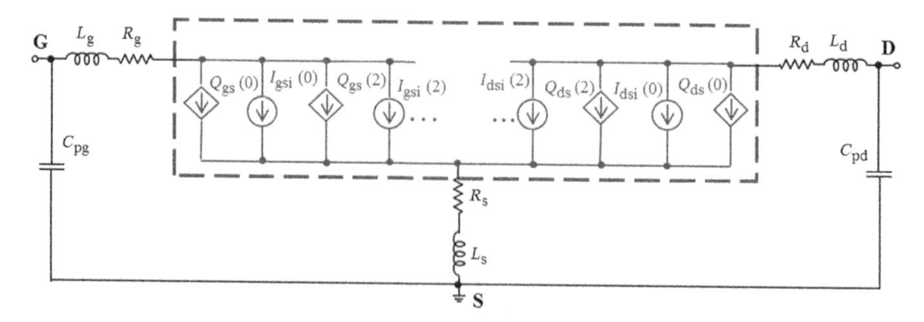

FIGURE 1.15

Non-quasi-static large-signal equivalent circuit for a microwave FET. The non-quasi-static effects are included by extending the number of nonlinear sources.

techniques have been developed to model successfully the dispersion effects affecting mostly the drain current of relatively immature technologies (e.g., gallium nitride (GaN) devices) [79–83].

As will be discussed in detail in Chapter 5, the two main approaches for implementing empirical large-signal models in commercial circuit simulators are based on analytical functions [83–87] and lookup tables [72] and [88–91]. The former representation offers the advantage of much better extrapolation capability, in both voltage and frequency domains, beyond the investigated operating conditions. Furthermore, the analytical functions allow achieving a compact representation based on a limited number of fitting parameters without using huge tables. In turn, the benefits of better extrapolation and a more compact representation enable shortening of the simulation time and minimizing convergence problems. Nevertheless, the determination of the analytical function is based on a challenging model formulation and a complex optimization process. This is a critical drawback because the identification of the appropriate function and the associated parameters are critically dependent on the specific technology. On the other hand, the lookup table approach allows overcoming this drawback with a straightforward model implementation based on loading the intrinsic elements in tables as functions of the intrinsic voltages.

1.4 From de-embedding to waveform engineering

The mathematical art of shifting the electrical reference planes is referred to as de-embedding. Its importance comes from the fact that the electrical characteristics are not always directly measurable at the desired reference planes because the terminals of interest can be inaccessible. The determination of the data at the desired electrical reference planes plays a crucial role for measurements, modeling, and design purposes. Depending on whether the reference planes are shifted closer to or farther away from the actual DUT, a distinction can be made between de-embedding and embedding, as depicted in Figure 1.16 [92–96]. The process of de-embedding consists in shifting the electrical reference planes closer to the actual DUT in order to subtract unwanted contributions. A typical application for on-wafer transistors is the subtraction of the extrinsic effects arising from the contact pads and access transmission lines. This step allows one to obtain the behavior of the actual transistor from the measurements performed on the whole device at the calibration planes. By contrast, the embedding consists of moving the electrical reference planes further away from the actual DUT to include additional contributions. In general, the embedding concept can be useful to analyze the tested device as if it were to be placed in a hypothetical network.

To de-embed properly, the data are a work of art that require knowledge and experience. This task becomes even more challenging by working at higher frequencies. Microwave de-embedding can be differentiated into two different cases depending on whether it is applied to linear or nonlinear data. In light of that, the

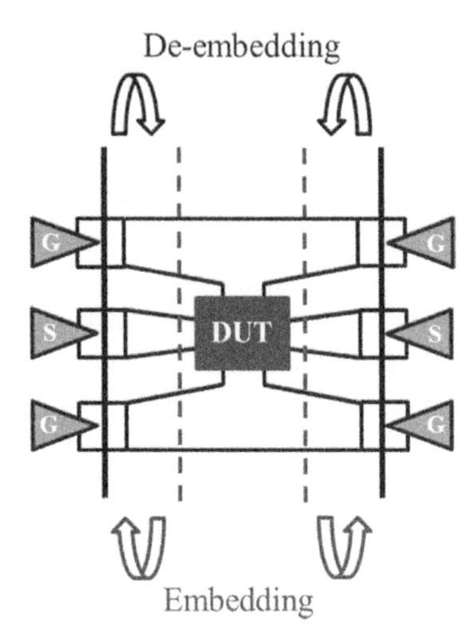

FIGURE 1.16

Illustration of the application of the de-embedding and embedding concepts to shift the input and output electrical reference planes from the whole to the actual device and vice versa for on-wafer characterization based on ground-signal-ground (GSG) probes.

remainder of this section is organized into two parts. The first subsection is devoted to the linear case, whereas the second subsection is focused on the nonlinear case. As will be discussed in the following two subsections, the de-embedding concept enables one to move the reference planes closer to the actual transistor, but its effect is different for linear and nonlinear data. The unwanted extrinsic contributions are mathematically fully removed in the linear case, whereas they should be included in the source and load impedances seen by the transistor at the new reference planes in the nonlinear case. This is because a full characterization of the linear behavior allows prediction of the device response for any external excitation and termination at the studied bias point because the small-signal behavior of a transistor depends only on its properties at the chosen bias condition. On the other hand, the large-signal behavior depends also on the time-varying port voltages and currents and, consequently, on the input and output terminations. It is worth pointing out that, as will be further elucidated in Chapter 9, the term "nonlinear" in the expression "nonlinear de-embedding" refers to the fact that nonlinear contributions should be de-embedded from the data.

1.4.1 De-embedding: from scattering to noise parameters

With the aim to achieve the *S* parameters at the desired reference planes, different strategies have been proposed to apply accurately the de-embedding concept. These procedures are based on matrix manipulations to subtract unwanted contributions.

A typical application consists of removing the extrinsic contributions to determine the behavior of the intrinsic transistor.

As an illustrative example, the following equation shows how to de-embed the contributions of the extrinsic ECPs when working with the circuit topology in Figure 1.9:

$$\mathbf{Z_{DUT\_Intr}} = [\mathbf{Y_{DUT}} - j\omega\mathbf{C_{Extr}}]^{-1} - \mathbf{R_{Extr}} - j\omega\mathbf{L_{Extr}} \tag{1.38}$$

where the three matrices $\mathbf{C_{Extr}}$, $\mathbf{R_{Extr}}$, and $\mathbf{L_{Extr}}$ representing the extrinsic effects are given by the following:

$$\mathbf{C_{Extr}} = \begin{bmatrix} C_{pg} & 0 \\ 0 & C_{pd} \end{bmatrix} \tag{1.39}$$

$$\mathbf{R_{Extr}} = \begin{bmatrix} R_g + R_s & R_s \\ R_s & R_d + R_s \end{bmatrix} \tag{1.40}$$

$$\mathbf{L_{Extr}} = \begin{bmatrix} L_g + L_s & L_s \\ L_s & L_d + L_s \end{bmatrix} \tag{1.41}$$

On the other hand, the knowledge of the S parameters of the test structures allows shifting the reference planes directly with matrix manipulations without having to extract the associated circuit representation. One of the most common de-embedding methods for on-wafer transistors is the two-step open-short de-embedding technique [48], which has been illustrated in Figure 1.11. By using Eqns (1.2) and (1.3), the matrix $\mathbf{Y_{DUT\text{-}ADOS}}$ representing the Y parameters of the transistor after applying the de-embedding based on open and short structures can be obtained straightforwardly as follows [48]:

$$\mathbf{Y_{DUT\text{-}ADOS}} = \left((\mathbf{Y_{DUT}} - \mathbf{Y_O})^{-1} - (\mathbf{Y_S} - \mathbf{Y_O})^{-1}\right)^{-1} \tag{1.42}$$

where $\mathbf{Y_{DUT}}$, $\mathbf{Y_O}$, and $\mathbf{Y_S}$ are the admittance matrices of the whole transistor, open, and short, respectively.

When the tested transistor is packaged, the two-step short-open method should be adopted [48]:

$$\mathbf{Y_{DUT\text{-}ADOS}} = (\mathbf{Z_{DUT}} - \mathbf{Z_S})^{-1} - (\mathbf{Z_O} - \mathbf{Z_S})^{-1} \tag{1.43}$$

where $\mathbf{Y_{DUT\text{-}ADSO}}$ is given by the admittance matrix after applying the de-embedding procedure based on short and open structures, whereas $\mathbf{Z_{DUT}}$, $\mathbf{Z_O}$, and $\mathbf{Z_S}$ represent the impedance matrices of the whole transistor, open, and short, respectively.

Nevertheless, the S parameters of the test structures, which can be determined with measurements or with electromagnetic simulations, can be used also to identify a lumped equivalent circuit network. Typically, $\mathbf{Y_O}$ and $(\mathbf{Y_S}-\mathbf{Y_O})^{-1}$ represent, respectively, the contributions given by $j\omega\mathbf{C_{Extr}}$ and $\mathbf{R_{Extr}} + j\omega\mathbf{L_{Extr}}$. When the corresponding circuit is determined, the extrinsic contributions can be de-embedded with formulas based on the same philosophy used in Eqns (1.38)–(1.41).

Regarding the four noise parameters, they represent completely the noise behavior of a linear two-port network at a given bias point by allowing prediction of the noise factor for any source impedance. Therefore, similar to the case of S parameters, the extrinsic effects can be completely removed from the noise parameters by using de-embedding methods based on simple matrix manipulations. Although the noise parameters and the noise correlation matrix are equivalent representations, the latter provides a more convenient description to apply the de-embedding concept to the noise characteristics. This is because the noise correlation matrix allows using Eqns (1.5)–(1.8), which are of great utility for performing the de-embedding and embedding processes. To accomplish this task, the noise correlation matrix of the transistor can be calculated by its noise parameters (see Eqn (1.10)), whereas the noise correlation matrix of a passive network such as the test structures can be determined with only thermal noise sources from its S parameters and temperature [2–5]:

$$\mathbf{C_Y} = 2kT\text{Re}(\mathbf{Y}) \tag{1.44}$$

$$\mathbf{C_Z} = 2kT\text{Re}(\mathbf{Z}) \tag{1.45}$$

It should be observed that, if the de-embedding concept is exploited for shifting the reference planes of the noise factor determined with one fixed source impedance (e.g., 50 Ω), the unwanted extrinsic contributions are not fully removed. This is because the noise factor determined with one fixed source impedance cannot be considered to be a full characterization of the noise behavior. Nevertheless, the unwanted extrinsic contributions should be included only in the source impedance, since Eqn (1.9) shows that the noise factor does not depend on the load impedance.

1.4.2 De-embedding: nonlinear time-domain waveforms

Contrary to the small-signal case, the transistor cannot be described with a linear matrix representation under large-signal condition. As a consequence, the electrical reference planes of the nonlinear data should be shifted by using the Kirchhoff laws relating voltages and currents before and after applying de-embedding. On the other hand, the test structures can be described with linear matrix representations because they are linear devices. As illustrative example, the two-step de-embedding based on open and short structures may be applied to nonlinear data at each frequency point as follows (see Figure 1.17) [37]:

$$i_{1\_\text{DUT-ADOS}} = i_{1\_\text{DUT}} - v_{1\_\text{DUT}}Y_1 - (v_{1\_\text{DUT}} - v_{2\_\text{DUT}})Y_3 \tag{1.46}$$

$$i_{2\_\text{DUT-ADOS}} = i_{2\_\text{DUT}} - v_{2\_\text{DUT}}Y_2 - (v_{2\_\text{DUT}} - v_{1\_\text{DUT}})Y_3 \tag{1.47}$$

$$v_{1\_\text{DUT-ADOS}} = v_{1\_\text{DUT}} - i_{1\_\text{DUT-ADOS}}(Z_1 + Z_3) - i_{2\_\text{DUT-ADOS}}Z_3 \tag{1.48}$$

$$v_{2\_\text{DUT-ADOS}} = v_{2\_\text{DUT}} - i_{2\_\text{DUT-ADOS}}(Z_2 + Z_3) - i_{1\_\text{DUT-ADOS}}Z_3 \tag{1.49}$$

where i_{1_DUT}, i_{2_DUT}, v_{1_DUT}, v_{2_DUT} and $i_{1\_\text{DUT-ADOS}}$, $i_{2\_\text{DUT-ADOS}}$, $v_{1\_\text{DUT-ADOS}}$, $v_{2\_\text{DUT-ADOS}}$ are, respectively, the nonlinear currents and voltages before and after applying the de-embedding, the admittances Y_1, Y_2, and Y_3 represent the

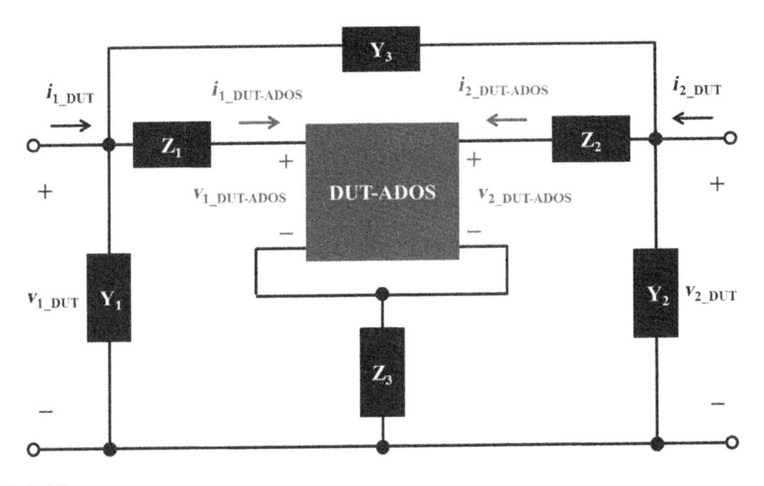

FIGURE 1.17

Illustration of open-short de-embedding for nonlinear data.

contributions associated to the open structure (i.e., $Y_1 = Y_{11\_O} + Y_{12\_O}$, $Y_2 = Y_{22\_O} + Y_{12\_O}$, and $Y_3 = -Y_{12\_O}$), and the impedances Z_1, Z_2, and Z_3 represent the contributions associated to the short structure after de-embedding of open structure effects (i.e., $Z_1 = Z_{11\_S\text{-}ADO} - Z_{12\_S\text{-}ADO}$, $Z_2 = Z_{22\_S\text{-}ADO} - Z_{12\_S\text{-}ADO}$, and $Z_3 = Z_{12\_S\text{-}ADO}$). Because both open and short structures are reciprocal devices, Y_{12_O} and $Z_{12\_S\text{-}ADO}$ can be considered to be equal to Y_{21_O} and $Z_{21\_S\text{-}ADO}$. As mentioned, Eqns (1.46)–(1.49) enable one to move the electrical reference planes, but the extrinsic contributions represented by the test structures should be included in the source and load impedances [3] and [50].

It should be pointed out that, if only linear contributions (e.g., Eqns (1.46)–(1.49)) are de-embedded from nonlinear data, the de-embedding itself should be considered to be linear. Actually, de-embedding can be considered to be nonlinear only when nonlinear contributions are de-embedded. A typical example of nonlinear de-embedding is given by its application to retrieve the transistor load-line at the current generator plane, which is essential for power amplifier design. To accomplish this goal, the current and voltage waveforms should be de-embedded from the contributions arising from both the linear extrinsic elements and the nonlinear intrinsic capacitances. This particular application of the de-embedding concept highlights its importance as a crucial prerequisite for the success of the waveform engineering design technique [97–102], which is aimed at shaping the current and voltage waveforms at the transistor current generator plane to optimize the device performance. Further details can be found in Chapter 9.

It should be highlighted that the application of the de-embedding concept to the measured load line is of great utility for designers when they need to engineer the shape of the time-domain waveforms for optimal performance. This is because

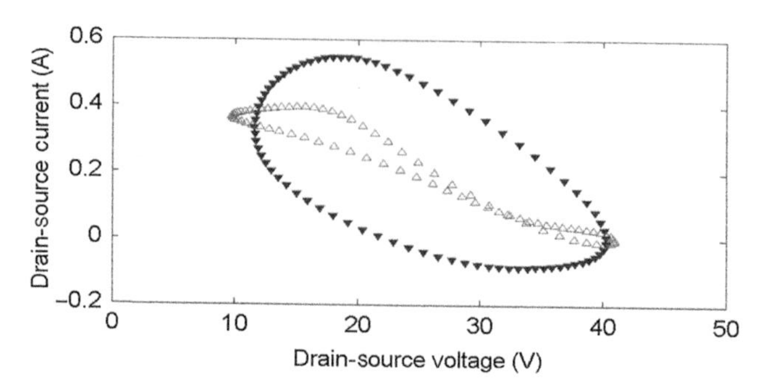

FIGURE 1.18

Comparison between the measured extrinsic load line (down triangles) and the load line at the current generator plane (up triangles) for a GaN HEMT ($0.7 \times 800\ \mu m^2$) at $V_{GS} = -2$ V, $V_{DS} = 25$ V, $f_0 = 4$ GHz, and $P_{out} = 1.55$ W.

the extrinsic contributions along with the intrinsic capacitive effects can significantly mask the shape of the load line at the current generator plane. As an illustrative example, Figure 1.18 shows the measured extrinsic load line and the corresponding load line at the current generator plane for a GaN HEMT.

1.5 De-embedding: experimental results

This section discusses experimental results illustrating the application of the de-embedding concept to the FinFET. The first subsection is devoted to presenting the basics of this advanced FET device, whereas the other two subsections show the experimental results for both linear and nonlinear cases.

1.5.1 Basics of FinFET

As well known, the aggressive downscaling of the conventional complementary metal oxide semiconductor (CMOS) technology is reaching its inherent limit, due to short channel effects. To go beyond this limit, the last three decades have witnessed the advent and the exponential growth of a plethora of multiple-gate structures, which are called multiple-gate field-effect transistors (MuGFETs) [103]. This revolutionary change in the device architecture consists of obtaining gate control over the transistor channel from more than one side, as opposed to the conventional planar MOSFET. Hence, based on the number of sides, MuGFETs can be differentiated into three main groups: double- or dual-gate, triple- or tri-gate, quadruple- or surrounding-gate or gate-all-around. By using multiple-gate structures, the better electrostatic control of the gate over the channel enables a reduction in the short-channel effects, which turns into a lower threshold voltage roll-off and a lower subthreshold slope with associated larger on–off current ratio.

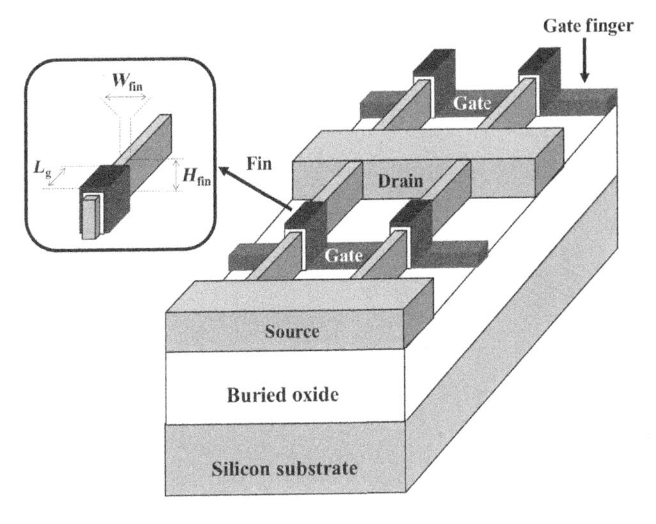

FIGURE 1.19

Illustrative three-dimensional schematic view of a triple-gate FinFET structure on SOI substrate.

Among the many types of multiple-gate structures proposed over the years, the FinFET stands out for being compatible with the conventional planar CMOS technology. This innovative transistor was introduced at the end of the second millennium with the name of folded-channel FET [104]. Subsequently, the term FinFET was coined to account for the fact that multiple gates are folded over the surface of a thin silicon body in such a way to look like the dorsal fin of a fish [105,106]. The FinFET has attracted worldwide attention going from theoretical research to practical applications. The latter can be seen in the fact that Intel has used 22 nm FinFET technology in Ivy Bridge, which is a new generation of processors that was launched in the market in Spring 2012 [107].

As illustrated in Figure 1.19, the total gate width associated with a triple-gate fin structure can be estimated as twice the fin height H_{fin} plus the fin width W_{fin}. The latter contribution should be disregarded in the case of a double-gate fin structure, namely when the top insulating layer is made sufficiently thick to electrically isolate the top gate. To enlarge the active channel, multifinger and multifin layouts are used (see Figure 1.20). Consequently, the total gate periphery W increases proportionally to the total number of fins, which is given by the product between the number of fingers N_{finger} and the number of fins for each finger N_{fin}:

$$W = N_{finger} N_{fin} \left(2H_{fin} + W_{fin} \right) \tag{1.50}$$

The multifinger layout allows the reduction of R_g, which is a critical circuit element that can have a strong impact on microwave characteristics, such as the noise performance and the maximum frequency of oscillation f_{max}. As a matter of

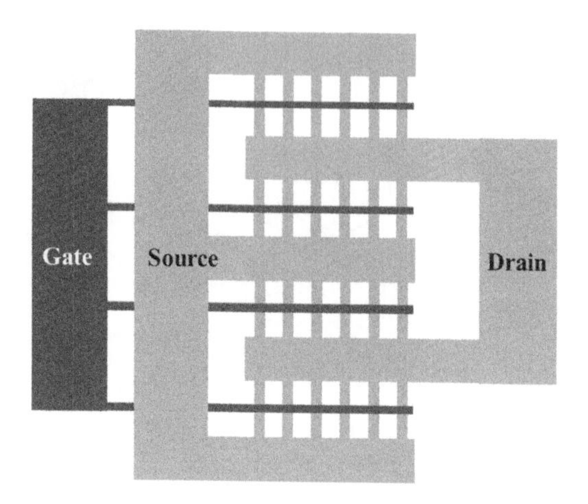

FIGURE 1.20

Illustrative schematic top view of a FinFET with four gate fingers and six fins for each finger.

fact, an increased number of fingers in parallel leads to a wider path, not only for the drain current but also for the gate current, which implies a decrease in R_g.

Typically, the FinFETs are built on silicon-on-insulator (SOI) substrate because its high resistivity allows one to minimize the substrate losses. Nevertheless, it should be highlighted that the interest in bulk FinFETs is increasing in relation to overcoming the drawbacks of the SOI substrate, namely high wafer cost, poor thermal conductivity, and high defect density [108,109].

The FinFET offers attractive properties mainly for digital applications. Its high-frequency performance is quite promising but further improvements are still required. However, the analysis of this technology at high frequencies plays a key role to target a successful integration of both digital and analog circuits on the same chip for mixed-mode applications.

In the present chapter, the FinFET is used as reference device to illustrate the application of the microwave de-embedding concept with experimental data achieved with on-wafer characterization (see Figure 1.21). Figure 1.22 illustrates the concept of de-embedding and embedding for the FinFET by moving the reference planes from the whole device terminals to the actual device and vice versa. The tested FinFET was fabricated at Imec (Belgium), as described in [110]. This device is a triple-gate nMOSFET fabricated on SOI substrate with a gate length of 60 nm, a total gate periphery of 45.6 μm, a fin height of 60 nm, a fin width of 32 nm, and 50 fingers, with each finger composed of six fins.

1.5.2 From scattering to noise parameters

Figure 1.23 shows the small-signal equivalent circuit used to model the whole FinFET at the input and output probe tip reference planes [50]. This circuit can be

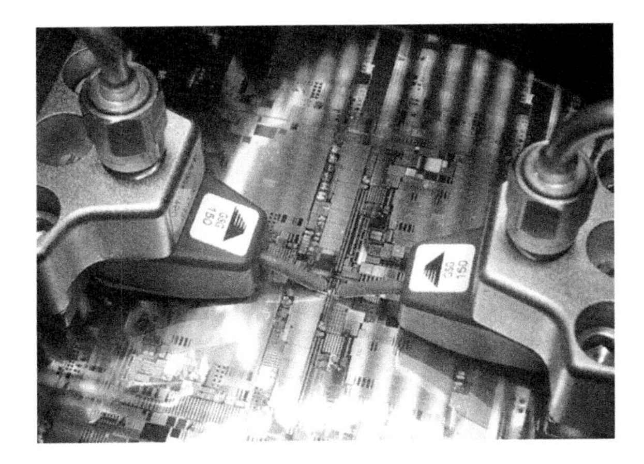

FIGURE 1.21

Photo of CPW probes contacting an on-wafer FinFET.

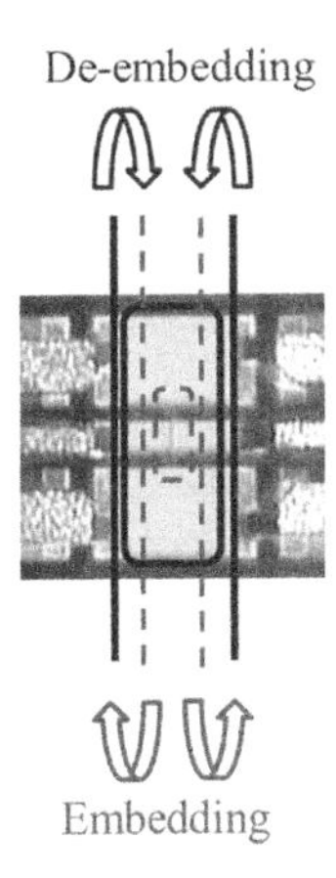

FIGURE 1.22

Photo of an on-wafer FinFET to illustrate the application of the de-embedding and embedding concepts to shift the input and output electrical reference planes from the whole to the actual device and vice versa.

broken into two parts: the circuit network used to model the actual transistor and the circuit network used to model the contact pads, the metal interconnections between the pads and the actual device, and the substrate. The latter network is composed of 16 extrinsic elements (i.e., C_{1x}, R_{1x}, C_{1y}, R_{1y}, C_{2x}, R_{2x}, C_{2y}, R_{2y}, C_{3x}, R_{3x}, C_{3y}, R_{3y}, L_1, L_2, L_3, R_3), which are determined from S parameters measured on open and short

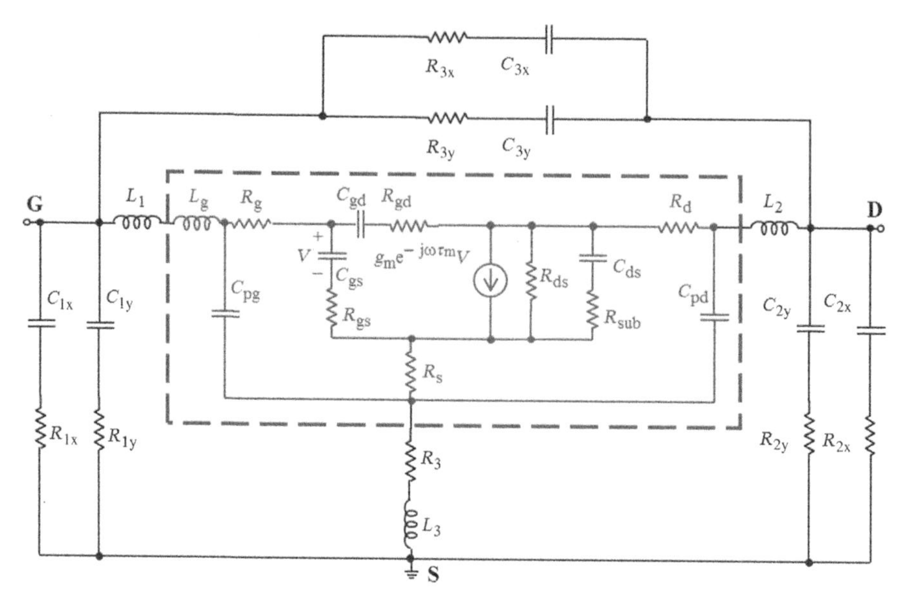

FIGURE 1.23

Small-signal equivalent circuit of a FinFET.

dummy structures. It should be pointed out that the tested transistor and the associated dummy structures have been accurately designed to have the same layout and they were also fabricated on the same die of the wafer. In particular, the three input, output, and feedback RC branches (i.e., R_{ix}–C_{ix} and R_{iy}–C_{iy}, with i being 1, 2, or 3) are determined in order to reproduce the measured S parameters of the open structure including the low-frequency kinks associated to the lossy substrate, whereas the other elements (i.e., L_1, L_2, L_3, R_3) are determined in order to reproduce the measured S parameters of the short structure. The circuit network for the actual transistor is composed of six extrinsic elements (i.e., C_{pg}, L_g, R_g, C_{pd}, R_d, R_s), which are extracted from S parameters measured on the cold transistor to model residual extrinsic contributions after applying the open-short de-embedding procedure, and nine intrinsic elements (i.e., C_{gs}, R_{gs}, C_{gd}, R_{gd}, g_m, τ_m, C_{ds}, R_{ds}, R_{sub}), which are calculated from the intrinsic Y parameters at each bias point.

The resistance R_{sub} has been included in the model to improve the simulated intrinsic Y_{22} at high frequencies by accounting for residual substrate losses. To extract the nine intrinsic elements from real and imaginary parts of the intrinsic four Y parameters, the resistance R_{ds} is determined at low frequencies by treating the RC series network (i.e., C_{ds} and R_{ds}) as an open circuit. Hence, the contribution of the extracted R_{ds} is subtracted before calculating analytically the other intrinsic elements.

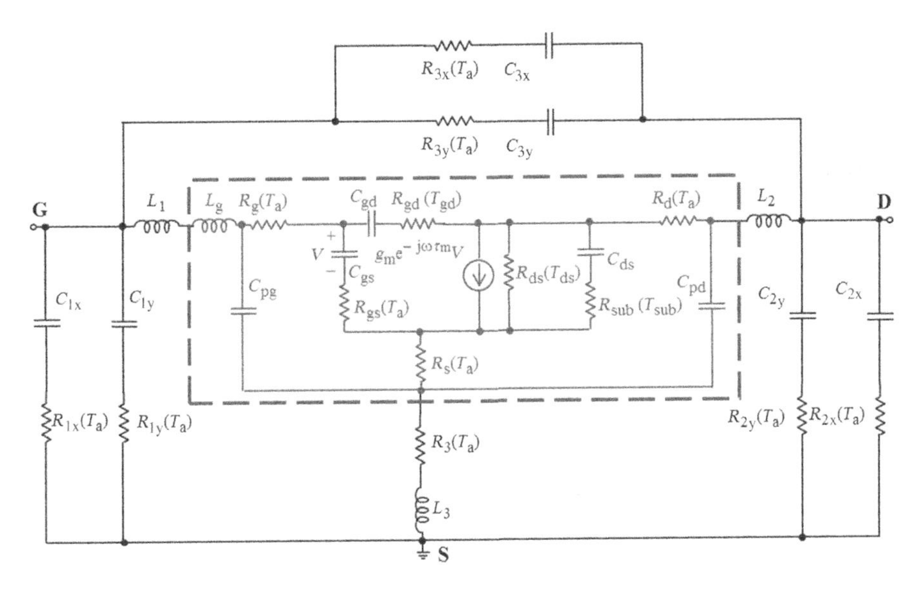

FIGURE 1.24

Noise equivalent circuit of a FinFET.

To obtain a noise model, equivalent temperatures are assigned to each resistor of the small-signal equivalent circuit (see Figure 1.24) [18]. In particular, the temperatures T_{gd}, T_{ds}, and T_{sub}, which are associated to the intrinsic output, feedback, and substrate resistances, are obtained by minimizing the difference between simulated and measured F_{50} up to 26.5 GHz, whereas the other temperatures are selected to be equal to the ambient temperature.

Figure 1.25 shows the comparison between measured and simulated S parameters before and after applying the open-short de-embedding procedure. As can be clearly observed, significant changes in the behavior of the S parameters are detected by de-embedding the contributions associated to the two dummy structures. In particular, the gain given by the magnitude of S_{21} is increased, the S parameters exhibit a smaller phase rotation, and the low frequency kinks disappear. The latter change is due to the subtraction of the RC networks determined from S-parameter measurements of the open structure.

Figure 1.26 shows the noise parameters simulated by the noise model with and without the contributions associated to the open and short structures to exploit their impact on the noise performance. As can be clearly seen, the extrinsic network contributions significantly affect the simulated noise characteristics. The observed differences can be roughly summarized in the fact that the absence of this extrinsic network allows lowering NF_{min}, reducing R_n at frequencies lower than 16.5 GHz, and removing the low-frequency kink in Γ_{opt}. The determination of the circuit

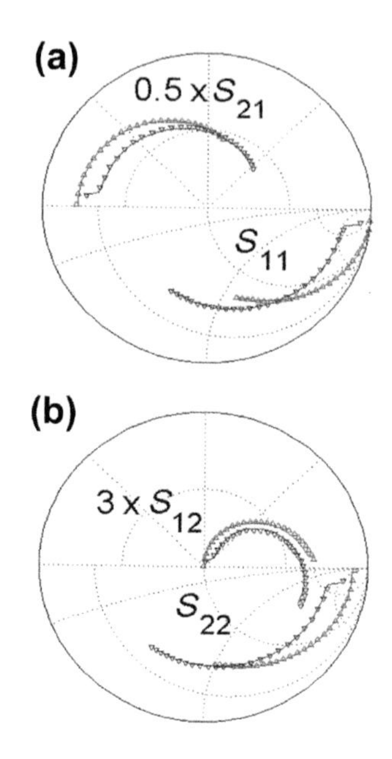

FIGURE 1.25

Measured (symbols) and simulated (lines) *S* parameters before (down triangles) and after (up triangles) applying open-short de-embedding procedure for a FinFET at $V_{GS} = 0.8$ V, $V_{DS} = 1.2$ V with frequency ranging from 0.3 GHz to 50 GHz.

representing the open and short structures allows determination of the origin of the differences observed by simulating the model with and without the effects of two such dummy structures. For instance, the change in Γ_{opt} after shifting the reference planes is mainly due to the removal of the input and feedback RC networks.

1.5.3 Nonlinear time-domain waveforms

The small-signal equivalent circuit in Figure 1.23 is used as cornerstone for building the corresponding large-signal equivalent circuit reported in Figure 1.27 [50] and [77]. This large-signal model is based on loading all intrinsic circuit elements (i.e., C_{gs}, C_{gd}, C_{ds}, C_m, I_{dsiDC}, I_{dsiRF}, R_{gs}, R_{gd}, R_{sub}) in lookup tables as a function of the intrinsic voltages.

Figure 1.28 shows the comparison between measured and simulated large-signal data before and after applying the open-short de-embedding procedure. The most significant difference is observed in the input locus showing that, as a consequence of the de-embedding process, the gate current exhibits much lower values and tends to be in quadrature rather than in phase with the gate voltage. This result is due to the substrate losses represented by the extrinsic RC networks because the de-embedding

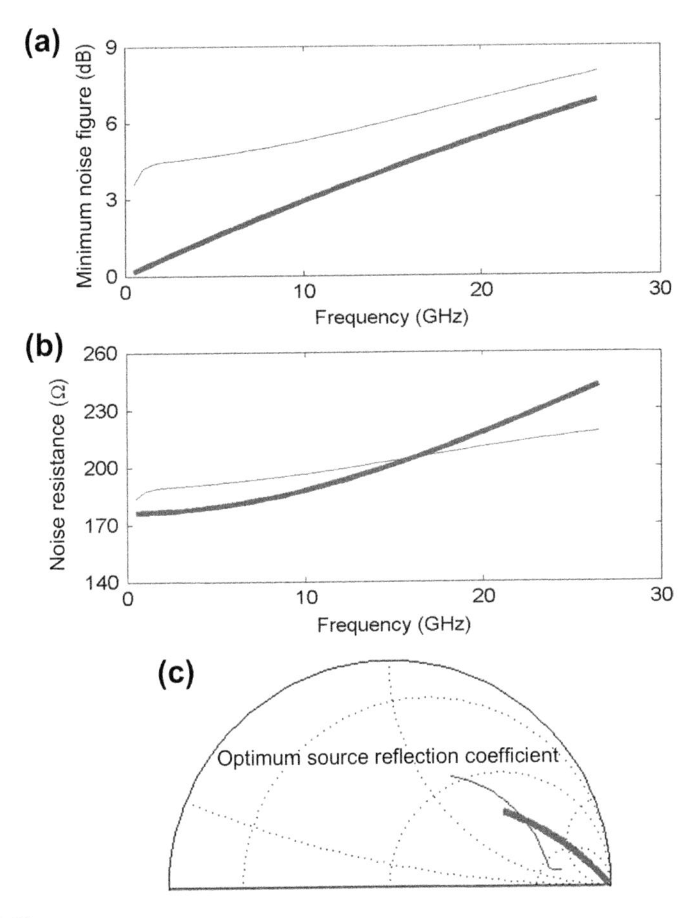

FIGURE 1.26

Model simulations of the noise parameters (a) NF_{min}, (b) R_n, and (c) Γ_{opt} from 0.5 GHz to 26.5 GHz for a FinFET at $V_{DS} = 1$ V and $V_{GS} = 0.8$ V, with (thin lines) and without (thick lines) the contributions of the external part of the equivalent circuit determined from *S* parameters measured on open and short structures.

process implies that the current contributions associated with the input and feedback extrinsic RC networks are subtracted from the gate current (see Eqn (1.46)).

It is worth a reminder that the application of de-embedding to nonlinear data implies the source and load impedances presented to the device at the new reference planes should include also the contributions of the access transmission lines. Nevertheless, once the large-signal model has been extracted and validated with measurements, the nonlinear device behavior can be predicted for any source and load terminations. On the other hand, this is not necessary in the linear case because

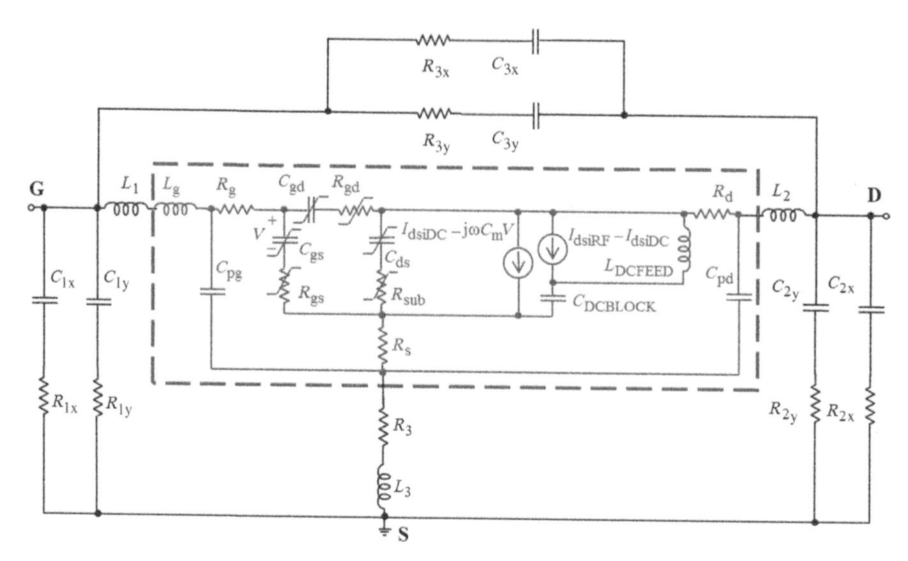

FIGURE 1.27

Large-signal equivalent circuit of a FinFET.

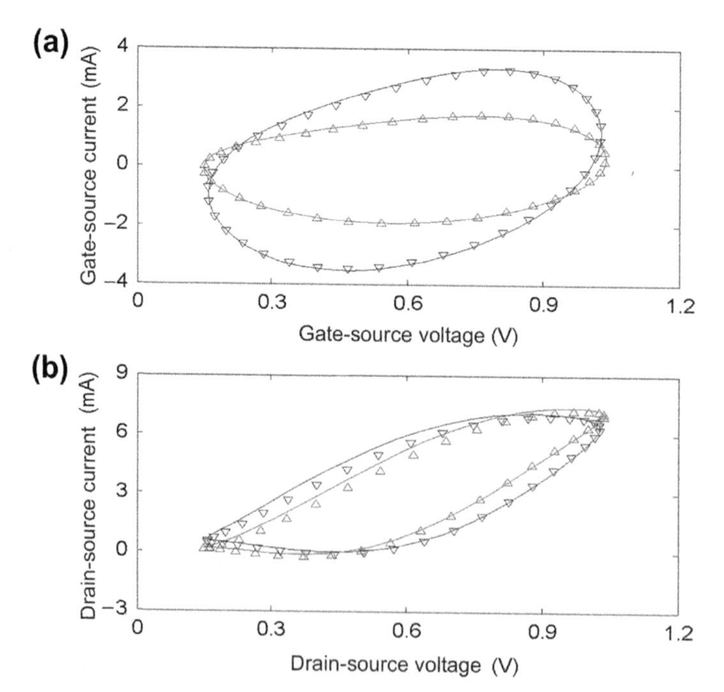

FIGURE 1.28

Measured (symbols) and simulated (lines) (a) input and (b) output loci before (down triangles) and after (up triangles) applying open-short de-embedding procedure for a FinFET at $V_{GS} = 0.6$ V, $V_{DS} = 0.6$ V, $f_0 = 15$ GHz, and $P_{in} = -1.7$ dB m.

the extrinsic source/output loads are mathematically replaced by a predefined impedance for *S*-parameter calculations (i.e., 50 ohm).

This introductory chapter concludes with the hope that, now, readers have all the ingredients required to easily read and comprehend the following extensive chapters.

Acknowledgments

This work was supported by the project PON 01_01322 PANREX with financial support by Italian MIUR, the KU Leuven GOA-project, and FWO-Vlaanderen. The authors would like to thank Dr Bertrand Parvais, Dr Morin Dehan, Dr Abdelkarim Mercha, Dr Stefaan Decoutere, Dr Wojciech Wiatr, Dr Zlatica Marinković, Prof. Vera Marković, Prof. Jean-Pierre Raskin, Dr Gustavo Avolio, Dr Antonio Raffo, Prof. Giorgio Vannini, and Prof. Iltcho Angelov for their valuable support and useful discussions.

References

[1] Rohde UL, Rudolph M. RF/microwave circuit design for wireless applications. (Hoboken, NJ, USA): John Wiley & Sons; 2012.

[2] Rohde UL, Poddar AK, Böck G. The design of modern microwave oscillators for wireless applications: theory and optimization. (Hoboken, NJ, USA): John Wiley & Sons; 2005.

[3] Fager C, Root DE, Rudolph M. Nonlinear transistor model parameter extraction techniques. (Cambridge, UK): Cambridge University Press; 2011.

[4] Hillbrand H, Russer P. An efficient method for computer aided noise analysis of linear amplifier networks. IEEE Trans Circuits Syst April 1976;23(4):235–8.

[5] Biber CE, Schmatz ML, Morf T, Lott U, Morifuji E, Bachtold W. Technology independent degradation of minimum noise figure due to pad parasitics. In: IEEE MTT-S Int Microwave Symp Dig (Baltimore, MD, USA), June 1998. pp. 145–8.

[6] Pucel RA, Struble W, Hallgren R, Rohde UL. A general noise de-embedding procedure for packaged two-port linear active devices. IEEE Trans Microw Theory Tech November 1992;40(11):2013–24.

[7] Golio M. The RF and microwave handbook. (Boca Raton, FL, USA): CRC Press; 2001.

[8] Collantes J-M, Pollard RD, Sayed M. Effects of DUT mismatch on the noise figure characterization: a comparative analysis of two Y-factor techniques. IEEE Trans Instrum Meas December 2002;51(6):1150–6.

[9] Fundamentals of RF and Microwave Noise Figure Measurements. Agilent Application Note 57-1, 2010.

[10] Wiatr W, Crupi G, Caddemi A, Mercha A, Schreurs DMM-P. Source-pull characterization of FinFET noise. In: Int Conf Mixed Design Integr Circuits Syst (Wrocław, Poland), June 2010. pp. 425–30.

[11] Davidson AC, Leake BW, Strid E. Accuracy improvements in microwave noise parameter measurements. IEEE Trans Microw Theory Tech December 1989;37(12): 1973–8.

[12] O'Callaghan JM, Mondal JP. A vector approach for noise parameter fitting and selection of source admittance. IEEE Trans Microw Theory Tech August 1991;39(8): 1376–82.

[13] Boudiaf A, Laporte M, Dangla J, Vernet G. Accuracy improvements in two port noise parameter extraction method. In: IEEE MTT-S Int Microwave Symp Dig (Albuquerque, NM, USA), June 1992. pp. 1569–72.

[14] Caddemi A, Martines G, Sannino M. HEMT for low noise microwaves: CAD oriented modeling. IEEE Trans Microw Theory Tech July 1992;40(7):1441–5.

[15] Dambrine G, Happy H, Danneville F, Cappy A. A new method for on wafer noise measurement. IEEE Trans Microw Theory Tech March 1993;41(3):375–81.

[16] Tasker PJ, Reinert W, Hughes B, Braunstein J, Schlechtweg M. Transistor noise parameter extraction using a 50-Ω measurement system. In: IEEE MTT-S Int Microwave Symp Dig (Atlanta, GA, USA), June 1993. pp. 1251–4.

[17] Lazaro A, Pradell L, O'Callaghan JM. FET noise-parameter determination using a novel technique based on a 50-Ω noise-figure measurements. IEEE Trans Microw Theory Tech March 1999;47(3):315–24.

[18] Crupi G, Caddemi A, Schreurs DMM-P, Wiatr W, Mercha A. Microwave noise modeling of FinFETs. Solid State Electron February 2011;56(1):18–22.

[19] Van Moer W, Rolain Y. A large-signal network analyzer: why is it needed? IEEE Microw Mag December 2006;7(6):46–62.

[20] Verspecht J. Large-signal network analysis. IEEE Microw Mag December 2005;6(4): 82–92.

[21] Verspecht J, Debie P, Barel A, Martens L. Accurate on wafer measurement of phase and amplitude of the spectral components of incident and scattered voltage waves at the signal ports of a nonlinear microwave device. In: IEEE MTT-S Int Microwave Symp Dig (Orlando, FL, USA), May 1995. pp. 1029–32.

[22] Blockley P, Gunyan D, Scott JB. Mixer-based, vector-corrected, vector signal/network analyzer offering 300 kHz-20 GHz bandwidth and traceable phase response. In: IEEE MTT-S Int Microwave Symp Dig (Long Beach, CA, USA), June 2005. pp. 1497–500.

[23] Schreurs DMM-P. Applications of vector non-linear microwave measurements. IET Microw Antennas Propag April 2010;4(4):421–5.

[24] Pailloncy G, Avolio G, Myslinski M, Rolain Y, Vanden Bossche M, Schreurs DMM-P. Large-signal network analysis including the baseband. IEEE Microw Mag April 2011;12(2):77–86.

[25] Crupi G, Avolio G, Schreurs D, Pailloncy G, Caddemi A, Nauwelaers B. Vector two-tone measurements for validation of nonlinear microwave FinFET model. Microelectron Eng October 2010;87(10):2008–13.

[26] Ghione G, Naldi CU, Filicori F. Physical modeling of GaAs MESFETs in an integrated CAD environment: from device technology to microwave circuit performance. IEEE Trans Microw Theory Tech March 1989;37(3):457–68.

[27] Snowden CM, Pantoja RR. GaAs MESFET physical models for process orientated design. IEEE Trans Microw Theory Tech July 1992;40(7):1401–9.

[28] Baccarani G, Reggiani S. A compact double-gate MOSFET model comprising quantum-mechanical and non-static effects. IEEE Trans Electron Dev August 1999;46(8):1656–66.

[29] Denis D, Snowden CM, Hunter IC. Coupled electrothermal, electromagnetic, and physical modeling of microwave power FETs. IEEE Trans Microw Theory Tech June 2006;54(6):2465–70.

[30] Bonani F, Guerrieri SD, Ghione G. Physics-based simulation techniques for small- and large-signal device noise analysis in RF applications. IEEE Trans Electron Dev March 2003;50(3):633–44.

[31] Schreurs DMM-P, Wood J, Tufillaro N, Barford L, Root DE. Construction of behavioural models for microwave devices from time-domain large-signal measurements to speed-up high-level design simulations. Int J RF Microw Comput-Aided Eng January 2003;13(1):54–61.

[32] Pedro JC, Maas SA. A comparative overview of microwave and wireless power amplifier behavioral modeling approaches. IEEE Trans Microw Theory Tech April 2005;53(4):1150–63.

[33] Root DE, Verspecht J, Sharrit D, Wood J, Cognata A. Broad-band poly-harmonic distortion behavioral models from fast automated simulations and large-signal vectorial network measurements. IEEE Trans Microw Theory Tech November 2005;53(11):3656–64.

[34] Qi H, Benedikt J, Tasker PJ. Nonlinear data utilization: from direct data lookup to behavioral modeling. IEEE Trans Microw Theory Tech June 2009;57(6): 1425–32.

[35] Marinković Z, Crupi G, Schreurs DMM-P, Caddemi A, Marković V. Microwave FinFET modeling based on artificial neural networks including lossy silicon substrate. Microelectron Eng October 2011;88(10):3158–63.

[36] Currás-Francos MC, Tasker P, Fernandez-Barciela M, Campos-Roca Y, Sanchez E. Direct extraction of nonlinear FET Q–V functions from time domain large signal measurements. IEEE Microw Guided Wave Lett December 2000;10(12):531–3.

[37] Crupi G, Schreurs DMM-P, Xiao D, Caddemi A, Parvais B, Mercha A, et al. Determination and validation of new nonlinear FinFET model based on lookup tables. IEEE Microw Wirel Compon Lett May 2007;17(5):361–3.

[38] Dambrine G, Cappy A, Heliodore F, Playez E. A new method for determining the FET small-signal equivalent circuit. IEEE Trans Microw Theory Tech July 1988; 36(7):1151–9.

[39] Rorsman N, Garcia M, Karlsson C, Zirath H. Accurate small-signal modeling of HFET's for millimeter-wave applications. IEEE Trans Microw Theory Tech March 1996;44(3):432–7.

[40] Wood J, Root DE. Bias-dependent linear scalable millimeter-wave FET model. IEEE Trans Microw Theory Tech December 2000;48(12):2352–60.

[41] Crupi G, Xiao D, Schreurs DMM-P, Limiti E, Caddemi A, De Raedt W, et al. Accurate multibias equivalent circuit extraction for GaN HEMTs. IEEE Trans Microw Theory Tech October 2006;54(7):3616–22.

[42] Crupi G, Schreurs DMM-P, Raffo A, Caddemi A, Vannini G. A new millimeter wave small-signal modeling approach for pHEMTs accounting for the output conductance time delay. IEEE Trans Microw Theory Tech April 2008;56(4):741–6.

[43] Brady RG, Oxley HCH, Brazil TJ. An improved small-signal parameter-extraction algorithm for GaN HEMT devices. IEEE Trans Microw Theory Tech July 2008; 56(7):1535–44.

[44] Crupi G, Schreurs DMM-P, Caddemi A. On the small signal modeling of advanced microwave FETs: a comparative study. Int J RF Microw Comput-Aided Eng September 2008;18(5):417–25.
[45] Zarate-de Landa A, Zuniga-Juarez JE, Loo-Yau JR, Reynoso-Hernandez JA, Maya-Sanchez MC, del Valle-Padilla JL. Advances in linear modeling of microwave transistors. IEEE Microw Mag April 2009;10(2). pp. 100, 102–11, 146.
[46] Chen CH, Deen MJ. High frequency noise of MOSFETs I modeling. Solid State Electron November 1998;42(11):2069–81.
[47] Vandamme EP, Schreurs DMM-P, van Dinther C. Improved three-step de-embedding method to accurately account for the influence of pad parasitics in silicon on-wafer RF test structures. IEEE Trans Electron Dev April 2001;48(4):737–42.
[48] Jamal Deen M, Fjeldly TA. CMOS RF modeling, characterization and applications. (Singapore): World Scientific Publishing; 2002.
[49] Liang Q, Cressler JD, Guofu N, Yuan L, Freeman G, Ahlgren DC, et al. A simple four-port parasitic deembedding methodology for high-frequency scattering parameter and noise characterization of SiGe HBTs. IEEE Trans Microw Theory Tech November 2003;51(11):2165–74.
[50] Crupi G, Schreurs DMM-P, Caddemi A. Accurate silicon dummy structure model for nonlinear microwave FinFET modeling. Microelectron J September 2010;41(9): 574–8.
[51] Cidronali A, Collodi G, Vannini G, Santarelli A, Manes G. A new approach to FET model scaling and MMIC design based on electromagnetic analysis. IEEE Trans Microw Theory Tech June 1999;47(6):900–7.
[52] Rautio JC. Deembedding the effect of a local ground plane in electromagnetic analysis. IEEE Trans Microw Theory Tech February 2005;53(2):770–6.
[53] Resca D, Santarelli A, Raffo A, Cignani R, Vannini G, Filicori F, et al. Scalable nonlinear FET model based on a distributed parasitic network description. IEEE Trans Microw Theory Tech April 2008;56(4):755–66.
[54] Resca D, Raffo A, Santarelli A, Vannini G, Filicori F. Scalable equivalent circuit FET model for MMIC design identified through FW-EM analyses. IEEE Trans Microw Theory Tech February 2009;57(2):245–53.
[55] Pozar DM. Microwave engineering. (New York, NY, USA): John Wiley and Sons; 1998.
[56] Berroth M, Bosch R. High frequency equivalent circuit of GaAs FET's for large-signal applications. IEEE Trans Microw Theory Tech February 1991;39(2):224–9.
[57] Root DE. Analysis and exact solution of relaxation-time differential equations describing non quasi-static large-signal FET models. In: Eur Microwave Conf (Cannes, France), September 1994. pp. 854–9.
[58] Barciela MF, Tasker PJ, Campos-Roca Y, Demmler M, Massler H, Sanchez E, et al. A simplified broad-band large-signal nonquasi-static table-based FET model. IEEE Trans Microw Theory Tech March 2000;48(3):395–405.
[59] Santarelli A, Di Giacomo V, Raffo A, Filicori F, Vannini G, Aubry R, et al. Nonquasi-static large-signal model of GaN FETs through an equivalent voltage approach. Int J RF Microw Comput-Aided Eng June 2008;18(6):507–16.
[60] Crupi G, Schreurs DMM-P, Caddemi A, Raffo A, Vannini G. Investigation on the non-quasi-static effect implementation for millimeter-wave FET models. Int J RF Microw Comput-Aided Eng January 2010;20(1):87–93.

[61] Crupi G, Schreurs DMM-P, Caddemi A. Theoretical and experimental determination of onset and scaling of non-quasi-static phenomena for interdigitated FinFETs. IET Circuits Dev Sys November 2010;5(6):531–8.

[62] Root DE, Hughes B. Principles of nonlinear active device modeling for circuit simulation. In: ARFTG Conf Dig (Tempe, AZ, USA), December 1988. pp. 1–24.

[63] Jen SH-M, Enz CC, Pehlke DR, Schroter M, Sheu BJ. Accurate modeling and parameter extraction for MOS transistors valid up to 10 GHz. IEEE Trans Electron Dev November 1999;46(11):2217–27.

[64] Van der Ziel A. Gate noise in field effect transistors at moderately high frequencies. Proc IRE March 1963;51(3):461–7.

[65] Statz H, Haus HA, Pucel RA. Noise characteristics of gallium arsenide field-effect transistor. IEEE Trans Electron Dev September 1974;21(9):549–62.

[66] Cappy A. Noise modeling and measurement techniques. IEEE Trans Microw Theory Tech January 1988;36(1):1–10.

[67] Pospieszalski MW. Modeling of noise parameters of MESFET's and MODFET's and their frequency and temperature dependence. IEEE Trans Microw Theory Tech September 1989;37(9):1340–50.

[68] Dambrine G, Raskin J-P, Danneville F, Vanhoenaker-Janvier D, Colinge J-P, Cappy A. High-frequency four noise parameters of silicon-on-insulator-based technology MOSFET for the design of low-noise RF integrated circuits. IEEE Trans Electron Dev August 1999;46(8):1733–41.

[69] Danneville F. Microwave noise and FET devices. IEEE Microw Mag October 2010;11(6):53–60.

[70] Felgentreff T, Olbrich G, Russer P. Noise parameter modeling of HEMTs with resistor temperature noise sources. In: IEEE MTT-S Int Microwave Symp Dig (San Diego, CA, USA), May 1994. pp. 853–6.

[71] Pascht A, Grözing M, Wiegner D, Berroth M. Small-signal and temperature noise model for MOSFETs. IEEE Trans Microw Theory Tech August 2002;50(8): 1927–34.

[72] Root DE, Fan S, Meyer J. Technology independent large signal non quasistatic FET models by direct construction from automatically characterized device data. In: Eur Microwave Conf (Stuttgart, Germany), September 1991. pp. 927–32.

[73] Schreurs DMM-P, Verspecht J, Vandenberghe S, Vandamme E. Straightforward and accurate nonlinear device model parameter-estimation method based on vectorial large-signal measurements. IEEE Trans Microw Theory Tech October 2002;50(10): 2315–9.

[74] Myslinski M, Schreurs DMM-P, Wiatr W. Development and verification of a nonlinear lookup table model for RF Silicon BJTs. In: Eur GaAs Other Semiconductors Appl Symp (Milano, Italy), September 2002. pp. 93–6.

[75] Wood J, Aaen PH, Bridges D, Lamey D, Guyonnet M, Chan DS, et al. A nonlinear electro-thermal scalable model for high-power RF LDMOS transistors. IEEE Trans Microw Theory Tech February 2009;57(2):282–92.

[76] Aaen PH, Plà JA, Wood J. Modeling and characterization of RF and microwave power FETs. (Cambridge, UK): Cambridge University Press; 2007.

[77] Crupi G, Schreurs DMM-P, Caddemi A, Angelov I, Homayouni M, Raffo A, et al. Purely analytical extraction of an improved nonlinear FinFET model including non-quasi-static effects. Microelectron Eng November 2009;86(11):2283–9.

[78] Homayouni M, Schreurs DMM-P, Crupi G, Nauwelaers B. Technology independent non-quasi-static table-based nonlinear model generation. IEEE Trans Microw Theory Tech December 2009;57(12):2845–52.

[79] Camacho-Penalosa C. Modeling frequency dependence of output impedance of a microwave MESFET at low frequencies. Electron Lett June 1985;21(12):528–9.

[80] Roff C, Benedikt J, Tasker PJ, Wallis DJ, Hilton KP, Maclean JO, et al. Analysis of DC-RF dispersion in AlGaN/GaN HFETs using RF waveform engineering. IEEE Trans Electron Dev January 2009;56(4):13–9.

[81] Raffo A, Vadalà V, Schreurs DMM-P, Crupi G, Avolio G, Caddemi A, et al. Nonlinear dispersive modeling of electron devices oriented to GaN power amplifier design. IEEE Trans Microw Theory Tech April 2010;58(4):710–8.

[82] Crupi G, Raffo A, Schreurs DMM-P, Avolio G, Vadalà V, Di Falco S, et al. Accurate GaN HEMT non-quasi-static large-signal model including dispersive effects. Microw Opt Tech Lett March 2011;53(3):692–7.

[83] Filicori F, Vannini G, Santarelli A, Sanchez AM, Tazon A, Newport Y. Empirical modeling of low-frequency dispersive effects due to traps and thermal phenomena in III-V FET's. IEEE Trans Microw Theory Tech December 1995;43(12):2972–81.

[84] Angelov I, Zirath H, Rorsman N. A new empirical nonlinear model for HEMT and MESFET devices. IEEE Trans Microw Theory Tech December 1992;40(12): 2258–66.

[85] Angelov I, Bengtsson L, Garcia M. Extensions of the Chalmers nonlinear HEMT and MESFET model. IEEE Trans Microw Theory Tech October 1996;44(10):1664–74.

[86] Siligaris A, Dambrine G, Schreurs DMM-P, Danneville F. A new empirical nonlinear model for sub-250 nm channel MOSFET. IEEE Microw Wirel Compon Lett October 2003;13(10):449–51.

[87] Avolio G, Schreurs DMM-P, Raffo A, Crupi G, Angelov I, Vannini G, et al. Identification technique of FET model based on vector nonlinear measurements. Electron Lett November 2011;47(24):1323–4.

[88] Vandamme EP, Schreurs DMM-P, van Dinther C, Badenes G, Deferm L. Development of a RF large signal MOSFET model, based on an equivalent circuit, and comparison with BSIM3v3 compact model. Solid State Electron March 2002;46(3): 353–60.

[89] Currás-Francos MC. Table-based nonlinear HEMT model extracted from time-domain large-signal measurements. IEEE Trans Microw Theory Tech May 2005; 53(5):1593–600.

[90] Orzati A, Schreurs DMM-P, Pergola L, Benedickter H, Robin F, Homan OJ, et al. A 110-GHz large-signal lookup-table model for InP HEMTs including impact ionization effects. IEEE Trans Microw Theory Tech February 2003;51(2):468–74.

[91] Crupi G, Raffo A, Sivverini G, Bosi G, Avolio G, Schreurs DMM-P, et al. Non-linear look-up table modeling of GaAs HEMTs for mixer application. In: IEEE Int Workshop on Nonlinear Microwave and Millimeter Wave Integr Circuits (Dublin, Ireland), September 2012. pp. 1–3.

[92] Elmore G. De-embedded measurements using the 8510 microwave network analyzer. In: Hewlett-Packard RF & Microwave Symp (Palo Alto, CA, USA), August 1985.

[93] Williams D. De-embedding and unterminating microwave fixtures with nonlinear least squares. IEEE Trans Microw Theory Tech June 1990;38(6):787–91.

[94] Embedding/de-embedding, Anritsu Application Note 11410-00278, 2001.

[95] Three and four port S-parameter measurements. Anritsu Application Note 11410-00279, 2002.

[96] De-embedding and embedding S-parameter networks using a vector network analyzer. Agilent Application Note 1364-1, 2004.

[97] Cripps SC. RF power amplifiers for wireless communication. (Norwood, MA, USA): Artech House; 1999.

[98] Maas SA. Nonlinear microwave and RF circuits. Norwood, MA: Artech House; 2003.

[99] Colantonio P, Giannini F, Limiti E. High efficiency RF and microwave solid state power amplifiers. (Chichester, UK): John Wiley & Sons; 2009.

[100] Tasker PJ. Practical waveform engineering. IEEE Microw Mag December 2009;10(7):65−76.

[101] Raffo A, Avolio G, Schreurs DMM-P, Di Falco S, Vadalà V, Scappaviva F, et al. On the evaluation of the high-frequency load line in active devices. Int J Microw Wirel Tech February 2011;3(1):19−24.

[102] Avolio G, Schreurs DMM-P, Raffo A, Crupi G, Vannini G, Nauwelaers B. Waveforms only based nonlinear de-embedding in active devices. IEEE Microw Wirel Compon Lett April 2012;22(4):215−7.

[103] Colinge J-P. Multiple-gate SOI MOSFETs. Microelectron Eng September 2007;84(9/10):2071−6.

[104] Hisamoto D, Lee WC, Kedzierski J, Anderson E, Takeuchi H, Asano K, et al. A folded-channel MOSFET for deep-sub-tenth micron era. In: Int Electron Devices Meeting Tech Dig (San Francisco, CA, USA), December 1998. pp. 1032−4.

[105] Huang X, Lee WC, Kuo C, Hisamoto D, Chang L, Kedzierski J, et al. Sub 50-nm FinFet: PMOS. In: Int Electron Devices Meet Tech Dig (Washington, DC, USA), December 1999. pp. 67−70.

[106] Crupi G, Schreurs DMM-P, Raskin J-P, Caddemi A. A comprehensive review on microwave FinFET modeling for progressing beyond the state of art. Solid State Electron February 2013;80(2):81−95.

[107] Bohr M, Mistry K. Intel's Revolutionary 22 nm Transistor Technology. Intel website; May 2011.

[108] Nawaz M, Decker S, Giles L-F, Molzer W, Schulz T. Evaluation of process parameter space of bulk FinFETs using 3D TCAD. Microelectron Eng July 2008;85(7):1529−39.

[109] de Andrade MGC, Martino JA, Aoulaiche M, Collaert N, Simoen E, Claeys C. Behavior of triple-gate bulk FinFETs with and without DTMOS operation. Solid State Electron May 2012;71:63−8.

[110] Crupi G, Schreurs DMM-P, Parvais B, Caddemi A, Mercha A, Decoutere S. Scalable and multibias high frequency modeling of multi fin FETs. Solid State Electron November/December 2006;50(10/11):1780−6.

CHAPTER

2 Millimeter-Wave Characterization of Silicon Devices under Small-Signal Regime: Instruments and Measurement Methodologies

Gilles Dambrine

IEMN, CNRS, University of Lille, France

2.1 Preliminary concepts

2.1.1 Advanced silicon technologies and mm-wave applications

The nanoelectronics industry will impact modern society in a wide range of applications: computing, communications, health care, security and defense, and environmental monitoring. In 2024, the technological node for metal oxide semiconductor field effect transistors (MOSFETs) is predicted to be 8 nm (currently 28 nm) and the cutoff frequency will be in the order of the THz (currently 0.35 THz) [1]. In the same manner, silicon germanium-based heterojunction bipolar transistors (HBTs, under BiCMOS) offer the possibility to build circuits and subsystems to compete with optical solutions in the submillimeter wave range [2].

The spectrum above 100 GHz, the so-called millimeter (<300 GHz) and submillimeter (300 GHz up to 3 THz) wave ranges, has been envied for several years. The III–V based technologies opened the way to address military and spatial applications. The performances of the silicon technologies, planned also for a long time, are today able to answer (technically and doubtlessly economically) the large-scale applications above 100 GHz (75–110 GHz: the W-band; 90–140 GHz: the D-band; 140–220 GHz: the G-band; 220–325 GHz: the H-band; 325–500 GHz: the Y-band, etc.). Contrary to centimeter waves, these frequency ranges are mainly free and their propagation properties are well suited for even large-scale civil applications, such as very short distance wireless ultra-high-speed communications, security, and health care (e.g., remote sensing systems, passive and active imaging, biological spectroscopy), among others.

However, at present, most compact models of silicon devices used for the design of circuits and systems are validated in limited frequency ranges. Generally, for small signal regimes, calculated scattering parameters (so-called scattering [*S*] parameters; see Section 2.1.2), and even noise figures (see Section 2.1.3) are compared with measurements up to 110 GHz or less.

Microwave De-embedding. http://dx.doi.org/10.1016/B978-0-12-401700-9.00002-1

Today, on-wafer probing systems are able to measure *S* parameters up to 700 GHz [3]. The challenge is now to maintain the quality and accuracy of measurements in the millimeter (mm) and submillimeter wave ranges. This task is much more challenging for advanced silicon devices, using characterizations that are useful for the validation of the electrical model at such frequencies.

2.1.2 *S*-parameter measurements: basics

The concept and associated calculation and measurement procedures of alternating current (AC) and voltages are commonly used to design and to measure low-frequency circuits. This concept becomes limited when the frequency increases while phase and magnitude of currents and voltages become not only time (or frequency) dependent but also spatially dependent, even along a uniform lossless conductor. On technicians' and engineers' experimental tables, AC multimeters and impedance meters are replaced by vectorial network analyzers (VNAs). A VNA measures (generally in the frequency domain) the phase and magnitude of wave quantities and it delivers *S* parameters.

The first report on *S* parameters appears to be an article concerning an electrical circuit (ideal transformer with resistive loads) [4]. Several definitions of *S* parameters have ensued, based on the principles of waveguide and microwave circuit theories and their analogies [5—11]. An important work [12] mentioned that such analogy between waveguide and electrical circuit can entail shortcuts and can be the cause of errors. A waveguide-based circuit is described in terms of travelling waves using a matrix based on wave quantity ratios. A linear electrical network is described by impedance (or admittance/chain) matrix characterized by current and voltage ratios. The coefficients of such matrices quantify the linear relations between input and output quantities. To define these coefficients related to a given node, we must simultaneously load the other nodes by specific ideal terminations. For instance, the coefficients of the impedance matrix of a given node are calculated by "opening" the others (ideal load impedance of infinite value). In the case of *S* parameters, the coefficients are defined by using ideal reflection-free terminations. These specific terminations make the reference (normalization) to ensure that the coefficients of a given matrix characterize the linear properties of the network independently of its output environment. For the definition of *S* parameters, this notion of reflection-free termination is generally associated with the theoretical characteristic impedance of the propagating mode, which is not easy to define or measure. For instance, in the case of a rectangular waveguide VNA setup, it is commonly said that the VNA is calibrated under 50 Ω and the propagating single mode in such a waveguide is TE10, where the wave impedance is frequency dependent [13]. Moreover, the commonly used analogy between a waveguide transmitting a single mode and a R, L, C, and G distributed equivalent electrical circuit permits a mathematical calculation of the *S*-parameter matrix from an impedance matrix. In this case, the *S* parameters are normalized by introducing an arbitrary reference impedance.

From an experimental point of view, the notion of an electrical circuit is well suited and practical for many cases of *S*-parameter measurements and associated calibration procedures, especially in the case of on-wafer characterizations. However, users of VNAs must keep in mind all the causes of errors engendered by this analogy and its domain of validity.

In the case of a two-port, the S-parameter matrix is defined as the linear relationship between reflective wave quantities (b_i) versus the incident ones (a_i) (Eqn (2.1)):

$$\begin{aligned} b_1 &= S_{11}a_1 + S_{12}a_2 \\ b_2 &= S_{21}a_1 + S_{22}a_2 \\ &\text{or} \\ \begin{bmatrix} b_1 \\ b_2 \end{bmatrix} &= \begin{bmatrix} S_{11} & S_{12} \\ S_{21} & S_{22} \end{bmatrix} \begin{bmatrix} a_1 \\ a_2 \end{bmatrix} \end{aligned} \tag{2.1}$$

$|a_i|^2$ and $|b_i|^2$ are respectively the power values of the incident and reflective waves; the unity is the watt. S_{11} and S_{22} are called *reflection coefficients*, while S_{12} and S_{21} are the *transmission coefficients*. The indices 1 and 2 correspond respectively to Port 1 (input) and Port 2 (output) of the two-port.

The S_{ii} and S_{ij} parameters are defined in two ways: the forward path and reverse one (see Figure 2.1). For the forward path, the incident wave a_1 stimulates Port 1, whereas Port 2 is loaded by a reflection-free termination ($a_2 = 0$); S_{11} and S_{21} are determined using Eqn (2.2). This definition can be extended to multiport concepts.

$$S_{11} = \left.\frac{b_1}{a_1}\right|_{a_2=0} ; \quad S_{21} = \left.\frac{b_2}{a_1}\right|_{a_2=0} \tag{2.2}$$

For the reverse path, the incident wave a_2 stimulates Port 2, whereas Port 1 is loaded by a reflection-free termination ($a_1 = 0$); S_{12} and S_{22} are determined using Eqn (2.3).

$$S_{12} = \left.\frac{b_1}{a_2}\right|_{a_1=0} ; \quad S_{22} = \left.\frac{b_2}{a_2}\right|_{a_1=0} \tag{2.3}$$

The instruments able to measure S parameters are classified into two categories, for which the principles are completely different:

1. The homodyne-based system uses a six-port reflectometer for reflection coefficient measurement and a six-port network analyzer measuring both reflection and transmission coefficients. The six-port reflectometer was developed by

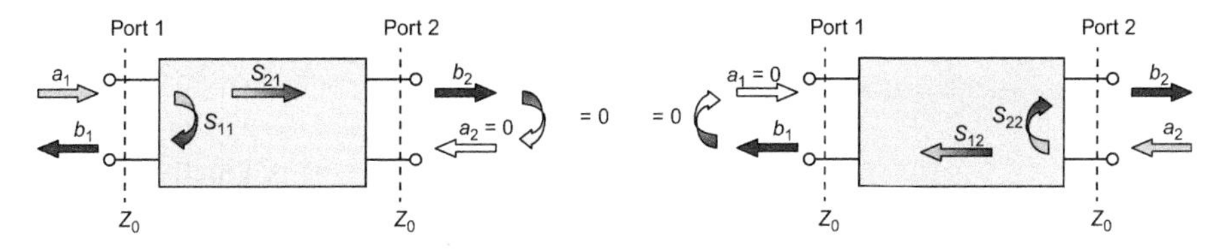

FIGURE 2.1

Schematic of a two-port under forward and reverse wave paths.

Glenn F. Engen [14] and his colleagues at the National Bureau of Standards (which today is called National Institute of Standards and Technology) in Boulder, Colorado, USA, starting in the early 1970s. This instrument measures only the power of waves, which can be traceable with primary standards; it does not need phase-locked signal sources.

2. The heterodyne-based system uses four-port reflectometer(s), heterodyne receivers, and a phase-locked source (or multiple synchronized phase-locked sources). The earliest automatic system was developed by K. Rytting Douglas [15] and commercialized by ® Hewlett-Packard in the 1960s. Currently, most of the commercially available VNAs are based on this principle.

The basic architecture of a heterodyne-based VNA is depicted in Figure 2.2. The principles and functionalities of such an instrument are detailed in the literature [16–18].

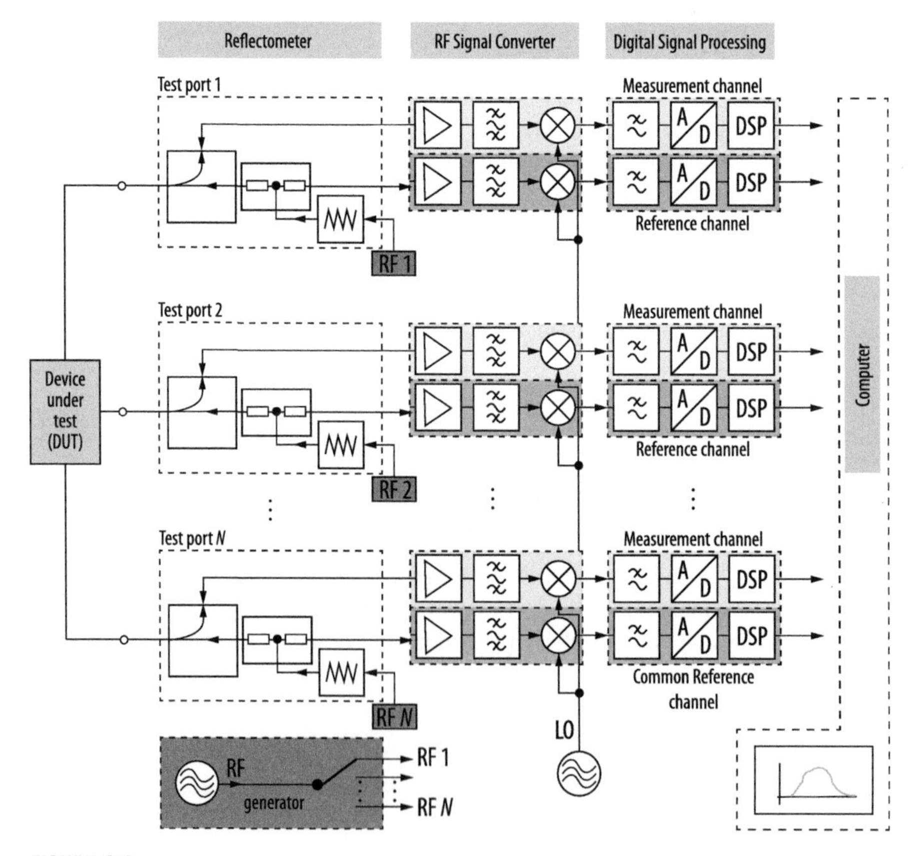

FIGURE 2.2

Basic architecture of a heterodyne-based VNA.

The *test set* separates the incident and reflected waves at the test port. Although many microwave devices (e.g., directional couplers, bridges, power splitters) constitute this test set, it can be represented as a four-port reflectometer.

The *generator* provides the radio frequency (RF) signal; one part of the signal flows to the active test port while the other flows to the reference channel. A source switch is commonly used to pass the stimulus signal to each of the test ports. The frequency, magnitude, and phase of the stimulus signal can be accurately tuned and controlled. For multiple-port VNAs, two synchronized generators are used.

Two receivers per test port are necessary to convert RF signals into intermediary frequency (IF) signals: one is used to downconvert the measured signal flowing from the device under test (DUT), while the other converts the reference one related to the stimulus signal. These receivers involve a low noise amplifier (LNA), RF band-pass filter, mixer, IF low-pass filter, analog/digital converter, and digital signal processor. A local oscillator (LO) is necessary to convert an RF signal into an IF signal. The LO and RF signals are synchronized; several phase locked loop-based architectures (involving a crystal-stabilized frequency reference) can be used. The dynamic range of the instrument depends mainly on the properties of the receiver and its frequency range. The sensitivity is related to the noise level and the IF bandwidth setting. Due to improvements in high-frequency, mixed-mode, and digital circuit technologies, modern VNAs are very sensitive; they can even be used for noise measurements. For instance, up to 26.5 GHz coaxial-based VNAs present a noise floor that is lower than −110 dBm for a 10 Hz resolution IF bandwidth.

The principle and procedures of VNA calibration consist of mathematically extracting the *S* parameters of the DUT from the measured data. In other words, one has to remove from the measured quantities all the parts that are not related to the DUT. This added quantity comes from all the instrument elements themselves (so-called systematic errors), as well as from stochastic (e.g., noise, uncontrolled system variations) and drift causes (e.g., environment conditions such as temperature, humidity, or degradation processes). "Mathematically" means that one has to model the instrument and its environment of use. The solver algorithms correspond mainly to the fields of linear equations or matrix analysis and sometimes of statistics. Nowadays, these solvers are integrated into VNA or specialized software (e.g., on-wafer calibration procedures).

With a minimum of training, nonexpert users are now able to measure accurate *S* parameters at least up to 50–60 GHz using modern instruments and suited calibration tools. Nevertheless, it is important to keep in mind that these commonly used VNA calibration procedures are based on straightforward assumptions:

1. The model used for calibration treats most of the imperfections of the overall measurement setup (systematic errors), not the stochastic causes.
2. This model has to remain constant during all steps of the calibration procedure as well as during the measurement of the DUT.

3. Although the parameters of the model are numerous, well-defined (or well-known) input values (standards) have to be fixed to solve the mathematical system of equations; the final solution depends directly on these input data.

The numerous kinds of calibration procedures are based on two categories of models. The first category is based on the physical properties of a reflectometer-based network analyzer. In the case of two reflectometers (two-port reflection and transmission measurements), the model, which was proposed for the first time in 1970s [19], involves 12 terms (vectors), as presented in Figure 2.3. Each vector (except for isolation vectors) is assumed to represent the main characteristics of the two reflectometers. For instance, in the case of the reflection coefficient measurement scheme in forward path, the vector modeling the directivity (E_{FD}) is added to the signal reflected by the DUT through the coupling path of the reflectometer (Meas1). Uncertainty for the determination of this term will impact the accuracy of S_{11A} and will be a limitation for the measurement of reflection coefficient of low magnitude. E_{FS} is assumed to model the source mismatch. This term makes a reverse loop with S_{11A} on the flow graph. The resulting signal, of traveling wave between the DUT (S_{11A}) and the source (E_{FD}), attenuated by the reflection tracking term (E_{FRT}), will flow also to port Meas1. This phenomenon will be more important for the high-magnitude reflection coefficient.

If we omit the two isolation vectors (E_{FX} and E_{RX}), the linear equations relating the S parameters of the DUT and measured quantities (a_1, b_1, a_2, b_2) can be solved by using at least six reflective one-port standards: three connected at Port 1 and three others at Port 2, as well as one two-port standard connected between Port 1 and Port 2.

The so-called short-open-load-thru (SOLT, or sometimes OSLT) calibration method is established on this 12-term error model. After calibration, the accuracy of reference vectorial planes definition (reflection and transmission) directly depends on the accuracy of knowledge of the overall standards in the whole frequency range. In reflection, the reference that fixes the magnitude and phase values of S_{11} and S_{22} is mainly defined by the reflective standards (SOL). In transmission, the thru standard is the key element.

The second category of calibration algorithms is based on a pure mathematical approach associated with eight [20] or seven [21] error term-based models in the case of two-port measurements. At least two distinct two-port standards are necessary to solve the mathematical problem. A third two-port standard is generally added. As depicted in Figure 2.4, the error model involves two two-ports determined by their respective [S] matrices as noted (A and B in [20]; G and H in [21]). This model assumes that the two test sets are alternately active. The notion of a forward and reverse path-based model (SOLT) is not required, but this model is suited for modern VNAs with at least four receivers. It is usually referred to as VNA self-calibration techniques.

Thru-Reflect-Line Method—The first calibration technique published in literature [20] is the thru-reflect-line (TRL) method. The *thru* standard, which corresponds to a

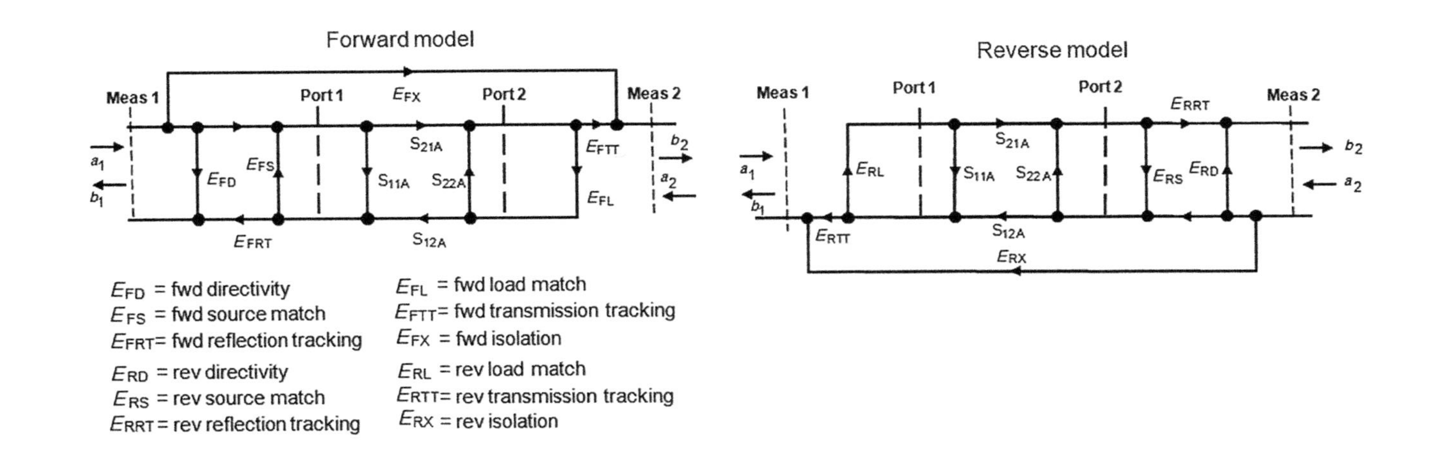

FIGURE 2.3

The 12-term based error model for forward and reverse paths in a two-port *S*-parameter calibration procedure.

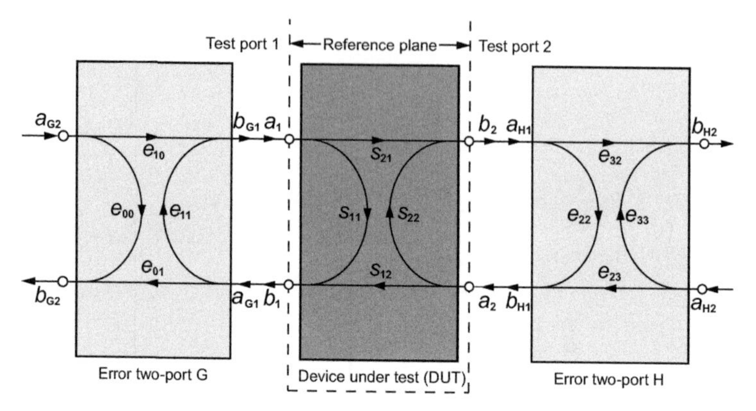

FIGURE 2.4

Synopsis of an eight or seven-term error model.

direct lossless connection between the two DUT ports, is assumed to be perfectly and fully known. Its [S_{Thru}] matrix is given in Eqn (2.4). The *line* standard is assumed to be partially known. S_{11Line} and S_{22Line} are fixed to 0 (see Eqn (2.5)). In other words, the travelling wave properties of the line have to be strictly similar to the thru properties. We also assume that the connection of any kind of two-port (thru, line, reflect, DUT) between the measurement planes does not change the travelling properties of the overall measurement system. Moreover, while these two standard matrices have to be distinct to avoid underdetermined solving procedures, the length of the line is generally chosen to be equal to quarter of the wave length at the frequency corresponding to the middle of the frequency range. In practice, only 60–80% of this frequency range is usable. This method is then by concept limited in terms of bandwidth. By using a set of several lines with different lengths (but with same $S_{11Lines}$ and $S_{22Lines}$), a broadband TRL-like calibration procedure is proposed in [22]. Finally, the *reflect* standard is assumed to be fully unknown, but $S_{11Reflect}$ has to be strictly equal to $S_{22Reflect}$ (noted as ρ in Eqn (2.6)); moreover, the magnitude of these reflection coefficients has to be distinct compared to the thru and line magnitudes. This standard allows one to overcome a phase indecision (0 or PI) of the roots of a quadratic equation system. When one converts S parameters into Z parameters, the reference impedance is related to the reflection coefficient of the thru and line standards.

From the metrology point of view, this technique makes sense because a line supporting a single traveling wave mode can be realistic and lead to accurate modeling in coaxial and even rectangular waveguide technologies.

$$[S_{Thru}] = \begin{bmatrix} 0 & e^{-\gamma \cdot 0} = 1 \\ e^{-\gamma \cdot 0} = 1 & 0 \end{bmatrix} \tag{2.4}$$

$$[S_{\text{Line}}] = \begin{bmatrix} 0 & e^{-\gamma l} \\ e^{-\gamma l} & 0 \end{bmatrix} \tag{2.5}$$

$$[S_{\text{Reflect}}] = \begin{bmatrix} \rho & x \\ x & \rho \end{bmatrix} \tag{2.6}$$

Thru-Reflect-Match Method—The thru-reflect-match (TRM) calibration procedure, which was introduced in 1988 [23], is similar to the TRL technique, although the line standard is replaced by the match standard. The match standard can be considered ideally as a nonreflective line having an infinite length. The reference impedance of Z parameters, which are needed to convert *S* parameters, is fixed by the reflection coefficient of the match standard. The main advantages of this technique are: (1) from a theoretical point of view, there is no frequency limitation, and therefore this technique is well suited for broadband *S*-parameter measurements; and (2) from an experimental point of view, similarly to the SOLT technique, the standards can be realized with "lumped" elements. A lumped element is assumed to have dimensions much smaller than the wavelength (typically $<\lambda_g/20$); its characteristics can be accurately modeled using the basic current–voltage concept and measured under static or quasi-static regimes.

This calibration technique and its derivative ones (e.g., line-reflect-match [LRM] and line-reflect-reflect-match [LRRM] [24]) are now largely used for on-wafer *S*-parameter measurements.

All of these calibration algorithms and associated procedures are available for users and integrated within VNA software or other specialized commercially available software, such as that dedicated for on-wafer measurements [25,26]. Nevertheless, users must have in mind the definition of the calibration procedures and *S*-parameters previously detailed, as well as their domains of validity and/or applicability. Moreover, they can modify some of the standard definitions, but they have to check if such a modification relates best to these concepts and the experimental setups. These usual precautions become indispensable for on-wafer *S*-parameter measurements, especially in the mm-wave range (see Section 2.2).

2.1.3 High-frequency noise measurements: basics

The high-frequency (HF) noise signal is naturally present in all passive and active components, circuits, and systems. This signal is stochastic and superposed to the other coherent ones (DC and AC). The causes of HF noise are multiple and are related to fundamental incoherent electronic mechanisms (thermal Brownian agitation [27,28], nonequilibrium stochastic electronic conduction/diffusion mechanisms [29], or thermal or tunneling assisted conduction mechanisms through a thin barrier of potential [30], quantum noise). This type of noise source is characterized to be quasi-frequency independent (at least up to few 100 GHz); therefore, such causes are called white-spectrum sources.

A noise source is quantified by its spectrum intensity. From the mathematical point of view, this spectrum intensity is obtained by the Fourier transformation of the autocorrelation function of the time-dependent noise signal (Wiener–Khinchin–Einstein theorem). For instance, the noise voltage spectrum intensity generated by a pure resistance is written as:

$$\overline{\delta v^2} = 4kT_{\text{amb}}R \tag{2.7}$$

where k is the Boltzmann constant, R is the resistance value, and T_{amb} is the ambient temperature.

The noise power is the mean value of the spectrum intensity in a defined bandwidth Δf (see Eqn (2.8)).

$$\overline{v^2} = 4kT_{\text{amb}}R\Delta f \tag{2.8}$$

For the calculation of noise properties in electrical circuits, we can use the same voltage–current concept as that used for linear AC calculations. Figure 2.5 represents a noisy linear two-port that can be fully described by its four linear electrical parameters (e.g., ABCD parameters) and two correlated noise sources. An important figure of merit to quantify the noise performance of a two-port is the noise figure (NF), which is generally given in decibels (dB). The noise figure translates the signal-to-noise ratio degradation through the two-port when the equivalent noise temperature of generator is fixed at 290 K (see Eqn (2.9)).

$$\text{NF(dB)} = 10\log\left(\frac{\text{Signal}_{\text{in}}/\text{Noise}_{\text{in}}}{\text{Signal}_{\text{out}}/\text{Noise}_{\text{out}}}\right) \quad \text{where Noise}_{\text{in}} \text{ is related to 290 K} \tag{2.9}$$

For noise circuit calculation using classical Kirchhoff rules, we first attribute an arbitrary phase to voltage and current noise sources: the voltage source arrows are oriented to the outside of the two-port while the current source arrows are oriented downward. Then, the noise properties of the two-port are completely described by four parameters (two real ones and one complex). We also note an efficient method to calculate the noise properties of a linear electrical network—the correlation matrix approach proposed in [31]. The noise matrix in chain representation of a noisy two-port (see Figure 2.5) is detailed in Eqn (2.10). The two noise sources and their cross-correlation vector are related to F_{min} (minimum noise figure

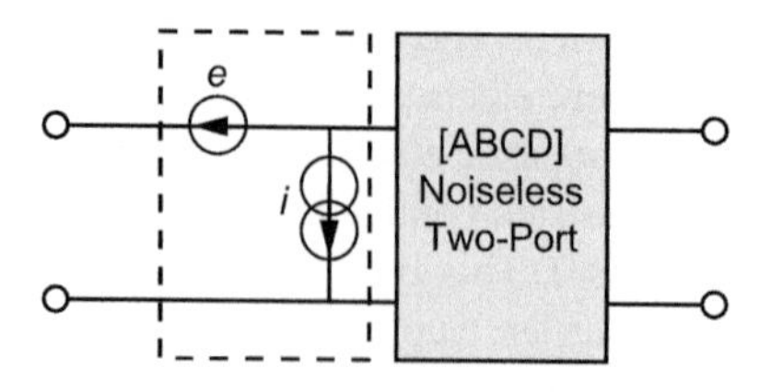

FIGURE 2.5

Electrical description of noisy two-port and its associated equivalent noise sources.

in linear units), R_n (equivalent noise resistance), and Y_{opt} (optimum generator admittance to reach F_{min}), which are the most commonly used noise parameters by designers and experimenters.

$$[C_A] = \begin{bmatrix} \overline{e^2} & \overline{ei^*} \\ \overline{e^*i} & \overline{i^2} \end{bmatrix} = 4kT_0\Delta f \begin{bmatrix} R_n & \frac{F_{min}-1}{2} - R_n Y_{opt}^* \\ \frac{F_{min}-1}{2} - R_n Y_{opt} & R_n \left| Y_{opt} \right|^2 \end{bmatrix} \tag{2.10}$$

where $T_0 = 290$ K.

From the experimental side, the intensity of HF noise is very low compared to coherent signals and necessitates highly sensitive instruments for its measurement. For instance, the thermal noise available power generated by pure resistance is close to -113.9 dBm at 293.5 K (20°C) for 1 MHz bandwidth. High-frequency noise measurements are generally carried out in the frequency domain either by using a dedicated instrument such as noise figure meter (NFM) or conventional spectrum analyzers. In any case, experimenters must have in mind that noise measurement is uniquely related to a power measurement. Then, one measures a mean value of the noise power in a given frequency bandwidth. Even if it is often used in the usual language, it is not rigorous to say "We measure a noise figure (figure of merit)". It is the same for the other usual parameters such as F_{min}, R_n, and Y_{opt} (only parameters of noise modeling). For noise power calculation or measurement, the notions of available power and available gain are generally preferred. The available power of a noise source (or generator) is defined as the power delivered to a noiseless ($T_{noise} = 0$ K) load, in which impedance is equal to the conjugated source impedance. In the case of a passive one-port at ambient temperature T_{amb} (e.g., thermal noise source), the available noise power is equal to $k\,T_{amb}\,\Delta f$. The available gain of a two-port is defined as:

$$G_{av\_DUT} = \frac{\left(1-\left|\Gamma_g\right|^2\right)\left|S_{21}\right|^2}{\left|1-\Gamma_g S_{11}\right|^2\left(1-\left|\Gamma_{out}\right|^2\right)} \quad \text{with } \Gamma_{out} = S_{22} + \frac{S_{12}S_{21}\Gamma_g}{1-S_{11}\Gamma_g} \tag{2.11}$$

where, S_{12}, S_{21} and S_{22} are the S parameters of the two-port and Γ_g is the reflection coefficient of the generator.

For the noise characterization of a two-port, we connect a noise generator at the two-port input and a noise power measurement instrument at the output. A simple description of the setup is represented in Figure 2.6.

The available noise power at the two-port output can be expressed as follows:

$$P_{av\_out} = kT_{in}G_{av\_DUT}\Delta f + N_{add\_out} \tag{2.12}$$

where N_{add_out} is the output available noise power added by the noisy two-port. T_{in} is the input equivalent noise temperature (noise source).

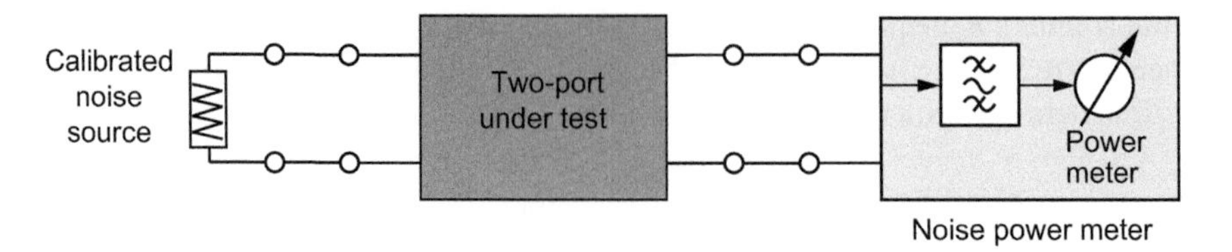

FIGURE 2.6

Basic synopsis of two-port noise measurement setup.

The actual power delivered to the instrument and corresponding to the actual measurement can be expressed as:

$$P_{meas} = P_{av\_out}\frac{\left(1-|\Gamma_{out}|^2\right)\left(1-|\Gamma_L|^2\right)}{|1-\Gamma_{out}\Gamma_L|^2} \tag{2.13}$$

From Eqns (2.11–2.13), we can note that the measured power is linearly dependent from the noise temperature of the generator T_{in}. If Γ_{out} and Γ_L are known (or measured), we can extract N_{add_out} (linear extrapolated value at $T_{in}=0$ K) and $kG_{av\_DUT}\Delta f$ (slope of the linear extrapolation) from two power measurements corresponding to two known distinct values of T_{in} (generally noted as T_{cold} and T_{hot}). By considering the initial definition (Eqn (2.9)) and Eqn (2.12), the noise figure can be expressed as follows:

$$F_{DUT} = 1+\frac{N_{add\_out}}{kG_{av\_DUT}T_0\Delta f} \quad \text{with } T_0 = 290\text{ K} \tag{2.14}$$

The first method to extract the noise figure from noise power measurements is called the Y-factor method. The Y factor is the ratio of two output available powers, noted as N_1 and N_2, corresponding respectively to T_{cold} and T_{hot}. The noise figure is expressed as a function of the Y factor, T_0, T_{cold}, and T_{hot} as follows:

$$F_{DUT} = \frac{(T_{hot}-T_{cold})}{T_0(Y-1)} \quad \text{with } Y = \frac{N_2}{N_1} \tag{2.15}$$

In Eqn (2.15), the term $\frac{(T_{hot}-T_{cold})}{T_0}$ is called the *excess noise ratio* of the noise generator.

Let us come back now to the actual measured noise power P_{meas} (Eqn (2.13)) and discuss the measurement conditions and requirements in terms of the power meter (or NFM, spectrum or receiver) sensitivity. First of all, the full chain of measurement has to be operated under the linear condition. If the gain of the chain is modified due to the DUT or to the power meter, the measured power has to be affected by the same quantity. This gain of chain may be very high and sensitive to nonlinearity. For instance, if we want to measure an output matched ($\Gamma_{out}=0$) DUT of NF = 5 dB and $G_{av}=20$ dB with a conventional 50 Ω ($\Gamma_L=0$) power meter (bolometer) with −40 dBm of sensitivity, preceded by a 100 MHz bandpass lossless filter and a hot/cold (300 K/77 K) 50 Ω ($\Gamma_g = 0$) noise source, we need to insert in the chain RF and/or baseband (or IF in case of mixer) amplifiers with a minimum total gain close to 30 dB (preferably adjustable up to 40 dB to ensure the accuracy of measurement). Such high gain can easily engender some nonlinearity if the DUT gain and/or the noise figure increases. For that reason, it is important to carefully control the overall gain of the measurement chain both in high frequency (RF part) and in baseband (IF, detected signal and data processing). Some instruments, such as a dedicated NFM or spectrum analyzer, include automatic gain control capability.

The sensitivity of the noise power meter (noise receiver) is also a crucial characteristic. The sensitivity is defined by the minimum detectable signal (MDS_{in} in dBm) at the input, defined as follows:

$$MDS_{in}(\mathrm{dBm}) = 10\log(kT_{amb}\Delta f) + NF_{receiver} + S/N_{min} \tag{2.16}$$

The noise meter for high-frequency noise measurements includes downconverters and the actual noise detection is performed generally around a few 10 MHz of IF frequency. Commercially available dedicated instruments are able to measure up to 26.5 GHz and even higher for spectrum analyzers with noise measurement capability. For noise measurements in the millimeter-wave frequency range, external downconverters are necessary. Such downconverters operate generally in the double side band (DSB).

Generally, the instrument passband (Δf) is low, typically around a few MHz, to minimize the effect of extra noise signals coming from the RF path through frequency converters. For a $\Delta f = 1$ MHz passband ideal system, the MDS_{in} with S/N_{in} assumed to be 3 dB is close to -111 dBm + NF_{dB} ($T_{amb} \sim 300$ K). Figure 2.7 represents the theoretical noise power (calculated from Eqn (2.13)) at the noise meter input versus the noise figure of the DUT for several values of DUT available gain. The equivalent noise temperature of the noise generator is fixed to 300 K (T_{cold}), the magnitude of the reflection coefficients at the output of the DUT (Γ_{out}) and at the power meter input (Γ_L) is fixed to 0 (best case), and the bandpass bandwidth is 1 MHz. If we assume that the power meter has a noise figure of 5 dB and that an S/N_{min} ratio of 3 dB is requested, the DUT has to present at least an available gain of 10 dB to ensure noise power detection for any values of NF_{DUT} (NF_{DUT} has to be higher than 3 dB for $G_{av\_DUT} = 5$ dB and $NF_{DUT} > 8$ dB for $G_{av\_DUT} = 0$ dB). Of course, this simple approach is questionable because the input signal is a noisy and not a coherent signal. In fact, a basic view of this problem of sensitivity for active DUT is that the product $G_{DUT} \cdot F_{DUT}$ (in linear operating conditions) has to be higher than the noise figure of the receiver. The sensitivity of receiver can be improved by placing an LNA in front of the receiver in order to reduce its noise figure. The output mismatch effect can be also reduced by using an isolator with at least 20 dB of isolation. The sensitivity of the receiver is a huge problem for on-wafer noise measurements in the mm-wave frequency range. At such frequencies, the gain of DUT is degraded by probe and waveguide losses; the DSB noise figure of mixers operating beyond 100 GHz needed for frequency conversion

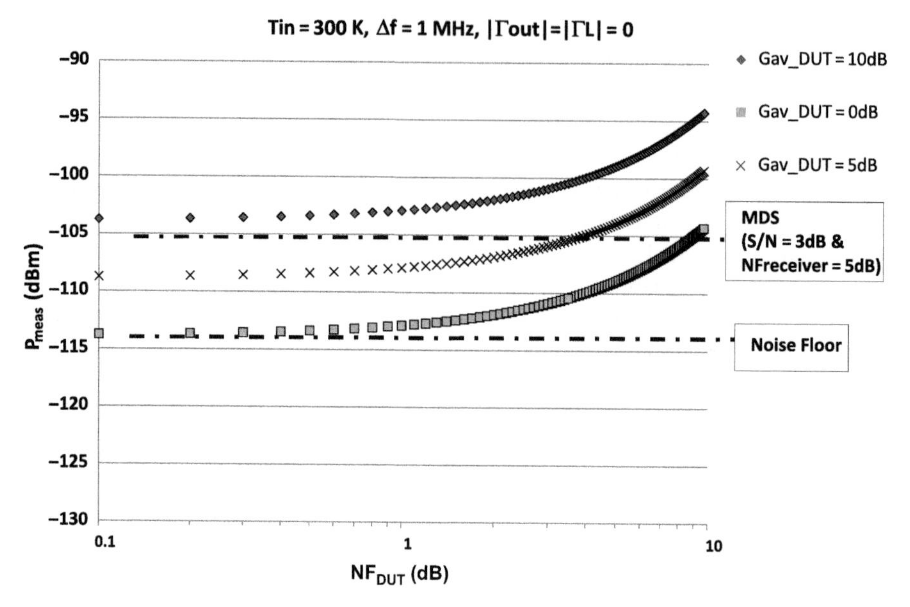

FIGURE 2.7

Sensitivity of an ideal noise receiver: Theoretical values of noise powers presented at the input of an ideal receiver ($NF_{receiver} = 5$ dB) versus NF_{DUT} for several values of a DUT's available gain.

is at least 5–7 dB, and moreover packaged LNAs operating at mm-wave frequencies are generally not available.

As mentioned previously, to design low noise circuits, it is necessary to elaborate on accurate noise modeling of transistors. F_{min}, R_n, and Y_{opt} (or Γ_{opt}, Z_{opt}) are generally the output parameters of such models that can be easily introduced in computer-aided design (CAD) software. From Eqn (2.10), we can express the noise figure of a two-port as function of these parameters as follows:

$$F(Y_g) = F_{min} + R_n \frac{|Y_g - Y_{opt}|^2}{\Re(Y_g)} \quad \text{where } Y_g \text{ is the generator admittance} \tag{2.17}$$

Equation (2.17) shows that the noise figure of a two-port depends on the four noise parameters and the admittance Y_g (or reflection coefficient, impedance) of the generator. This equation is also the basis of experimental extraction techniques. Indeed, if we present at least four different values of Y_g (with defined equivalent noise temperature) at this input, it is mathematically possible to solve Eqn (2.17) and to extract from it the four characteristic noise parameters of the DUT [32]. In practical terms, the variation of the admittance of the noise generator is realized by placing an impedance tuner in front of the DUT. Precise automatic mechanical-based tuners are preferred, which are commercially available in a broad frequency range (even above 100 GHz). With these setups, we can either measure several noise figures (hot/cold method) or several noise powers (cold method), corresponding to several values of the noise generator admittance. Didactic application notes are available in [33,34]. In the case of the cold method or multiple-impedances method [35], the equivalent noise temperature of the generator is equal to the ambient temperature of the tuner. Equation (2.18) details the expression of the measured noise power (see Figure 2.7) as a function of the DUT and noise power meter (or noise receiver) properties.

$$P_{\mathrm{meas}} = \frac{4k\Delta f \Re(Y_{\mathrm{out}})\Re(Y_{\mathrm{L}})}{|Y_{\mathrm{out}} - Y_{\mathrm{L}}|^2}\left[T_{\mathrm{e\_receiver}} + G_{\mathrm{av\_DUT}}\left(T_{\mathrm{e\_generator}} + T_{\mathrm{e\_DUT}}\right)\right] \quad (2.18)$$

where $T_{\mathrm{e\_receiver}}$, $T_{\mathrm{e\_DUT}}$, and $T_{\mathrm{e\_generator}}$ are respectively the input equivalent noise temperatures of the receiver and of the DUT, and $T_{\mathrm{e\_generator}}$ is the equivalent noise temperature of the generator, which is equal to the ambient temperature in the cold noise measurement method.

A photograph of a 5–40 GHz on-wafer noise measurement setup, including an automatic tuner, is represented in Figure 2.8.

2.2 On-wafer mm-wave instruments and setup

Two important discoveries have revolutionized the world of HF measurements: the automatic VNA and the microwave probe. For instance, in microelectronics, these two tools have permitted extending conventional DC parametric setups up to microwave and now millimeter wave ranges for industrial online tests. Before the

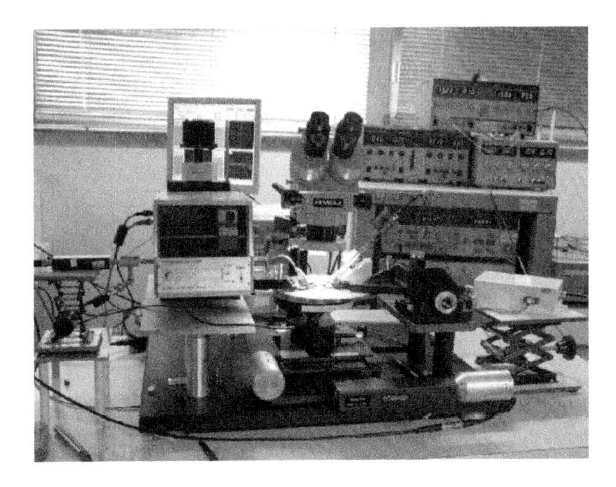

FIGURE 2.8

On-wafer noise measurement setup, including an input automatic tuner for the extraction of device noise parameters.

availability of HF probes, technicians and engineers had to insert the device or circuit into specific microwave test fixtures (or packages) to perform the measurement of *S* parameters or other HF quantities. This technique was very invasive because it necessitated wafer dicing, chip brazing, and bonding steps. Moreover, these historic solutions were not well suited for accurate calibration procedures. Although many tentative works were carried out in the 1970s, the first modern microwave probe was proposed by E.W. Strid and K.R. Gleason and published in the 1980s [36,37].

2.2.1 On-wafer station and setup

A graph view of typical manual on-wafer stations for HF measurements is depicted in Figure 2.9. It should be noted that the automatic version of such on-wafer stations is more dedicated for continuous inline testing in industrial environments. The wafer size capability is up to 300 mm, and its mechanical accuracy and repeatability of positioning are close to few microns in XY (resolution: 0.5 μm) and 1 micron in Z (resolution: 0.25 μm) [38].

A typical station contains four main parts:

1. The mechanical structure
2. The XYZ positioners (HF and DC probe holder)
3. The sample holder (or chuck)
4. The optical systems.

The overall setup also needs several accessories, such as an antivibration table, probes, cables or waveguide, and contact and calibration substrates. Up to 110 GHz, the probes are connected to the VNA through semirigid or flexible cables, which are generally around 10–100 cm long. The choice of high-phase stability cables is primordial. The temperature control of the environment is also very important to avoid phase drift due to thermal dilatation mechanisms. For systems operating above 110 GHz, the millimeter wave frequency extensions of VNA are directly fixed on the positioner (Figure 2.9, right) and connected to the probes through dedicated waveguides. Because the commonly used HF probe contains at least three contacts

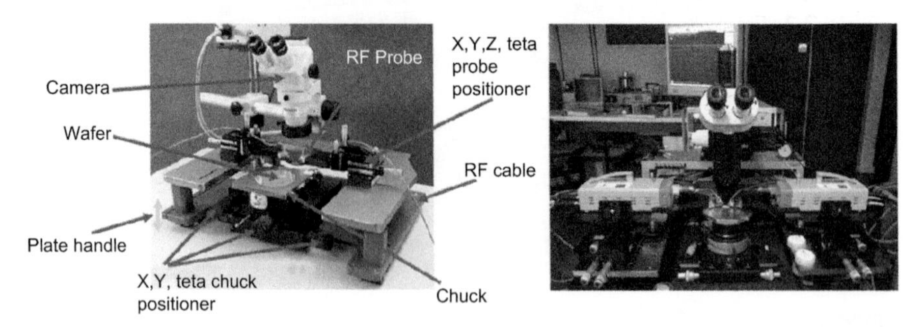

FIGURE 2.9

On-wafer setups. (left: conventional prober; right: millimeter-wave prober).

called ground-signal-ground (GSG), the positioners have to include a precise and stable tilt control. Finally, a high-quality microscope is mandatory to ensure high performance in terms of zooming and focusing capabilities.

2.2.2 Millimeter-wave vectorial network analyzer: principle and architecture

Because by definition mm-waves cover the 30–300 GHz frequency range, we have to distinguish between several types of VNAs. We can list three types of modern mm-wave VNAs:

1. The full coaxial-based VNA up to 67 GHz; all the RF test-set components (reflectometers, etc.) in this instrument are coaxial (e.g., 1.85 mm for a 67 GHz system)
2. The quasi-coaxial VNA, which operates in the full range up to 110 GHz; such an instrument is based on a 67 GHz coaxial (1.85 mm) two-port or four-port VNA, to which is added a 67–110 GHz waveguide (WR10) head extension and a combiner or diplexer (WR10 + 1.85 mm) with coaxial (1.0 mm) output
3. The waveguide (from WR15 to WR1.0)-based VNA from 50 GHz to 1100 GHz; this system is based on a standard coaxial two-port or four-port VNA improved by waveguide mm and sub-mm wave measurement heads. The frequency ranges of such frequency extenders, given in Table 2.1, are fixed by the waveguide sizes (TE10).

A basic architecture corresponding to a mm-wave extender is represented in Figure 2.10. The mm-wave extender contains a waveguide reflectometer (dual directive couplers), input $RF_{\_IN}$ and $LO_{\_IN}$ amplifiers, frequency multipliers, and harmonic mixers to combine and convert the RF and LO signals into IF one. Some of them contain also an RF source attenuator to control its level and flatness over the frequency range. The overall system is based on four-port VNA to benefit its input/output (test-ports and receiver inputs) versatility as well as its two generators. These generators are

Table 2.1 Frequency ranges of millimeter and submillimeter wave VNA extenders

Waveguide Flange	Frequency (GHz)	Band Designation
WR15	50–75	V
WR10	75–110	W
WR8	90–140	F
WR6	110–170	D
WR5	140–220	G
WR3	220–325	H
WR2.2	325–500	Y
WR1.5	500–750	
WR1.0	750–1100	

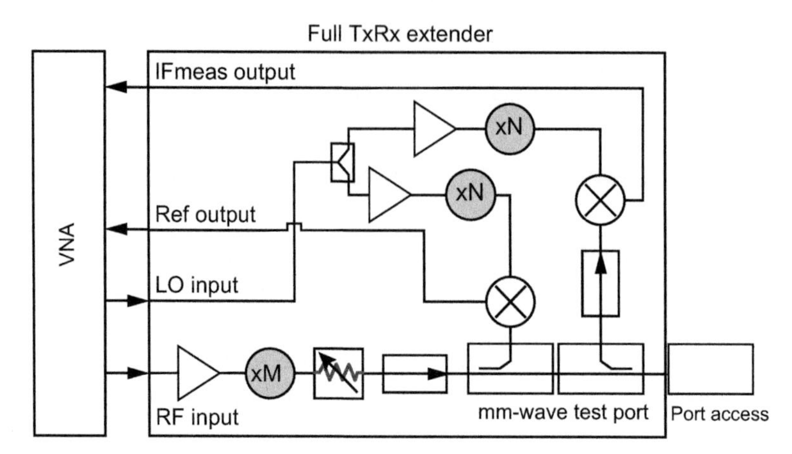

FIGURE 2.10

Schematic of a VNA millimeter-wave extender.

able to generate, through the VNA test ports, the $LO_{\_IN}$ and $RF_{\_IN}$ signals at the inputs of mm-wave extenders; these signals can be tuned separately both in frequency and output power. This last point is important and useful to optimize the up- and downfrequency conversion properties of the mm-wave extender by choosing the optimum rank of harmonic and optimal input power. The output IF signals of the mm-wave extenders are connected to unused receiver inputs of the VNA.

A four-port VNA (26.5 GHz or sometimes 50 GHz) commonly constitutes the basis of the mm-wave system that necessitates frequency converters with large rank of harmonics. For instance, for generated input LO and RF ($LO_{\_IN}$ and $RF_{\_IN}$) being in the 8–20 GHz frequency band, the total rank of harmonics (multiplier + H-mixer) is from 6 (W-band) to 30 (Y-band) for RF and 8 (W-band) to 24 for LO [39,40]. The input LO and RF powers range typically between 5 and 13 dBm. The IF output signal is less than 400 MHz. The RF power level and its flatness, as well as the dynamic range, are important features of mm-wave VNAs; these parameters depend on the frequency range (see Table 2.2).

2.2.3 Millimeter-wave noise receivers: principle and architecture

An example of the architecture of D-band noise receivers is presented in Figure 2.11. Similarly to a VNA frequency extender, a mm-wave noise receiver is generally assembled with a microwave NFM (or spectrum analyzer), to which is added a mm-wave down-/up-converter and solid-state noise source. For noise measurements below 60–70 GHz, high spectral purity microwave synthesizers are available. In such a case, a fundamental mode downconverter is preferred over a harmonic downconverter. At frequencies beyond, the primary LO frequency has to be multiplied (by N). This implies that its phase noise increases by N^2. Such noise is converted through the mixer and increases the noise level at IF output.

Table 2.2 Typical values of RF power levels and dynamic range of millimeter-wave VNAs [32]

Frequency (GHz)	RF Power Level Typ. Value (dBm/mW)	Dynamic Range Typ. Value (dB) (RBW = 10 Hz)
75–110	4.5/2.8	>110
90–140	4/2.5	>100
110–170	−3/0.5	>90
140–220	−8/0.16	>90
220–325	−15/0.032	>90
325–500	−23/0.005	>65

Consequently, it is crucial to choose a synthesizer as primary LO with a low noise phase option. To moderate high multiplication rank of the LO, the downconverter can be constituted with subharmonically pumped mixers, which generally operate with LO = 1/2RF. Such mixers present quite good noise performance, with DSB equivalent noise temperatures close to 600 K (NF = 5 dB) in D-band, 750 K (NF = 5.5 dB) in G-band, and 900 K (NF = 6.1 dB) in H-band. It is important to note that such mm-wave mixers are very sensitive to electrostatic discharges and have to be manipulated with caution. In particular, each component of the setup and the user have to be connected to ground; in addition, during the mounting, the IF port must never be connected to an uncontrolled load (e.g., an open-ended 1-m long IF cable not previously discharged).

The total DSB noise figure of the W-band and D-band receivers is plotted in Figure 2.12. The mixer losses (and thus NF) are optimized at each frequency point by adjusting the subharmonic LO power.

The solid state (avalanche diode) noise source is also a key element of the noise measurement bench. Such sources are commercially available up to 170 GHz. The important feature is the ENR (excess noise ratio), which fixes the reference hot equivalent noise temperature (see Section 2.1.3). Accurate and traceable values of ENR as a function of frequency have to be supplied with the source. The typical ENR value of noise source, operating above 100 GHz, is 12 dB (15 dB maximum). The second important characteristic of a noise source concerns its reflection coefficient, which has to be as low as possible and very similar for both on- and off-states. Generally, an isolator is integrated (or connected) within the source. For measurements higher than 170 GHz, the noise source can be realized by using an efficient absorber placed alternately into liquid nitrogen (77 K) and in an ambient environment (~300 K) to perform hot/cold noise measurements (Y-factor technique). The thermal noise radiated from the absorber is collected by a directive waveguide horn. A photograph of such a technique is presented in Figure 2.13.

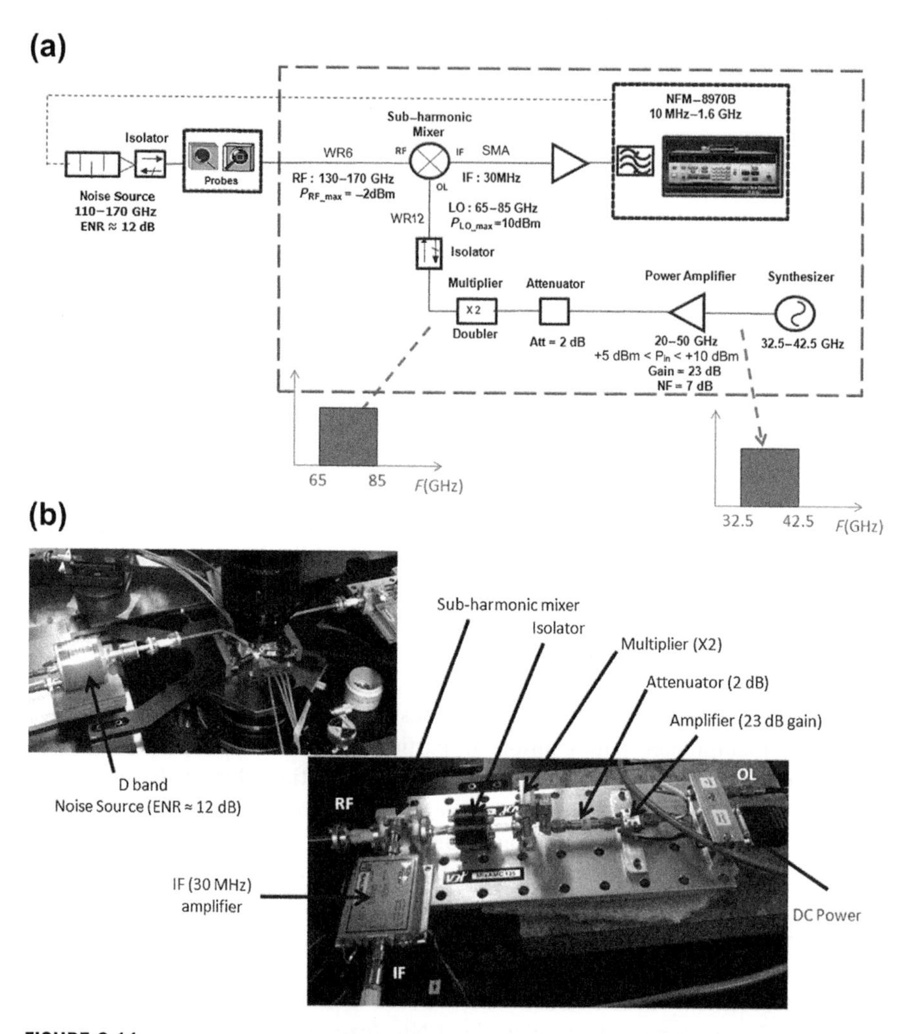

FIGURE 2.11

Schematic (a) and photographs (b) of on-wafer D-band noise measurement setup.

2.3 Calibration and de-embedding procedures for on-wafer measurements

Figure 2.14 shows the principle of calibration steps for *S*-parameters for on-wafer measurements. The main target is to accurately define the reference planes BB′ of the DUT using calibration procedures described in Section 2.1.2. It should be noted that the topology of the DUT's embedded access transmission lines must be compatible with the probe (pitch and tip size), which generally necessitates three GSG access pads.

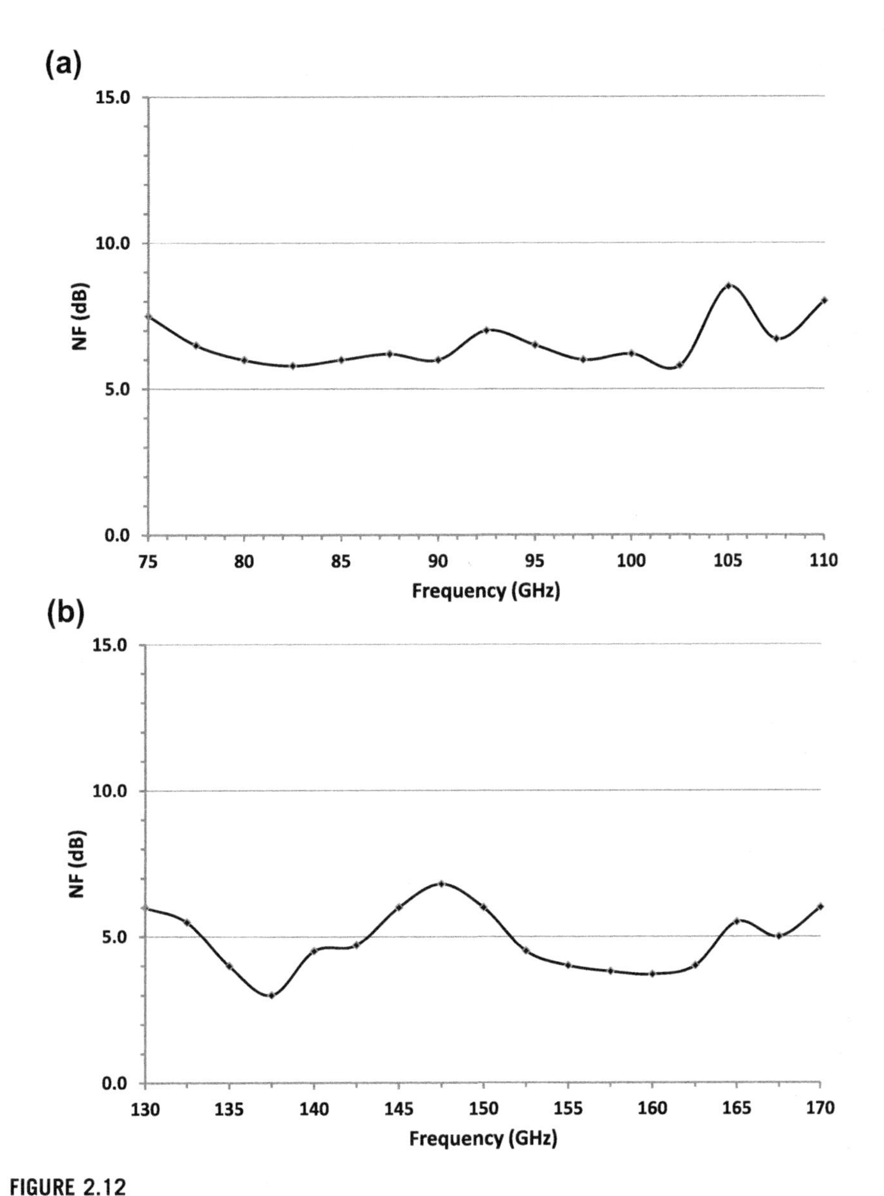

FIGURE 2.12

Total noise figure of W-band (a) and D-band (b) noise receivers.

Two methods or concepts can be considered. The first one consists of fabricating several on-wafer standards (e.g., line, open, short, load) directly embedded into reference planes BB′. The second method is carried out in two steps: a first calibration is performed at the probe tips (reference planes AA′), in which case the calibration standards can be fabricated on other substrates, whereas the second

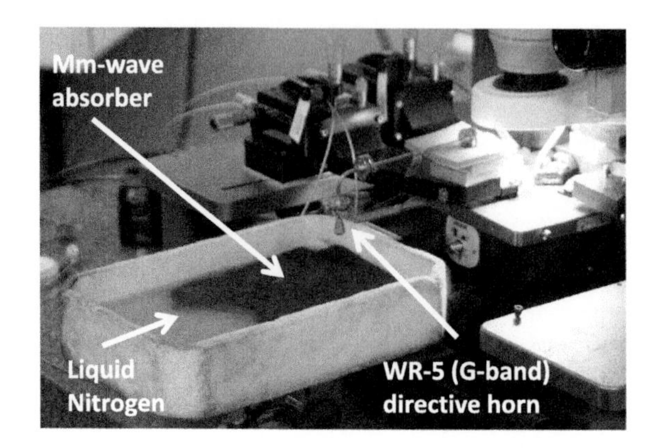

FIGURE 2.13

Hot/cold noise measurement bench operating in G-band (140–220 GHz).

calibration is commonly called the "de-embedding procedure." The latter consists of removing the access transmission lines of the DUT using specific dummy structures generally associated with an equivalent electrical model. These two approaches merit a preliminary discussion. By considering the theory of S parameters (see Section 2.1.2), the first method might be the best one. It would be enough to insert the standards directly between reference planes B and B′. For example, the ideal case would be to insert perfectly defined transmission lines and to apply a calibration method such as LRL or multiline LRL. Nevertheless, such an approach suffers from many drawbacks: (1) the dimensions and electrical (or electromagnetic [EM]) properties of the inner connections of the DUT (e.g., a transistor) are rarely compatible

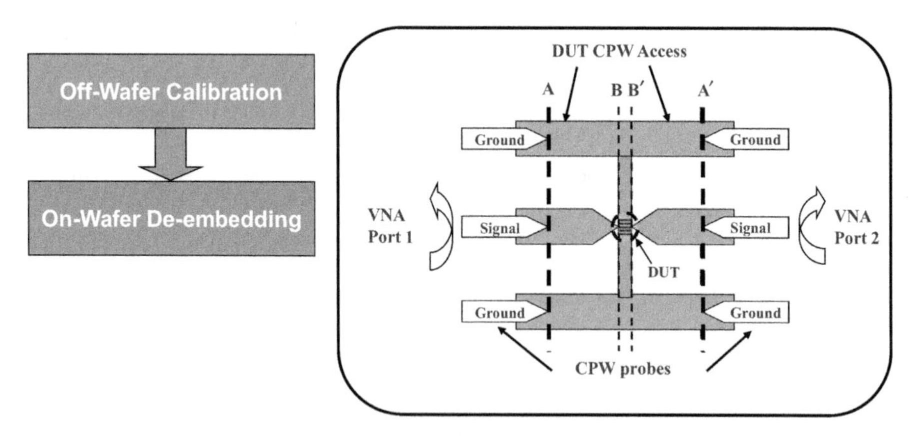

FIGURE 2.14

Schematic of the calibration procedure and steps.

with well-defined standards such as transmission lines (without dimensional discontinuities) or other types of standards; (2) transmission lines fabricated in advanced silicon technologies can present complex properties that render modelling difficult especially in the mm-wave frequency range; (3) overall standards have to be fabricated directly on-wafer, which necessitates a non-negligible area of the semiconductor.

The two-step calibration method is most commonly used. However, this method generally accumulates the imperfections of two off- and on-wafer calibration steps. For the off-wafer procedure, the main causes of uncertainty result from the actual standard properties compared to their respective definition, but also from the control of probe contact (electrical quality and repeatability) and probe position on the pad. For the on-wafer de-embedding procedure, we need dummy structures such as pad, open, short, and line standards with identical topology, which are assumed to have the same electrical behaviors as the DUT's access transmission lines. The technological process permits the fabrication of dummy structures with strictly similar topology. The second assumption, related to having electric (or EM) properties that are perfectly similar to those of the DUT environment, is more difficult to achieve. Despite these difficulties, this method, with its numerous variants, is widely used for *S*-parameter measurements of transistors and circuits, even for advanced silicon technologies. This method, which is relatively easy to perform, is very satisfactory up to 60–70 GHz. The main problems appear for higher frequency ranges, as will be explained in the following.

2.3.1 Off-wafer calibration procedures: principle and calibration kit

2.3.1.1 Mm-wave probes

Before going into detail, we must elaborate on microwave and mm-wave probe technologies and properties. First, a probe could be considered as a guided travelling wave support. All the components of the probe are designed to favor the excitation and to couple EM energy on a single mode in a given frequency range. Therefore, a probe can also be considered as a filter of modes in the case of multiple-modes conditions. A probe is composed of three main parts: (1) a coaxial connector (down to 1 mm standard, maximum frequency: 110 GHz) or a waveguide (with available flanges from WR15 to WR1.5; maximum frequency: 750 GHz); (2) a miniaturized propagation structure (mini-coaxial, CPW (coplanar wave guide), or microstrip lines); and (3) GSG (or more contacts) coplanar-like tips (e.g., metal bolds, miniature metallic blades, metallic needles, other shapes).

Some waveguide-based probes include an integrated bias tee, which is generally indispensable for active device measurements. This biasing network presents a low-pass frequency response by providing a direct low resistance DC path for supplying sufficient DC current (e.g., up to 1.5 A) to a device under test and by presenting resistive impedance below the cutoff frequency of the waveguide. Indeed, active DUTs present very high gain, which can lead to instabilities due to the strong reactive impedance of the waveguide under the low cutoff frequency of TE10 mode. The

DC output of this integrated bias is terminated generally by a 50 Ω microwave coaxial connector (e.g., 2.92 mm). To avoid instabilities in the low frequency range, it is recommended to connect a second external performing bias tee loaded to a broadband coaxial 50 Ω termination. Table 2.3 summarizes the typical best performances (insertion and return losses) of commercially available mm-wave probes.

2.3.1.2 Impedance standard substrates

The typical elements for calibrating an on-wafer measurement system comprise opens, shorts, matched loads, and CPW lines (see Figure 2.15). All these elements are fabricated on high-quality dielectric substrate (e.g., alumina); each element is generally repeated in column. For measurements above 100 GHz, the substrate thickness is reduced and an absorbing holder is required to avoid the propagation of parasitic modes.

Any set of standards (open, short, thru, and load) is of two-port topology and consists of two identical elements spaced at 150–200 μm from each other. The space between two sets of standards is close to half a millimeter to reduce coupling effects between the probe tips' back and other metallic areas. The material of the pad metallization is gold that is a few microns thick. The contact resistance between the probe tips' metal (BeCu, W, or Ni) and gold pads is between 10^{-2} and 10^{-1} Ω. For on-wafer measurements on silicon technology, the metal of test pads is aluminum. In this case, tungsten or, even better, nickel-based tips ensure a low contact resistance.

The more identical the standards are, the more accurate the calibration process becomes, which in turn results in accurate on-wafer testing. For instance, the coplanar load is the key element for LRM, LRRM, and SOLT calibration procedures (see Section 2.1.2) because it fixes the reference impedance. This standard consists of two parallel 100 Ω metallic (NiCr) resistors embedded in CPW-like pads. The value of these resistors is trimmed to be 50 Ω ± 1%. It is also important to note that the total area of standards is less than 200 μm of length × 600 μm

Table 2.3 Typical values of insertion and return losses of commercially available waveguide-based probes

Frequency (GHz)	Insertion Loss (dB)	Return Loss (dB)
75–110	<0.8	<15
90–140	<1.5	<15
140–220	<2	<15
220–325	<2.5	<15
325–500	4–6	<15
500–750	6–8	<10

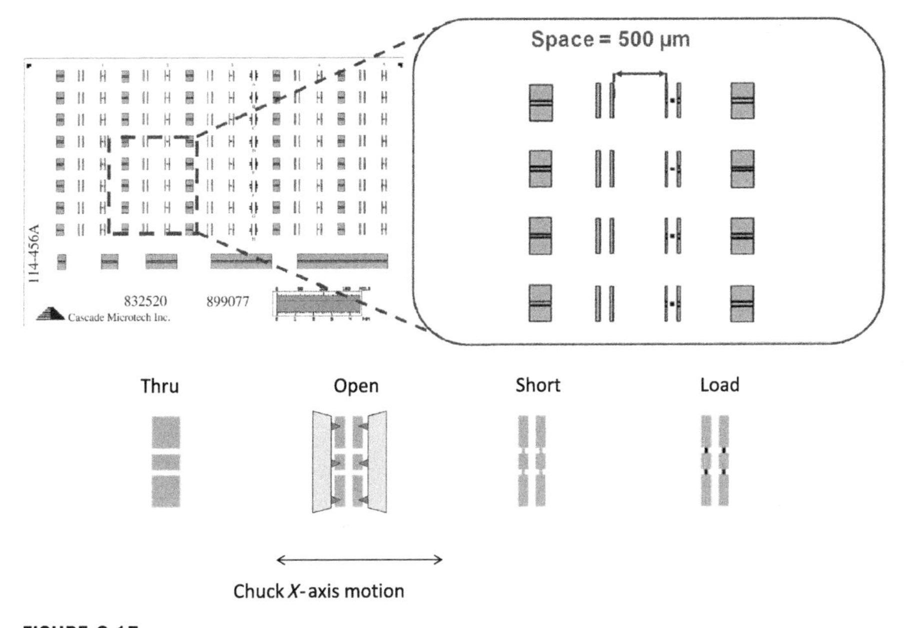

FIGURE 2.15

Layout of impedance standard substrate (ISS ®Cascade Microtech) with details of the standard layout.

of width. The pad size is between 25 and 50 μm, which can be considered as a lumped element up to 250 GHz (λ_g ~0.5 mm for CPW on alumina).

As mentioned in Section 2.1.2, standards, calibration procedures (in the case of SOLT), and self-calibration techniques must be completely defined. These standards are characterized using electrical-based models. The set of element values of these equivalent circuits is called a "cal-kit." Before the calibration sequence starts, the cal-kit file is stored in the VNA or specific external software. These reference values depend partly on the probe technology (type of propagation structure, tip shapes, and probe pitch). More generally, these values depend on the probe's EM environment when probes are in contact on the standard pads (or sometimes "open," in the air). The probe position on the contact pad is also an important cause of standard equivalent element deviation. An initial set of reference values (cal-kit) is provided by the probe supplier. It is important to note here that, contrary to high-frequency traceable standards, such as TEM 7 mm coaxial system or even precise waveguide-based standards, the impedance standard substrate (ISS) integrates only planar elements, of which the complete HF properties are not directly to metrological primary standards.

The standard thru is characterized by an ideal lossless line defined by its characteristic impedance and its delay (between 0.5 and 1 ps). This delay must in theory define the reference planes at the probe tips. We have to note that the insertion losses of CPW lines fabricated in low-loss dielectric are close to 1 dB/mm at 200 GHz. We can estimate that a 150-μm long thru standard could present 0.1−0.2 dB of insertion losses, which becomes non-negligible.

The standard load is modeled by a resistance in series with an inductance. The physical origin of this inductance is complex. The only controllable parameter by the user is the placement of the tips on the pad. The value of the inductance varies between approximately 1 and 1.5 pH ($Z_{Inductance}$ ~ j1.25 to j2 Ω at 200 GHz) for 10 μm of contact position deviation. Compared to XY automatic prober accuracy around a few micrometers, 10 μm is relatively important but the contact position depends also on the overlap of the tips (sliding effect) controlled by the Z-axis of the positioner to ensure a low resistive contact. An optimal overlap is 15−30 μm and depends on the nature of both the tips (e.g., shape morphology, contact area, type of metallization, cleanliness) and the pad (e.g., type of metallization, thickness, native oxide, morphology after several contacts).

The standard open is modeled by one pure capacitance or one equivalent capacitance including frequency dispersive effects (e.g., polynomial-based model) or even by an open-ended line offset. This standard can be of different natures, such as an open-air probe, probe on dielectric, or probe on dedicated open structure.

Finally, the standard short is modeled by a pure or equivalent frequency-dependent inductance or short-ended line offset. The topology of this standard consists of short-circuiting the GSG pads.

The ISS substrate also integrates multiple-length CPW lines with similar transversal topology (dimensions of signal and ground strips and slot) to perform LRL (line reflect line) or multiline LRL. To optimize the area, the ground strips are transversally enlarged by a resistive metallization layer to cause the electric field lines to vanish. These lines do not integrate air bridges, which can cause multimodal phenomena.

To summarize, these "standards" are characterized by numerous unknown, even uncontrolled values. In light of these difficulties, especially for mm-wave measurements, the experimenter must choose the calibration procedures that are the least sensitive to the standards' definitions. Self-calibration techniques are certainly the best candidates because they are less (or not at all) sensitive to the "reflect" standards' definitions. Multiline LRL is certainly the most suitable for metrology. The main drawback of this technique is in the practical sense because one needs to perform a lot of probe contacts, and ISS and probes are expensive, especially for mm-wave experiments. For measurements in an industrial environment, the best tradeoff between performance and practical easiness is definitively the LRM or LRRM techniques. For on-wafer measurements below 50−60 GHz, the "battle" between SOLT and self-calibration defenders is not necessary because one can obtain quite similar results if the usual good precautions are taken for each of them.

2.3.1.3 Experimental methodology

Highly successful equipment and well-likened calibration procedures are not enough. The quality of on-wafer measurements also depends on well-established experimental practices.

It would be too difficult to detail here all of the experience needed to perform good measurements, but we will highlight the main stages. First of all, many application notes and videos are available on the web sites of wafer prober suppliers. Secondly, probes are very fragile, so their longevity and their quality require highly checked mechanical adjustments.

The main objective that the user must achieve is to ensure the best contact of the probe tips on the DUT pads. "Best contact" means morphologic uniformity but also optimal DC and HF electrical properties. Prior to beginning the probe contact, it is important to carry out a visual inspection both of the probe tips (e.g., track of wear, contact deformation, dielectric or metallic dust) and of the pads' surface (presence or not of uniform probe marks).

Because HF probes own at least three contacts, the plane of the contact fingers has to be perfectly parallel to the plane of the bottom side of the DUT's pads to ensure uniform contacts. For instance, a nonperfect ground contact can produce parasitic effects, resulting in noncontrolled resonances in *S*-parameter frequency responses. In the case of planarity defects, we recommend making fine adjustments of the probe tilt using a dedicated contact substrate with uniform metallization in terms of thickness and roughness. Defects of probe tips or nonuniformity of pad metallization can also produce nonuniform contacts despite a right planarity adjustment. The electrical quality of the contacts is related at the first order to the contact resistance (close to a few 10 mΩ in the best case).

Because the contact area is close to $10 \times 10\ \mu m^2$ or even less, the contact mechanisms at this scale are complex and related to elastic contact properties, with dynamic Coulomb friction versus the perpendicular force inducing sliding displacement of the contact over several micrometers (at least 15–20 μm). The electrical contact quality can be estimated by connecting a DC multimeter through the bias tee and by contacting all the probe contacts on the same metallization (or on a short standard). Of course, this measurement is not accurate enough to extract the contact resistance while the bias tee resistance is close to a few ohms but can highlight an important defect. *S* parameters (e.g., in reflection), especially in mm-waves, are sensitive to the contact quality. The check of the reflection coefficient over the broadband frequency range is of primary importance. In normal cases, the frequency response of S_{11} (and S_{22}), when the two probes are connected on a short circuit, must be free of resonance and must remain constant over a long period. For instance, at 100 GHz, variations of magnitude and phase less than ±0.05 dB and ±0.5°, respectively, are acceptable after few minutes of contact.

When these preliminary and important checks are achieved, the calibration procedure can be started. Before contacting the several standards on the ISS, substrate alignment is carried out using chuck rotation adjustments, and the offset distance between the two probes related to the standard pads has to be fixed (the contact overlap previously optimized has to be taken into account). The *Y*-axis of the chuck must be adjusted to obtain the same distance between the two ground contacts and the end sides of the standard. It imposes the use of a dedicated standard layout adapted to the pitch of the probes.

Once the prober is configured, the different standards are automatically or manually measured by moving the X-axis of the chuck and by raising and by lowering the probes between every sample. In the case of mm-wave measurements, for instance up to 325 GHz, this calibration procedure has generally to be repeated on different prober systems supporting the right VNA frequency extender heads. Although it is time consuming, the measurement methodology previously described must be applied with the same precautions.

2.3.2 On-wafer de-embedding procedures: principle and de-embedding test structures

2.3.2.1 Principles and specifics in mm-wave

As mentioned in Section 2.3, the principle of the on-wafer de-embedding procedure consists basically of moving the complex reference planes of the S parameters at the probe tips to the ones at the DUT. This principle is illustrated in Figure 2.16. Moving the reference planes means to change—mathematically speaking—the reference of the complex planes by using linear relationships between each of them. This set of relationships is generally related to an electrical model assumed to represent the properties of the DUT accesses. For this concept, we often take a shortcut, which makes the mixture between reference planes dependent on electrical measurements and geometrical planes dependent on the physical nature of the system. This mixture becomes less and less credible for nanoscale devices operating in the mm-wave frequency range.

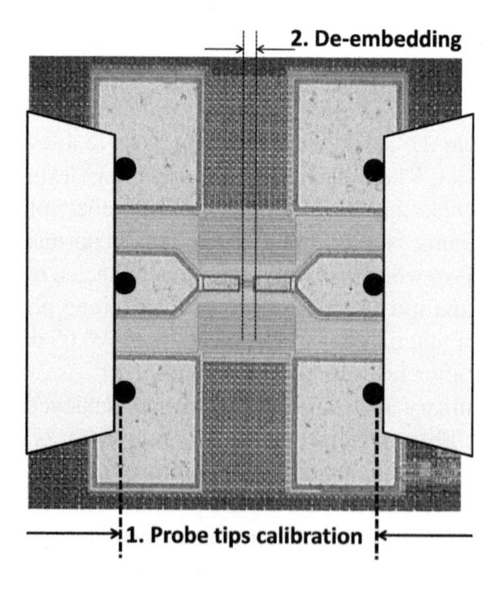

FIGURE 2.16

On-wafer de-embedding procedure.

First of all, we have to consider mm-wave measurements and associated de-embedding techniques on advanced silicon technologies (CMOS, BiCMOS) as the most complex situation compared to other mm-wave technologies (e.g., III–V high-electron-mobility transistor [HEMT], HBT). There are two main reasons:

1. Silicon is a semiconductor material presenting naturally low resistivity (several ohm centimeters, sometimes less), whereas III–V such as gallium arsenide (GaAs) is a semi-insulating material and presents a resistivity six orders of magnitude higher than that of silicon. Therefore, the commonly used silicon wafer is not the best material as substrate for high-frequency propagation structures.
2. The back-end (BEOL) of advanced silicon technologies is much more complex compared to the GaAs one. The silicon CMOS or BiCMOS roadmap is driven by powerful integration trends to address both digital and analog circuits and systems. The recent silicon process includes in a few microns thickness at least seven levels of metal layers separated by thin layers of silicon nitride or silicon oxides, whereas in the case of the GaAs microwave monolithic integrated circuit process, only two to three metallic layers are commonly available. This complex BEOL of silicon process induces many complex electrical parasitics (e.g., wire and via resistive, inductive effects and pads, wires and substrate capacitive couplings), which are to be accurately extracted (or modeled) during the de-embedding procedure.

To illustrate the main difficulty of de-embedding on advanced silicon process, Figure 2.17 represents the measured and simulated S_{11} (and S_{22} in the case of HEMT) up to 220 GHz. In both cases, the simulated S_{11} is calculated using the usual small signal equivalent circuit (insert in right picture) deduced from broadband S-parameter measurements (typically up to 50–60 GHz) [42]. The off-wafer calibration at the probe tips is LRRM. The on-wafer de-embedding in the case of III–V HEMT is simply performed by removing an ideal line delay ($\sim$0.9 ps) at each CPW access. For silicon n-MOSFETs, the de-embedding consists of mathematically removing the thin-film microstrip (TFMS) transistor's accesses by using short and open structures (pad-short, pad-open, complete-short, and complete-open using the technique detailed in [43]).

Despite a much more complex de-embedding method used in the silicon case, Figure 2.17 (right) shows strong resonance phenomena both in the W-band and G-band, and one can expect the same in between these two bands.

2.3.2.2 Description of de-embedding procedures based on silicon dummy structures

Before making more in-depth explanations of these phenomena, we have to present the main de-embedding techniques commonly used for characterization of advanced silicon devices. A typical layout of HF transistor test structure is represented in Figure 2.18, along with a schematic describing each part (pads, access lines, top-down interconnections).

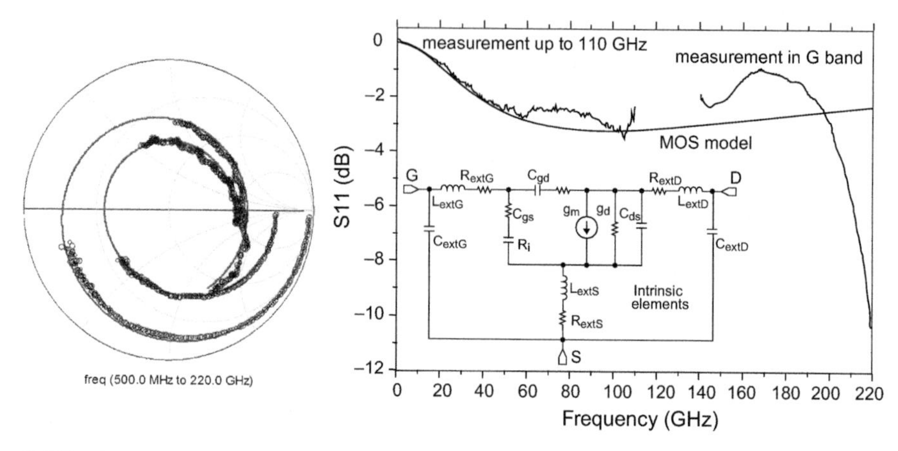

FIGURE 2.17

Comparison of measured and calculated reflection coefficients of mm-wave transistors (left: 70-nm metamorphic HEMT process; right: 65-nm *n*-MOSFET–CMOS process) [41].

Complex procedures (several dummy structures: pad-open and pad-short, pad-open, pad-short, complete-open, complete-short, thru-line, etc.) can provide appropriate subtraction of parasitics. Based on these dummy structures, several methods that allow high-accuracy correction, in the centimeter wave range, are available in literature [41,43–49]. Similar techniques with specific variants were then extended to the mm-wave range [50–53].

For instance, we describe below the pad-open-short-short method detailed in Ref. [43] and extended up to 220 GHz in Ref. [50]. Four dummy structures are

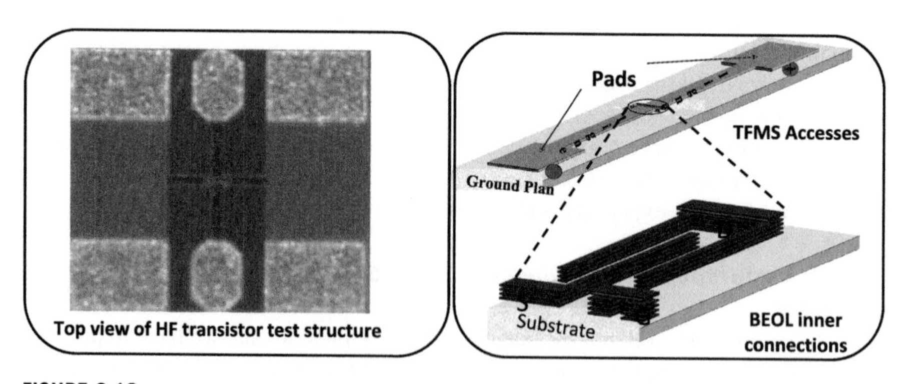

FIGURE 2.18

Top view of silicon transistor high-frequency test structure (left); schematic views (right).

necessary: pad-open, complete-open (composed of pad, line, top-down via holes, and comb), pad-short (composed of pad, line, top-down via holes connected to the ground at the end of the line), and complete-short structures (composed of pad, line, top-down via holes connected to the ground at each side of the comb) (see Figure 2.19). This method is based on an electrical equivalent circuit presented in Figure 2.19, including parallel and series one-ports. The nature of the series elements is mainly inductive and the parallel elements are mainly capacitive. As discussed previously, we assume implicitly, using such a model, that the transistor's accesses present quasi-lumped behavior. The following equations are used for this procedure of de-embedding:

$$
\begin{aligned}
[Z_S] &= \begin{bmatrix} Z_{S1} + Z_{S2} & Z_{S2} \\ Z_{S2} & Z_{S3} + Z_{S2} \end{bmatrix} \\
&= \left(\left[Y_{\text{Pad\_Short}}\right] - \left[Y_{\text{Pad\_Open}}\right]\right)^{-1} \\
&\quad + \frac{\left\{\left(\left[Y_{\text{Complete\_Short}}\right] - \left[Y_{\text{Pad\_Open}}\right]\right)^{-1} - \left(\left[Y_{\text{Pad\_Short}}\right] - \left[Y_{\text{Pad\_Open}}\right]\right)^{-1}\right\}}{3}
\end{aligned}
\tag{2.19}
$$

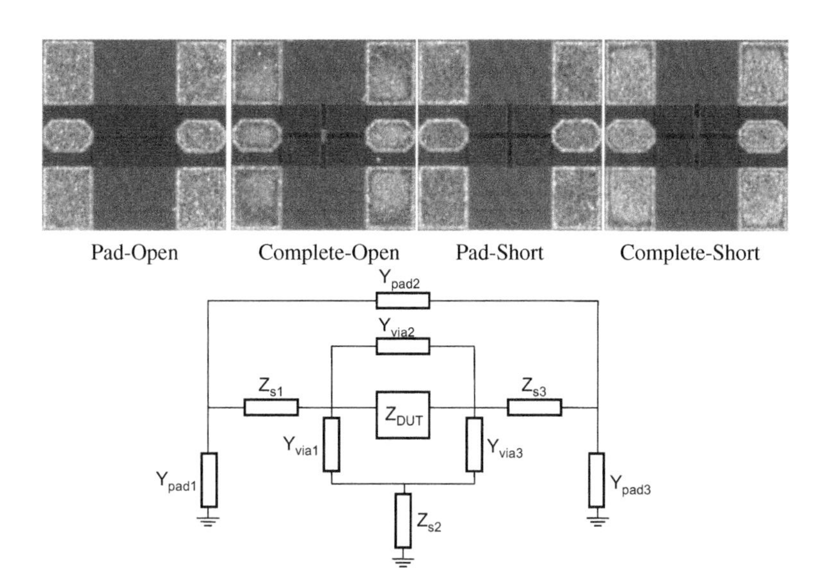

FIGURE 2.19

Photograph of dummy structures and access equivalent circuit used for on-wafer silicon de-embedding method described in ref. [43].

$$[Y_{\text{via}}] = \begin{bmatrix} Y_{\text{via1}} + Y_{\text{via2}} & -Y_{\text{via2}} \\ -Y_{\text{via2}} & Y_{\text{via3}} + Y_{\text{via2}} \end{bmatrix} = \left\{ \left([Y_{\text{Complete\_Open}}] - [Y_{\text{Pad\_Open}}] \right)^{-1} - [Z_S] \right\}^{-1} \tag{2.20}$$

$$[Z_{\text{DUT}}] = \begin{bmatrix} Z_{\text{DUT\_11}} & Z_{\text{DUT\_12}} \\ Z_{\text{DUT\_21}} & Z_{\text{DUT\_22}} \end{bmatrix} = \left\{ \left\{ \left([Y_{\text{Meas}}] - [Y_{\text{Pad\_Open}}] \right)^{-1} - [Z_S] \right\}^{-1} - [Y_{\text{via}}] \right\}^{-1} \tag{2.21}$$

In Eqn (2.19), the factor 3 for $[Z_S]$ takes into account the distribution of series parasitic of the metallic bar, which connects to the transistor. The matrix $[Y_{\text{via}}]$ (Eqn (2.20)) is calculated to evaluate the top-down via holes contribution near the transistor. Finally, $[Z_{\text{DUT}}]$ represents the intrinsic parameters of the transistor. The de-embedding accuracy depends on the measurement of dummy structures used and the amount of parasitics that are taken into account, as well as their electrical (or EM) pertinences.

A successful de-embedding procedure can be obtained by basic verifications. For instance, the intrinsic parameters of the transistor's equivalent circuit should be frequency independent. Simpler procedures based also on lumped equivalent circuit models, such as pad-open or pad-open/pad-short de-embedding applied in the mm-wave range, lead to resonances and fluctuation of transistor model parameters due to wrong subtraction of access parasitics. Figure 2.20 illustrates the limitation of simple de-embedding techniques using only two open and short dummy structures [49] compared to the more complex one detailed previously [50].

Other on-wafer de-embedding methods combine lumped element circuit models and transmission line theory [53–56]. The minimum dimensions (area of pads and length of access line) of test structures are related to probe pitch, probe contact overlap, and prober positioning accuracy. Typically, the total length (pad + line) of mm-wave silicon transistor access is from 50 to 100 μm; if we assume a propagation time delay per millimeter of TFMS line close to 6–7 ps/mm, the basic limit of $\lambda_g/20$ is achieved around 150 GHz for $L_{\text{access}} = 50$ μm. Therefore, beyond 100–150 GHz, distributed circuit based de-embedding procedures have to be employed. Because advanced silicon transistors (CMOS and HBT) will present, as scheduled in ITRS (international technology roadmap for semiconductors), a cutoff frequency close to 1 THz, the access dimensions will have to be strongly reduced. For example, pads with 25-μm pitch have been recently designed to characterize 32-nm indium phosphide (InP) HEMTs at 700 GHz [57].

This indispensable reduction of the pad size is not really compatible with what is foreseen for silicon metrology in the ITRS roadmap. For instance, the RF pad area used for probe contact will remain around 650 μm^2 for the 8-nm node, which induces huge dimensional and electrical mismatches with nanodevices under test. A drastic reduction of test pad area by one or two orders of magnitude is straightforward from a technological point of view but revolutionary in respect of the instrumentation

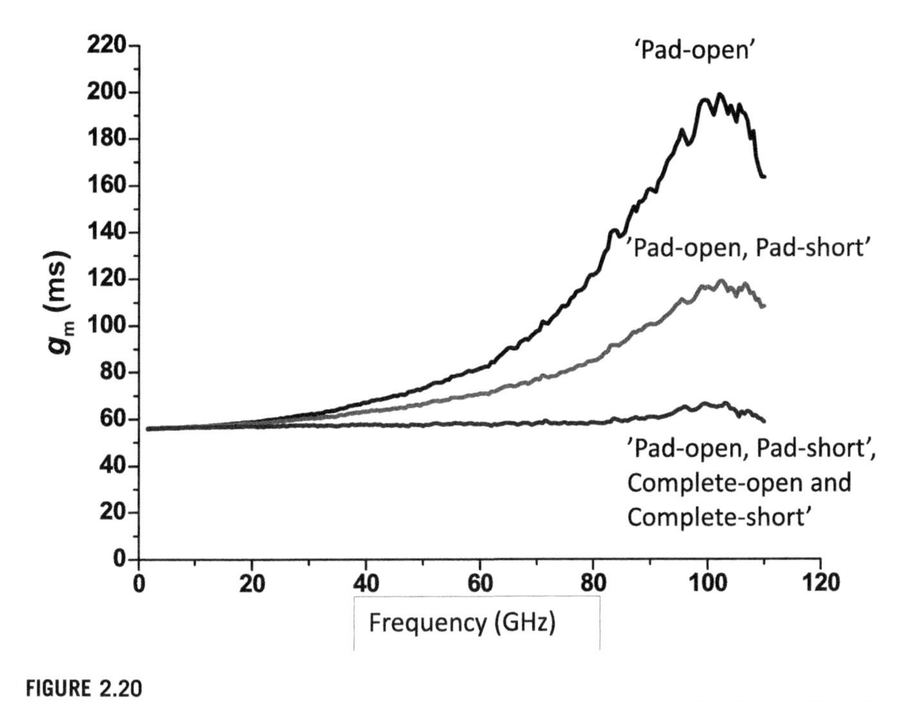

FIGURE 2.20

Example of intrinsic transconductance (g_m) of 65 nm *n*-MOSFET versus frequency (up to 110 GHz) extracted using three de-embedding procedures.

required (prober). In such a vision of integration, the prober requirements (accuracy of alignment, positioning, repeatability) should be much more severe, as is foreseen by ITRS. Moreover the sub-10 μm pitch probe does not exist.

2.3.2.3 Coupling on-wafer measurement phenomena

At the mm-wave range, other parasitic coupling paths appear between probes, test structures, and adjacent structures (see Figure 2.21). These parasitic coupling paths are different when DUT or dummy structures are measured due to the change of the measurement area near the structure under test (dummy structures or transistors), generating de-embedding errors (see Figure 2.17 right). On the other hand, automatic on-wafer measurements for in-line characterization set some constraints for the design of the test structures (e.g., the minimum distance required between probe pads). Moreover, the silicon consumption for the structures has to be reduced as much as possible.

A simple experience to highlight the influence of the probe environment in the mm-wave range consists of separating one test structure from the others (by dicing process) and reporting it at the fringe of a pure dielectric support. In this new environment, the back of the probe contacting the device is in the air. Figure 2.22 shows measurement results of a pad-open structure before and after it was cut. The air

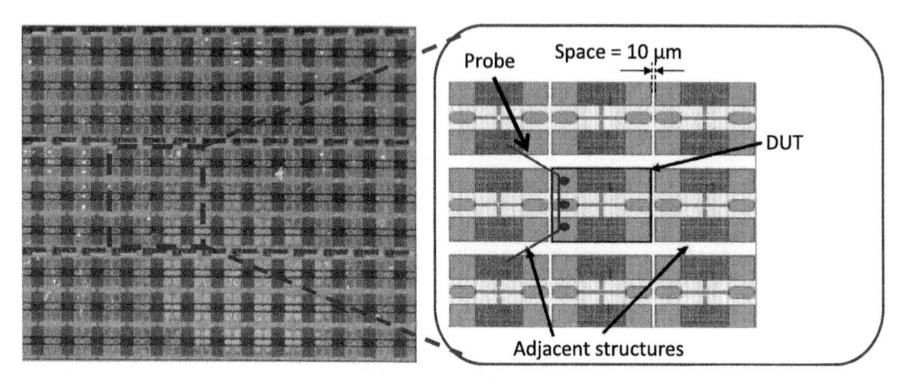

FIGURE 2.21

Layout of high-frequency test structures including transistors and dummy elements (CMOS 65 nm).

space back to the probe (isolated chip) attenuates the resonance phenomena at frequencies close to 60 and 90 GHz.

We can explain coupling effects between the probe and its environment by using three-dimensional EM simulations based on the layout of the test structures and probe architecture similar to the real one. For the sake of simplicity, the purpose of the

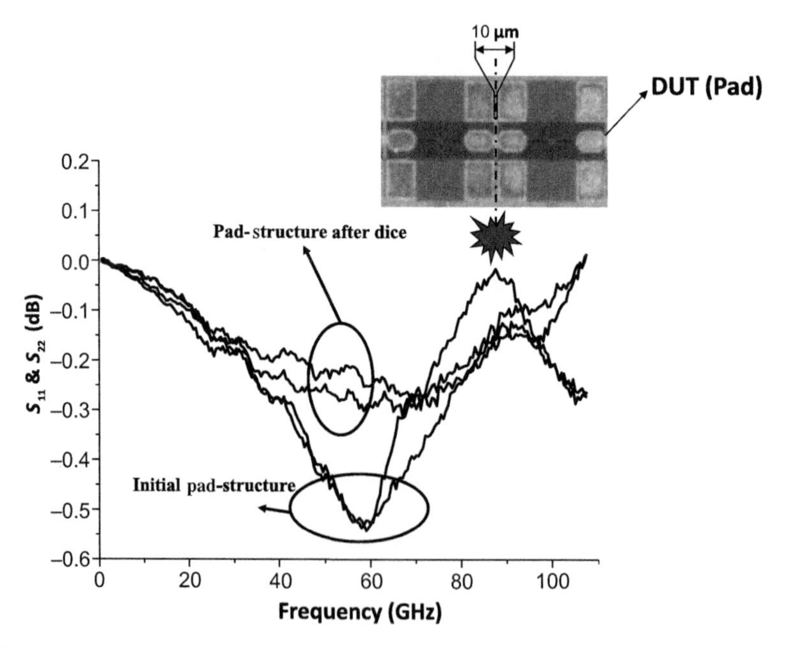

FIGURE 2.22

Magnitude of S_{11} and S_{22} parameters of dummy pad-open structures versus frequency Before and After Cut.

simulation structure (particularly for the probe) is to point out the environmental influence on DUT characteristics, even if the resulting simulation data might be slightly different than the real ones. The approached topology of the probe is created from actual pictures of ®Cascade Microtech 220 GHz Infinity probes taken by scanning electron microscope (SEM). Figure 2.23 gives an art view of the drawing used for EM simulations, describing the DUT and the probe topology. In this case, one adjacent device is placed under the probe with 10 μm of space between each device.

All the results of this study are detailed in Ref. [58]. The conclusion of this study is that one of the causes of resonance phenomena in the mm-wave frequency range is due to back probe environment. This induces error first during off-wafer calibration where the back probe environment is alumina (the adjacent standards are spaced out sufficiently), and secondly during the on-wafer measurements (dummy structures and transistors). In this last case, the back probe environment is complex (silicon and mainly metallization) and strongly modifies the frequency response of the DUT's S parameters. One solution to reduce these coupling effects is to change the layout rules by separating adjacent structures with a larger gap or better to realize quincunx test structures and to reduce if possible its number in order to keep as small as possible the used silicon area for the HF testing.

To summarize, these on-wafer de-embedding techniques based on lumped or distributed circuit representations of the transistor's accesses have been validated with quite good success up to the G-band. We can reasonably expect that they should operate up to 300 GHz by slightly scaling down the HF test structure layout. For sub-mm wave characterization, the problem of electrical and EM behaviors of such test structures becomes predominant and will be a huge hindrance for accurate measurements and parameter extractions. Unless a lot of research and economical

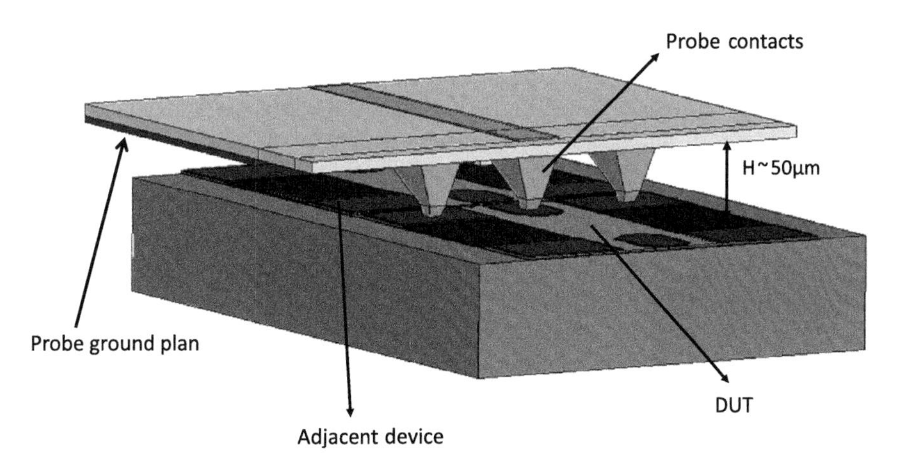

FIGURE 2.23

Art view of microstrip type mm-wave probe contacting a silicon HF structure with an adjacent device under the probe (a similar structure with real dimensions is used for 3D EM simulation).

investments are carried out on measurements tools (prober) in order to make possible viable industrial tests of very small test structures (e.g., with $10 \times 10\ \mu m^2$; pads or less), other approaches have to be investigated. One solution should consist of efficiently using the parasitic effects of HF test structures as a matching network. In this case, it is not a single transistor that would be measured but perhaps a simple amplifier. This idea is the base of the mm-wave built-in self-test (BIST).

2.3.2.4 Noise calibration and de-embedding procedures

One of the most used techniques consists of extracting the four noise parameters (e.g., NF_{min}, R_n, and Y_{opt}) of every part (assumed like two-port) constituting the overall noise setup. The mathematical procedure is based on the concept of noise correlation matrices (see Section 2.1.3). For instance, it is straightforward to calculate the noise correlation matrix (in chain form) of a DUT embedding into input and output networks using Eqn (2.22).

$$[C_A]_{DUT} = \left([A]_{input}\right)^{(-1)} \Big\{[C_A]_{total} - [C_A]_{input} - [A]_{input}[A]_{DUT}[C_A]_{output}[A]^{+}_{DUT}[A]^{+}_{input}\Big\}\left([A]^{+}_{input}\right)^{(-1)} \qquad (2.22)$$

where $[A]_{xx}$ are the usual chain matrices (coherent electrical signals) and $[C_A]_{xx}$ are the noise correlation matrices in chain form; exponents "(-1)" indicates a reversed matrix and "$+$" indicates a transposed and conjugated matrix.

The correlation matrix (impedance, admittance, or chain configurations) of the passive parts of the setup can be directly calculated from S parameters and temperature. The physical temperature of these passive elements is assumed to be uniform and has to be carefully controlled and measured. The S parameters of these passive two-ports are obtained using off- and on-wafer de-embedding procedures detailed previously. For active two-ports (e.g., transistor, noise receiver), the noise correlation matrix has to be determined from multiple noise power (or noise figure) measurements using a numerical procedure (see Section 2.1.3). Today, this method is implemented in specific software dedicated for noise measurements; it is widely used and validated for the characterization of transistors in the microwave range.

In the mm-wave range (above the W-band), the multiple impedance technique using external tuners is possible but limited in the case of on-wafer measurements, especially to accurately determine the noise behaviors of transistors. From a practical point of view, we mention again that only the noise powers or the noise figures are the measured quantities. At high frequency, it is not easy to measure directly and accurately the noise sources (elements of the noise correlation matrices) of the systems. Then, the natural calibration procedure consists of obtaining the actual output noise power or noise figure of the DUT by removing the noise contribution of each element (e.g., receiver, cable, waveguide) of the setup as well as the on-wafer embedded structures (e.g., pad, line). The simplest method consists of first calibrating the noise receiver using a calibrated noise source. Second, the noise figure of each passive two-port (external components and on-wafer structures) is calculated from de-embedded S parameters and its physical temperature. Finally, the noise figure or output noise power of the DUT is deduced from measurements using the well-known Friis equation. Although easy to implement and well adapted for on-wafer noise measurements in mm-wave, this method is far from being complete. In particular, all of the mismatch aspects between the various parts of the bench must be taken into account.

2.4 Applications: advanced silicon MOSFETs and HBTs

In microelectronics, DC and HF measurements are the core for the elaboration of accurate electrical and associated layout models of active and passive devices. These models are implemented as a library (design kit) within CAD platforms.

For active devices such as transistors, we can list two types of models. First, compact models are elaborated with multiparameter physical concepts based on semiconductor, electrokinetic, electromagnetic, and thermal equations. We can cite the BSIM (Berkeley short-channel IGFET (insulated-gate field-effect transistor) model) [59] for MOSFET (CMOS process) and the HICUM (high current model) for bipolar transistors [60] for HBT (BiCMOS process). These models are elaborated for a given technology node and are scalable versus technological parameters, bias, and thermal conditions. Second, we consider lumped electrical circuit-based models, of which parameters or elements are extracted from HF measurements (S parameters, noise and/or nonlinear power measurements). These models are more basic compared to the compact ones. They are strongly correlated to a limited set of transistor topologies, bias, and thermal conditions.

These two types of models are together useful particularly during the period of maturation of a technology node, but also for mm-wave. Indeed, with a basic lumped element-based model, it is possible to report feedback in real time for technological process and transistor architecture optimizations. Such a model is also well suited to carry out preliminary circuit designs, especially at mm-wave frequencies. On the other hand, compact models (even in beta versions) which are more connected to the physical behaviors of the complete process, are of precious help for the elaboration of lumped element-based models. Here, we are interested only in the basic lumped element-based linear model, including HF noise and their validities in mm-wave.

2.4.1 Small-signal modeling and validation in mm-wave frequency range

Conventional methodology (initially for III–V field-effect transistors [42] and [61]) for MOSFET [62–65] equivalent circuit extraction consists of the following steps: (1) off-wafer calibration using a classic method such as SOLT, LRM, or LRRM procedures, in which the reference plane is assumed to be at the probe tips (see Section 2.3.1); (2) on-wafer measurements of a set of dummy structures to move the S-parameter reference planes close to the transistor (see Section 2.3.2); and (3) measurement of the MOSFET in different bias conditions (V_{gs}, V_{ds}).

The equivalent lumped circuit, according to physical properties of the transistor, is composed of an extrinsic (assumed to be constant with bias conditions) and intrinsic part. The extraction of each of the model elements from broadband S-parameter measurements is realized under different biasing points. The extrinsic parameters are generally extracted at zero drain voltage using special techniques for variation of gate overdrive ($V_{gs} - V_{th}$) as a function of frequency [62,63]. The independence of the intrinsic parameters to frequency is considered as an extraction quality factor. Figure 2.24 (left side) represents an example of intrinsic elements as a function of frequency in the case of a 60 × 1 μm total gate width 65 nm n-MOSFET [64]. This example shows a very flat frequency response with element variations less than 10% and up to 110 GHz. To achieve such a result, we have to mention that all steps of the measurement procedure as detailed previously (e.g., off- and on-wafer calibration, determination of extrinsic field-effect transistor parameters) must be carefully checked. Figure 2.24 (right) shows the comparison between measured S parameters and calculated ones from this equivalent circuit up to 110 GHz. Moreover, Figure 2.25 shows that such a simple equivalent circuit remains valid at mm-wave frequencies at least up to 220 GHz. Some resonances in S_{11} and S_{22} in G-band are observable, although they are strongly attenuated thanks to a well-adapted on-wafer de-embedding procedure [50].

2.4.2 Noise parameter extraction techniques and validation in the mm-wave frequency range

As mentioned in Section 2.1.3, the HF noise model of a transistor and its associated parameters (NF_{min}, R_n, and Γ_{opt}) are, together with S parameters, indispensable data for the design of low-noise circuits and systems.

Because noise measurements at mm-wave frequencies (above 60 GHz) are not easy and necessitate specific equipment (see Section 2.2.3), the elaboration of transistor noise models, starting from noise power or noise figure measurements, is first carried out in the microwave frequency range (typically up to 40–60 GHz). Because the noise properties of a transistor are strongly related to its HF electrical behaviors, the noise characterization must start by DC and accurate broadband S-parameter measurements. The accuracy of the noise model will depend of the accuracy of such preliminary measured data. Then, all techniques and discussions concerning off-wafer and on-wafer calibrations mentioned previously (see Section 2.3) are the base of on-wafer noise measurements and the quality of the noise model. This is the reason that the prober station must be equipped both with noise instruments (noise source, receiver, and tuner) and a VNA.

There are two main approaches to elaborate a transistor noise model:

1. The first measurement strategy is based on Eqn (2.17) (see Section 2.1.3), which relates the noise figure (or output noise power, see Eqn (2.18)) of a two-port preceded by a given impedance (or admittance, reflection coefficient) of the generator as a function of NF_{min}, R_n, and Z_{opt} (or Y_{opt}, Γ_{opt}). This method remains

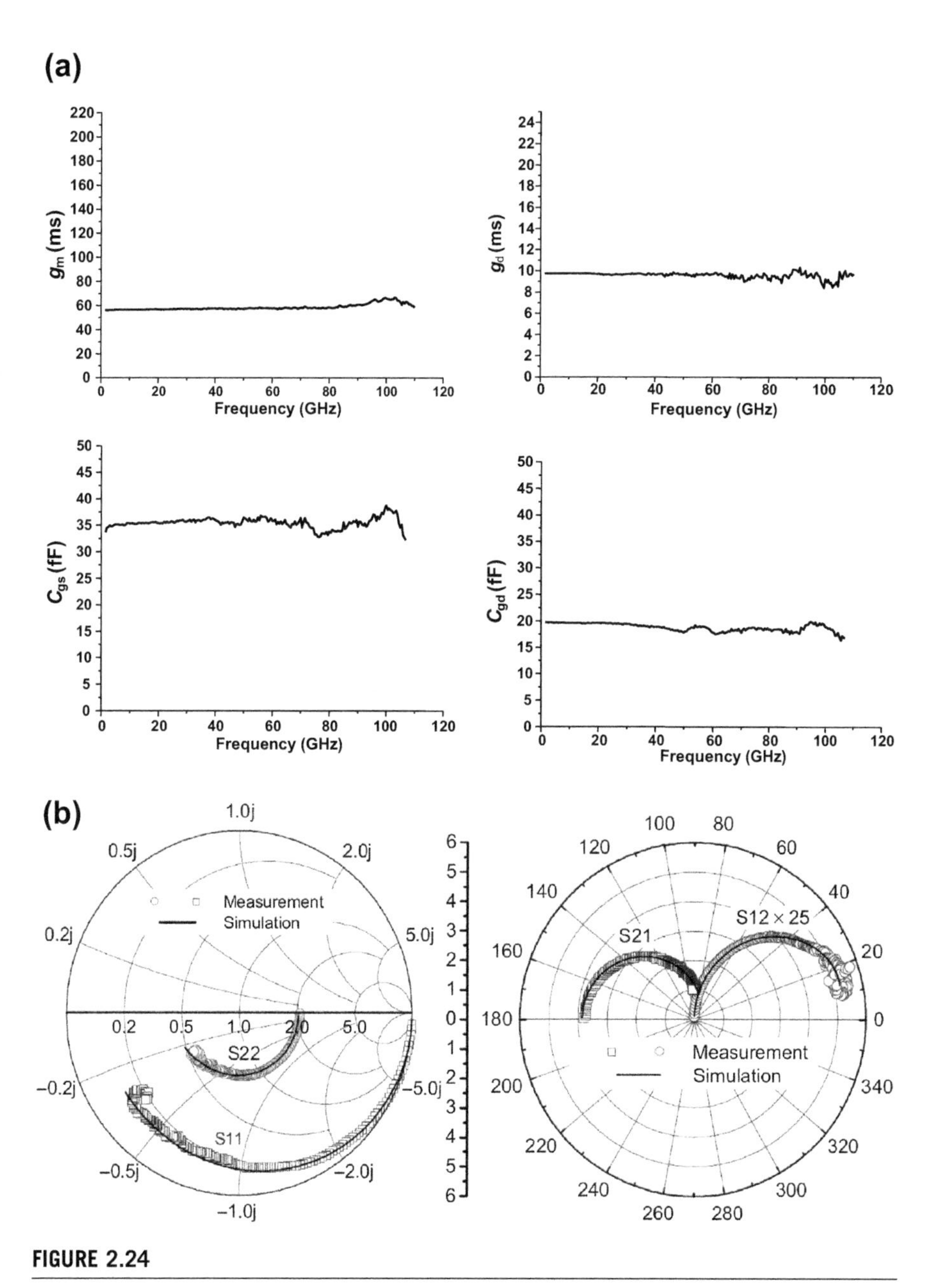

FIGURE 2.24

Left: intrinsic g_m, g_d, C_{gs}, and C_{gd} of a 65-nm *n*-MOSFET versus frequency (up to 110 GHz); right: comparison of measured and calculated *S* parameters (up to 110 GHz).

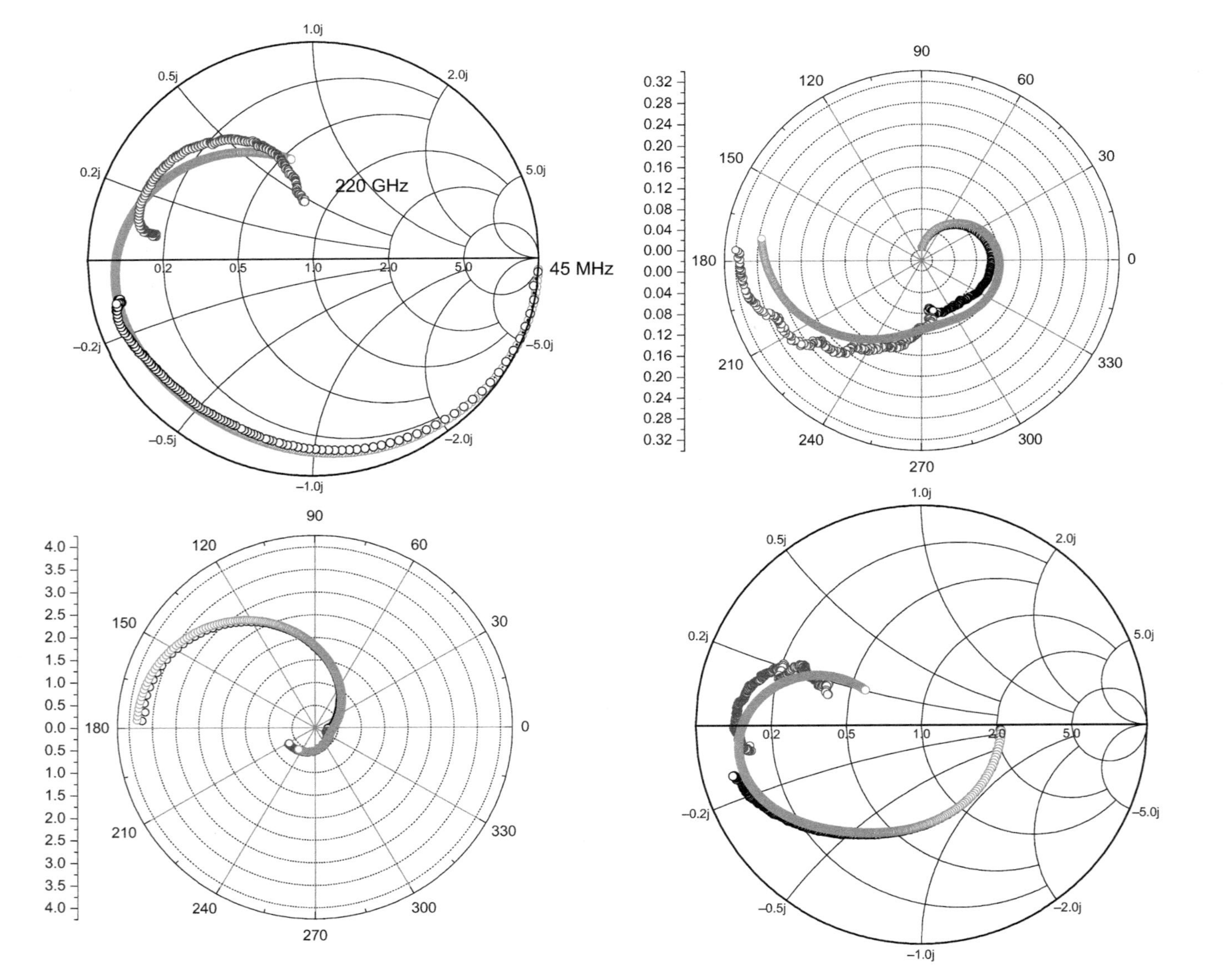

FIGURE 2.25

Comparison of measured and calculated S parameters of a 65-nm n-MOSFET (up to 220 GHz).

valid for any kind of two-port under a linear regime. From an experimental point of view, this method requires a system (tuner) able to modify (synthesize) the impedance of the DUT generator. The four noise parameters (or related noise sources; see Eqn (2.10)) are extracted by mathematically based solvers. It is noted that Eqn (2.17) is an elliptic paraboloid in the complex plan of generator impedance and it is completely solved by a set of four equations. In experimental procedures, seven or more noise measurements are necessary to improve the efficiency of such a numeric approach. Many methods based on this concept have been proposed. Some of them are built on noise figure measurements corresponding to several distinct impedances of the generator [32] and [66] (such a method is also called the multi-impedance hot/cold method), whereas other ones are based on multi-impedance power noise measurements [67] (called the multi-impedance cold method). The main advantage of the second method is that one does not need an active noise source. The input noise equivalent temperature is the ambient temperature of the impedance presented to the DUT, which supposes that the tuner is purely passive.

2. The second measurement approach consists of extracting the noise parameters of a transistor from a physical-based noise model. This technique is appropriate for field-effect transistors (FETs), and numerous theoretical studies were carried out from the 1960s to the 1990s [29] and [68–73]. The main principle is that HF noise (thermal, diffusion origins) sources in FETs are driven by specific properties and that less than four parameters (three parameters in the noise model described in [29], [68] and [70], two in [71], and only one in [69]), are sufficient to entirely calculate the noise properties of FETs. Using these theories, the noise measurement approach called F50 [72] consists of making noise figure measurements over a broad frequency range for a single known impedance of the generator (e.g., the natural impedance of the noise source close to 50 Ω). This method was a striking success because it was easy to implement, particularly in industrial testing.

In the case of HBTs, such a method is not applicable because the HF noise sources are of different natures (shot noise). Moreover, the electrical mechanisms are also fundamentally different compared to FETs. Nevertheless, there also exists a bipolar noise model established on an equivalent circuit and two correlated shot noise sources based on pioneering work by A. Van der Ziel [29], which was extended later (see Ref. [74]). The quadratic mean values of base-emitter and base-collector noise sources are related to DC currents (shot noise). Such a simple model can describe accurately the HF noise properties of HBT in the microwave range if the elements of the linear equivalent circuit, such as extrinsic (base, emitter, and collector) and intrinsic (base-emitter) access resistances, intrinsic transit times [75], and base-collector capacitance, are accurately extracted from broadband S parameters measured at different bias conditions. Nevertheless, such a simple model is much less efficient in the mm-wave frequency range, where the noise behavior depends strongly on the cross-correlation function between the noise sources, which is not straightforward to accurately determine from experiments.

For elaboration or validation of noise models of MOSFETs or HBTs at mm-wave frequencies, the general methodology consists of first extracting the four noise parameters (or noise sources) from multi-impedance techniques in broadband (typically up to 40–50 GHz) or using the F50 method in the case of MOSFETs. The tuner used for such characterization is a precise mechanical system placed in front of the input probe. With such a tuner, we can expect to present to the DUT a maximum magnitude of reflection coefficient close to 0.7 (by taking into account the probe and cables/connectors losses), even at 40–50 GHz. The value is nevertheless sufficiently high to cover the interesting part of the input's Smith chart area in order to accurately extract the noise parameters of many kinds of transistors. The noise properties are then calculated and extrapolated to a higher frequency range. Finally, the noise figures or noise powers of the DUT with or without tuners are measured using millimeter wave noise test sets, which for instance operate in W-band and even in D-band, and then are compared with the calculated ones.

An example of results concerning a 65-nm *n*-MOSFET (60 × 1 μm total gate width) is presented in Figure 2.26. The model is elaborated using the F50 method (up to 40 GHz) and the equivalent circuit has been determined by broadband (up to 60 GHz) *S*-parameter measurements. In this figure, the 8–40 GHz measured data are obtained using the multi-impedance technique and a mechanical 8–50 GHz tuner. Finally, the 75–110 GHz measured data are extracted using the

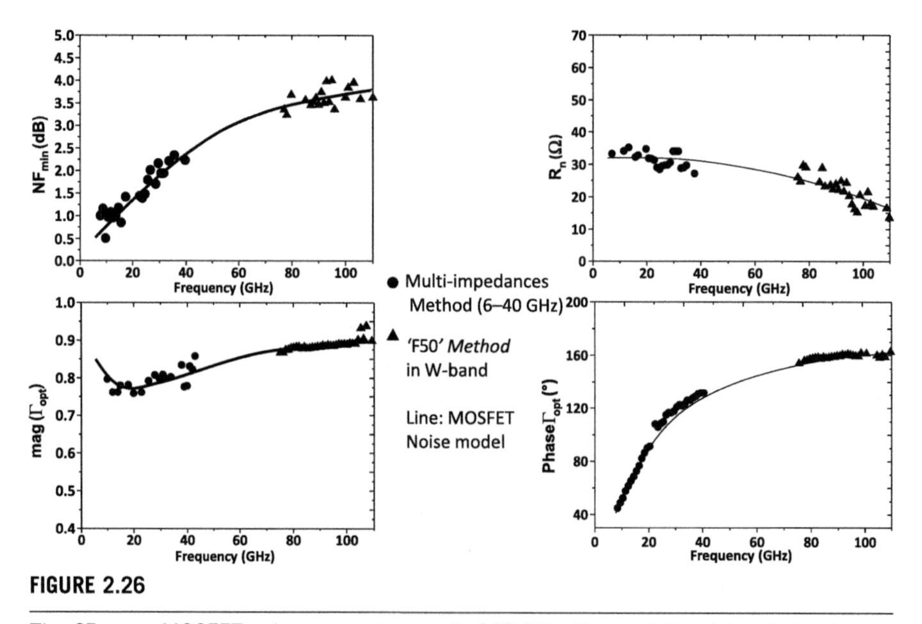

FIGURE 2.26

The 65-nm *n*-MOSFET noise parameters up to 110 GHz; the model is elaborated using the F50 method (up to 40 GHz). The 6–40 GHz measured data are obtained using the multi-impedance technique, and 75–110 GHz measured data are extracted using the F50 method.

F50 method from noise figure measurements in W-band. This example proves that all of these experimental techniques, setups, and associated theories are particularly efficient events in the mm-wave range. Such techniques are now largely used for inline industrial HF characterizations.

Beyond the W-band, the noise characterization of transistors using these conventional techniques becomes complex and even questionable. There are many reasons: (1) the gain of DUT is generally not high enough and the noise receiver's sensitivity becomes insufficient to obtain accurate noise power measurements—an effect that becomes more and more important due to the losses of waveguides and probes; (2) the commercially available solid-state noise sources are limited to 170 GHz; and (3) the influence of transistors' access parasitics (e.g., pads, line, top-down interconnections) strongly modifies the noise properties, and the noise parameters (or noise model) of an intrinsic transistor become unusable for CAD.

Due to these difficulties in accurately determining the noise parameters of transistors in the mm-wave frequency range, new concepts appeared. The main idea is to integrate (partly or integrally) some elements of the noise test set close to the DUT, such as the tuner, LNA, noise receiver, or even noise source. The studies concerning such in-situ test sets belong to the field of BIST systems.

A first basic approach of this principle is detailed in Ref. [76]. Here, the method consists of validating the mm-wave linear model (including noise) of 65 nm *n*-MOSFETs by using narrow-band prematched transistors designed at several frequencies covering the W-band and partly the D-band (77, 89, 105, and 150 GHz). The objective is to reduce the noise figure and increase the gain in order to increase

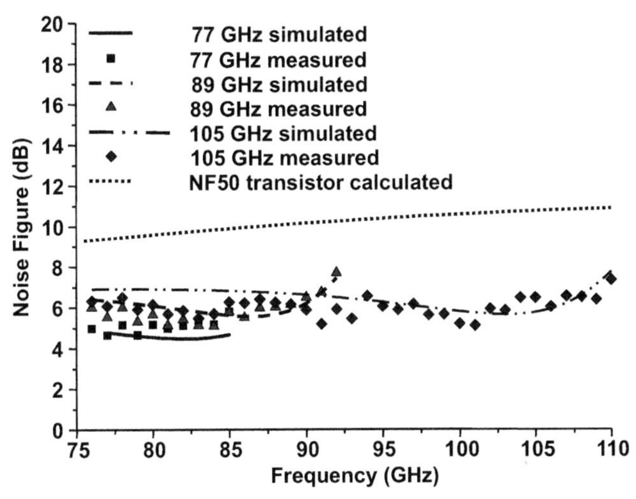

FIGURE 2.27

Measured noise figures of 77, 89, and 105 GHz prematched 65-nm *n*-MOSFET comparison with modeling.

the accuracy of noise measurements. An example of results is depicted in Figure 2.27 for 65-nm MOSFET technology, where the reduction of noise figures of the prematched structures compared to the 50 Ω noise figure of the unique transistor is clearly shown. Other approaches consist of integrating more parts of the noise measurement setup. For instance, in Ref. [77], a fully integrated measurement system is realized to extract the noise parameters of advanced HBT in W- and D-bands. This circuit integrates around the DUT an input passive tuner, an output LNA, and even for the D-band system, a complete receiver involving DSB mixer, VCO (voltage controlled oscillator), and LO and IF amplifiers. On the same chip, a strictly similar noise measurement setup without DUT is used for the calibration procedure. The main interest of such an integrated measurement system is to reduce the losses of passive elements, which are necessary in a conventional external bench (e.g., waveguides, probes) and the parasitic effects of the transistor accesses (pads, lines, top-down interconnections). Of course, such a circuit becomes a silicon area consuming with respect to its objective, which consists of characterizing only one DUT.

An intermediary solution between these two first examples of the BIST approach is proposed in Ref. [78]. In this case, only the tuner is integrated in front of the DUT. The integrated setup is designed for the noise parameters' extraction in the W-band of advanced mm-wave SiGe-based HBTs (BiCMOS process). Figure 2.28 (left) shows the photograph of the integrated tuner involving a cold FET for resistive impedance variations in parallel with a varactor for reactive ones. This tuner presents a large range of impedances over the W-band. The impedance constellation covered by the tuner at 90 GHz is presented in Figure 2.28 (right). Figure 2.29 represents the

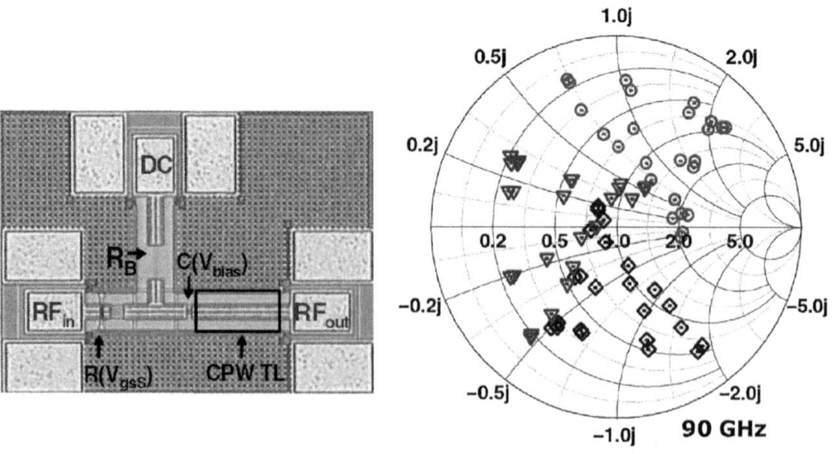

FIGURE 2.28

In-situ W-band tuner layout (left) and impedance constellation at 90 GHz (right).

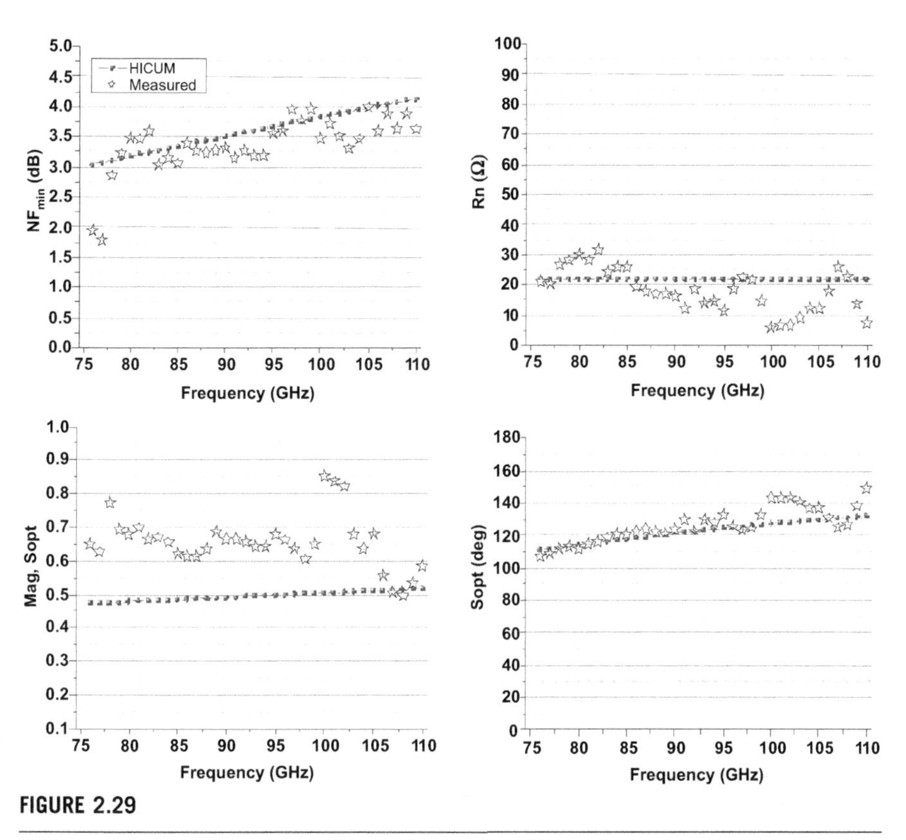

FIGURE 2.29

Four noise parameters of HBT in the W-band: comparison between experimental values extracted from in-situ tuner based system and calculated ones.

four noise parameters of an HBT in the W-band extracted using this in-situ tuner based system.

Acknowledgments

I would like to first acknowledge my Ph.D. students who have strongly contributed to the many experimental results presented in this chapter: Dr Nicolas Waldhoff, Dr Yoann Tagro, Dr Frédéric Gianesello, Dr Samuel Boret, Dr Laurent Poulain, and Marina Deng. Numerous results on silicon have been obtained in the framework of a common laboratory between ST Microelectronics, Crolles, and IEMN—CNRS—University of Lille; I would like particularly to thank Dr Daniel Gloria, who is in charge of the advanced characterization and modeling group at ST Microelectronics. Finally, I acknowledge my colleagues: Prof. François Danneville, Prof. Alain Cappy, Dr Lionel Buchaillot, Sylvie Lepilliet, and Dr Damien Ducatteau. Some content of this chapter has benefited from the facilities of the ExCELSiOR (ANR-11-EQPX-0015)—Nanoscience Characterization Center (www.excelsior-ncc.eu).

References

[1] ITRS Edition. RF and analog/mixed-signal technologies (RFAMS). http://www.itrs.net/Links/2011ITRS/Home2011.htm; 2011.

[2] Heinemann B, Barth R, Bolze D, Drews J, Fischer GG, Fox A, et al. SiGe HBT technology with fT/f max of 300GHz/500GHz and 2.0 ps CML gate delay. In: Proc Int Electron Device Meet (San Francisco, CA, USA), December 2010. pp. 688−769.

[3] Reck TJ, Chen L, Zhang C, Arsenovic A, Groppi C, Lichtenberger A, et al. Micromachined probes for submillimeter-wave on-wafer measurements—Part II: RF design and characterization. IEEE Trans Terahertz Sci Technol November 2011;1(2):357−63.

[4] Campbell GA, Foster RM. Maximum output networks for telephone substation and repeater circuits. Trans AIEE (Am Inst Electr Eng) 1920;39:231−80 [Available within IEEE Xplore©].

[5] Montgomery CG, Dicke RH, Purcell EM. Principles of microwave circuits. (New York, NY, USA): McGraw-Hill; 1948.

[6] Marcuvitz N. Waveguide handbook. (New York, NY, USA): McGraw-Hill; 1951.

[7] Carlin H. The scattering matrix in network theory circuit theory. Circuit theory, IRE Trans June 1956;3(2):88−97.

[8] Collin RE. Foundations for microwave engineering. (New York, NY, USA): McGraw-Hill; 1966.

[9] Kerns DM, Beatty RW. Basic theory of waveguide junctions and introductory microwave network analysis. (Oxford, UK): Pergamon Press; 1967.

[10] Youla DC. On scattering matrices normalized to complex port numbers. Proc IRE 1961;49(7):1221.

[11] Kurokawa K. Power waves and the scattering matrix. IEEE Trans Microwave Theory Tech March 1965;MTT-13(3):194−202.

[12] Marks RB, Williams DF. A general waveguide circuit theory. J NIST September−October 1992;97(5).

[13] Williams DF, Alpert BK. Characteristic impedance, power, and causality. IEEE Microwave Guided Wave Lett May 1999;9(5):165−71.

[14] Engen GF. The six-port reflectometer: an alternative network analyzer. IEEE Trans Microwave Theory Tech December 1977;MTT-25:1075−80.

[15] http://www.ieeeghn.org/wiki/index.php/Douglas_K._Rytting.

[16] Hiebel M. Fundamentals of vector network analysis. 3rd ed. Rohde & Schwarz, ISBN 978-3-939837-06-0; 2008.

[17] http://cp.literature.agilent.com/litweb/pdf/5965-7917E.pdf.

[18] http://www.anritsu.com/en-US/Downloads/Application-Notes/ApplicationNote/DWL7959.aspx.

[19] Rehnmark S. On the calibration process of automatic network analyzer systems. IEEE Trans Microwave Theory Tech April 1974;22(4):457−8.

[20] Engen GF, Hoer CA. Thru-reflect-line: an improved technique for calibrating the dual 6-port automatic network analyzer. IEEE Trans Microwave Theory Tech December 1979;MTT-27(12):983−7.

[21] Eul HJ, Schiek B. A generalized theory and new calibration procedures for network analyzer self-calibration. IEEE Trans Microwave Theory Tech April 1991;39:724−31.

[22] Marks RB. A multiline method of network analyzer calibration. IEEE Trans Microwave Theory Tech July 1991;MTT-39(7):1205−15.

[23] Eul H-J, Schiek B. Thru-match-reflect: one result of a rigorous theory for de-embedding and network analyzer calibration. In: Proc 18th Euro Microwave Conf (Amsterdam, The Netherlands), September 1988. pp. 909–14.

[24] Davidson A, Jones K, Strid E. LRM and LRRM calibrations with automatic determination of load inductance. In: IEEE Fall ARFTG Conf Dig (Monterrey, CA, USA), November 1990. pp. 57–63.

[25] http://www.nist.gov/pml/electromagnetics/rf_electronics/on_wafer_calibration_measurement_and_measurement_verification.cfm.

[26] http://www.cascademicrotech.com/products/probes/wincal-xe.

[27] Johnson J. Thermal agitation of electricity in conductors. Phys Rev July 1928;32:97–109.

[28] Nyquist H. Thermal agitation of electric charge in conductors. Phys Rev July 1928;32:110–13.

[29] Van Der Ziel A. Noise in solid-state devices and lasers. Proc IEEE August 1970;58(8):1178–206.

[30] Schottky W. Über spontane Stromschwankungen in verschiedenen Elektrizitätsleitern. Annalen der Physik 1918;57(362):541–67.

[31] Hillbrand H, Russer P. An efficient method for computer-aided noise analysis of linear amplifier networks. IEEE Trans Circuits Syst April 1976;CAS-23:235–8.

[32] Lane RQ. The determination of device noise parameters. Proc IEEE August 1969;57(8):1461–2.

[33] http://www.focus-microwaves.com/.

[34] http://www.maurymw.com/MW_RF/Application_Notes_Library.php.

[35] Adamian V, Uhlir U. A novel procedure for receiver noise characterization. IEEE Trans Instrum Meas June 1973;IM-22(2):181–2.

[36] Strid E, Gleason K. A microstrip probe for microwave measurements on GaAs FET and IC wafers. In: GaAs IC Symp Dig (Las Vegas, NV, USA), November 1980. Paper 31.

[37] Strid EW, Gleason KR. A DC-12 GHz monolithic GaAsFET distributed amplifier. IEEE Trans Electron Devices August 1982;29(7):1065–71.

[38] http://www.cmicro.com/products/probe-systems/300mm-wafer.

[39] Rhode & Schwarz Application Note: http://www2.rohde-schwarz.com/file_18442/ZVA-Zxx_dat-sw_en.pdf.

[40] Agilent Technologies N5261A and N5262A User's and Service Guide: http://cp.literature.agilent.com/litweb/pdf/N5262-90001.pdf.

[41] Chen CH, Deen MJ. A general noise and S-parameter deembedding procedure for on-wafer high-frequency noise measurements of MOSFETs. IEEE Trans Microwave Theory Tech May 2001;49(5):1004.

[42] Dambrine G, Cappy A, Heliodore F, Playez E. A new method for determining the FET small-signal equivalent circuit. IEEE Trans Microwave Theory Tech July 1988;36(7):1151–9.

[43] Carbonero JL, Joly R, Morin G, Cabon B. On-wafer high frequency measurement improvements. In: Proc IEEE Int Microelectron Test Struct Conf, vol. 7 (San Diego, CA, USA), March 1994. pp. 168–73.

[44] Tiemeijer LF, Havens RJ, Jansman ABM, Bouttement Y. Comparison of the 'pad-open-short' and 'open-short-load' de-embedding techniques for accurate on-wafer RF characterization of high quality passives. IEEE Trans Microwave Theory Tech February 2005;53(2):723–9.

[45] Tiemeijer LF, Pijper RMT, van Steenwijk JA, van der Heijden E. A new 12-term open-short-load de-embedding method for accurate on-wafer characterization of RF MOSFET structures. IEEE Trans Microwave Theory Tech February 2010; 58:419–33.

[46] Cho H, Burk DE. A three-step method for the de-embedding of high-frequency S-parameter measurements. IEEE Trans Electron Devices June 1991;38(6):1371.

[47] Kolding TE. A four-step method for de-embedding gigahertz on-wafer CMOS measurements. IEEE Trans Electron Devices April 2000;47(4):734–40.

[48] van Wijnen PJ, Claessen HR, Wolsheimer EA. A new straightforward calibration and correction procedure for 'on wafer' high frequency S-parameter measurements (45 MHz–18 GHz). In: Proc IEEE Bipolar Circuits Tech Meet (Santa Barbara, CA, USA), June 1987. pp. 70–3.

[49] Koolen MCAM, Geelen JAM, Versleijen MPJG. An improved de-embedding technique for on-wafer high-frequency characterization. In: Proc IEEE Bipolar Circuits Tech Meet (Minneapolis, MN, USA), May 1991. pp. 188–91.

[50] Waldhoff N, Andrei C, Gloria D, Lepilliet S, Danneville F, Dambrine G. Improved characterization methodology for MOSFETs up to 220 GHz. IEEE Trans Microwave Theory Tech May 2009;57(5):1237–43.

[51] Raya C, Derrier N, Chevalier P, Gloria D, Pruvost S, Celi D. From measurement to intrinsic device characteristics: test structures and parasitic determination. In: Proc IEEE Bipolar/BiCMOS Circuits Tech Meet (BCTM) (Monterrey, CA, USA), October 2008. pp. 232–9.

[52] Yau KHK, Dacquay E, Sarkas I, Voinigescu S. Device and IC characterization above 100GHz. IEEE Microwave Magazine January–February 2012;13(1):30–54.

[53] Andrei C, Gloria D, Danneville F, Dambrine G. Efficient de-embedding technique for 110 GHz deep-channel-MOSFET characterization. IEEE Microwave Wireless Components Lett April 2007;17(4):301–3.

[54] Mangan AM, Voinigescu SP, Yang M-T, Tazlauanu M. De-embedding transmission line measurements for accurate modeling of IC designs. IEEE Trans Electron Devices February 2006;53(2):235–41.

[55] Cho M-H, Huang G-W, Wang Y-H, Wu L-K. A scalable noise de-embedding technique for on-wafer microwave device characterization. IEEE Microwave Wireless Comp Lett October 2005;15(10):649–51.

[56] Yau KHK, Mangan AM, Chevalier P, Schvan P, Voinigescu SP. A transmission-line based technique for de-embedding noise parameters. In: Proc IEEE Int Conf Microelectron Test Struct (ICMTS) (Tokyo, Japan), March 2007. pp. 237–42.

[57] Deal W, Leong K, Mei XB, Sarkozy S, Radisic V, Lee J, et al. Scaling of InP HEMT cascade integrated circuits to THz frequencies. In: IEEE Comp Semiconductor IC Symp (CSICS) Digest (Monterey, CA, USA), October 2010. pp. 195–8.

[58] Andrei C, Gloria D, Danneville F, Scheer P, Dambrine G. Coupling on-wafer measurement errors and their impact on calibration and de-embedding up to 110 GHz for CMOS millimeter wave characterizations. In: Proc IEEE Int Conf Microelectron Test Struct (ICMTS) (Tokyo, Japan), March 2007. pp. 253–6.

[59] http://www-device.eecs.berkeley.edu/bsim/.

[60] http://www.iee.et.tu-dresden.de/iee/eb/hic_new/hic_start.html.

[61] Berroth M, Bosch R. Broadband determination of the FET small-signal equivalent circuit. IEEE Trans Microwave Theory Tech July 1990;38(7):891–5.

[62] Raskin JP, Dambrine G, Gillon R. Direct extraction of the series equivalent circuit parameters for the small-signal model of SOI MOSFET's. IEEE Microwave Guided Wave Lett December 1997;7(12):408–10.

[63] Bracale A, Pasquet D, Gautier JL, Fel N, Pelloie JL. A new approach for SOI devices small signal parameters extraction. In: Processing Analog Integr Circuits Signal Processing. (Norwell, MA, USA): Kluwer Academic Publisher; November 2000.

[64] Waldhoff N, Andrei C, Gloria D, Danneville F, Dambrine G. Small signal and noise equivalent circuit for CMOS 65 nm up to 110 GHz. In: Proc Euro Microwave Conf (Amsterdam, The Netherlands), October 2008. pp. 321–4.

[65] Poulain L, Waldhoff N, Gloria D, Danneville F, Dambrine G. Small signal and HF noise performance of 45 nm CMOS technology in mmW range. In: Proc IEEE Radio Freq Integr Circuits Symp (RFIC) (Baltimore, MD, USA), June 2011. pp. 1–4.

[66] Boudiaf A, Laporte M, Dangla J, Vernet G. Accuracy improvements in two-port noise parameter extraction method. In: IEEE MTT-s Int Microwave Symp Dig, 1, vol. 3 (Albuquerque, NM, USA), June 1992. pp. 1569–72.

[67] Adamian V, Uhlir A. A novel procedure for receiver noise characterization. IEEE Trans Instrum Meas June 1973;IM-22:181–2.

[68] Pucel RA, Haus HA. Signal and noise properties of gallium arsenide microwave field effect transistors. In: Adv Electro Electron Phys, vol. 38 (New York, NY, USA): Academic Press; 1975. pp. 195–265.

[69] Gupta MS, Greiling PT. Microwave noise characterization of GaAs MESFET's: determination of extrinsic noise parameters. IEEE Trans Microwave Theory Tech April 1988;36(4):745–51.

[70] Cappy A. Noise modeling and measurement techniques. Invited paper. IEEE Trans Microwave Theory Tech January 1988;36(1):1–10.

[71] Pospieszalski MW. Modeling of noise parameters of MESFET's and MODFET's and their frequency and temperature dependence. IEEE Trans Microwave Theory Tech September 1989;37(9):1340–50.

[72] Dambrine G, Happy H, Danneville F, Cappy A. A new method for on-wafer noise measurement. IEEE Trans Microwave Theory Tech March 1993;41(3):375–81.

[73] Danneville F, Dambrine G, Happy H, Tadyszak P, Cappy A. Influence of the gate leakage current on the noise performance of MESFETs and MODFETs. Solid-State Electronics May 1995;38(5):1081–7.

[74] Hawkins R. Limitations of Nielsen's and related noise equations applied to microwave bipolar transistors, and a new expression for the frequency and current dependent noise figure. Solid-State Electronics March 1977;20(3):191–6.

[75] Zerounian N. Transit times of SiGe:C HBTs using nonselective base epitaxy. Solid-State Electronics October 2004;48(10–11):1993–9.

[76] Waldhoff N, Tagro Y, Gloria D, Gianesello F, Danneville F, Dambrine G. Validation of the 2 temperatures noise model using pre-matched transistors in W-band for sub-65 nm technology. IEEE Microwave Wireless Components Lett May 2010;20(5):274–6.

[77] Yau KHK, Chevalier P, Chantre A, Voinigescu SP. Characterization of the noise parameters of SiGe HBTs in the 70–170 GHz Range. IEEE Trans Microwave Theory Tech August 2011;59(8):1983–2000.

[78] Tagro Y, Gloria D, Boret S, Morandini Y, Dambrine G. In-situ silicon integrated tuner for automated on-wafer MMW noise parameters extraction using multi-impedance method for transistor characterization. In: Proc IEEE Int Conf Microelectron Test Struct (ICMTS) (Oxnard, CA, USA), March–April 2009. pp. 184–8.

CHAPTER

3 Characterization and Modeling of High-Frequency Active Devices Oriented to High-Sensitivity Subsystems Design

Ernesto Limiti, Walter Ciccognani, Sergio Colangeli

Dipartimento di Ingegneria Elettronica, University of Roma "Tor Vergata", Roma, Italy

3.1 Introduction

Taking noise measurements is notoriously a very challenging task, since sources of errors are numerous and difficult to avoid, to correct, or even to figure out. Therefore, measurements should be taken by skillful, highly experienced personnel, especially if critical applications are involved or very low-noise devices are to be tested. This is in contrast with the often encountered opinion that characterization of linear devices is basically a simple task, for which the effects of the microwave test bench and the shift of its reference planes may be performed by well-defined and stabilized procedures. For the case of noise characterization and device modeling, however, considering the effects of the test bench and its complexity, together with the necessity to properly de-embed its influence on the performed measurements, such opinion is clearly not applicable.

On the other hand, today's trends are indeed oriented toward more and more advanced technologies, so that such expertise gets increasingly necessary—in fact, as the technological progress moves forward, measurement techniques and instrumentation need to be refined and made more sophisticated. Very high-frequency processes are being developed that represent a major challenge to those concerned with their measurement, while extremely low-noise or cryogenic devices are approaching the limits themselves beyond which classical models fail to be valid, and some common assumptions cease to hold.

The reason why such a rush toward the extreme possibilities of microwave devices is being pursued lies in the fact that new advanced systems are being addressed for applications in security, defense, and even communication markets. This is the case for passive imaging systems for security applications (e.g., millimeter-wave scanners [1−3]) where the increase in operating frequency range lies well within the millimeter-frequency bands (often touching the terahertz region), and expected sensitivity of the resulting systems should be kept as high as possible [4]. Regarding the advanced scientific applications, radio-astronomy

Microwave De-embedding. http://dx.doi.org/10.1016/B978-0-12-401700-9.00003-3

receivers [5–7] are nowadays approaching the multi-pixel camera [8] or the focal-plane array configuration [9], thus demanding compact, high-frequency, and even cryogenic operation [10]. Finally, even in the mass-market applications related to cellular base stations, a high-sensitivity receiving subsystem actually allows a larger unit cell size, thus leading to a major reduction of the cell number and hence system complexity. Such increase in sensitivity even led to the adoption of liquid nitrogen cryogenic cooling of the receiving section, requiring dedicated approaches and devices [11].

Another modern trend resides in the adoption of GaN devices also for low-noise amplifier (LNA) stages, thanks to the "robustness" of the resulting module to impinging input signals [12]. Such adoption is however strictly related to the capability of effectively characterizing and modeling GaN HEMTs not only for power applications but also if noise performance is concerned.

In this chapter, quite a broad overview of noise characterization-related topics will be offered to the reader, with different depth levels. Most of the attention, however, will be paid to the practical side of noise measurements and the subsequent steps of noise extraction and modeling, as well as to some advanced design methodologies. A major concern will be in the procedures that are necessary to effectively de-embed the measurements from the contribution of the test bench and the adopted methodologies. The scope of the discussion will cover a well-assessed theory concerning linear devices operated in the frequency range from a few megahertz to some 100 GHz, and at physical temperatures above some tens of kelvins. In these conditions, $1/f$ noise can be neglected and Johnson (thermal) noise is approximately independent of frequency; as a consequence, thermal and, possibly, shot noise of elemental noise sources add up to yield a white power spectrum, which can be conveniently described in terms of "equivalent" thermal noise.

The second part of the chapter will be devoted to the application of the device noise models in the proper design of single- and multistage low noise amplifiers, including a mixed technique that actually employs characterization techniques directly in the amplifier design.

Since the focus throughout the whole chapter will often fall on the concepts of "calibration" and "embedding" (or "de-embedding"), it will now be specified how those terms are used in this contribution. The term *calibration* refers to the operations that allow adjusting an instrument so that systematic effects are removed from measurements; examples of this are vector network analyzer (VNA) calibrations, such as SOLT and TRL, and noise receiver calibrations, such as through the Y-factor method (see Section 3.3.1). Especially—but not exclusively—when the calibration is simple enough that it is accomplished by plainly adding or subtracting a term, the word *correction* is often used (for instance, a measurement correction is accomplished when subtracting the receiver's noise temperature from the total noise temperature at its input section). Notice that the term *calibration* could also refer to the process of measuring some parameter of interest of a piece of equipment, to be subsequently used as a reference; this is the case of tuner calibrations (see Section 3.2.1).

It is worthwhile noting that, as far as high-frequency measurements are concerned, a calibration sets one or more reference planes, such as the probe-tips in a SOLT calibration, or the receiver's input section in noise calibrations. Depending on the particular case, these reference planes may or may not be those corresponding to the device under test (DUT)—which, notice, may not be directly accessible. In such cases, what is actually measured is a network in which the DUT is included, or "embedded"; therefore, the operations required to remove from measurements the unwanted contributions is referred to as "de-embedding," while the opposite action is obviously labeled "embedding." As a consequence, (de-)embedding is fundamentally a shift of reference planes that, in simple cases—e.g., (de-)embedding of linear two-ports—is readily accomplished by means of matrix multiplications (divisions).

It should therefore be apparent that the concepts of calibration and de-embedding are very strictly entangled and may even coincide in some cases. For instance, a TRL calibration can be thought of as a special case of de-embedding, where what is removed is the contribution of two error boxes representing the systematic effects of the VNA.

3.2 High-frequency noise measurement benches

3.2.1 Typical instrumentation

It is here taken for granted that the typical instrumentation of a high-frequency test laboratory is available and familiar to the reader: the phrase "typical instrumentation" refers here to low-level components, such as cables, adapters, bias-tees, RF probes, RF switches, and so on, as well as to power supplies, meters, and VNAs.

In addition to the above, noise measurements typically require more specific hardware, such as solid-state noise sources and noise figure meters. Very often ferrite isolators and LNAs are necessary, and in some cases even RF electromechanical tuners; as the test bench complexity grows, a number of RF switches may become necessary, as well as relays to properly actuate them and a personal computer to control and manage the setup along all the measurement phases.

As to the solid-state noise source, this component is fundamentally a matched load that can generate two different levels of broadband noise power, serving the function of reference levels. Since the source's noise is white, it is *equivalent* to that of a passive noisy resistor (which only generates thermal noise) whose physical temperature can be switched, upon request, between two values: a cold (ambient) temperature T_C and a hot temperature T_H. Notice that the physical temperature of the noise source is almost unaffected by the state being chosen, since the additional noise when in the hot state is commonly obtained by reverse-biasing a PN diode. Of course, the reverse current flowing in the diode could lead to a slight increase in T_C as compared to the room temperature, which can be easily detected by means of a temperature sensor thermally connected to the noise source connector or case.

It is noteworthy remarking that, if primary standards were to be used for calibration, T_C may be, for instance, a cryogenic temperature obtained from a liquid nitrogen bath, and T_H the ambient temperature of a matched termination. Therefore the terms *cold* and *hot* should be intended according to the specific context; in general, they simply refer to a pair of different temperatures, one (T_C) lower than the other (T_H). Of course, primary standards are not common if not in metrology laboratories; therefore only solid-state noise sources will be considered in the following.

With particular regard to such noise sources, notice that due to their working principle, it is difficult to guarantee a high return loss over the whole frequency range of operation and—in particular—to minimize the variations of their reflection coefficient between the cold (or "off") and the hot (or "on") states. To overcome this issue, solid-state noise sources typically embed an attenuator pad; unfortunately, however, the pad attenuation cannot be arbitrarily high, since this would result in lowering the T_H value too close to T_C and hence making the noise source useless for practical purposes. The distance between T_C and T_H for a noise source is usually quantified by means of its excess noise ratio (ENR), defined as:

$$\begin{aligned} \mathrm{ENR} &= \frac{T_H - T_C}{T_0} \\ \mathrm{ENR_{dB}} &= 10 \cdot \log_{10}(\mathrm{ENR}) \end{aligned} \tag{3.1}$$

where $T_0 = 290$ K is the standard noise temperature. The ENR of a commercial noise source is usually expressed in decibels and tabulated as function of frequency under specified environmental conditions: in particular, $T_C = T_0$ is usually assumed. Notice that in solid-state noise sources, T_H comes from the sum of thermal noise, which is the same as for T_C, and an "excess" noise due to diode current; hence the difference $T_H - T_C$, as well as ENR, may be safely assumed to be a constant quantity, while T_H should be computed as:

$$T_H = T_C + T_0 \cdot \mathrm{ENR} \tag{3.2}$$

Notice that the T_H value, which can be computed from the noise source's ENR as provided on its data sheets, is not really an *available* noise temperature but rather the *absorbed* noise temperature on a matched (50 Ω) load. If the noise source is also very well matched ($\Gamma_{out,NS} \cong 0$), the difference between these two quantities is negligible, otherwise the data sheet value should be corrected (normalized) by a factor $1 - |\Gamma_{out,NS}|^2$. No correction is required, on the contrary, on T_C, which is simply the noise source's physical temperature.

Common values for ENR are 6 dB or 15 dB; notice that the internal architecture of either type of noise source is identical, the only difference being in the attenuator pad. Hence low-ENR sources typically feature better characteristics in terms of return loss, as well as of reflection coefficient stability between off and on states. Nevertheless, the noise source selection also depends on other factors, as will be described in the following sections.

Together with the ENR, manufacturers provide its associated uncertainty, which is usually in the order of ± 0.1 dB, although varying with frequency. Such information is necessary in order to assess the global uncertainty of a noise measurement.

As to the actual noise measurements, a simple power meter is clearly unsuited—although noise measurements are a sophisticated form of power measurements. Indeed, other functionalities than mere power detection are needed in practice, i.e., filtering and down-conversion. The noise figure meter is the instrument specifically designed to integrate all these features, as well as routines for managing a complete 50 Ω noise factor (F_{50}) measurement, all corrections included. In most cases, LNA with good linear performance (matching, gain, and noise figure) can be measured with remarkable accuracy simply by means of a calibrated noise source and a noise figure meter.

However, a noise figure meter could also be replaced by a spectrum analyzer. The possible drawbacks concern the instrument noise figure, which is usually higher than in a noise figure meter, and the lack of preprogrammed routines for automated measurements. Both of the above, on the other hand, can be easily worked around: the former by using an external LNA as a preamplifier, the latter by remotely controlling the instrument via a PC running custom measurement and correction routines. Moreover, if a remote PC is used, a modern spectrum analyzer offers far more versatility than any standard noise figure meter, and in fact, due to its superior filtering capabilities, is even better suited than the latter when measurements over complex source terminations are needed, as it will clearly appear in Section 3.4.3.

Moreover, some of the latest models of commercial network analyzers also exploit internal receivers for noise measurements. In such cases, S parameters and noise figure can be measured simultaneously thanks to complex, turnkey systems essentially made up of a single instrument. Nevertheless, at present, these all-in-one systems are limited to noise measurements up to K-band; beyond this boundary, an additional, external receiver is therefore needed.

For the sake of generality, the instrument selected to measure noise power—whether a noise figure meter, a spectrum analyzer, or a last-generation network analyzer—will be in the following referred to as a *receiver*. In other words, the term will be used with no reference to the actual instrument architecture but simply to its functional use within the test bench.

The last component that will be encountered along the discussion is the microwave tuner. Although active realizations of these components exist, mainly for applications in power test benches, only passive, electromechanical tuners will be considered here (see, for instance, Figure 3.1(a)). The purpose of such two-port components is to synthesize arbitrary (passive) reflection coefficients at one port when the other port is terminated on a matched load; however, they can be inserted in a critical section of a two-port's cascade to improve the matching level.

In general, passive microwave tuners consist of a 50 Ω slabline and a probe, referred to as the "slug," which can be moved both along the slabline (x-direction) and vertically (y-direction). If port 1 (port 2) is terminated on a matched load, the reflection coefficient shown by the tuner at port 2 (port 1) moves along a curve line

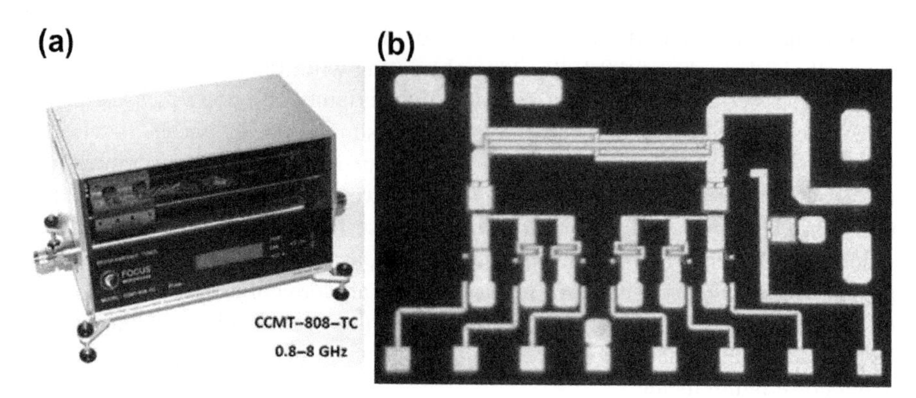

FIGURE 3.1

Examples of passive tuning devices for microwave applications. (a) Electromechanical tuner produced by focus microwaves. (b) Monolithic variable load (2.5×1.3 mm^2).

from the origin of the Smith chart to the edge as the slug, from a totally retracted position, approaches the inner conductor. On the other hand, for a fixed y-position of the slug, it is possible to vary the reflection coefficient's phase by acting on the x-position.

Before being used in a test bench, the tuner needs to be calibrated on a set of different positions—corresponding to as many reflection coefficients. A polar grid of points is commonly obtained on the Smith chart, which can be further refined by interpolation. When the tuner is used as a two-port (rather than a variable termination), more complex calibration schemes become necessary, since the whole scattering matrix of the tuner is needed.

For reasons of test bench automation and repeatability, the slug must be moved by stepped motors so tuner positioning can be performed by a computer controller (hence the denomination "electromechanical"). In any case, tuner repeatability is traditionally a major concern for its practical usability, making calibration a delicate (and time-consuming) step, to be periodically performed.

With the aim of overcoming repeatability issues, as well as avoiding the high costs of electromechanical tuners, fully electronic and monolithic realizations of variable loads have also been proposed [7], which allow a digital control of the synthesized impedance. The active devices used in such circuits—shown in Figure 3.1(b)—operate in "cold-FET" mode, and can therefore be thought as sources of thermal noise only. As a consequence, in much the same way as electromechanical tuners, they are fully described by their scattering parameters and physical temperature.

3.2.2 Measuring noise power

Electrical noise is typically thought of as an unwanted disturbance, whose power level should be kept as low as possible with respect to signal power. Precisely this broadband noise feature of presenting a very weak power level makes it very difficult to accurately measure it. Spurious signals in the receiver at the measurement frequencies should always be checked out and avoided, as well as electromagnetic interferences in the test laboratory: PCs should be shielded, wireless hot spots disabled, mobile phones turned off, and so on. For the same reason, shielded cables with carefully cleaned, threaded connectors should be preferred.

In order to raise the power level incident on the power sensor and increase accuracy, the receiver bandwidth should be kept as wide as possible, as long as the receiver itself does not risk becoming nonlinear by overdriving. Other limitations are represented by the device's own bandwidth—which of course cannot be exceeded—or, in source-pull measurements, by the speed at which the source termination moves along the Smith chart in the selected bandwidth (see Section 3.4.3). When measuring F_{50} of naked devices or LNAs from a few gigahertz up, a resolution bandwidth up to 5 MHz is common practice, since both scattering and noise parameters can be assumed to remain almost constant.

Since noise is basically a fluctuation of electric quantities, what is measured and displayed by the receiver is a series of random values characterized by statistical parameters such as mean value and standard deviation. Then the fluctuation, or jitter, can be reduced by a factor $\sqrt{N}$ by averaging N readings. Notice that, since the mean of power measurements is sought, each reading, if expressed in logarithmic form, should be converted in the corresponding value as a ratio before averaging; at which point, the so obtained average value could be converted back, if desired, to logarithmic. Not all instruments implement this kind of averaging; therefore one ought to check out the implemented routine on their receiver's manual or, if such information cannot be found, to execute the proper procedure off-line.

Quite alike, jitter can be reduced by widening the video bandwidth, that is, the pass-band of the low-pass filter following the power detector. Obviously, both averaging and filtering lead to an increase in measurement time, which can be unacceptable not only in itself but also because it exacerbates the effects of all kinds of drifts in the measurement chain, from room temperature variation to receiver stability, especially in terms of gain. Ideally, each measurement should be immediately followed by a receiver calibration, perhaps by means of controlled RF switches, with the exception of particular cases—i.e., when, as well known, the gain-noise factor product of the DUT is high enough to make the measurement not very sensitive to the receiver characteristics.

Finally, it is well known that low-noise receiver preamplification can be beneficial to measurement accuracy when its native noise figure is very high. A possible way to exemplify this effect is to consider that noise readings of a receiver featured by an equivalent temperature $T_{e,REC} = 1000$ K (NF$\cong$6.5 dB), an available gain $G_{av,REC} = 1$, and input-terminated on a load at a noise temperature T_S, are:

$$N_S = T_S \cdot kBG_{av,REC} = (T_S + T_{e,REC}) \cdot kBG_{av,REC} \tag{3.3}$$

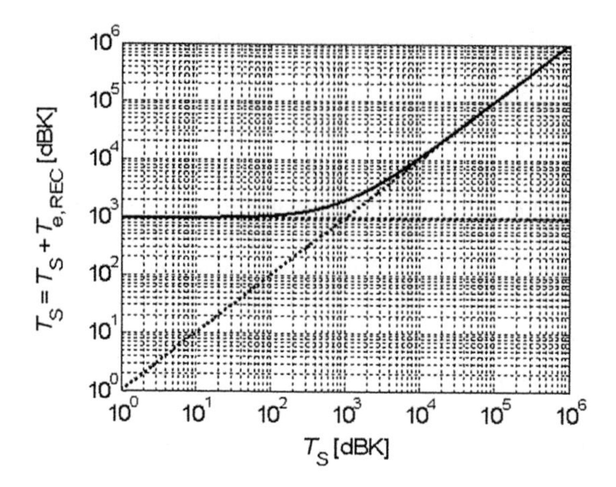

FIGURE 3.2

Normalized noise temperatures at a receiver's input section, $T_S = T_S + T_{e,REC}$. The dashed lines represent limit behaviors of T_S for $T_S \ll T_{e,REC}$ (horizontal line) and for $T_S \gg T_{e,REC}$ (inclined line).

where k is Boltzmann's constant and B is the receiver's effective bandwidth. The shape of $T_S + T_{e,REC}$ as a function of T_S, plotted in Figure 3.2 in a log–log scale after normalizing all temperatures to 1 K, shows that the source noise temperature gets more and more difficult to distinguish from the receiver's noise floor as it becomes smaller with respect to $T_{e,REC}$. Although a sharp limit cannot be assessed, the best readings occur for $T_S \geq T_{e,REC}$, while measurements of $T_S \leq \frac{1}{10} T_{e,REC}$ should not be attempted if high accuracy on T_S is desired. On the other hand, notice that, for a fixed uncertainty on $N_{S,dBm}$, both absolute and percentage uncertainties on T_S are reduced for low values of $T_{e,REC}$ and of T_S itself.

However, what—quite more subtly—affects the measurement of low T_S's is the presence of unknown spurious signals emerging from the receiver's noise floor. Moreover, carelessly using a preamplifier in an attempt to lower the receiver noise figure may turn out to be very detrimental to the measurement results—that is, when the added gain in front of the receiver causes the latter to operate beyond linearity conditions. As it will be shown in Section 3.4.2, even the slightest level of nonlinearity can affect the measurement accuracy, not to mention the worsening in gain stability deriving from arbitrarily cascading amplification stages.

3.2.3 Typical test bench architectures

Several architectures can be devised for noise test benches, each reflecting a particular measurement technique. Although a number of techniques have been proposed [13], only a couple of them have proved to be sufficiently accurate to be employed on high-performance devices and functional blocks nowadays available—namely, the "Y-factor" method and the "cold source" method.

The Y-factor method (sometimes also referred to as "hot-cold source" method, by analogy with the phrase "cold source") is the earlier of the two and is very well suited to measuring the noise factor on 50 Ω of non-frequency-converting devices. The cold source method is more recent and solves a number of problems that concern Y-factor measurements. Besides letting one measure correctly frequency-converting devices, which are not the subject of this work, it makes it feasible to implement source pull measurements, as far as an electromechanical tuner is available. Indeed, source pull is theoretically realizable also in a Y-factor test bench, but the resulting position of the tuner (i.e., between the noise source and the DUT) would produce huge issues in terms of tuner pre-characterization and repeatability, as well as negatively affecting the ENR presented to the DUT.

Either method, in its most essential form, can provide F_{50} measurements simply by means of a VNA (only off-line measurements will be considered from now on), a noise source, and a receiver. In both methods the receiver must be calibrated by means of a direct connection to the noise source, which is then switched on and off. However, a realistic setup also uses a matching network between the DUT and the receiver, so that the latter is always terminated on the same source reflection coefficient, both during the measurement and the calibration phase. Such a matching network is usually realized by means of an isolator, but it could also be an attenuator or an electromechanical tuner, each solution showing peculiar strong and weak points. Notice that this network could even be eliminated, at least in the measurement phase, if the receiver were fully pre-characterized and no doubts on its stability were to be nourished.

The reference setup for calibration, Y-factor measurement, and cold source pull measurement are schematically represented in Figure 3.3. Of course, the input load in Figure 3.3(c) may be replaced with a fixed termination, and a simple F_{50} cold source measurement obtained.

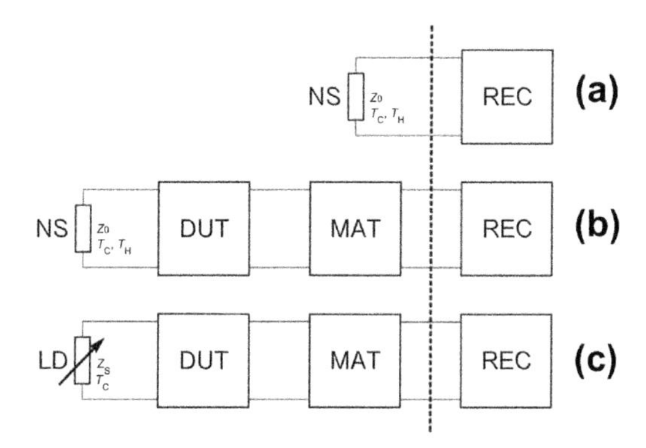

FIGURE 3.3

General test bench setups. (a) Receiver calibration. (b) Y-factor method. (c) Cold source pull method. Abbreviations have the following meanings: NS = noise source, LD = load, DUT = device under test, MAT = matching network, REC = receiver.

Notice that the architectures just shown are deliberately simplified and aim at providing a functional representation of the measurement chain. Of course, any practical implementation of a noise test bench requires a complex connection of hardware and the relevant controlling software routines; this is especially true whenever a high level of automation is required [14].

In a realistic case, and considering on-wafer measurements, each of the networks depicted in Figure 3.3 is made up of several components. For instance, network MAT could be the cascade of the DUT's output probe, a bias-tee, and an isolator, while network REC may include an isolator, an LNA, and a spectrum analyzer. Similarly, network LD may be the chain of a tuner, a coaxial cable, a bias-tee, and the DUT's input probe. In all these cases, the simplified architectures of Figure 3.3 are well suited to represent the actual setup, as long as the small-signal parameters of each functional block are computed by cascading those of its components.

Particular care is only required when applying the simplified representations of Figure 3.3 to network NS, which can stand for an actual noise source (Figure 3.3(a) and (b)) as well as the cascade of a noise source and an isolator/attenuator (Figure 3.3(a)), or of a noise source and an RF probe (Figure 3.3(b)), depending on the setup. More specifically, whenever the actual noise source is embedded in a more complex network by cascading passive components to it, not only the overall reflection coefficient must be computed but also the ENR must be shifted to the new reference plane. This shift from the noise source's native value, ENR, to the "effective" or "embedded" value, ENR', is easily accomplished by applying the following formula:

$$\mathrm{ENR}' = \mathrm{ENR} \cdot G_{\mathrm{av,NET}} \tag{3.4}$$

where $G_{\mathrm{av,NET}}$ is the available gain of the added passive network, NET. Equation (3.4) is derived under the (reasonable) assumption that the physical temperature both of the noise source and of the passive two-port equals T_{C}. More in general, ENR' can be computed from the shifted cold and hot noise temperatures:

$$\begin{aligned} T'_{\mathrm{C}} &= \left(T_{\mathrm{C}} + T_{\mathrm{e,NET,c}}\right) \cdot G_{\mathrm{av,NET,c}} \\ T'_{\mathrm{H}} &= \left(T_H + T_{\mathrm{e,NET,h}}\right) \cdot G_{\mathrm{av,NET,h}} \end{aligned} \tag{3.5}$$

which does not rely on that assumption, or on the statement that $G_{\mathrm{av,NET}}$ be constant for either source state; on the contrary, two different values are used both for $T_{\mathrm{e,NET}}$ and $G_{\mathrm{av,NET}}$, corresponding to the cold (subscript c) and hot (subscript h) states.

3.3 From noise power to noise parameters computation and modeling

3.3.1 Noise factor computation

In Section 3.2.3 the test bench setups required to implement Y-factor or cold source noise measurements were described. In this section, the detailed measurement procedures will be discussed step by step. In the following it will be assumed that the best accuracy is sought, and therefore all reflection coefficients and available gains needed to carry out the noise measurements are intended to be derived from two-port scattering parameters, obtained via a calibrated VNA; notice, in particular, that the receiver available gain does not need to be known, since either it cancels out in the final formulae or it appears in terms that can be found as a whole.

Notice that the content of this section equally applies to coaxial and on-wafer measurements, provided that the actual components of the test bench are combined as described at the end of Section 3.2.3. To perform such operations, the *S* parameters of each component or chain of components must be known. In the case of on-wafer measurements, of course this applies also to the RF probes, whose parameters may be determined through well-known methods, such as the so-called "adapter-removal" technique [15].

To help the reader to univocally identify the meaning of the symbols in the following equations, some conventions have been adopted. According to Figure 3.3, each two-port and bipole in the calibration and measurement phase is assigned a two- or three-letter capitalized label, respectively, which appears as subscript (e.g., F_{REC}); one-letter labels denote, on the other hand, constant values (e.g., T_C). Lower-case c and h refer to "cold" and "hot" steps, and add to previous subscripts, while superscripts *cal* and *meas* refer to "calibration" and "measurement," respectively, as in $T_{e,REC,c}^{cal}$ (effective noise temperature of the receiver in the cold step of calibration). For the sake of better readability, however, symbols will be simplified whenever possible.

As a further convention—however widely accepted—the term noise factor (F) refers in this contribution to the linear value of the quantity (namely, as a ratio), while its expression in decibels is referred to as the noise figure (NF). These quantities, as well as noise equivalent temperature, are not unique for a certain device; on the contrary, they depend on frequency, bias point, physical temperature, and source termination. If the first three variables are fixed, the noise behavior of a linear device can be described through a set of four real parameters, or more typically two real numbers and a complex number. For instance, one of the most common representations is based on the following quantities: noise resistance (R_n), minimum noise factor (F_{min}), and optimum noise admittance ($Y_{opt} = G_{opt} + jB_{opt}$)—alternatively, the last parameter is often expressed as a reflection coefficient on a certain normalization impedance, Γ_{opt}, so that F is written as:

$$F(\Gamma_S) = F_{min} + \frac{4R_n}{Z_0} \cdot \frac{\left|\Gamma_S - \Gamma_{opt}\right|^2}{\left|1 + \Gamma_{opt}\right|^2 \cdot \left(1 - \left|\Gamma_S\right|^2\right)} \tag{3.6}$$

Obviously, F_{min} is the minimum value of F over the passive Smith chart, which occurs for $\Gamma_S = \Gamma_{opt}$. As to R_n, it provides a measure of how sensitive F will be with respect to deviation of Γ_S from Γ_{opt}. As a consequence, when measuring the noise factor of a device, it is necessary to refer the measurement to the particular source

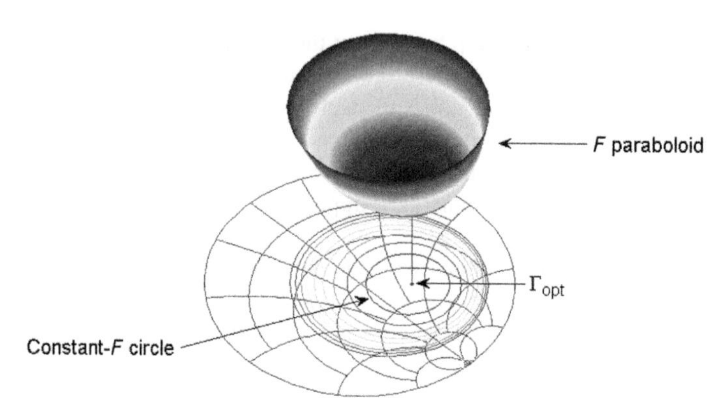

FIGURE 3.4

Example of noise paraboloid.

reflection coefficient. In the following, this is not done explicitly, so as to simplify the notation, but the reader should not overlook this dependence (Figure 3.4).

First, the receiver calibration will be addressed, since it is needed for both the Y-factor and cold-source method, and it is carried out in the same way. In this phase, the noise factor of the receiver F_{REC} is easily obtained from two readings, one corresponding to the noise source being off (or cold), one to the source being on (or hot):

$$\begin{aligned} N_{\mathrm{c}}^{\mathrm{cal}} &= \left(T_{\mathrm{C}} + T_{\mathrm{e,REC,c}}^{\mathrm{cal}}\right) \cdot kBG_{\mathrm{av,REC,c}}^{\mathrm{cal}} \\ N_{\mathrm{h}}^{\mathrm{cal}} &= \left(T_{\mathrm{H}} + T_{\mathrm{e,REC,h}}^{\mathrm{cal}}\right) \cdot kBG_{\mathrm{av,REC,h}}^{\mathrm{cal}} \end{aligned} \tag{3.7}$$

where the subscripts c in $T_{\mathrm{e,REC,c}}^{\mathrm{cal}}$ and in $G_{\mathrm{av,REC,c}}^{\mathrm{cal}}$ and h in $T_{\mathrm{e,REC,h}}^{\mathrm{cal}}$ and in $G_{\mathrm{av,REC,h}}^{\mathrm{cal}}$ account for possible variations of these quantities between the cold and hot steps, due to slightly different values of the noise source's reflection coefficient when off and on, or also to drifts in the receiver. Both effects, however, will be neglected for the moment (a discussion on this point can be found in Section 3.4.1) so that Eqn (3.7) is rewritten as:

$$\begin{aligned} N_{\mathrm{c}}^{\mathrm{cal}} &= \left(T_{\mathrm{C}} + T_{\mathrm{e,REC}}^{\mathrm{cal}}\right) \cdot kBG_{\mathrm{av,REC}}^{\mathrm{cal}} \\ N_{\mathrm{h}}^{\mathrm{cal}} &= \left(T_{\mathrm{H}} + T_{\mathrm{e,REC}}^{\mathrm{cal}}\right) \cdot kBG_{\mathrm{av,REC}}^{\mathrm{cal}} \end{aligned} \tag{3.8}$$

From Eqn (3.8) the "Y" factor in calibration is defined as:

$$Y^{\mathrm{cal}} \triangleq \frac{N_{\mathrm{h}}^{\mathrm{cal}}}{N_{\mathrm{c}}^{\mathrm{cal}}} = \frac{T_{\mathrm{H}} + T_{\mathrm{e,REC}}^{\mathrm{cal}}}{T_{\mathrm{C}} + T_{\mathrm{e,REC}}^{\mathrm{cal}}} \tag{3.9}$$

Hence:

$$\begin{aligned} T_{\mathrm{e,REC}}^{\mathrm{cal}} &= \frac{T_{\mathrm{H}} - Y^{\mathrm{cal}} T_{\mathrm{C}}}{Y^{\mathrm{cal}} - 1} \\ F_{\mathrm{REC}}^{\mathrm{cal}} &= 1 + \frac{T_{\mathrm{e,REC}}^{\mathrm{cal}}}{T_0} = \frac{\mathrm{ENR}}{Y^{\mathrm{cal}} - 1} - \left(\frac{T_{\mathrm{C}}}{T_0} - 1\right) \end{aligned} \tag{3.10}$$

In the measurement phase, similar computations can be worked out, starting from two new cold and hot readings:

$$\begin{aligned} N_{\mathrm{c}}^{\mathrm{meas}} &= \left(T_{\mathrm{C}} + T_{\mathrm{e,TOT,c}}^{\mathrm{meas}}\right) \cdot kBG_{\mathrm{av,TOT,c}}^{\mathrm{meas}} \\ N_{\mathrm{h}}^{\mathrm{meas}} &= \left(T_{\mathrm{H}} + T_{\mathrm{e,TOT,h}}^{\mathrm{meas}}\right) \cdot kBG_{\mathrm{av,TOT,h}}^{\mathrm{meas}} \end{aligned} \tag{3.11}$$

where the subscript TOT refers to the cascade composed by the DUT, MAT, and REC.

If variations of the source reflection coefficients in the measurement chain are again neglected,

$$\begin{aligned} N_{\mathrm{c}}^{\mathrm{meas}} &= \left(T_{\mathrm{C}} + T_{\mathrm{e,TOT}}^{\mathrm{meas}}\right) \cdot kBG_{\mathrm{av,TOT}}^{\mathrm{meas}} \\ N_{\mathrm{h}}^{\mathrm{meas}} &= \left(T_{\mathrm{H}} + T_{\mathrm{e,TOT}}^{\mathrm{meas}}\right) \cdot kBG_{\mathrm{av,TOT}}^{\mathrm{meas}} \end{aligned} \tag{3.12}$$

and the Y factor in measurement is defined as:

$$Y^{\mathrm{meas}} \triangleq \frac{N_{\mathrm{h}}^{\mathrm{meas}}}{N_{\mathrm{c}}^{\mathrm{meas}}} = \frac{T_{\mathrm{H}} + T_{\mathrm{e,TOT}}^{\mathrm{meas}}}{T_{\mathrm{C}} + T_{\mathrm{e,TOT}}^{\mathrm{meas}}} \tag{3.13}$$

Finally,

$$\begin{aligned} T_{\mathrm{e,TOT}}^{\mathrm{meas}} &= \frac{T_{\mathrm{H}} - Y^{\mathrm{meas}} T_{\mathrm{C}}}{Y^{\mathrm{meas}} - 1} \\ F_{\mathrm{TOT}}^{\mathrm{meas}} &= 1 + \frac{T_{\mathrm{e,TOT}}^{\mathrm{meas}}}{T_0} = \frac{\mathrm{ENR}}{Y^{\mathrm{cal}} - 1} - \left(\frac{T_{\mathrm{C}}}{T_0} - 1\right) \end{aligned} \tag{3.14}$$

From the cascade noise factor formula (often referred to as the Friis formula), however, the following must hold:

$$F_{\mathrm{TOT}} = F_{\mathrm{DUT}} + \frac{F_{\mathrm{MAT}} - 1}{G_{\mathrm{av,DUT}}} + \frac{F_{\mathrm{REC}} - 1}{G_{\mathrm{av,DUT}} G_{\mathrm{av,MAT}}} \tag{3.15}$$

and therefore the device noise factor is computed as:

$$F_{\mathrm{DUT}} = F_{\mathrm{TOT}} - \frac{F_{\mathrm{MAT}} - 1}{G_{\mathrm{av,DUT}}} - \frac{F_{\mathrm{REC}} - 1}{G_{\mathrm{av,DUT}} G_{\mathrm{av,MAT}}} \tag{3.16}$$

where all quantities are known from previous scattering parameters characterizations (i.e., $G_{av,DUT}$, $G_{av,MAT}$, and F_{MAT}), from the calibration phase (i.e., F_{REC}, which is assumed unchanged) or from the measurement itself (i.e., F_{TOT}). In particular, since the matching network is passive, its noise factor is a well-known function of its available gain and physical temperature (assumed equal to the ambient temperature, T_A):

$$T_{e,MAT} = T_A \cdot \left(\frac{1}{G_{av,MAT}} - 1 \right) \tag{3.17}$$

Notice, however, that in general there will be some difference between the receiver's noise factor during the measurement, F_{REC}^{meas}, and the value obtained during calibration, F_{REC}^{cal}. This difference must be minimized, taking the following precautions:

- The auto-ranging capability of the receiver, if present, must be disabled when switching from calibration to measurement (or vice versa).
- All measurement steps should be performed minimizing undue delays, in order to avoid drifts in the receiver and the whole chain.
- The receiver source termination should remain unchanged in calibration and measurement. If the noise source and the isolator are very well matched on 50 Ω, the setups in Figure 3.3 are sufficient. If this is not the case, one may think to use the isolator also during the calibration phase, as part of a new, "effective" noise source featured by novel noise temperatures (computed through Eqn (3.10)). If this option is preferred, a very accurate small-signal characterization of the isolator is mandatory, since any uncertainty on its available gain reflects directly on the ENR uncertainty of the effective noise source.

While the first point is simple to put into practice, the last two points require quite some work. However, they are worth the effort since they relate to sources of uncertainty that cannot be estimated (if not conservatively) but only reduced.

To review the steps necessary to perform a cold-source measurement, let Eqn (3.8) be applied to the measurement of the cascade represented in Figure 3.3(c):

$$N_M = T_M \cdot kBG_{av,REC} = \left(T_M + T_{e,REC}\right) \cdot kBG_{av,REC} \tag{3.18}$$

where T_M represents the noise temperature of the bipole at the receiver left, that is the cascade of a passive load at ambient temperature, the DUT, and a matching network:

$$T_M = \left[\left(T_A + T_{e,DUT}\right) \cdot G_{av,DUT} + T_{e,MAT}\right] \cdot G_{av,MAT} \tag{3.19}$$

Therefore, if the matching network is passive and at the physical temperature T_A, the device noise temperature, $T_{e,DUT}$, is expressed as:

$$T_{e,DUT} = \left[T_M - T_A \cdot \left(1 - G_{av,MAT}\right)\right] \cdot \frac{1}{G_{av,DUT} G_{av,MAT}} - T_A \tag{3.20}$$

The only quantity that cannot be derived from scattering parameters and temperature measurements is here T_M. Nevertheless, recalling Eqn (3.18) yields:

$$T_M = \frac{N_M}{kBG_{av,REC}} - T_{e,REC} \tag{3.21}$$

where $T_{e,REC}$ is known from calibration, following Eqn (3.10). As to $kBG_{av,REC}$, on the other hand, Eqn (3.8) allows computing it (under the same hypothesis of receiver gain stability) as:

$$kBG^{cal}_{av,REC} = \frac{N^{cal}_h - N^{cal}_c}{T_H - T_C} = \frac{N^{cal}_h - N^{cal}_c}{T_0 \cdot \mathrm{ENR}} \tag{3.22}$$

Alternatively, following Eqn (3.8):

$$kBG^{cal}_{av,REC} = \frac{N^{cal}_c}{T_C + T^{cal}_{e,REC}} = \frac{N^{cal}_h}{T_H + T^{cal}_{e,REC}} \tag{3.23}$$

Notice that all these expressions for $kBG^{cal}_{av,REC}$ are equivalent and always provide the same result. Therefore testing their equivalence cannot be used to double-check if something in the calibration phase went wrong (for instance, receiver compression).

Merging Eqns (3.21) and (3.22) finally results in:

$$T_M = \frac{N_m}{N^{cal}_h - N^{cal}_c} \cdot (T_H - T_C) - T_{e,REC} = \frac{N_m}{N^{cal}_h - N^{cal}_c} \cdot \mathrm{ENR} \cdot T_0 - T_{e,REC} \tag{3.24}$$

On the other hand, substituting Eqn (3.23) in (3.21) results in:

$$T_M = \frac{N_m}{N^{cal}_c} \cdot (T_C + T_{e,REC}) - T_{e,REC} = \frac{N_m}{N^{cal}_c} \cdot T_C + \left(\frac{N_m}{N^{cal}_c} - 1\right) \cdot T_{e,REC} \tag{3.25}$$

Following the steps described thus far, both procedures easily allow measuring the device's noise factor when its source termination is 50 Ω. More precisely, in the Y-factor method such termination is determined by the solid-state noise source used during the measurement phase and it is very difficult to change arbitrarily: the effect is an uncertainty on the F_{50} real value, which cannot be exactly quantified without a complete knowledge of the device noise parameters—but these parameters, of course, are supposed to be the result, not the basis, of the measurement. For this reason, it is advisable to declare the true source reflection coefficient (with respect to which this kind of uncertainty is not present) rather than refer the measurement to 50 Ω and use a conservative value for the respective uncertainty; this is especially true when noise measurements are taken on naked devices for modeling on the basis of equivalent circuit techniques (see Section 3.3.2).

On the other hand, the device source termination in the cold-source method is determined by a passive load, which could even be a calibration standard—but it could also be the cascade of a low-quality termination, a bias-tee, and an

RF probe. In the latter case, considerations similar to those in the previous paragraph hold. Nevertheless, loading the device with source terminations different from 50 Ω represents in some cases a deliberate choice, namely when a source pull technique is adopted. The cold-source method naturally lends itself to implement such technique, due to its prerequisite of a single-source temperature in the measurement phase. With reference to Figure 3.3, all that is needed in addition to a basic F_{50} setup is an input tuner, while the procedure to take each measurement is exactly the same as already described: the same formulae hold, provided the proper reflection coefficients are used when computing available gains.

3.3.2 Full noise characterization and modeling

3.3.2.1 Equivalent circuit approach

It is desirable that a two-port noise model is determined in the same way as a small-signal equivalent circuit (SSEC) model. It is also desirable that the model embrace a small number of frequency independent parameters, allowing extension of the model validity outside the measurement range. A remarkable number of references [16–22] show that under certain conditions and for a particular family of devices (HEMT and MESFET) it is possible to obtain a model fulfilling both conditions.

In a very popular work by Pospieszalski [18], the author shows that for an intrinsic device, two constant values, T_{gs} and T_{ds}, need to be known in addition to the elements of the equivalent circuit to predict all four noise parameters at any frequency. Such device-specific, frequency-independent values can be defined in the frequency range in which $1/f$ noise is negligible (and below the frequency limit at which the white noise approximation fails). As their symbols suggest, T_{gs} and T_{ds} are associated, respectively, to the gate-source resistance R_{gs} and the drain-source resistance R_{ds} with the meaning of equivalent noise temperatures.

These equivalent temperatures are associated to a voltage source, e_{gs}, in series with the gate resistance and a current noise source, i_{ds}, in parallel with the drain

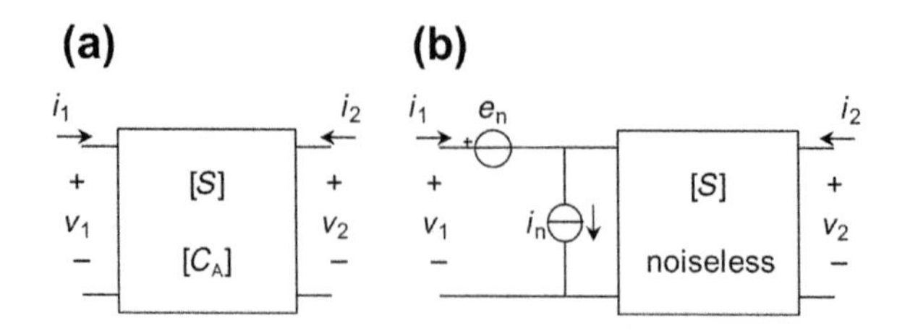

FIGURE 3.5

Representations of a linear, noisy two-port (a) and its equivalent circuit with external noise sources (b).

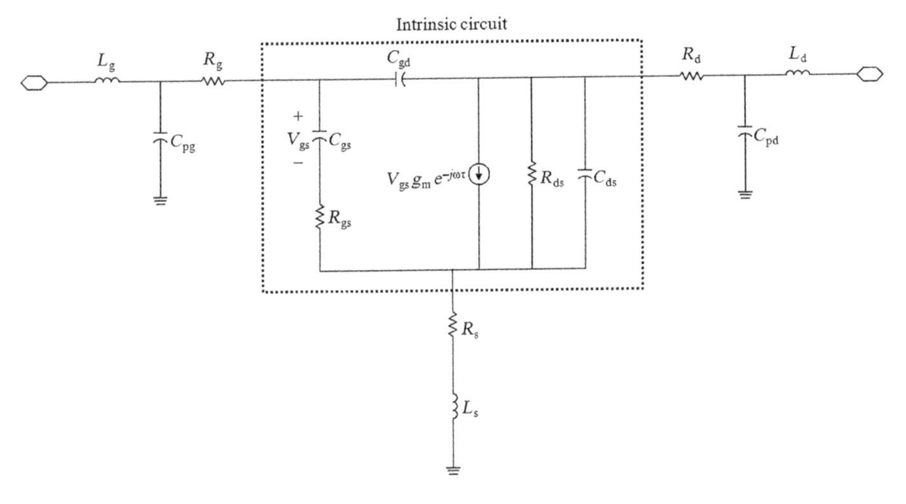

FIGURE 3.6

13-Elements equivalent circuit, 7 for the intrinsic part (5 scalars and 1 complex) and 6 for the extrinsic part.

terminal (see Figure 3.6). The root mean square voltage and current of these stochastic processes are determined by:

$$e_{gs} = \sqrt{\overline{e_{gs}e_{gs}^*}} = \sqrt{\overline{|e_{gs}|^2}} = \sqrt{4kT_{gs}r_{gs}B} \tag{3.26}$$

$$i_{ds} = \sqrt{\overline{i_{gs}i_{gs}^*}} = \sqrt{\overline{|i_{ds}|^2}} = \sqrt{4kT_{ds}g_{ds}B} \tag{3.27}$$

where the bar denotes the mean value and the asterisk stands for the conjugate operator. It is important to remark that Pospieszalski assumes no correlation between these sources. Furthermore, parasitic (extrinsic) resistances are considered to introduce thermal noise only, so their contribution can be easily taken into account if their values and the ambient temperature T_A are known. The extrinsic device can be modeled starting from the intrinsic device and including the effects of parasitic resistors, capacitors, and inductors.

The most straightforward way to embed parasitics is by means of Hillbrand and Russer's matrix formulation of parallel, series, and cascade connection of linear, noisy two-ports [23]. The noisy two-ports are represented by noise-free *Y* (admittance), *Z* (impedance), *H* (hybrid), or *ABCD* (chain) matrices, linked, respectively, to the appropriate correlation matrix C_Y, C_Z, C_H, C_A.

Each of the latter represents noise by means of two noise sources and the correlation between them, when the corresponding representation is selected for the two-port network. If for instance the *ABCD* representation of the correlation matrix is considered, a noisy two-port with small-signal parameters *S* (see Figure 3.5(a)) can be represented as the cascade of two partially correlated noise sources, e_n and

i_n, and a noiseless two-port with the same small-signal parameters, S (see Figure 3.5(b)). The terms of the correlation matrix are given by the auto- and cross correlations of the noise generators: the former are computed as $\overline{e_n e_n^*}$ and $\overline{i_n i_n^*}$, the latter as $\overline{e_n i_n^*}$ and $\overline{e_n^* i_n}$.

Similarly, in the other representations, a couple of noise generators are used of appropriate types; for instance, two noise current generators, $i_{n,1}$ and $i_{n,2}$, are associated to the first and second port of the noisy network, respectively, in the admittance representation, while in the impedance representation, two noise voltage generators are used, $e_{n,1}$ and $e_{n,2}$, respectively.

More especially, the correlation matrix in admittance representation is defined as:

$$C_Y = \begin{bmatrix} \overline{i_{n,1} i_{n,1}^*} & \overline{i_{n,1} i_{n,2}^*} \\ \overline{i_{n,2} i_{n,1}^*} & \overline{i_{n,2} i_{n,2}^*} \end{bmatrix} = \begin{bmatrix} \overline{|i_{n,1}|^2} & \overline{i_{n,1} i_{n,2}^*} \\ \overline{i_{n,2} i_{n,1}^*} & \overline{|i_{n,2}|^2} \end{bmatrix} \tag{3.28}$$

while for the impedance network representation the correlation matrix is

$$C_Z = \begin{bmatrix} \overline{e_{n,1} e_{n,1}^*} & \overline{e_{n,1} e_{n,2}^*} \\ \overline{e_{n,2} e_{n,1}^*} & \overline{e_{n,2} e_{n,2}^*} \end{bmatrix} = \begin{bmatrix} \overline{|e_{n,1}|^2} & \overline{e_{n,1} e_{n,2}^*} \\ \overline{e_{n,2} e_{n,1}^*} & \overline{|e_{n,2}|^2} \end{bmatrix} \tag{3.29}$$

Notice that the correlation matrix is Hermitian in each representation (terms on the secondary diagonal only differ by a conjugate operation); moreover, terms on the principal diagonal are real and positive, so that, as expected, also the correlation matrix representation of a noisy two-port accounts for four real parameters. Furthermore, it should be stressed that other definitions could be accepted for the correlation matrix, only differing for a normalization factor; in the following, for instance, a normalization on $4kBT_0$ will be adopted, leading to simpler formulae.

In this contribution, the procedure used for the embedding of parasitics is considered with reference to noise parameters, while the linear parameters' counterpart is taken for granted.

To add the contribution of parallel elements, first the noise parameters (i.e., R_n, F_{min}, and Y_{opt}) of the starting embedded network (or the one resulting from the previous embedding step) are transformed into the admittance correlation matrix $C_{Y,e}$ (the topic of converting representations will be addressed shortly) and, then, the resulting correlation matrix C_Y is computed as:

$$C_Y = C_{Y,e} + C_{Y,p} \tag{3.30}$$

where $C_{Y,p}$ represents the correlation matrix of the "parallel" parasitic (Π-network).

On the other hand, to add the contribution of series elements, first the noise parameters of the starting network are transformed into the impedance correlation matrix, and then the resulting correlation matrix is computed:

$$C_Z = C_{Z,e} + C_{Z,s} \tag{3.31}$$

where C_Z is the resulting matrix, $C_{Z,e}$ the correlation matrix of the embedded network, and $C_{Z,s}$ the one related to series parasitics.

In a similar way, combination of noisy two-ports represented by their hybrid matrices is obtained by a simple sum:

$$C_H = C_{H,e} + C_{H,h} \tag{3.32}$$

where C_H is the overall correlation matrix while $C_{H,e}$ and $C_{H,h}$ are those of the component networks.

On the contrary, to combine the noise properties of two cascaded two-port networks—assuming that network "e" precedes network "c"—is slightly more complicated:

$$C_A = C_{A,e} + A_e C_{A,c} A_e^H \tag{3.33}$$

where C_A is the overall correlation matrix, $C_{A,e}$ and $C_{A,c}$ are those of the two networks, and A_e is the $ABCD$ matrix of the first network.

The pending discussion on how to switch between the classical noise parameters is now dealt with. Only the conversion formulae from the set of parameters R_n, F_min, and Y_opt to the chain correlation matrix C_A is described, and vice versa:

$$C_A = 4kBT_0 \cdot \begin{bmatrix} R_\mathrm{n} & \dfrac{F_\mathrm{min}-1}{2} - R_\mathrm{n} Y_\mathrm{opt}^* \\ \dfrac{F_\mathrm{min}-1}{2} - R_\mathrm{n} Y_\mathrm{opt} & R_\mathrm{n} \left| Y_\mathrm{opt} \right|^2 \end{bmatrix} = 4kBT_0 \cdot \hat{C}_A \tag{3.34}$$

$$\begin{cases} R_\mathrm{n} = \hat{c}_{11} \\ Y_\mathrm{opt} = \sqrt{\dfrac{\hat{c}_{22}}{\hat{c}_{11}} - \left(\dfrac{\Im\{\hat{c}_{12}\}}{\hat{c}_{11}} \right)^2} + j \dfrac{\Im\{\hat{c}_{12}\}}{\hat{c}_{11}} \\ F_\mathrm{min} = 1 + 2\left(\hat{c}_{12} + \hat{c}_{11} Y_\mathrm{opt}^* \right) \end{cases} \tag{3.35}$$

where the generic $\hat{c}_{ij}$ is clearly the term of $\hat{C}_A$ in position (i, j), and the cap denotes the normalization on $4kBT_0$. Nevertheless, notice that, once one form of the correlation matrix is known (C), it is possible to compute, if needed, any other form (C') by applying the following matrix relation:

$$C' = PCP^H \tag{3.36}$$

where P is a suitable conversion matrix from Table 3.1 and superscript H denotes the Hermitian transposition.

The counterpart of Eqn (3.34) for passive networks is expressed by the following equations:

$$C_Y = 4kBT_\mathrm{A} \cdot \Re\{\mathbf{Y}\} = \frac{T_\mathrm{A}}{T_0} \cdot \hat{C}_Y \tag{3.37}$$

Table 3.1 Conversion matrices

Resulting representation		Original Representation			
		Admittance	**Impedance**	**Transmission**	**Hybrid**
	Admittance	$\begin{pmatrix} 1 & 0 \\ 0 & 1 \end{pmatrix}$	$P_{ZY} = \begin{pmatrix} y_{11} & y_{12} \\ y_{21} & y_{22} \end{pmatrix}$	$P_{AY} = \begin{pmatrix} -y_{11} & 1 \\ -y_{21} & 0 \end{pmatrix}$	$P_{HY} = \begin{pmatrix} -y_{11} & 0 \\ -y_{21} & 1 \end{pmatrix}$
	Impedance	$P_{YZ} = \begin{pmatrix} z_{11} & z_{12} \\ z_{21} & z_{22} \end{pmatrix}$	$\begin{pmatrix} 1 & 0 \\ 0 & 1 \end{pmatrix}$	$P_{AZ} = \begin{pmatrix} 1 & -z_{11} \\ 0 & -z_{21} \end{pmatrix}$	$P_{HZ} = \begin{pmatrix} 1 & -z_{12} \\ 0 & -z_{22} \end{pmatrix}$
	Transmission	$P_{YA} = \begin{pmatrix} 0 & B \\ 1 & D \end{pmatrix}$	$P_{ZA} = \begin{pmatrix} 1 & -A \\ 0 & -C \end{pmatrix}$	$\begin{pmatrix} 1 & 0 \\ 0 & 1 \end{pmatrix}$	$P_{HA} = \begin{pmatrix} 1 & B \\ 0 & D \end{pmatrix}$
	Hybrid	$P_{YH} = \begin{pmatrix} -h_{11} & 0 \\ -h_{21} & 1 \end{pmatrix}$	$P_{ZH} = \begin{pmatrix} 1 & -h_{12} \\ 0 & -h_{22} \end{pmatrix}$	$P_{AH} = \begin{pmatrix} 1 & -h_{11} \\ 0 & -h_{21} \end{pmatrix}$	$\begin{pmatrix} 1 & 0 \\ 0 & 1 \end{pmatrix}$

$$C_Z = 4kBT_A \cdot \Re\{\mathbf{Z}\} = \frac{T_A}{T_0} \cdot \hat{C}_Z \tag{3.38}$$

$$C_H = 4kBT_A \cdot \Re\{\mathbf{H}\} = \frac{T_A}{T_0} \cdot \hat{C}_H \tag{3.39}$$

For a purely reactive component, each element of the correlation matrix is zero, and thus only the linear parameters have to be known for the embedding; for a purely resistive network (as the series parasitic resistors), also the physical temperature of the parasitic(s) has to be known for the determination of the correlation matrix expressed by Eqns (3.37) and (3.38).

The first step necessary to the model extraction consists in performing the S parameter measurements for each bias point in which the model is to be extracted and in cold-FET conditions for the extraction of the parasitic elements [24–28]. Such a set of measurements allows the determination of the intrinsic and extrinsic element values of the SSEC.

The second step consists in performing a suitable number of noise factor measurements on the active device. For each bias point, the resulting system of independent equations can be solved for the unknown equivalent temperature variables.

Pospieszalski suggests that such measurements be performed with the input of the device terminated on 50 Ω, considering at least two frequency points to verify the requirement on the minimum number of measurements that allow solving the system. Such measurements are usually well suited for 50 Ω test systems.

Once the SSEC is determined and the noise factor measurements are performed, it is possible to define the Pospieszalski equivalent temperature noise model by assigning the proper value to the gate and drain equivalent temperatures.

The system of equations is based on one of the following expressions:

$$F(Y_S) = 1 + \frac{y^H \hat{C}_A y}{\Re\{Y_S\}} \quad y = \begin{pmatrix} Y_S \\ 1 \end{pmatrix} \tag{3.40}$$

$$F(Z_S) = 1 + \frac{z^H \hat{C}_A z}{\Re\{Z_S\}} \quad z = \begin{pmatrix} Z_S \\ 1 \end{pmatrix} \tag{3.41}$$

By rewriting Eqn (3.41) in terms of the chain matrix coefficients $\hat{c}_{ij}$ with explicit dependence on frequency, the following result holds:

$$F(Z_S, f, T_{gs}, T_{ds}) = 1 + \frac{Z_S^2 \hat{c}_{11}(f, T_{gs}, T_{ds}) + 2\Re\{Z_S \hat{c}_{12}(f, T_{gs}, T_{ds})\} + \hat{c}_{22}(f, T_{gs}, T_{ds})}{\Re\{Z_S\}} \tag{3.42}$$

As shown in the following, the system is linear with respect to its variables T_{gs}, T_{ds}, and, therefore, it is directly inverted if the number of measurements equals the number of unknowns. Typically, with the aim of lowering the uncertainty in the determination of the equivalent temperatures, more than two measurements are considered. In this case the Moore–Penrose pseudo-inverse can be used to compute a best fit (least squares) solution as follows:

$$x = A^{+}y \tag{3.43}$$

where

$$A^{+} = \left(A^{T}A\right)^{-1}A^{T} \tag{3.44}$$

is the pseudo-inverse matrix of A, defined if A is an $m \times n$ matrix with $m > n$ (notice that if $m = n$ the pseudo-inverse matrix is identical to the usual inverse matrix). This solution corresponds to finding the unique x vector that minimizes the Euclidean distance:

$$e(x) = \|Ax - y\|^{2} \tag{3.45}$$

Equation (3.43) is thus written in terms of the $n = 2$ unknowns T_{gs} and T_{ds} and the m noise factor measurements $F_1,\ldots,F_m$

$$\begin{bmatrix} T_{gs} \\ T_{ds} \end{bmatrix} = A_M \cdot \begin{bmatrix} F_1 \\ F_2 \\ \cdots \\ F_{m-1} \\ F_m \end{bmatrix} \tag{3.46}$$

in which the coefficients matrix A_M is determined by Eqn (3.42) and Eqn (3.44).

T_{gs} and T_{ds} may be treated as fitting factors or parameters with a physical meaning; what is important from the modeling point of view is that their values be determined once the products $R_{gs}T_{gs}$ and $G_{ds}T_{ds}$—involved in the computation of the noise parameters—have already been ascertained. As a consequence, the accuracy on T_{gs} and T_{ds} is directly correlated with the accuracy on the value of R_{gs} and G_{ds}, besides the obvious dependence on the accuracy of the noise factor measurements. Usually R_{gs} can be estimated with lower accuracy if compared to G_{ds}. This suggests that T_{gs} be considered as an unknown variable even though both the typical values extracted for common devices and the physical considerations on the nature of the gate noise contributors allow one to suppose that it must be close to the device physical temperature. In this way an error in the estimation of R_{gs} can be compensated by T_{gs} in the product $R_{gs}T_{gs}$ (for a more extensive explanation, refer to [18]).

As can be easily understood from such considerations, Pospieszalski's approach for noise modeling relies on a preliminary, very accurate extraction of the device's SSEC. This very critical, time-consuming step is, however, necessary when a complete technology assessment is desired [29], but may be conveniently avoided, whenever "noise-only" modeling is of interest [14] and [30–32], by recurring to source-pull techniques, such as those described in the following sections.

3.3.2.2 Direct four-parameter extraction

In established theory of linear noisy networks (see, for instance, [33]), two-ports can be represented at a given frequency as in Figure 3.5(a), where S is the two-port scattering matrix and C_A is its noise correlation matrix. It has already been shown that the noise properties of the two-port can be de-embedded so as to obtain an equivalent representation, as in Figure 3.5(b). The relationship between the noise sources e_n and i_n in Figure 3.5(b) and the correlation matrix (here considered in *ABCD* form) is as follows:

$$C_A = \begin{bmatrix} c_{11} & c_{12} \\ c_{21} & c_{22} \end{bmatrix} = \begin{bmatrix} c_{11} & c_{12} \\ c_{12}^* & c_{22} \end{bmatrix} = \begin{bmatrix} \overline{e_n e_n^*} & \overline{e_n i_n^*} \\ \overline{e_n^* i_n} & \overline{i_n i_n^*} \end{bmatrix} \tag{3.47}$$

where the same notation is used as in the preceding section. Furthermore, the correlation parameters can be linked to classic noise parameters by the following expressions (equivalent to the formerly shown relations (3.30)):

$$R_n = \frac{c_{11}}{4kBT_0} = \hat{c}_{11} \tag{3.48}$$

$$Y_\gamma = \frac{c_{12}^*}{c_{11}} = \frac{\hat{c}_{12}^*}{\hat{c}_{11}} \tag{3.49}$$

$$G_u = \frac{c_{22} - c_{11} \cdot |Y_\gamma|^2}{4kBT_0} = \hat{c}_{22} - \hat{c}_{11} \cdot |Y_\gamma|^2 \tag{3.50}$$

By means of these parameters, the two-port noise factor can be expressed as a function of the source admittance $Y_S = G_S + jB_S$:

$$F(Y_S) = 1 + \frac{G_u + R_n \cdot |Y_S - Y_\gamma|^2}{G_S} \tag{3.51}$$

Note that all representations of the two-port noise properties are equivalent and account for four real parameters (two real numbers and a complex one in the case of Eqns (3.48)–(3.50)). The source-pull technique applied to noise characterization allows extracting such parameters at a given frequency by means of at least four noise factor measurements, each on a different input termination. In order to avoid an ill-conditioned system of equations and to reduce extraction uncertainties, a much higher number of measurements is usually performed in practice, and a least-squares minimization is then performed on an over-dimensioned system.

However, a direct application of a least-squares algorithm on a nonlinear function of unknowns, as in Eqn (3.51), may result in multiple and/or unphysical solutions, unless a good initial guess is provided. Various solutions have been proposed to this problem (see, for instance, [34]), the most used being the one proposed by Lane [35]; his work starts from an alternative expression of F as a nonlinear function of F_{min}, R_n, and Y_{opt}, which is then transformed into a linear function of four new, factitious unknowns A, B, C, and D:

$$F(Y_S) = A + B \cdot \left(G_S + \frac{B_S^2}{G_S} \right) + C \cdot \frac{1}{G_S} + D \cdot \frac{B_S}{G_S} \tag{3.52}$$

These parameters—related to the conventional noise parameters through simple, yet nonlinear, relationships—must not be confused with the terms of the *ABCD* matrix.

As a final note, it should be stressed that, the number of noise measurements being equal, an important role is played by the particular pattern of the selected source reflection coefficients [36,37].

3.3.2.3 Advanced approaches

3.3.2.3.1 Single-frequency extraction

In [38] a novel extraction technique is presented, featured by the following peculiarities: first, it can be shown to be mathematically equivalent, at single frequency, to Lane's algorithm, and, second, it can be easily extended in order to be applied to multifrequency, source-pull noise data. These aspects are the subject of the present and the following sections, respectively. Finally, in Section 3.3.2.3.3 a real example of noise extraction is reported and discussed.

In this approach only classic, physical parameters are used. In particular, using Eqns (3.48)–(3.50) allows, after some math, to rearrange Eqn (3.51) as follows:

$$(F(Y_S)-1)\cdot G_S = |Y_S|^2\cdot\hat{c}_{11} + 2G_S\cdot\Re\{\hat{c}_{12}\} - 2B_S\cdot\Im\{\hat{c}_{12}\} + \hat{c}_{22} \tag{3.53}$$

where the left-hand side is clearly linear with respect to $\hat{c}_{11}$, $\Re\{\hat{c}_{12}\}$, $\Im\{\hat{c}_{12}\}$, and $\hat{c}_{22}$. It can be shown, by using Eqns (3.48)–(3.50) and the definition of Lane's parameters (not reported here), that Eqn (3.53) is equivalent to (3.52).

After taking a number n of F measurements (where n is adequate to the uncertainty estimated for the particular setup), a linear system of equations is set up as follows:

$$b = A\cdot x \tag{3.54}$$

where:

$$b = \begin{bmatrix} (F(Y_{S,1})-1)\cdot G_{S,1} \\ (F(Y_{S,2})-1)\cdot G_{S,2} \\ \vdots \\ (F(Y_{S,n})-1)\cdot G_{S,n} \end{bmatrix} \tag{3.55}$$

$$A = \begin{bmatrix} |Y_{S,1}|^2 & 2G_{S,1} & -2B_{S,1} & 1 \\ |Y_{S,2}|^2 & 2G_{S,2} & -2B_{S,2} & 1 \\ \vdots & \vdots & \vdots & \vdots \\ |Y_{S,n}|^2 & 2G_{S,n} & -2B_{S,n} & 1 \end{bmatrix} \tag{3.56}$$

$$x = \begin{bmatrix} \hat{c}_{11} \\ \Re\{\hat{c}_{12}\} \\ \Im\{\hat{c}_{12}\} \\ \hat{c}_{22} \end{bmatrix} \tag{3.57}$$

and simply solved, for instance, by means of a pseudo-inversion (see end of Section 3.3.2.1). Note, however, that Eqn (3.53) can be rearranged in order to assign a different weighting to system rows, or, instead, an explicit weighting function may be devised to reduce the influence of rows believed to be less accurate.

3.3.2.3.2 Multifrequency extraction

Provided non-white noise generation mechanisms are neglected, it is empirically found that, for a typical high-frequency active device, the four terms of the correlation matrix can be easily fitted by Taylor polynomials of (at most) the third degree, over a frequency band at least up to f_x, which is the frequency at which the optimum noise match becomes purely resistive. As an example, in Figure 3.7 the correlation

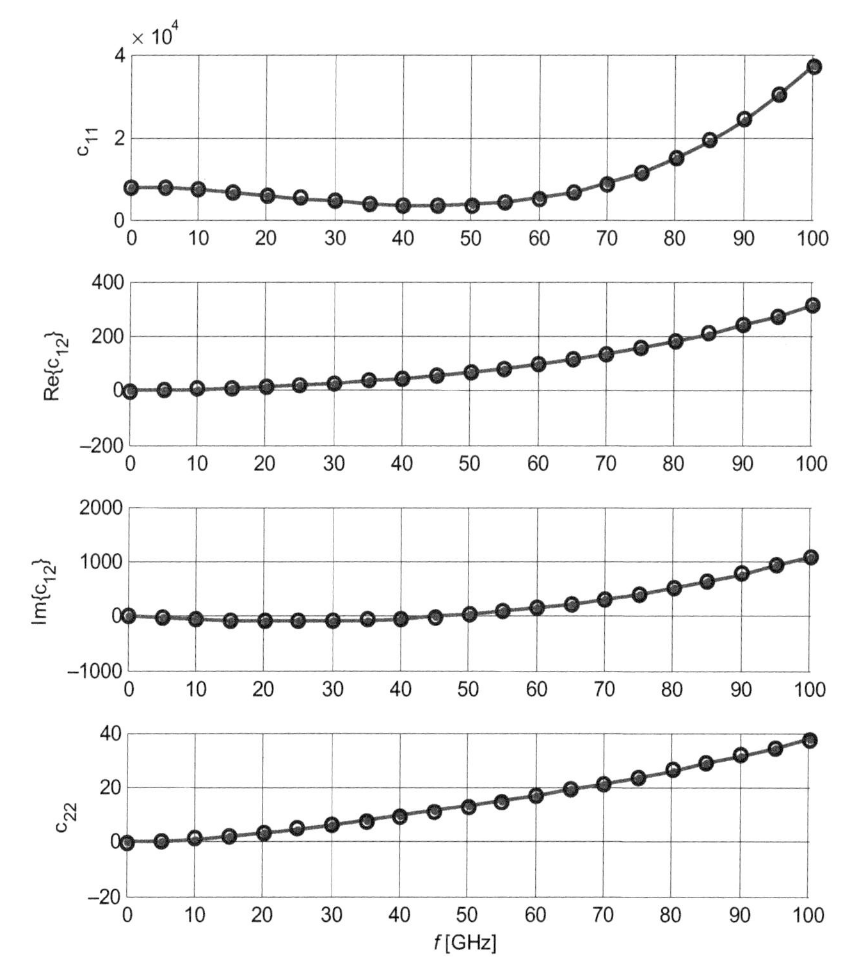

FIGURE 3.7

Noise correlation matrix elements of a GaN HEMT over frequency. Circle markers: actual values. Continuous line: polynomial fitting.

matrix of an ideal HEMT (whose parameters are based on equivalent circuit extractions [29]) are plotted: the computed behavior (continuous trace) is excellently fitted by third-degree polynomials (round markers) well beyond f_x, which equals 58 GHz for the considered HEMT device.

However, in view of physical considerations (in particular with respect to the first derivatives of the classical noise parameters at zero frequency), some from the resulting grand total of 16 coefficients can be zeroed. The following 10-coefficient model results:

$$\begin{array}{rcllll} \hat{c}_{11} & = & +\alpha_0 & & +\alpha_2 f^2 & +\alpha_3 f^3 \\ \Re\{\hat{c}_{12}\} & = & & & +\beta_2 f^2 & +\beta_3 f^3 \\ \Im\{\hat{c}_{12}\} & = & & +\gamma_1 f & +\gamma_2 f^2 & +\gamma_3 f^3 \\ \hat{c}_{22} & = & & & +\delta_2 f^2 & +\delta_3 f^3 \end{array} \tag{3.58}$$

where f is the frequency. In some practical cases (see Section 3.3.2.3.3), other simplifications are possible.

Two possibilities now arise as far as the determination of such coefficients is concerned:

Treating each of them as fitting parameters of the relevant $\hat{c}_{ij}$ term;
Substituting Eqn (3.58) in Eqns (3.54)–(3.57), so as to obtain a new system of the form Eqn (3.54) but in the unknowns $\alpha_0, \alpha_2 \ldots \delta_3$. The vector b will remain unchanged, while the characteristic matrix rows and the vector of unknowns will have the following forms, respectively:

$$\left[|Y_{S,i}|^2 \begin{bmatrix} 1 \\ f^2 \\ f^3 \end{bmatrix}^T \; 2G_{S,i} \begin{bmatrix} f^2 \\ f^3 \end{bmatrix}^T \; -2B_{S,i} \begin{bmatrix} f \\ f^2 \\ f^3 \end{bmatrix}^T \; \begin{bmatrix} f^2 \\ f^3 \end{bmatrix}^T \right] \tag{3.59}$$

$$\begin{bmatrix} \alpha_0 & \alpha_2 & \cdots & \delta_3 \end{bmatrix}^T \tag{3.60}$$

Method (a) is based on a previous extraction of each $\hat{c}_{ij}$ term versus measurement frequencies by repeatedly applying Eqns (3.54)–(3.57); as a consequence, the determination of the correlation matrix is independent for every frequency, and the subsequent fitting process is independent for each $\hat{c}_{ij}$ term. On the contrary, approach (b) is expected to be somewhat more robust since all unknowns are computed through one single-step extraction; in other words, measurements at a given frequency influence the resulting model at every frequency.

However, both methods are feasible and typically lead to good results even when applied to measurements affected by random errors. As an example, Figure 3.8 shows:

The ideal terms of the correlation matrix—black trace;
The frequency-by-frequency extraction—round markers;
The extraction following approach (a)—cross markers;
The extraction following approach (b)—triangle markers.

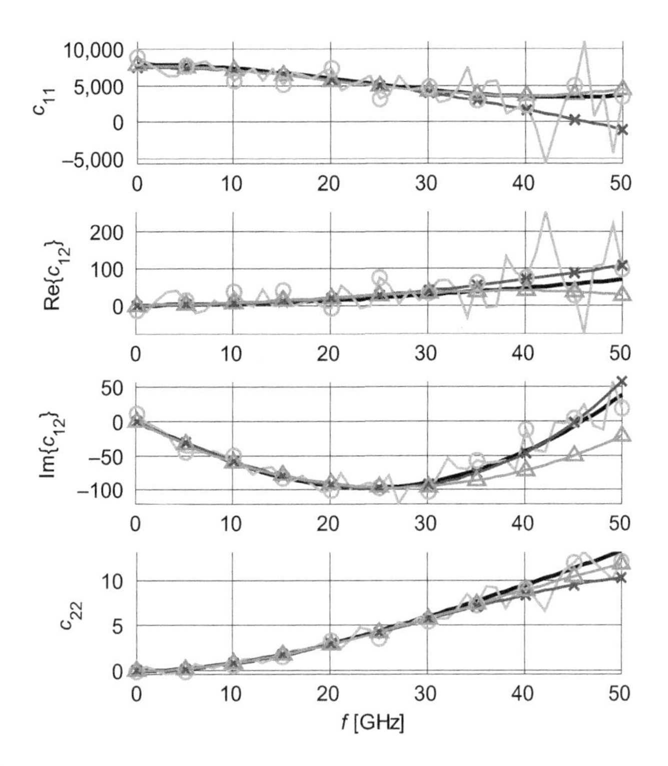

FIGURE 3.8

Example of extraction using the proposed methodologies on an ideal HEMT. Continuous line: true values. Circle markers: single-frequency extractions. Cross markers: extraction based on approach (a). Triangle markers: extraction based on approach (b). Notice that extraction frequencies are from 0.1 GHz to 50.1 GHz, step 1 GHz, but not all points are marked, to enhance readability.

All extractions were accomplished on simulated source pull measurements of noise figures from 8 to 32 GHz, to which a randomly generated error was purposely added as the product of a Gaussian variable (described by $\mu = 0$ and $\sigma = 0.025$) multiplied by NF. Also, a shrink of the Smith chart region of synthesizable input loads at higher frequencies was taken into account while generating the simulated measurements. These intentional non-idealities explain why the trace with round markers (frequency-by-frequency extraction) in Figure 3.8 does not perfectly overlap the black one (true values); on the other hand, the traces with cross markers (approach (a)) and triangle markers (approach (b)) well replicate the ideal correlation matrix also outside the extraction frequency range.

As a final remark, notice that a model as simple as that in Eqn (3.58) would be difficult to justify if one attempted a direct approximation of Lane's parameters, since their relationships to classic noise parameters are less familiar than those of the correlation matrix terms. Therefore, although it is maybe feasible to extend approach (b) directly to Lane's parameters, this would require a specific work, which was not addressed by the authors. However, one may use Lane's algorithm to determine the classical noise parameters, then compute the $\hat{c}_{ij}$ terms, and finally apply approach (a).

3.3.2.3.3 Application to a real device

In Section 3.3.2.3.2 a method was described that allows modeling of the noise behavior of a high-frequency active device over a broad frequency range by means of 10 constant coefficients only. It was also anticipated that in some cases even simpler models are possible. In particular a good approximation can often be achieved by neglecting coefficients β_2 and β_3, since the imaginary part of $\hat{c}_{12}$ is typically predominant on the real part, and γ_3, whose contribution does not typically play a significant role if not at high frequencies.

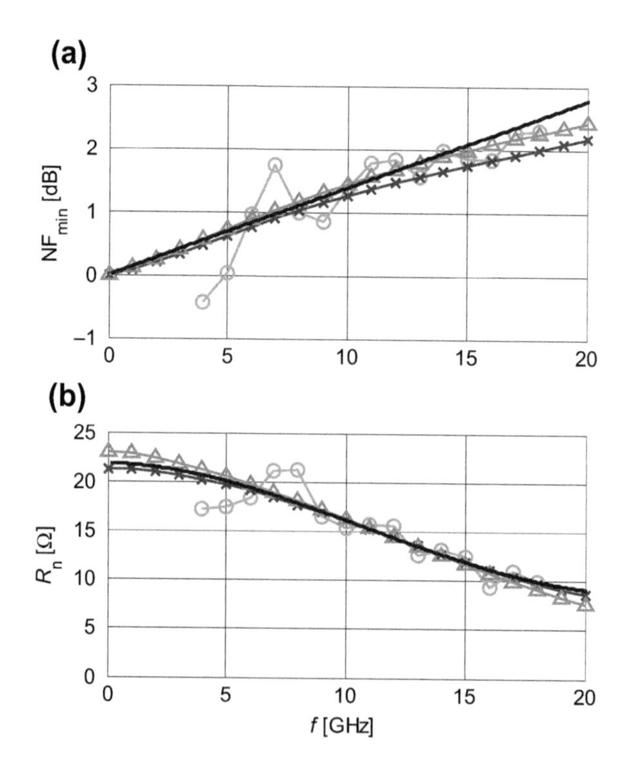

FIGURE 3.9

Example of extraction using the proposed methodologies on a real HEMT: minimum noise figure and noise resistance. Continuous line: reference values. Circle markers: single-frequency extractions. Cross markers: extraction based on approach (a). Triangle markers: extraction based on approach (b).

The seven-coefficient model thus obtained was used on a family of four-finger HEMT active devices with increasing finger lengths ($4 \times 25\ \mu m$, $4 \times 50\ \mu m$, $4 \times 75\ \mu m$). These devices, realized by Selex-SI on a GaN/SiC, 5-μm gate length technology, were characterized from 4 to 18 GHz with a 1-GHz step. As an example of extraction, the $4 \times 50\ \mu m$ device noise performance is reported in Figure 3.9 (minimum noise figure and noise resistance) and in Figure 3.10 (optimum input termination). In these figures, again, the round markers represent the sequence of single-frequency, source-pull extractions, while cross and triangle markers indicate results of approaches (a) and (b) in Section 3.3.2.3.2, respectively. Note that in this case real measurements are considered, so the true value of parameters is not known; instead, a thick trace is plotted representing the same parameters extracted by a reference alternative method, whose validity is discussed in [22]. A good agreement is achieved between the proposed methods and the reference over the whole frequency range considered.

Note how, as in the ideal case discussed in Section 3.3.2.3.3, a direct extraction applied distinctly at each frequency results in irregular fluctuations of noise parameters, while the proposed approaches smooth such undesired behavior and allow obtaining a characterization quite in agreement with the reference model.

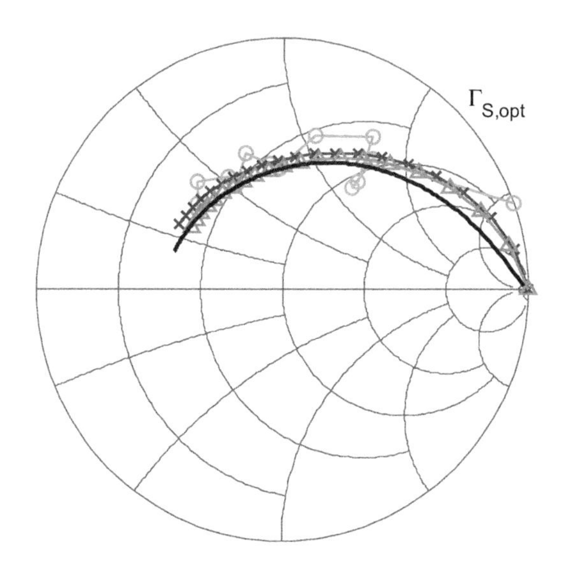

FIGURE 3.10

Example of extraction using the proposed methodologies on a real HEMT: optimum input termination. Continuous line: reference values. Circle markers: single-frequency extractions. Cross markers: extraction based on approach (a). Triangle markers: extraction based on approach (b).

3.4 Measurement errors

3.4.1 Noise factor uncertainty

Literature regarding noise measurement uncertainty is quite broad, especially as far as the Y-factor method is concerned. For this reason, and considering that the cold-source method is somewhat more general purpose (easily enabling source-pull measurements), in this section some insights will be provided only on the latter. Nevertheless, a certain degree of inaccuracy is found, unfortunately, in the most common references, in particular application notes and technical material, about which the reader should be warned in advance. To get a correct sense of the matter, refer for instance to [39], although such work is particularly focused on error correction. On the contrary, a simpler approach to the topic (oriented only to uncertainty evaluation) will be provided hereafter, based on the following hypotheses:

1. Variations in the reflection coefficient of the noise source, which is only used for receiver calibration, are negligible. This is not at all a restrictive assumption if a 6-dB noise source is used—or if an "effective" source is obtained by cascading the commercial component and an isolator.
2. The receiver source termination during measurement is the same as during calibration. Using the same isolator in the two steps is a way to guarantee this condition.
3. The three measurement phases are taken shortly after each other, so all system characteristic quantities can be assumed as constant (stable), although not exactly known, over the duration of each measurement.
4. The noise parameters of the DUT are almost constant over the range of temperature variations that it may experience during the measurement campaign.
5. Moreover, the nominal value of the DUT's physical temperature is not considered as an input variable to the system but rather as a predetermined operating condition; as such, it plainly represents a part of the measurement.

The previous assumptions—which, although implicitly, are often taken for granted—allow a practical handling of the topic and lead to closed-form expressions.

Having this background fixed, and with regard to Figure 3.3(a) and (c), recall now Eqn (3.20). Computing the total differential of this equation allows to link variations of the directly measured quantities, in the right-hand side, to variations of $T_{\mathrm{e,DUT}}$, yielding:

$$\begin{aligned} \mathrm{d}T_{\mathrm{e,DUT}} = {} & \frac{1}{G_{\mathrm{av,DUT}}G_{\mathrm{av,MAT}}}\cdot\mathrm{d}T_{\mathrm{M}} + \left(\frac{G_{\mathrm{av,MAT}}-1}{G_{\mathrm{av,DUT}}G_{\mathrm{av,MAT}}}-1\right)\cdot\mathrm{d}T_{\mathrm{A}} \\ & + \frac{-T_{\mathrm{A}}\cdot\left(1-G_{\mathrm{av,MAT}}\right)}{G_{\mathrm{av,DUT}}G_{\mathrm{av,MAT}}}\cdot\mathrm{d}G_{\mathrm{av,DUT}} \\ & + \left[\frac{T_{\mathrm{A}}}{G_{\mathrm{av,DUT}}G_{\mathrm{av,MAT}}}+\frac{T_{\mathrm{A}}\cdot\left(1-G_{\mathrm{av,MAT}}\right)}{G_{\mathrm{av,DUT}}G_{\mathrm{av,MAT}}^{2}}\right]\cdot\mathrm{d}G_{\mathrm{av,MAT}} \end{aligned} \tag{3.61}$$

Following a commonly accepted methodology [40], suppose now that all uncertainties come from random measurement errors, with normal probability distributions, and that such errors are not correlated (notice from Eqn (3.21), in particular, that T_{M} is independent of T_{A}, $G_{\mathrm{av,DUT}}$, and $G_{\mathrm{av,MAT}}$). The mean value of these distributions is assumed to be the measured one, while the standard deviation is associated to the measurement uncertainty, through a multiplicative factor depending on the desired confidence level; this factor will be considered to be unity for simplicity (so as to use the phraseology "measurement uncertainty" and "standard deviation" interchangeably), however, the presented results are independent of this choice.

Under these premises, Eqn (3.61) allows one to compute the uncertainty on $T_{\mathrm{e,DUT}}$, denoted by $\delta T_{\mathrm{e,DUT}}$, as the root of the sum of squares of the individual component uncertainties, δT_{M}, δT_{A}, $\delta G_{\mathrm{av,DUT}}$, and $\delta G_{\mathrm{av,MAT}}$, taken with their coefficients:

$$\begin{aligned}\delta T_{\mathrm{e,DUT}} = \Bigg\{ & \left(\frac{1}{G_{\mathrm{av,DUT}}G_{\mathrm{av,MAT}}}\right)^2 \cdot (\delta T_{\mathrm{M}})^2 \\ & + \left(\frac{G_{\mathrm{av,MAT}}-1}{G_{\mathrm{av,DUT}}G_{\mathrm{av,MAT}}} - 1\right)^2 \cdot (\delta T_{\mathrm{A}})^2 \\ & + \left[\frac{-T_{\mathrm{A}}\cdot(1-G_{\mathrm{av,MAT}})}{G_{\mathrm{av,DUT}}G_{\mathrm{av,MAT}}}\right]^2 \cdot (\delta G_{\mathrm{av,DUT}})^2 \\ & + \left[\frac{T_{\mathrm{A}}}{G_{\mathrm{av,DUT}}G_{\mathrm{av,MAT}}} + \frac{T_{\mathrm{A}}\cdot(1-G_{\mathrm{av,MAT}})}{G_{\mathrm{av,DUT}}G_{\mathrm{av,MAT}}^2}\right]^2 \cdot (\delta G_{\mathrm{av,MAT}})^2 \Bigg\}^{1/2}\end{aligned} \tag{3.62}$$

The problem results thus split into smaller ones, and it can be easily handled by determining separately δT_{M}, δT_{A}, $\delta G_{\mathrm{av,DUT}}$, and $\delta G_{\mathrm{av,MAT}}$. As to the latter two, they depend heavily on the measurement setup and VNA calibration, and should be computed following the specific guidelines given in the VNA manual. Fortunately, some vendors also offer software tools to compute S parameters' accuracy, from which $\delta G_{\mathrm{av,DUT}}$ and $\delta G_{\mathrm{av,MAT}}$ can be straightforwardly estimated, for instance by means of simple Monte Carlo simulations. Term δT_{A}, which represents the uncertainty on the ambient temperature, essentially depends on the temperature sensor; therefore it is easily determined by following the relevant documentation.

The only term left to determine is now δT_{M}, which requires some more work. First of all, substituting Eqns (3.10) and (3.22) in (3.21) yields:

$$T_{\mathrm{M}} = T_0 \cdot \mathrm{ENR} \cdot \frac{N_{\mathrm{m}}^{\mathrm{meas}} - N_{\mathrm{c}}^{\mathrm{cal}}}{N_{\mathrm{h}}^{\mathrm{cal}} - N_{\mathrm{c}}^{\mathrm{cal}}} + T_{\mathrm{C}} \tag{3.63}$$

Notice that T_{C} is the noise source's physical temperature, not to be confused with T_{A}, the physical temperature of the measurement chain. Indeed, since the noise source may be prone to heating due to the bias current used during the "hot" state, it is customary to measure T_{C} with a second temperature sensor, which results in T_{A} and T_{C} being uncorrelated (under hypothesis above). If this were not the case (i.e., if

T_A and T_C were equal), one should substitute Eqn (3.63) into (3.61), then collect dT_A before computing $\delta T_{e,DUT}$, obtaining an expression different from Eqn (3.62).

Once again, starting by differentiating Eqn (3.63):

$$\begin{aligned} dT_M = & T_0 \cdot \frac{N_m^{meas} - N_c^{cal}}{N_h^{cal} - N_c^{cal}} \cdot d\text{ENR} + T_0 \cdot \text{ENR} \cdot \frac{1}{N_h^{cal} - N_c^{cal}} \cdot dN_m^{meas} \\ & + T_0 \cdot \text{ENR} \cdot \frac{N_m^{meas} - N_h^{cal}}{\left(N_h^{cal} - N_c^{cal}\right)^2} \cdot dN_c^{cal} + T_0 \cdot \text{ENR} \cdot \frac{N_c^{cal} - N_m^{meas}}{\left(N_h^{cal} - N_c^{cal}\right)^2} \cdot dN_h^{cal} + dT_C \end{aligned} \tag{3.64}$$

the uncertainty on the quantity of interest can be expressed as a function of component uncertainties (i.e., δENR, δN_m^{meas}, δN_c^{cal}, δN_h^{cal}, and δT_C).

All the information needed at this point can be found on the basis of the instrumentation manuals (i.e., the noise source's, the receiver's, and the temperature sensor's data sheets). Notice, however, that the quantities appearing in Eqn (3.64) are linear, while uncertainties related to power levels, such as δENR and δN, are usually given in logarithmic form. A conversion is then required in such cases, remembering that:

$$\delta Q = \frac{\ln 10}{10} Q \cdot \delta Q_{dB} \tag{3.65}$$

which relates the uncertainty of a generic quantity represented logarithmically, δQ_{dB}, to the expression of the uncertainty of the same quantity as a ratio, δQ, where $Q_{dB} = 10 \log Q$.

Once all the above computations have been carried out, the uncertainty on $T_{e,DUT}$ is known. Outside the radio-astronomy community, however, it is normal practice to refer to the concepts of noise factor and, more commonly, noise figure. To obtain the noise factor representation, the following relationship is exploited:

$$F_{DUT} = 1 + \frac{T_{e,DUT}}{T_0} \tag{3.66}$$

from which it is obviously obtained that:

$$\delta F_{DUT} = \frac{1}{T_0} \delta T_{e,DUT} \tag{3.67}$$

The uncertainty on the DUT noise figure, NF_{DUT}, is now derived by applying Eqn (3.65) to δF_{DUT}.

A final remark is necessary concerning Eqn (3.64): terms dN_m^{meas}, dN_c^{cal}, and dN_h^{cal} in this equation were treated as a mere consequence of instrument errors in reading absolute power levels, of jitter and similar phenomena. Nevertheless, other issues may be involved causing errors on noise power readings, and in particular changes in the receiver's gain and/or noise figure. As observed previously, these and other possible inconveniences were not taken into account in order to obtain closed formulae, yet for this analysis to be valid it is vital that the hypotheses at the beginning of this section be fulfilled.

3.4.2 Effects of receiver compression

The aim of this section is to warn the reader against the often underestimated issue of receiver compression. As for Section 3.4.1, the focus is on the cold-source method.

To clearly state the problem, first Eqn (3.63) is recast as follows:

$$T_M = T_0 \cdot \mathrm{ENR} \cdot \frac{Y_m - 1}{Y_c - 1} + T_C \tag{3.68}$$

where, for the sake of notation clarity, the symbol Y_m was used to represent the ratio $\frac{N_m^{meas}}{N_c^{cal}}$ and Y_c to represent $Y^{cal} = \frac{N_h^{cal}}{N_c^{cal}}$.

Now suppose that a certain level of receiver gain compression, $-l_{dB}$, is present during the measurement phase. It is here assumed that the noise source has been chosen accordingly to the receiver spurious-free dynamic range (SFDR), so that the problem does not appear during the calibration phase.

The effect of such compression is clearly an underestimation of N_m^{meas} and therefore an erroneous evaluation of T_M, which may be denoted as $T_{M,comp}$:

$$T_{M,comp} = T_0 \cdot \mathrm{ENR} \cdot \frac{l \cdot Y_m - 1}{Y_c - 1} + T_C \tag{3.69}$$

where

$$l = 10^{-l_{dB}/10} \quad \in \quad (0 \;\; 1) \tag{3.70}$$

Hence the absolute error on T_M is:

$$T_{M,comp} - T_M = T_0 \cdot \mathrm{ENR} \cdot \frac{Y_m}{Y_c - 1} \cdot (l - 1) = T_0 \cdot \mathrm{ENR} \cdot \frac{Y_{m,comp}}{Y_c - 1} \cdot \left(1 - \frac{1}{l}\right) \tag{3.71}$$

where $Y_{m,comp} = l \cdot Y_m$. Equation (3.71) shows that, as expected, $T_{M,comp}$ is less than T_M; moreover, it could provide a way to compensate for compression a posteriori, if only l were known. Of course, compression should not be compensated but simply avoided; consequently, Eqn (3.71) is only useful to confirm the hypothesis of compression when a measurement gives unexpected results.

Figure 3.11 shows, in a sample situation ($\mathrm{ENR}_{dB} = 6$ dB, $T_M = 4T_H$, $\mathrm{NF}_{REC} = 3$ dB, 6 dB, 9 dB), the relative error on T_M, computed as $\frac{T_{M,comp} - T_M}{T_M}$, as a function of receiver gain compression.

Notice that, unlike the sources of error discussed in Section 3.4.1, receiver compression cannot be treated as a Gaussian variable. On the contrary, compression is peculiar in that it always causes, if present, underestimations of T_M, and then of $T_{e,DUT}$. Therefore, it produces a shift of the uncertainty bars toward the bottom, making them asymmetrical with respect to the nominal values.

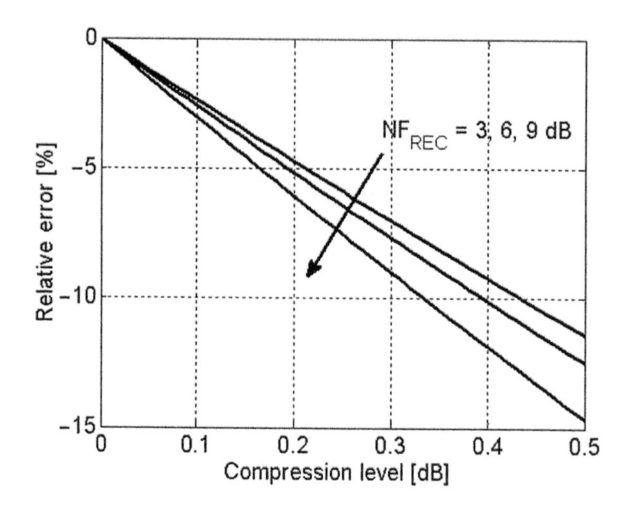

FIGURE 3.11

Relative error on T_M in a sample situation, as a function of receiver compression.

Notice that the previous discussion is greatly simplified with respect to the phenomena actually under way. In particular, whereas the attention has only focused on the receiver's gain, the effects of compression on its noise factor have been neglected. A more precise description of the receiver's behavior under compression should, on the contrary, leave aside fundamentally linear concepts such as noise factor and gain, and start from a low-level, nonlinear representation of the receiver. Although this issue is not very present in the literature, it is formally equivalent to the noise characterization of active devices operated in the nonlinear region—a topic that has been recently rediscovered by several authors [41–44], but whose mathematical foundations have been long known [45–47].

Due to the complexity of the matter, simplified models are assumed for the nonlinearly operated devices, such as a memoryless cubic model. From this assumption, the power spectral densities can be computed at the input and output port of the nonlinear amplifier, which allows evaluating the deviation of the measured noise factor (following the standard Y-factor method) from its small-signal, or linear, value [42] with a greater mathematical correctness than previously described; notice in particular that, depending on the receiver [44], the influence of compression on the noise power spectral density may be more significant than on the gain itself.

With a greater focus on device characterization, on the other hand, notice that this topic has not achieved yet a standard, well-established treatment; for instance, novel definitions have been proposed, but are still not commonly accepted [43], and only a little experimental validation can be found at present [44]. The interested reader may want to explore more in depth this subject—which is beyond the scope of this chapter—starting from the above suggested references.

3.4.3 Considerations on receiver bandwidth

As pointed out in [37], a not very well-known source of errors in noise factor measurements—especially when the source pull technique is involved—is the receiver's finite bandwidth. Indeed, although the noise factor is a point function of frequency, whose definition is based on noise power densities, a real receiver can only measure integral power levels over an effective bandwidth, B, greater than zero. As long as B is small enough that all linear parameters of the measurement (or calibration) chain can be supposed to be constant, noise power density is approximated as the ratio of noise integral power to effective bandwidth. In a number of cases (see, for instance, Eqn (3.9)), the exact value of B need not even be known, since it naturally cancels out; nevertheless, the "small bandwidth" hypothesis still must hold for the measurement to be correct.

In practice, a trade-off between narrow bandwidth (for close approximation of noise power densities, not to mention compression issues) and wide bandwidth (for measurement accuracy) is necessary; therefore some useful insights into this matter will be given in the following.

Figure 3.12 shows the typical behavior over a 500-MHz span of a high-magnitude source reflection coefficient synthesized by cascading a matched load, an electromechanical tuner, and a coaxial cable about 1.3 feet long. As it can be noted, the magnitude of the reflection coefficient can be considered almost constant, while the phase varies very rapidly over frequency—in this example, at a rate of −2.42 degrees per MHz. If a cold-source noise measurement were to be taken with a 5-MHz bandwidth, the DUT would see a source reflection coefficient varying about 12 degrees in phase! In this condition, one could hardly argue that the DUT available gain keeps constant, although its scattering and noise parameters are expected to be so.

In general, following [37], a tuner-synthesized reflection coefficient can be sufficiently well represented, for frequencies f close to a center frequency f_0, as:

$$\Gamma_{S,DUT}(f) = \Gamma_{S,DUT}(f_0) \cdot e^{-j \cdot 2\pi \cdot (f - f_0) \cdot 2\tau} \tag{3.72}$$

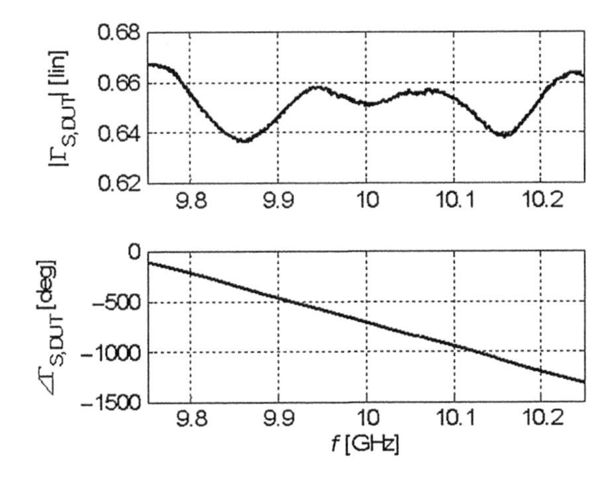

FIGURE 3.12

Example of reflection coefficient presented to a DUT in a source pull setup.

where τ is the (effective) delay time of the tuner (or of the source tuning network, however complex), evaluated at f_0:

$$\tau = -\frac{1}{2\pi}\cdot\frac{\partial\phi}{\partial f}\bigg|_{f_0} = -\frac{1}{4\pi}\cdot\frac{\partial\left(\angle\Gamma_{S,DUT}\right)}{\partial f}\bigg|_{f_0} \tag{3.73}$$

although it may be approximately regarded as a constant in the considered case (notice that the coaxial cable is nondispersive, but the tuner's reflection coefficient is in general an unknown function of frequency). From Eqn (3.72), the maximum phase deviation from $\angle\Gamma_{S,DUT}(f_0)$ is clearly $\pm 2\pi\cdot B\cdot\tau$ radians; therefore two ways to minimize finite-bandwidth errors can be immediately devised, namely reducing B or reducing τ.

While the former option undergoes constraints that should now be clear, the latter relates to avoiding long cables between the tuner and the DUT, which in fact are common for on-wafer measurements. Even the position of the tuner slug may be optimally chosen (i.e., positions closer to the DUT should be preferred), the reflection coefficient at f_0 being equal (this choice also limits losses in the tuner, and allows one to synthesize higher-magnitude terminations).

Observing Eqn (3.72), one can notice that bandwidth errors vanish when $\left|\Gamma_{S,DUT}(f_0)\right|$ approaches zero, while they increase when such magnitude approaches unity. On the one hand, this explains why F_{50} measurement is virtually unaffected by this kind of error. On the other hand, this means that, if the product $B\cdot\tau$ cannot be reduced to acceptably low values, taking noise measurements on high-magnitude source reflection coefficients should be avoided: as a reference case, the condition $\left|\Gamma_{S,DUT}(f_0)\right| < 0.7$ should typically be fulfilled, according to [37], to avoid the largest errors.

Related to finite-bandwidth issues as just described is another kind of error, which depends on the accuracy with which the receiver is able to tune at the measurement frequency. As it can be understood by considering again Eqn (3.72) in the case of an arbitrarily narrow measurement bandwidth, a noise factor measurement on a source termination different than desired is obtained if f_0 is shifted. The practical effect is approximately a rotation of the measurement reflection coefficients.

Unfortunately, the effect of receiver bandwidth on measurements cannot be accurately estimated without knowing the peculiar features of all the chain components, in particular the tuning network and the DUT. Furthermore, even if all needed parameters were available, closed formulae could hardly be derived; this is especially true if one is interested not as much in the accuracy of the single measurement on a particular reflection coefficient but rather in the accuracy of the four-parameter extraction as a whole—in other words, after applying complicated minimization algorithms such as those described in Sections 3.3.2.2 and 3.3.2.3, which also add a dependency of the result on the chosen source termination pattern.

The only reasonable way to overcome these difficulties is probably setting up a Monte Carlo simulation of the entire test bench. The real operation, taking into account finite-bandwidth issues as well as all other possible sources of error (provided they can be modeled), can straightforwardly—yet, at the cost of laborious programming—be emulated and compared to the ideal result for a number of representative DUTs.

3.5 High-sensitivity subsystems design

3.5.1 Traditional approaches

LNA is typically the first active stage of a microwave receiver system. Its noise-gain performance affects the overall receiver's noise figure, and therefore great care must be used during the design phase to optimize it. Ideally, the LNA's active device should be input-matched to achieve the optimum noise condition so the amplifier exhibits its minimum noise figure.

A schematic single-stage amplifier is depicted in Figure 3.13. The *S* parameter block represents an active device (biased FET) where some feedback may have been introduced to ease the trade-off between otherwise contrasting goals (noise, gain, input matching). Input and output matching networks (IMN and OMN, respectively) are the pair of passive two-port networks in charge of matching the active device's terminals.

The typical LNA design flow is as follows: the source termination is selected to trade off the amplifier's noise, gain, and input matching performance when the output is considered perfectly matched. This can be accomplished by the corresponding constant noise, constant available gain, and constant input matching family of circles in the Γ_S plane. The intersection (assuming that it exists) of those three families of circles is drawn on a Smith chart, representing the region of source complex loads that allows the fulfillment of all the specifications. Conversely, if a trade-off between the input and output matching is a target in the specification table—i.e., a better input match although not with a perfect match at the output—a blind optimization step has to be considered, unless more sophisticated approaches (such as those detailed in the following sections) are known to the designer.

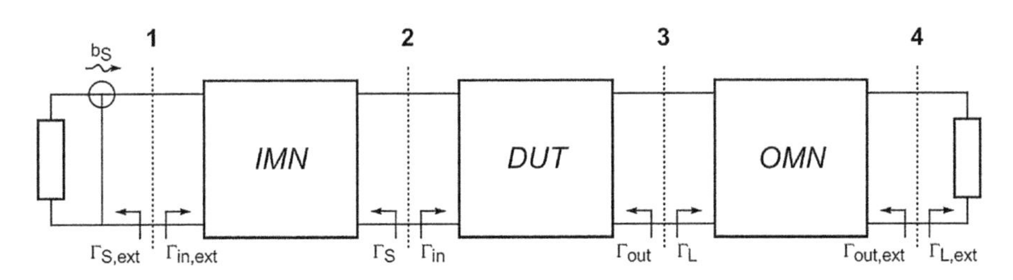

FIGURE 3.13

Typical microwave amplifier block diagram with the corresponding reflection coefficients.

3.5.2 A systematic approach utilizing constant-mismatch circles, degenerative feedback

The method reported in this section is based on a design chart where input return loss (IRL), output return loss (ORL), and the transducer gain (G_T) can be simultaneously visualized, and therefore the fulfillment of the design specifications can be directly verified [48].

The chart in Figure 3.13 is obtained by fixing the source termination to the optimum noise value (therefore minimizing the amplifier's noise figure) or in general to a given source termination resulting from stability, technological, and noise constraints. The value of the load impedance is directly obtained after considering constant input and output mismatch circles.

3.5.2.1 Constant mismatch circles

Let Γ_S be set, as stated before, equal to a certain value, either due to structural constraints, technical reasons, or any other electrical specifications.

With reference to Figure 3.13, the goal is to realize a design chart where the best trade-off between the amplifier's input and output return losses, IRL and ORL, can be instantaneously established. Notice that the following definitions are adopted for return losses:

$$\mathrm{IRL} = -20\cdot\log_{10}\left|\Gamma_{\mathrm{in,ext}}\right| \tag{3.74}$$

$$\mathrm{ORL} = -20\cdot\log_{10}\left|\Gamma_{\mathrm{out,ext}}\right| \tag{3.75}$$

The reflection coefficient at the device input (section-2 in Figure 3.13) is given by:

$$\Gamma_{\mathrm{in}} = f(S,\Gamma_L) = \frac{s_{11} - \Delta\cdot\Gamma_L}{1 - s_{22}\cdot\Gamma_L} \tag{3.76}$$

where Δ is the determinant of the matrix S and Γ_L is the termination at the device output port (section-3 in Figure 3.13). As well known, such relationship is a conformal mapping between the Γ_L and the Γ_{in} planes, so that a circle in the Γ_L plane is transformed into another circle in the Γ_{in} plane.

The mismatch occurring at section-2 in Figure 3.13 between the values of Γ_{in} and the preselected Γ_S is given by Eqn (3.77). Clearly, from Eqn (3.76), Γ_{in} is in turn a function of S and Γ_L, and therefore $|\mathrm{MM_{in}}|$, defined as:

$$|\mathrm{MM_{in}}| = f([S],\Gamma_L) = \left|\frac{\Gamma_{\mathrm{in}} - \Gamma_S^*}{1 - \Gamma_S\Gamma_{\mathrm{in}}}\right| \tag{3.77}$$

is a function of the same variables.

Incidentally, the expression in Eqn (3.77), for fixed values of the input mismatch, represents a family of circles in the Γ_{in} plane for fixed Γ_S. The center and radius of each circle depend on the selected value of $|\mathrm{MM_{in}}|$ and the predetermined position of Γ_S.

If IMN is synthesized through ideal reactive elements, then the mismatch occurring at section-2 equals the mismatch occurring at section-1 of the cascade. This leads to the fact that once the amplifier's input return loss is specified, $|MM_{in}|$ (in decibels) is automatically computed and equals the opposite of the amplifier's input return loss, i.e.:

$$|MM_{in}|[dB] = |\Gamma_{in,ext}|[dB] = 10^{\left(\frac{-IRL}{220}\right)} dB \tag{3.78}$$

Therefore, through Eqn (3.78), one of the design specifications regarding the overall amplifier's input behavior can be directly related to Γ_L.

As already stated, Eqn (3.77) represents a family of circles in the Γ_{in} plane parameterized through the value of $|MM_{in}|$. Through (3.76), that is a conformal mapping between the Γ_{in} plane and Γ_L plane, such family of circles can be mapped in a one-to-one relationship: for a given $|MM_{in}|$ value, a single circle is obtained in the Γ_L plane. Such circle represents all the possible Γ_L values yielding the required input return loss.

The best possible value of ORL for a given IRL and fixed Γ_S has now to be determined. Once again, this condition is related to the mismatch at section-3 in Figure 3.13, $|MM_{out}|$, defined similarly as in Eqn (3.77):

$$|MM_{out}| = f(S, \Gamma_L) = \left|\frac{\Gamma_L - \Gamma_{out}^*}{1 - \Gamma_L \cdot \Gamma_{out}}\right| \tag{3.79}$$

Where the device output reflection coefficient Γ_{out}, here constant since S and Γ_S are fixed, is given by

$$\Gamma_{out} = f(S, \Gamma_S) = \frac{s_{22} - \Delta \cdot \Gamma_S}{1 - s_{11} \cdot \Gamma_S} \tag{3.80}$$

Through the assumption that the OMN is synthesized by reactive elements, then $|MM_{out}|$ equals the mismatch occurring at section-4 of the cascade. Once again $|MM_{out}|$ (in decibels) is automatically computed and equals the opposite of the amplifier's ORL.

Equation (3.79) represents, for fixed values of the output mismatch, a set of circles in the Γ_L plane. The center and radius of each circle depend on the selected value of $|MM_{out}|$. The value of Γ_L that, for a given IRL, maximizes ORL, is geometrically attainable as the tangent point between the circles expressed in Eqn (3.79) and that is expressed in Eqn (3.77) after the latter has been appropriately transferred in the Γ_L plane through Eqn (3.76). Analytically, ORL can be found by stating that the distance between the centers of the two circles, when varying the value of $|MM_{out}|$, must equal the sum of their radiuses (when each circle is totally outside the other one) or the difference of their radiuses (when one circle is totally inside the other one).

An example CAD implementation of the above concepts is depicted in Figure 3.14 in the real case of a conditionally stable transistor. The selected device is a NEC ne321000, a commercially available *n*-channel HJFET whose linear and noise parameters at 10 GHz are listed in Table 3.2.

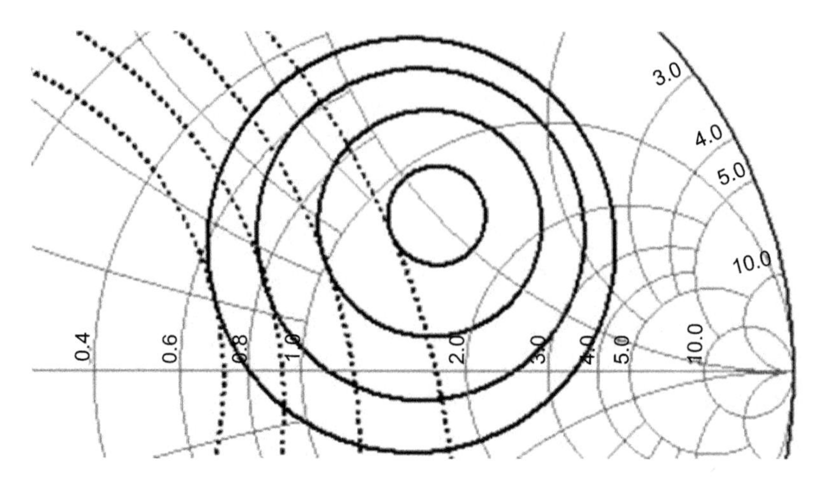

FIGURE 3.14

IRL circles (dashed) and ORL circles (solid) in the Γ_L plane for fixed Γ_S, [S] and degenerative source inductance (0.35 nH).

Table 3.2 NEC ne321000 Scattering and noise parameters at 10 GHz

	magnitude (S_{ij})	angle (S_{ij})
S_{11}	0.90	−68.6
S_{21}	4.39	129.9
S_{12}	0.09	46.7
S_{22}	0.56	−52.6
NF_{min} [dB]		0.28
$\Gamma_{opt,noise}$	mag	0.73
	ang	24.8
R_n		0.32

As clarified later (in the following section), an 0.35 nH inductor has been inserted between the FET's source and ground terminals, thus applying a series feedback degeneration. Dashed curves in Figure 3.14 represent the circles expressed in Eqn (3.77), that is to say, the load terminations realizing a given value of IRL (8, 10, 12, and 14 dB). Solid line circles, expressed in Eqn (3.79), represent the load terminations yielding a constant ORL. The tangency point between two circles simultaneously fulfills the two conditions and is selected as the optimum trade-off between such contrasting goals. The larger the circle, the smaller the value of the corresponding return loss.

The values of the optimum ORL for a given IRL, as depicted in Figure 3.14, are reported in Table 3.3.

Table 3.3 Values of IRL and ORL for the curves in Figure 3.14

IRL [dB]	14	10	12	8
ORL [dB]	6.3	8.1	11.3	18.4

3.5.2.2 The role of inductive feedback

Conditions on IRL and ORL often cannot be simultaneously fulfilled for a certain set active device's *S* parameters. Therefore the latter are modified by means of feedback. A method extensively described in the literature consists in inserting a series inductor between the FET's common terminal (source) and ground. This technique is typically used in LNA design since it eases the ORL—NF trade-off that has to be performed at the active device's input. The procedure becomes particularly helpful when the device is conditionally stable and therefore an adequate simultaneous input—output matching cannot be realized without the risk of triggering oscillation phenomena. The drawback of this method is clearly the reduction of the active device's maximum available gain. Therefore the amount of feedback (i.e., the inductor's value) must be carefully selected so the active device can be more simply matched at its terminals and to avoid an excessive impact on the active device's gain.

For this reason it is typical to observe, in multistage LNA designs, a gradual decrease in source inductance, since a higher feedback is required in the first stages to ease the simultaneous noise and signal matching, although at the expense of gain. A microphotograph of an example MMIC LNA for space-borne applications [49] is given in Figure 3.15.

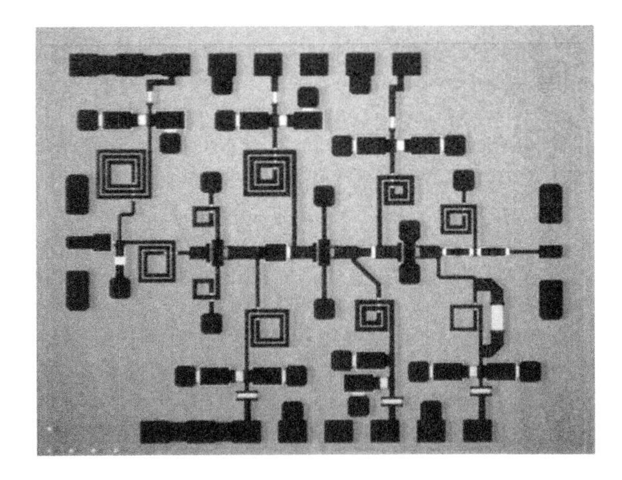

FIGURE 3.15

Microphotograph of a three-stage, low-noise amplifier for space-borne applications, based on a GaN-SiC process by Selex-SI.

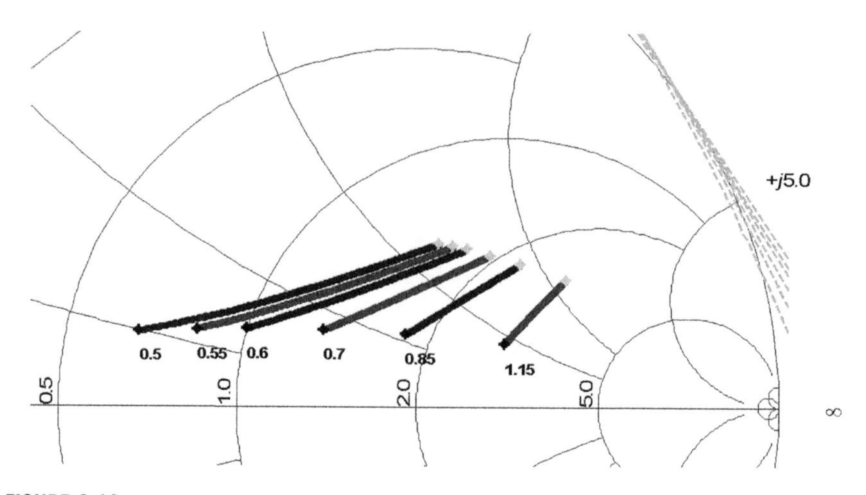

FIGURE 3.16

$\Gamma_{L,opt}$ curves that realize the most favorable ORL, for a given IRL and fixed Γ_S, for six different values of degenerative source inductance. Output stability circles are also plotted.

Figure 3.16 depicts the curves in the Γ_L plane implementing the most favorable ORL, for a given IRL, for six different values of degenerative source inductance, utilizing the approach described in the previous section. It has to be stressed that for each value of degenerative source inductance, both the scattering and noise parameters of the active device are modified accordingly, thus automatically yielding a different optimum source termination Γ_S.

One end of each curve in Figure 3.16 represents the load termination that matches the amplifier's output to 50 Ω (maximum power transfer condition), while the other end of each curve represents the load termination yielding conjugate matching condition at the device input. The other points of the curve correspond to intermediate trade-off between IRL and ORL.

When acting on the feedback inductance value, the curves tend to move and change in length, thus relaxing or stiffening the obtainable IRL−ORL trade-off. The dashed curves represent the values of Γ_L yielding $|\Gamma_{in}|$ greater than unity (output stability circles).

3.5.2.3 Design chart

A design chart can now be introduced where the optimum trade-off between IRL and ORL is plotted as a function of the feedback inductor value, together with the corresponding transducer gain. As a general rule, a simultaneous acceptable input−output matching becomes easier as the value of the source inductor increases to the detriment of the active device's gain. This feature is generally minor since extra gain can be obtained by cascading additional stages; on the other hand, IRL and ORL heavily depend on the single stage's source and load impedances and are therefore very sensitive to the active device's stability factor.

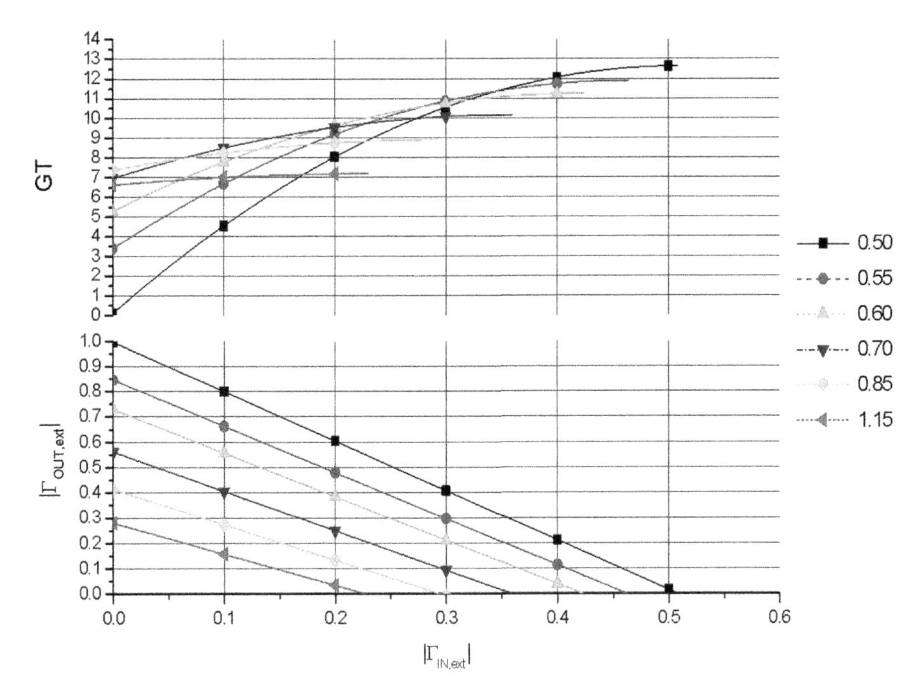

FIGURE 3.17

I/O match optimum trade-off and corresponding transducer gain (GT) for six different values of degenerative source inductance.

$|\Gamma_{\text{in,ext}}|$ and $|\Gamma_{\text{out,ext}}|$ (instead of IRL and ORL) are plotted in Figure 3.17 so that the axes can be limited between 0 and 1; 0 represents the perfect 50 Ω match while 1 represents a reactive termination (no active power can be transferred). The value of the series/series feedback inductance that fulfills the design constraints can be selected by simply inspecting the proposed chart. Once the appropriate value of feedback inductance has been selected, Γ_S is simply determined by the active devices' noise parameters while the corresponding value of Γ_L can be synthesized by analyzing a chart similar to that in Figure 3.14.

As an example, the two larger inductor values (0.85 nH and 1.15 nH) in Figure 3.17 can be used to design an amplifier having both $|\Gamma_{\text{in,ext}}|$ and $|\Gamma_{\text{out,ext}}|$ at the same time smaller than 0.2 ($\approx$ 14 dB return loss). An appropriate choice of the degenerative source inductance's value would be such that the line intersects the (0.2, 0.2) point—in this case, approximately 0.75 nH. In this way, the conditions on IRL and ORL would be simultaneously fulfilled with minimum gain reduction.

With the proposed design chart, the selection of the optimum trade-off among IRL, ORL, and transducer gain becomes readily available, as opposed to the typical iterative design process, in which the feedback amount is selected after a lengthy procedure.

3.5.3 Noise measure-based design for multistage amplifiers

The strategy described in the Section 3.5.1 is very well suited to design narrow-band, single-stage LNAs; in such cases the designer only has to keep under control three figures of merit depending on the source termination—namely, noise figure, available gain, and input return loss—since the graphical instruments used assume, implicitly, that a perfect matching is realized at the output port. Unfortunately, when a multistage design is involved, only the last stage's output section is expected to show a conjugate matching, while some degree of mismatch will be present in general at the intermediate sections; as a consequence, the constant-IRL circles cannot rigorously be used, although constant-F and constant-G_{av} circles are still valid, since they depend, for each stage, only on the source termination. In conclusion, generalizing the graphical methodology of Section 3.5.2 to multistage designs, via iteration in a left-to-right fashion, is not straightforward; it requires, in the best case, a lot of "trial and error," which, unfortunately, cannot be made an automatic task for a computer to handle.

In this section an alternative methodology [50–52] will be outlined that allows design of a multistage LNA, taking into account the issue of input matching. Given the scattering and noise parameters of the active stages, which are considered unconditionally stable, the result is the optimum lossless realization of the matching and interstage networks. "Optimality" here relates to global noise figure and matching, while as to gain it is assumed that there is enough freedom in choosing the number of stages to fulfill the design requirements. As a final remark, the presented methodology is suited to automatic implementations in numerical environments, such as MATLAB®.

First of all, some concepts will be recalled upon which the subsequent steps are based. One is the definition of mismatch level, ML, at any given reference plane along a chain of two-ports; in general, if we call Γ_l and Γ_r the reflection coefficients of the considered section looking to the left and to the right, respectively, the mismatch level is the ratio of the available power from the left side, $P_{av,l}$, to the power absorbed by the right side, P_r. Its expression is:

$$\mathrm{ML} = \frac{P_{av,l}}{P_r} = \frac{|1-\Gamma_l\Gamma_r|^2}{\left(1-|\Gamma_l|^2\right)\left(1-|\Gamma_r|^2\right)} \tag{3.81}$$

In the following, only unconditionally stable stages will be considered, therefore $|\Gamma_l| \leq 1$, $|\Gamma_r| \leq 1$, and $\mathrm{ML} \geq 1$, where the equality holds when conjugate matching is achieved.

Notice also that mismatch level is invariant under any cascading of lossless elements. Hence, if Γ_S and Γ_L denote the source and load reflection coefficients of a lossless two-port, and Γ_{out} and Γ_{in} the corresponding ones at the output and input port, respectively, it is possible to write:

$$\frac{|1-\Gamma_S\Gamma_{in}|^2}{\left(1-|\Gamma_S|^2\right)\left(1-|\Gamma_{in}|^2\right)} = \mathrm{ML} = \frac{|1-\Gamma_{out}\Gamma_L|^2}{\left(1-|\Gamma_{out}|^2\right)\left(1-|\Gamma_L|^2\right)} \tag{3.82}$$

It is thus apparent that, once three reflection coefficients out of four have been chosen in Eqn (3.82), the fourth is constrained to lie on some constant-mismatch circle. For instance, if Γ_{out} and Γ_L (or, equivalently, ML) and Γ_S have been fixed, then Γ_{in} will lie on a circle whose center and radius are as follows:

$$\begin{aligned} C(\Gamma_S, \mathrm{ML}) &= \frac{\Gamma_S^*}{\mathrm{ML}\cdot\left(1-|\Gamma_S|^2\right)+|\Gamma_S|^2} \\ R(\Gamma_S, \mathrm{ML}) &= \frac{\left(1-|\Gamma_S|^2\right)\cdot\sqrt{\mathrm{ML}\cdot(\mathrm{ML}-1)}}{\mathrm{ML}\cdot\left(1-|\Gamma_S|^2\right)+|\Gamma_S|^2} \end{aligned} \tag{3.83}$$

The other important concept is that of noise measure of a two-port, M, first defined by Haus and Adler [53] as:

$$M = \frac{F-1}{1-\frac{1}{G_{av}}} \tag{3.84}$$

The minimum value of this parameter, M_{min}, is invariant under any lossless transformation of the two-port, including feedback. Moreover, the minimum noise measure of a cascade of stages, each featured by a noise measure M_{STi}, is obtained when these are sorted by increasing values of M_{STi}; if, now, the overall available gain is much greater than unity, also the cascade noise factor is approximately minimum.

All of the previous observations lead to devising a procedure made of two parts. Supposing for example a three-stage design, in the first part the following steps are to be carried out:

- Choose a viable geometry, bias, and stabilization network for each stage's active device. The first two stages are the main ones responsible for overall noise behavior, so they are expected to show low values of minimum noise figure (reactive feedback can be used with no drawback on minimum noise measurement). The third stage, on the other hand, should be optimized with respect to gain and power requirements.
- Sort the active stages by increasing values of their *minimum* noise measure, then assign to the first and second stages the optimum source termination for noise measure, i.e., $\Gamma_{S,ST1} \doteq \Gamma^M_{S,opt,ST1}$ and $\Gamma_{S,ST2} \doteq \Gamma^M_{S,opt,ST2}$. Assign to the third stage the optimum source termination for available gain, i.e., $\Gamma_{S,ST3} \doteq \Gamma^G_{S,opt,ST3}$. Notice that at this point also $\Gamma_{out,ST3}$, $\Gamma_{out,ST3}$, and $\Gamma_{out,ST3}$ are univocally determined.
- Since the output matching network is lossless, the output mismatch level of the third stage and of the LNA are the same. The latter is constrained by the specifications to a maximum value, so $\Gamma_{out,LNA}$ is now to be chosen anywhere inside a particular circle centered on the origin of the Smith chart.

Notice that, thus far, neither the input, nor the interstage, nor the output matching network have been designed. It has simply been taken for granted that lossless networks realizing the desired reflection coefficients exist; indeed, an infinite number of such networks does exist. Furthermore, these simple steps already ensure that the chain shows the desired output return loss and a noise factor equal to:

$$F = M \cdot \left(1 - \frac{1}{G_{av}}\right) + 1 \cong M + 1 \cong M_{min} + 1 \tag{3.85}$$

regardless of how the lossless networks will be implemented. The two approximations in Eqn (3.85) follow from considering $G_{av,ST1} \cdot G_{av,ST2} \cdot G_{av,ST3} \gg 1$ and $G_{av,ST1} \cdot G_{av,ST2} \gg 1$, respectively. Incidentally, it may be pointed out that the available gain of a lossless network is unity; therefore such gains are dropped in the formulae.

In the second part of the proposed methodology, this will proceed backward, from the right to the left:

- Since $\Gamma_{out,ST3}$, $\Gamma_{out,LNA}$, and $\Gamma_{L,ST3}$ have now been fixed and the output matching network is lossless, $\Gamma_{out,ST3}$ must lie, by construction, on a constant-mismatch circle, with an expression similar to Eqn (3.83). Then, let $\Gamma_{L,ST3}$ be selected from the points of such circle; in doing so, $\Gamma_{in,ST3}$ also gets determined.
- The situation at point 0 now occurs again between the second and the third stage. Therefore the circle of possible values for $\Gamma_{L,ST2}$ can be evaluated. Let $\Gamma_{L,ST2}$ be pinpointed on that circle, and $\Gamma_{in,ST2}$ evaluated.
- As in the previous points, now choose $\Gamma_{L,ST1}$ and compute $\Gamma_{in,ST1}$.
- Considering that both $\Gamma_{S,ST1}$ and $\Gamma_{in,ST1}$ are now fixed, the input mismatch level of the first stage—and, consequently, of the LNA—can be computed. If this mismatch level is satisfactory, choose $\Gamma_{in,LNA}$ as in points 0 to 0. Otherwise, repeat the algorithm starting from point 0, choosing other values each time.

After applying this algorithm in its entirety, the designer knows a full set of suitable reflection coefficients, at each section of the LNA, but they are still left with the problem of how to realize the lossless networks synthesizing those reflection coefficients. If the focus is restricted to three-element topologies, and in particular to T- and Π-networks, simple formulae can be derived to compute their circuit parameters. For instance, if the symbols X_1, X_2, and X_3 are used to denote the input, central and output reactance of a T-network, it can be verified that:

$$\begin{aligned} X_1 &= q \pm \sqrt{k} \\ X_2 &= \mp\sqrt{k} \\ X_3 &= p \pm \sqrt{k} \end{aligned} \tag{3.86}$$

where the following auxiliary variables have been defined:

$$\begin{aligned} q &= R_S \frac{R_L X_{in} + R_{in} X_{out}}{R_S R_L - R_{in} R_{out}} \\ p &= R_L \frac{R_{out} X_{in} + R_S X_{out}}{R_S R_L - R_{in} R_{out}} \\ k &= R_{out} \frac{R_S^2 + q^2}{R_S} = R_{in} \frac{R_L^2 + p^2}{R_L} \end{aligned} \tag{3.87}$$

with: R_S and R_L source and load resistive terminations of the network; $R_{in} + jX_{in}$ and $R_{out} + jX_{out}$ desired input and output impedances. Notice that reactive components in the terminations, if present, are included in X_1 and X_3, respectively: in other words, Eqns (3.86) and (3.87) must be applied between two reference planes, outside that are kept the resistive parts and inside where there are all and only the reactive parts.

As to X_2, its sign can be chosen arbitrarily, thus determining an inductive or capacitive behavior in the shunt element. Therefore, there are in general two possible three-element T-networks that satisfy Eqn (3.82), as well as two corresponding Π-networks: instead of deriving the relevant formulae, it is noticed that their Y-matrix representations can be easily derived by inverting the Z-matrix representations of the T-networks. Closed formulae of topologies including also distributed components have been found by the authors, but are not reported here.

Although the methodology detailed herein is valid for narrow-band designs, the existence of several realizations of the lossless networks gives the designer the possibility of choosing the best combination of them. Moreover, some topologies will turn out to be more feasible (according to the physical values of the components in them) or more convenient (if their components can serve other functions, such as biasing). In conclusion, this section should be regarded not so much as a rigid set of rules but rather as a general guide, highly susceptible to personal modifications by the experienced designer.

3.5.4 Postproduction design

In [54], an alternative LNA design flow that addresses some of the typical issues connected with MMIC realization is given. The role of the input matching network is highlighted and investigated for low noise applications.

MMIC technology is usually guaranteed to fulfill several statistical properties rather than realize exact element values; this is even more evident if a prototypical advanced technology is considered. Yield-oriented design methodologies [55,56] represent a conservative way to approach this issue. In this case, smaller sensitivity to process variations is typically obtained to the detriment of peak performance. The MMIC, although exhibiting a satisfactory RF behavior, does not exploit the technology to its limits.

Another drawback of MMIC technology is that designers must usually rely on model accuracy and precision. However, while passive elements modeling is relatively simple, active modeling is more complex, since a single model has to describe, over a wide frequency range, many device geometries (different numbers of fingers, various gate widths) and several bias regions (ohmic, pinch-off, low noise, high-gain) [57,58]. In addition, the corresponding electrical model [59] cannot easily take into account the physical temperature of the active component. Especially for cryogenic applications [60,61] the difference between room temperature (where the model is typically extracted and validated) and the environment of operation could be significant, possibly leading to large dissimilarities between the MMIC's actual behavior and the predicted one.

Together with the active device, the input matching network (IMN) is probably the most crucial "functional block" in the LNA design. It significantly affects the LNA's overall noise behavior, gain, and input return loss. The LNA measured performance may be worse than simulated, both if the FET optimum noise termination is different from the modeled one and if the realized IMN does not agree perfectly with the corresponding simulated schematic.

A less conservative way to approach the problem consists in designing the IMN after the MMIC realization and characterization is done. Initially the MMIC section of the overall LNA (right part in Figure 3.18(a)), not including the IMN, is designed and realized. Subsequently, the MMIC is fully characterized through a set of linear and noise measurements. Finally, the off-chip input matching network (OC-IMN, the furthest left part in Figure 3.18(a)) is designed, considering the MMIC section as an active noisy two-port network described by its four noise parameters and scattering parameters. The proposed design flow must not be confused with the "classical" technique, which consists in designing the IMN in a hybrid form to reduce the resistive losses. In the classical case the LNA and IMN designs are concurrent and therefore unable to address all the issues connected with MMIC realization. In this approach the overall LNA behavior is less dependent on process drift and on discrepancies between the realized component and its simulated schematic, since these phenomena can be characterized and evaluated in the linear and noise characterization step, and corrected during the OC-IMN design step. In this way the OC-IMN design is based on the overall measured MMIC electrical behavior and not on the equivalent active and passive models.

In the following, the design and performance of a C-band (4–8 GHz) LNA using the proposed design flow are reported. The design of the MMIC section was performed adopting the Northrop Grumman Space Technology (NGST) ultra low-noise InP 0.1 μm technology. The active device periphery and bias point were

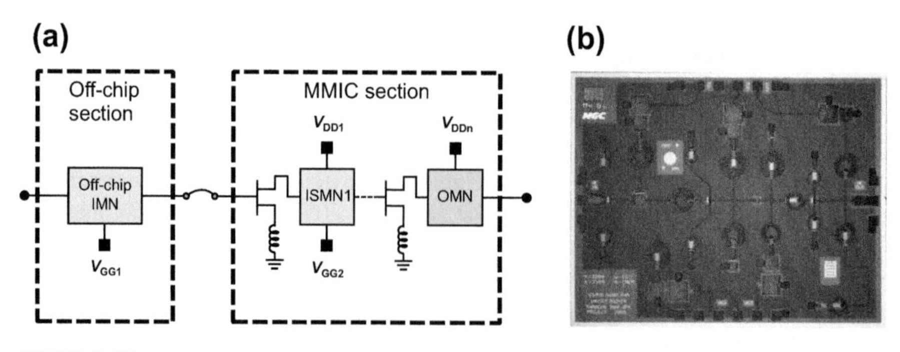

FIGURE 3.18

(a) Overall LNA, obtained by cascading an off-chip section with an MMIC one and (b) MMIC LNA, chip size 3.2 × 2.5 mm^2.

selected for noise–gain trade-off and minimum power dissipation (corresponding to a rather small drain to source voltage, $V_{DS} = 1$ V, and a limited drain current, $I_{DS} = 18$ mA). A properly sized series inductor was inserted between the FET's common terminal (source) and ground to ease the overall IRL–NF trade-off of the LNA, as clarified in Section 3.5.2.2. Finally, the interstage and output matching networks were designed. The MMIC microphotograph is reported in Figure 3.18(b).

Subsequent to the manufacturing, the MMIC section was measured to gain knowledge of its linear and noise behavior in the nominal bias condition. *S* parameters were measured on-wafer by an HP 8510C VNA, while on-wafer noise measurements were done via a tuner test bench using the cold source pull method described in Section 3.2.3. The final noise parameters extraction was carried out following Lane's methodology [35], which is outlined in Section 3.3.2.2.

In Figure 3.19 the source reflection coefficients (Γ_S) realized at 6 GHz are represented by round, gray markers. As can be noted, tuner losses do not allow the synthesis of magnitude close to unity. The black circle corresponds to the source reflection coefficient exhibiting the lowest measured noise figure; the black triangle is the predicted Γ_{opt} resulting from the application of the Lane methodology on the measured data. The characterized NF is around 0.3 dB, while R_n is close to 11 Ω.

The OC-IMN was designed on a low-loss Duroid substrate ($H = 254$ μm, $T = 5$ μm, $\varepsilon_r = 2.2$, $\tan\,\delta = 9 \times 10^{-4}$). A distributed-elements topology with short-circuited stubs was adopted both to minimize resistive losses and to employ

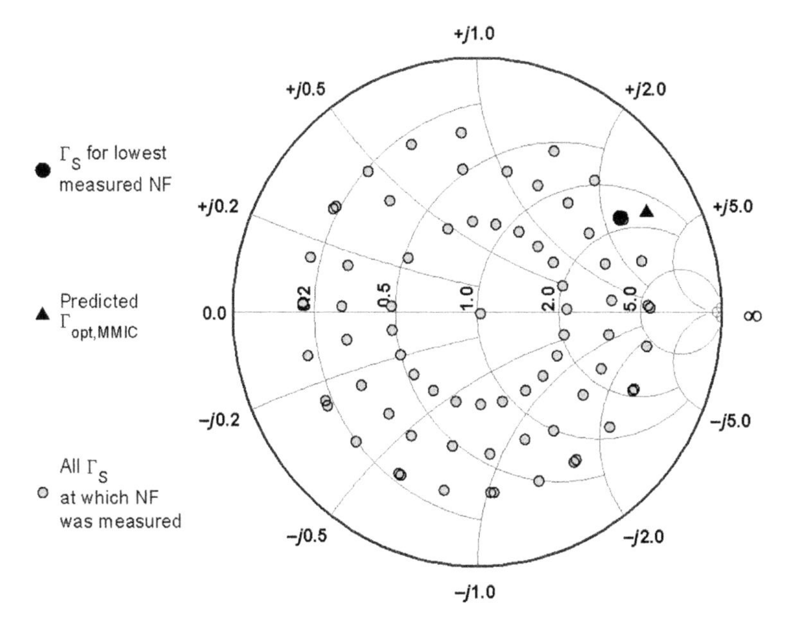

FIGURE 3.19

Source mapping of the MMIC LNA for the Lane methodology.

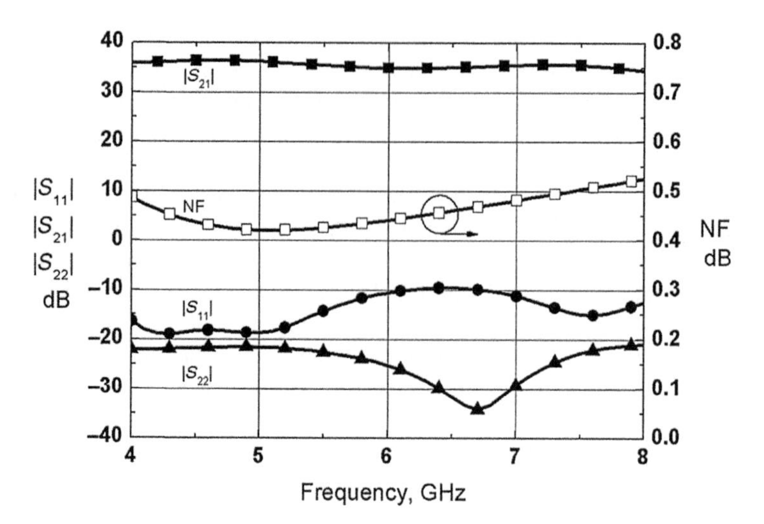

FIGURE 3.20

Overall LNA behavior versus frequency (■ gain, □ noise figure (right axis), ● input match, ▲ output match).

the network for gate bias injection. A series resistor, $R = 1.5\ \text{k}\Omega$, was inserted to further decouple the DC bias path. The bonding wire's inductive effect was also carefully taken into account during the off-chip matching network design step. The OC-IMN design was oriented toward obtaining the lowest possible overall NF with at least 33 dB small-signal gain and IRL = 10 dB over the targeted bandwidth.

Figure 3.20 depicts the linear and noise behavior of the overall (see Figure 3.18 (a)) LNA. The gain (■) results in being higher than 35 dB, with an associated noise figure (□, right axis) better than 0.55 dB over the whole C-band. In the same bandwidth the IRL (●) is better than 10 dB, while the ORL (▲) is better than 20 dB.

References

[1] Appleby R, Anderton RN. Millimeter-wave and submillimeter-wave imaging for security and surveillance. Proc IEEE October 2007;95(8):1683–90.

[2] Alexander N, Frijlink P, Hendricks J, Limiti E, Löffler S, Macdonald C, et al. IMAGINE project: a low cost, high performance, monolithic passive mm-wave imager front-end. In: Proc SPIE, Millimetre Wave Terahertz Sensors Tech V, vol. 8544 (Edinburgh, UK), September 2012. p. 854404.

[3] Xiang G, Chao L, Shengming G, Guangyou F. Study of a new millimeter-wave imaging scheme suitable for fast personal screening. IEEE Antennas Wirel Propag Lett June 2012;11(7):787–90.

[4] Dicker SR, Abrahams JA, Ade PAR, Ames TJ, Benford DJ, Chen TC, et al. A 90-GHz bolometer array for the Green Bank Telescope. In: Proc SPIE, Millimeter Submillimeter Detectors Instrumentation Astronomy III, vol. 6275 (Orlando, FL, USA), June 2006 p. 62751B.

[5] Ciccognani W, Giannini F, Limiti E, Longhi PE. Full W-Band high-gain LNA in mHEMT MMIC technology. In: Proc Eur Microwave Integr Circuit Conf (EuMIC) (Amsterdam, The Netherlands), October 2008. pp. 314–17.

[6] Ciccognani W, Limiti E, Longhi PE, Renvoise M. MMIC LNAs for radioastronomy applications using advanced industrial 70 nm metamorphic technology. IEEE J Solid-State Circuits October 2010;45(10):2008–15.

[7] Colangeli S, Ciccognani W, Bentini A, Scucchia L, Limiti E. A monolithic variable load for application in source-pull noise measurements. In: Proc 7th Eur Microwave Integr Circuits Conf (Amsterdam, The Netherlands), October 2012 pp. 544–7.

[8] Jankowski M, Limiti E, Palombini D. Q-band down-converting module for multi pixel camera receivers. In: Proc Integr Nonlinear Microwave Millimetre-wave Circuits (INMMIC), workshop on (Vienna, Austria), April 2011. pp. 1–4.

[9] Hay SG, O'Sullivan JD, Kot JS, Granet C, Grancea A, Forsyth AR, et al. Focal plane array development for ASKAP (Australian SKA Pathfinder). In: Proc Second Eur Conf Antennas Propagation (EuCAP) (Edinburgh, UK), November 2007. pp. 1–5.

[10] Glynn D, Nesti R, de Vaate J, Roddis N, Limiti E. Realization of a focal plane array receiver system for radio astronomy applications. In: Proc Eur Microwave Conf (EuMC) (Rome, Italy), September–October 2009. pp. 922–25.

[11] Cassinese A, Barra M, Ciccognani W, Cirillo M, De Dominicis M, Limiti E, et al. Miniaturized superconducting filter realized by using dual-mode and stepped resonators. IEEE Trans Microw Theory Tech January 2004;52(1):97–104.

[12] Ciccognani W, Limiti E, Longhi E, Mitrano C, Peroni M, Nanni A. An ultra-broadband robust LNA for defence applications in AlGaN/GaN technology. In: Proc of IEEE MTT-S Int Microwave Symp Dig (MTT) (Anaheim, CA, USA), May 2010. pp. 1–5.

[13] Agilent-Technologies. Fundamentals of RF and microwave noise figure measurements. Application note 57–1; August 2010.

[14] Bentini A, Ciccognani W, Colangeli S, Scucchia L, Limiti E. Highly reliable characterization approaches oriented to active device noise modeling. In: Int Symp Microwave Optical Tech (ISMOT) (Prague, Czech Republic), June 2011.

[15] Randa J, Wiatr W, Billinger RL. Comparison of methods for adapter characterization. In: Proc IEEE MTT-S Int Microwave Symp Dig, vol. 4 (Anaheim, CA, USA), June 1999. pp. 1881–4.

[16] Pucel RA, Statz H, Haus HA. Signal and noise properties of gallium arsenide microwave field-effect transistors. Adv Electron Electron Phys 1975;38:195–265.

[17] Gupta MS, Greiling PT. Microwave noise characterization of GaAs MESFET's: determination of extrinsic noise parameters. IEEE Trans Microw Theory Tech April 1988;36(4):745–51.

[18] Pospieszalski MW. Modeling of noise parameters of MESFETs and MODFETs and their frequency and temperature dependence. IEEE Trans Microw Theory Tech September 1989;37(9):1340–50.

[19] Tasker PJ, Reinert W, Hughes B, Braunstein J, Schlechtweg M. Transistor noise parameter extraction using a 50 Omega measurement system. In: Proc IEEE MTT-S Int Microwave Symp Dig, vol. 3 (Atlanta, GA, USA), June 1993. pp. 1251–4.
[20] Dambrine G, Belquin JM, Danneville F, Cappy A. A new extrinsic equivalent circuit of HEMT's including noise for millimeter-wave circuit design. IEEE Trans Microw Theory Tech September 1998;46(9):1231–6.
[21] De Dominicis M, Giannini F, Limiti E, Serino A. Direct noise characterization of microwave FET using 50 Ω noise figure and Y-parameter measurements. Microw Opt Technol Lett February 2005;44(6):565–9.
[22] Ciccognani W, Giannini F, Limiti E, Nanni A, Serino A, Lanzieri C, et al. Extraction of microwave FET noise parameters using frequency-dependent equivalent noise temperatures. In: Proc Microwave Optoelectronics Conf (IMOC) (Salvador, Brazil), October–November 2007. pp. 856–60.
[23] Hillbrand H, Russer P. An efficient method for computer aided noise analysis of linear amplifier networks. IEEE Trans Circuits Syst April 1976;23(4):235–8.
[24] Fukui H. Determination of the basic device parameters of a GaAs MESFET. Bell Syst Tech J March 1979;58(3):771–97.
[25] Dambrine G, Cappy A, Heliodore F, Playez E. A new method for determining the FET small-signal equivalent circuit. IEEE Trans Microw Theory Tech July 1988;36(7): 1151–9.
[26] Berroth M, Bosch R. Broad-band determination of the FET small-signal equivalent circuit. IEEE Trans Microw Theory Tech July 1990;38(7):891–5.
[27] Anholt R, Swirhun S. Equivalent-circuit parameter extraction for cold GaAs MESFET's. IEEE Trans Microw Theory Tech July 1991;39(7):1243–7.
[28] Berroth M, Bosch R. High-frequency equivalent circuit of GaAs FETs for large-signal applications. IEEE Trans Microw Theory Tech February 1991;39(2):224–9.
[29] Reveyrand T, Ciccognani W, Ghione G, Jardel O, Limiti E, Serino A, et al. GaN transistor characterization and modeling activities performed within the frame of the KorriGaN project. Int J Microw Wirel Tech March 2010;2(01):51–61.
[30] Peroni M, Nanni A, Romanin P, Dominijanni D, Notargiacomo A, Giovine E, et al. Low noise performances of scalable sub-quarter-micron GaN HEMT with field plate technology. In: Proc Eur Microwave Integr Circuits Conf (EuMIC) (Manchester, UK), October 2011. pp. 344–47.
[31] Limiti E, Colangeli S, Bentini A, Nanni A. Characterization and modeling of low-cost, high-performance GaN-Si technology. In: Proc 19th Int Conf Microwave Radar Wireless Communications (MIKON) (Warsaw, Poland), May 2012. pp. 599–604.
[32] Colangeli S, Bentini A, Ciccognani W, Palombini D, Palomba M, Limiti E. Noise characterization of 0.5 μm HEMTs fabricated in GaN-on-silicon technology, ESA Microwave Technologies & Techniques Workshop, (Noordwijk, The Netherlands), May 2012.
[33] Engberg J, Larsen T. Noise theory of linear and nonlinear circuits. (Chichester, UK): John Wiley & Sons Ltd; August 1995.
[34] Boudiaf A, Laporte M. An accurate and repeatable technique for noise parameter measurements. IEEE Trans Instrum Meas April 1993;42(2):532–7.
[35] Lane RQ. The determination of device noise parameters. Proc IEEE August 1969;57(8):1461–2.
[36] De Dominicis M, Giannini F, Limiti E, Saggio G. A novel impedance pattern for fast noise measurements. IEEE Trans Instrum Meas June 2002;51(3):560–4.

[37] Wiatr W, Adamson D. Finite bandwidth related errors in noise parameter determination of PHEMTs. In: AIP Conf Proc, vol. 780 (Salamanca, Spain), September 2005. pp. 685–8; no. 1.
[38] Colangeli S, Ciccognani W, Palomba M, Limiti E. Automated determination of device noise parameters using multi-frequency, source-pull data. In: Proceedings of EuMIC (Amsterdam, The Netherlands), October–November 2012. pp. 163–66.
[39] Otegi N, Collantes JM, Sayed M. Cold-source measurements for noise figure calculation in spectrum analyzers. In: Proc 67th ARFTG Conf, (San Francisco, CA, USA), June 2006. pp. 223–8.
[40] JCGM. Evaluation of measurement data—guide to the expression of uncertainty in measurement, vol. 100. JCGM; September 2008. Available at: http://www.bipm.org/en/publications/guides/gum.html.
[41] Cappy A, Dannevielle F, Dambrine G. Noise analysis in devices under non-linear operation. In: Proc 27th Eur Solid-State Device Res Conf (Stuttgart, Germany), September 1997. pp. 117–24.
[42] Geens A, Rolain Y. Noise figure measurements on nonlinear devices. IEEE Trans Instrum Meas August 2001;50(4):971–5.
[43] Lavrador PM, de Carvalho NB, Pedro JC. Noise and distortion figure—an extension of noise figure definition for nonlinear devices. In: Proc IEEE MTT-S Int Microwave Symp Dig, vol. 3 (Philadelphia, PA, USA), June 2003. pp. 2137–40.
[44] Escotte L, Gonneau E, Chambon C, Graffeuil J. Noise behavior of microwave amplifiers operating under nonlinear conditions. IEEE Trans Microw Theory Tech December 2005;53(12):3704–11.
[45] Rice SO. Mathematical analysis of random noise. Bell Syst Tech J July 1944;23(3):282–332.
[46] Rice SO. Mathematical analysis of random noise. Bell Syst Tech J January 1945;24(1):46–156.
[47] Rowe HE. Memoryless nonlinearities with gaussian inputs: elementary results. Bell Syst Tech J September 1982;61(7):1519–25.
[48] Ciccognani W, Giannini F, Limiti E, Longhi P, Serino A. Determining optimum load impedance for a noisy active 2-port network. In: Proc Eur Microwave Week (Munich, Germany), October 2007. pp. 1393–6.
[49] Barigelli A, Ciccognani W, Colangeli S, Colantonio P, Feudale M, Giannini F, et al. Development of GaN based MMIC for next generation X-band space SAR T/R module. In: Proc 7th Eur Microwave Integr Circuits Conf (Amsterdam, The Netherlands), October 2012. pp. 369–72.
[50] Ciccognani W, Colangeli S, Limiti E, Longhi P. A novel design methodology for simultaneously matched LNAs based on noise measure. In: Proc Eur Microwave Conf (EuMIC) (Rome, Italy), September 2009. pp. 1808–11.
[51] Ciccognani W, Colangeli S, Limiti E, Longhi P. A novel design methodology for simultaneously matched LNAs based on noise measure. In: Proc Eur Microwave Integr Circuits Conf (EuMIC) (Rome, Italy), September 2009. pp. 455–8.
[52] Ciccognani W, Colangeli S, Limiti E, Longhi P. Noise measure-based design methodology for simultaneously matched multi-stage low-noise amplifiers. IET Circuits Devices Syst January 2012;6(1):63–70.
[53] Haus HA, Adler R. Circuit theory of linear noisy networks (New York, NY, USA): J Wiley & Sons; 1959.

[54] Ciccognani W, Cremonini A, Limiti E, Longhi P, Orfei A, Scucchia L. LNA performance optimisation using post-production noise characterisation. In: Eur Microwave Integr Circuits Conf (EuMIC) (Rome, Italy), September 2009. pp. 234–7.
[55] Suffolk JR, Buck BJ, Tombs PN, Charlton RW. A production oriented GaAs monolithic low noise amplifier for phased array radar systems. In: Proc IEE Colloquium on Solid State Components Radar (London, UK), February 1988.
[56] Nieuwoudt A, Ragheb T, Nejati H, Massoud Y. Increasing manufacturing yield for wideband RF CMOS LNAs in the presence of process variations. In: Proc 8th Int Symp Quality Electronic Design (ISQED) (San Jose, CA, USA), March 2007 pp. 801–6.
[57] Heymann P, Prinzler H. Improved noise model for MESFETS and HEMTS in lower gigahertz frequency-range. Electron Lett March 1992;28(7):611–2.
[58] Klepser B-UH, Bergamaschi C, Schefer M, Diskus CG, Patrick W, Bachtold W. Analytical bias dependent noise model for InP HEMTs. IEEE Trans Electron Devices November 1995;42(11):1882–9.
[59] Caddemi A, Donato N, Sannino M. Low-noise device and circuit characterization at cryogenic temperatures for high sensitivity microwave receivers. In: Proc 8th IEEE Int Symp High Performance Electron Devices Microwave Optoelectronic Appl (Glasgow, UK), November 2000. pp. 89–94.
[60] Pospieszalski MW, Weinreb S, Norrod RD, Harris R. FET's and HEMT's at cryogenic temperatures—their properties and use in low-noise amplifiers. IEEE Trans Microw Theory Tech March 1988;36(3):552–60.
[61] Ciccognani W, Di Paolo F, Giannini F, Limiti E, Longhi PE, Serino A. GaAs cryo-cooled LNA for C-band radioastronomy applications. Electron Lett April 2006; 42(8):471–2.

CHAPTER

4

High-Frequency and Microwave Electromagnetic Analysis Calibration and De-embedding

James C. Rautio

Sonnet Software, Inc., North Syracuse, NY, USA

4.1 Introduction

Just as physical microwave measurements must be de-embedded from test fixtures by means of calibration, electromagnetic (EM) analysis must likewise be calibrated so the results may be de-embedded. EM analysis includes all physical effects. Thus radiation, higher order transmission line modes (propagating and evanescent), loss, and anisotropy (depending on the EM analysis sophistication) contribute to the analysis results. For typical microwave circuit design, radiation should be negligible, and higher order transmission line modes on port connecting transmission lines should be highly evanescent.

A port in an EM analysis can take many forms. The form we discuss here is that of an infinitesimal gap across which a voltage source is impressed. Of course, for the purpose of illustration, the infinitesimal gap ports are actually drawn in this chapter with a small gap. The voltage source excites all modes on a transmission line. If the port connecting transmission lines is properly designed, a single port can excite only one propagating mode. Energy inserted into the propagating mode propagates to the circuit being analyzed. The circuit reflects a portion of that energy back to the originating port. The amount (magnitude and phase) of this reflected energy is the information we seek.

Unfortunately, the gap port also excites all other transmission line modes, which are evanescent (non-propagating) with proper transmission line design. Since these modes are non-propagating, energy is immediately reflected and appears as additional inductive or capacitive reactance in the port input impedance. To make matters more complicated, if the evanescent modes are lossy (due to either metal or dielectric loss), then the evanescent modes absorb some of the incident energy, modifying the real part of the port input impedance.

The port calibration problem in EM analysis is analogous to a coaxial connector in a test fixture. The connector is a small capacitive or inductive, as well as a resistive, discontinuity. That discontinuity must be removed from a physical measurement. Likewise, the reactive-resistive component of the EM port input impedance that is due to the evanescent higher-order modes on the port connecting transmission line must be removed from the EM analysis in order to realize high-accuracy results.

Microwave De-embedding. http://dx.doi.org/10.1016/B978-0-12-401700-9.00004-5

The sum of the higher-order evanescent modes excited by the port discontinuity can also be viewed as fringing fields surrounding the port discontinuity. With this analogy, the EM port de-embedding problem is equivalent to removing the effect of fringing fields. In order to compare a physical measurement to EM analysis results, we must remove the connector discontinuity from the physical measurement, and we must also remove the EM port discontinuity from the EM analysis results. Determination of the discontinuity is called calibration. Removing the discontinuity is called de-embedding. De-embedding may also optionally remove a user-specified length of port connecting transmission line. This is called a reference plane shift. If there is some point on the device under test (DUT) that can be considered the "input" or "output," then the reference plane is usually shifted to that point. When given EM analysis data, or physical measurements, one should always be sure that the reference plane is explicitly specified. When the resulting data are used in other circuit analysis software, connections are then made to that point on the device.

This chapter describes how we do this for planar EM analysis. We introduce techniques for single ports and then extend to multiple coupled ports. Next we show how to provide a similar level of calibration for groups of coupled internal ports. We also provide substantial detail on error evaluation. Finally, we illustrate a few applications.

4.2 Double-delay calibration

4.2.1 Overview of double-delay

There is a wide range of techniques for calibrating an EM analysis. For example, one might use "wave" ports, which assume a specific transmission line mode coming from infinity incident on the port. In this case, there is no port discontinuity and calibration is not needed. However, the fields of the incident wave must be determined, and, for most practical cases, the characteristic impedance of these modes (based on transverse fields) is not unique. In addition, internal ports and multiple coupled ports become difficult.

Another class of techniques is related to the old "slotted-line" measurements. A length of uniform transmission line is added to the port and the standing wave, caused by the wave reflected from the circuit connected to the port, is observed. *S* parameters may be inferred from the standing wave. While this technique is intuitive and straightforward, it requires a length of line that is a reasonable fraction of a wavelength (difficult at low frequency), and the determination of characteristic impedance is still not unique. Internal ports of this type are generally not possible because there is simply no place to put the extra length of transmission line. *N*-coupled port calibration has yet to be accomplished using these techniques.

For physical measurements, slotted lines represent "ancient microwave measurement." Modern (physical) microwave measurement/calibration techniques (first developed in the late 1970s) measure multiple known standards to calibrate the measurement equipment. In this section, we describe the EM analysis calibration technique called "double-delay" [1]. This is one of the first modern EM analysis calibration techniques and was directly inspired by the physical through-reflect-line (TRL) calibration algorithm. Using TRL, the three namesake standards, with precisely known characteristics, are measured. The (precise) correct answer for the measurement of each standard is known. This is compared to the actual measurement result. If the characteristic impedance of the through and the line are known, then the errors introduced by the measurement equipment are determined and can be precisely removed and 50 ohm normalized S parameters are determined.

The double-delay EM calibration algorithm uses two standards, an L-length through line and a $2L$-length through line. Both of these standards are EM analyzed. The results of the EM analyses are then used to characterize the port discontinuity. In addition, the port connecting transmission line characteristic impedance, Z_0, is determined. Determination of Z_0 is a side effect of calibration. The calibration itself is not dependent in any way on knowing what the Z_0 is. The calibration returns precise 50 ohm normalized S parameters, regardless of what the Z_0 is of the two transmission line standards.

Determination of Z_0 is a major difference from all other calibration techniques, both EM and physical. The reason for this is in the port discontinuity. As described later, the EM port can be designed so the port discontinuity has no series component. There is only a shunt component. A physical port will typically always have a series inductive component and a shunt capacitive component. Imagine a "T" or a "Pi" network. If the EM port has no series component, then determination of, rather than dependence on, port Z_0 now becomes possible.

Given exactly zero series component to the port discontinuity, exact (to within numerical precision) port calibration becomes possible. The only error that remains is error due to discretization of the structure being analyzed. Exact calibration is important, for example, when analyzing a large filter by cutting it in half, analyzing the two pieces (which might be the same due to symmetry), and then connecting the halves together to see the response for the entire filter.

Double-delay (and TRL, and all other calibration algorithms) can give bad results. Nearly all failure mechanisms are in one way or another related to one critical assumption: "Double-delay assumes all port connecting lines are not over moded." There are several ways in which this assumption can be violated, as described in the following section. The competent designer will have a full and complete understanding of these failure mechanisms and will always be alert for their unwelcome presence in both EM analysis and in physical measurement.

4.2.2 Theory of double-delay

Double-delay [1] assumes that the port discontinuity is a pure shunt admittance as illustrated in Figure 4.1 with a capacitor. While our illustrations show only a capacitive port discontinuity for simplicity, in general, there may also be a shunt conductance. The two standards to be EM analyzed are an L-length through line and a $2L$-length through line, also illustrated in Figure 4.1.

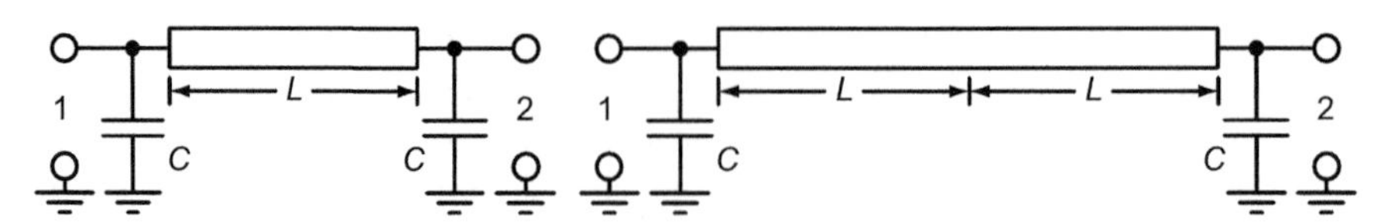

FIGURE 4.1

The equivalent circuit models of the two double-delay calibration standards, assuming a pure shunt port discontinuity.

The object is to determine the port discontinuity capacitance based only on EM analysis of the two standards. This is done using cascading parameters, a common form of which is referred to as the ABCD parameter. A two-port is represented by a 2×2 matrix. Given two two-ports, each represented by an ABCD matrix, the ABCD matrix of a cascade of the two two-ports corresponds to multiplying the two ABCD matrices. With a few exceptions, the order of multiplication matters.

If we obtain the ABCD matrix of the port discontinuity, then we can de-embed the EM analysis by inverting the port discontinuity ABCD matrix and multiplying it by the ABCD matrix of the circuit being analyzed. Consistent with physical measurement calibration and de-embedding, we refer to the circuit being analyzed as a DUT.

Our final objective is to obtain the S parameters of the DUT. We obtain the ABCD parameters of the DUT by multiplying the EM analysis ABCD parameters by the inverted ABCD parameters of the port discontinuity. That result is then converted from ABCD parameters to S parameters and the problem is solved.

Figure 4.2 illustrates how this is done. First, EM analysis is used to calculate the ABCD matrix for the L-length through and the $2L$-length through. Next, we invert the L-length through ABCD matrix. This gives us the ABCD matrix of a negative L-length through with a negative port discontinuity on each port. The negative port discontinuity corresponds to a negative capacitance.

Next, we pre- and post-multiply the $2L$-length through the ABCD matrix by the inverted L-length through the ABCD matrix. Notice that a negative shunt

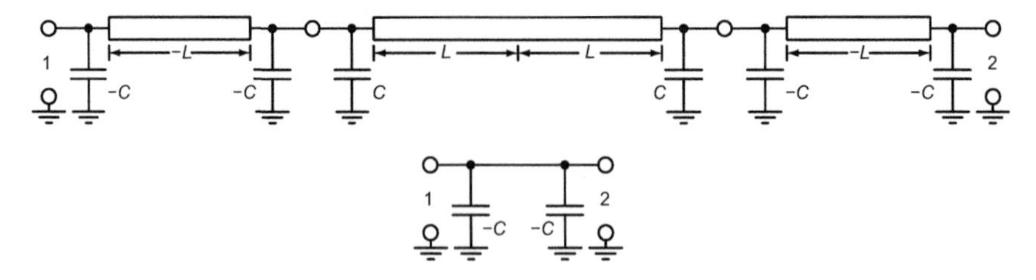

FIGURE 4.2

Method of determining the double port discontinuity. A combination of inverting and cascading ABCD matrices, illustrated schematically (above), yields the inverted double port discontinuity (below).

capacitance cascaded with an equal but positive shunt capacitance yields zero shunt capacitance. A negative L-length through cascaded with a $2L$-length through yields an L-length through. When we cascade that L-length through with a second negative L-length through, we now have no transmission line at all. The calibration process is represented by

$$\begin{bmatrix} A & B \\ C & D \end{bmatrix}^{-1}_{\text{DoublePort}} = \begin{bmatrix} A & B \\ C & D \end{bmatrix}^{-1}_{L\ \text{Length}} \begin{bmatrix} A & B \\ C & D \end{bmatrix}_{2L\ \text{Length}} \begin{bmatrix} A & B \\ C & D \end{bmatrix}^{-1}_{L\ \text{Length}}. \tag{4.1}$$

The net result of these matrix manipulations is a double negative port discontinuity, (see Figure 4.2, bottom). "Negative" means the port discontinuity shunt admittance is of opposite sign. The ABCD matrix for the negative discontinuity is the inverse of the positive port discontinuity matrix. In order to de-embed our DUT, we need a single negative port discontinuity. Mathematically, this corresponds to taking a matrix square root of the double negative port discontinuity. In general, the matrix square root is nonunique; there are many possible solutions. This is indeed the case here if the port discontinuity has any series inductance. When we restrict the port discontinuity to a pure shunt admittance, the ABCD matrix takes the form of

$$\begin{bmatrix} A & B \\ C & D \end{bmatrix}_{\text{Shunt}} = \begin{bmatrix} 1 & 0 \\ Y_{\text{Shunt}} & 1 \end{bmatrix}. \tag{4.2}$$

The negative double port discontinuity has double the desired port discontinuity shunt admittance. Thus, the desired single negative port discontinuity matrix is obtained by simply dividing the C element of the double negative port discontinuity matrix by two. This represents the unique matrix square root of Eqn (4.2).

The assumption of no series impedance in the port discontinuity is easily verified by checking that the A, B, and D elements of the calculated port discontinuity matrix take the anticipated values of 1, 0, and 1, respectively.

If port 2 of the DUT is different (e.g., Z_0) from port 1, a second negative single port discontinuity matrix is evaluated. With a negative single port discontinuity ABCD matrix in hand for each port, the EM analysis is calibrated. We can now de-embed the DUT, as shown in Figure 4.3:

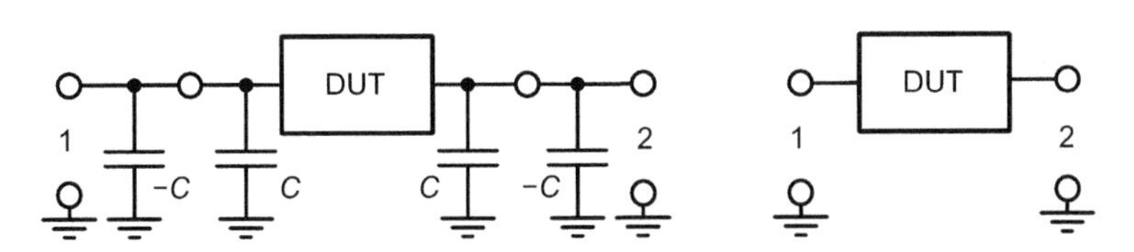

FIGURE 4.3

Schematic illustration of de-embedding an EM analysis (DUT) result (left) and the de-embedded DUT (right).

$$\begin{bmatrix} A & B \\ C & D \end{bmatrix}_{\text{DebedDUT}} = \begin{bmatrix} A & B \\ C & D \end{bmatrix}_{\text{Port1}}^{-1} \begin{bmatrix} A & B \\ C & D \end{bmatrix}_{\text{DUT}} \begin{bmatrix} A & B \\ C & D \end{bmatrix}_{\text{Port2}}^{-1}. \tag{4.3}$$

Note that if we place the calculated ABCD parameters of the L-length through calibration standard in Eqn (4.3) as the DUT, we have the de-embedded ABCD parameters of the L-length through. The ABCD parameters of an ideal L-length transmission line are

$$\begin{bmatrix} A & B \\ C & D \end{bmatrix}_{\text{IdealTLine}} = \begin{bmatrix} \cos(\beta L) & jZ_0\sin(\beta L) \\ j\dfrac{\sin(\beta L)}{Z_0} & \cos(\beta L) \end{bmatrix}. \tag{4.4}$$

Given calculated values for Eqn (4.4), the characteristic impedance of the L-length through is evaluated as the square root of the ratio of B over C. The sign of the square root is set so the real part is positive. Designers sometimes forget that when there is any loss of any kind the characteristic impedance is complex. Neglecting the imaginary part of the characteristic impedance necessarily inserts error into an analysis. We refer to this definition of characteristic impedance as "circuit theory equivalent" because this is the characteristic impedance that we must insert into our circuit theory analyses (which uses Eqn (4.4) or equivalent) in order to get the same result as our EM analysis. Sometimes it is called "TEM equivalent" because our circuit theory analyses might assume a TEM (transverse electromagnetic) transmission line [2].

Once we have the ABCD parameters of the L-length through line, we can invert the matrix and use it to remove an L-length of port connecting line from the DUT EM analysis result. This is referred to as "shifting the reference plane."

Note that the double-delay port calibration and the reference plane shift are in no way dependent on the aforementioned determination of characteristic impedance. The calculation of characteristic impedance is only to satisfy our curiosity—and to use it in circuit theory programs. Once we have both Z_0 and the propagation constant, β, we no longer need the EM analysis. We can get exactly the same result for any length transmission line with the same cross-sectional geometry by using circuit theory.

Since de-embedded S parameters are desired, we convert the de-embedded ABCD parameters to S parameters by first converting to Y parameters.

$$Y = \begin{bmatrix} Y_{11} & Y_{12} \\ Y_{21} & Y_{22} \end{bmatrix} = \begin{bmatrix} DB^{-1} & C - DB^{-1}A \\ B^{-1} & B^{-1}A \end{bmatrix} \tag{4.5}$$

$$S = (\mathbf{Y}_0 - Y)(\mathbf{Y}_0 + Y)^{-1}. \tag{4.6}$$

The $\mathbf{Y}_0$ matrix is a diagonal matrix with the normalizing characteristic admittances for the resulting S parameters. These are usually set to 1/50 so that the S parameters are normalized to 50 ohm, as they are for nearly all physical measurements. Sometimes S parameters resulting from EM analysis are presented in the form of "generalized" S parameters, which means that the S parameters are normalized to the (typically unknown) characteristic impedance of the port connecting lines. Unless the generalized S parameters just happen to correspond to a structure with 50 ohm port connecting lines, generalized S parameters and 50 ohm normalized S parameters should never be combined in the same circuit analysis. In addition, generalized S parameters do not exist for circuits that lack port connecting transmission lines, as with lumped models.

Since the DUT EM analysis results are usually provided in the form of S parameters, we also need to convert from S parameters to ABCD parameters. We do this by first converting to Y parameters:

$$Y = \mathbf{Y}_0(1 - S)(1 + S)^{-1}. \tag{4.7}$$

$$\begin{bmatrix} A & B \\ C & D \end{bmatrix} = \begin{bmatrix} -Y_{21}^{-1}Y_{22} & -Y_{21}^{-1} \\ Y_{12} - Y_{11}Y_{21}^{-1}Y_{22} & -Y_{11}Y_{21}^{-1} \end{bmatrix} \tag{4.8}$$

As mentioned before, this technique is critically dependent on the port discontinuity having no series component. This is checked by making sure the port discontinuity matrix takes the form of Eqn (4.2). A physical understanding of how there can be no series component is helpful. In Figure 4.4 we view the circuit from above with voltage sources connected between the perfectly conducting sidewall (negative terminal) and the subsections of the immediately adjacent port connecting transmission line. The voltage sources are connected across an infinitesimal gap.

The perfectly conducting sidewall represents a perfect ground reference. This is important. If we have multiple ports with different ground references, in certain cases substantial error can be introduced as described in subsequent sections. This can be the case even if there is only a small amount of resistance separating the two port ground references.

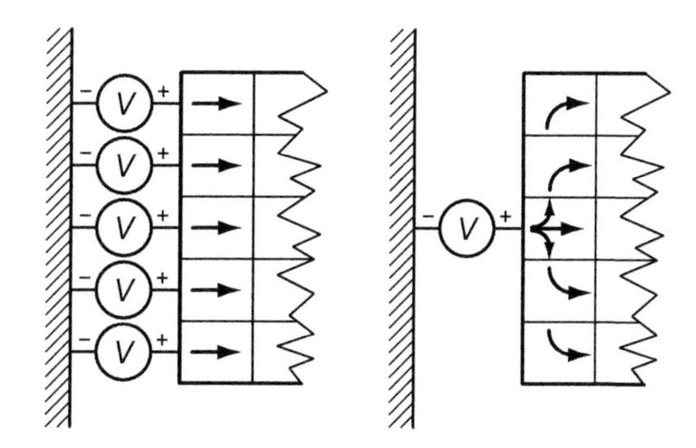

FIGURE 4.4

On the left, all port subsections are at the same potential and no transverse current can flow. The port discontinuity is pure shunt capacitance. On the right, the voltage source excitation is applied to only one subsection, and transverse current must flow yielding a port discontinuity with series inductance.

For the purposes of EM analysis, the planar port connecting transmission line is meshed into a number of subsections across the width of the line. If we connect the same voltage across each of these gaps for each of these subsections, see Figure 4.4 (left), then there can be no voltage between the port subsections. Thus there is no transverse current flow. Exactly at the port, all current flow is into the transmission line, not across it. Thus, there can be no port discontinuity inductance.

In contrast, if we connect a voltage source across the infinitesimal gap at only one subsection, see Figure 4.4 (right), then current must flow transversely to the edge of the line. This introduces series inductance into the port discontinuity.

The above qualitative discussion is confirmed numerically by simply viewing the calculated single port discontinuity ABCD parameters and verifying that they take the form of Eqn (4.2).

4.2.3 Failure mechanisms of double-delay

Under no circumstances does a competent microwave designer assume that the results of an EM analysis (or of a measurement) are correct. They do not even assume that the results might be wrong. The competent designer assumes the results are in fact wrong, and takes on the job of estimating by just how much the results are wrong. In both this section and the next, we consider this task.

First, we discuss violations of the fundamental assumption for calibration/de-embedding. These violations can occur in either physical measurements or in EM analysis with equal ease. In situations where all ports share exactly the same ground reference, these violations are typically related to violation of the requirement that the port connecting lines not be over-moded.

In addition to the perfect ground reference, if there is only one conductor in the port connecting transmission line, then there should be only one propagating mode. If there are two conductors (a coupled line), then there should be only two possible propagating modes. If there are three conductors, there should be only three propagating modes.

Over-moding can take the form of higher-order transmission line modes. For example, if the width of the planar line or the substrate thickness (i.e., the distance to the ground return conductor) is a significant fraction of a wavelength, then higher-order modes can start propagating. It should be emphasized that this can happen in both EM analysis and in measurement. If we build a circuit with large transverse transmission line dimensions, we risk failure due to higher-order mode propagation.

As a specific example, a 0.635 mm (0.025 in)-wide line on a 0.635 mm-thick alumina substrate (relative dielectric constant 9.8) used at frequencies higher than about 15 GHz allows multiple propagating modes. Any design or EM analysis that assumes the line is not over-moded will fail.

Grounded coplanar waveguide (CPW) is a particularly insidious transmission line geometry for the unwary. The correct CPW current flow is shown in Figure 4.5 (top). The input current flows on the center line and the ground return current is split equally between the two ground strips on the right and left sides.

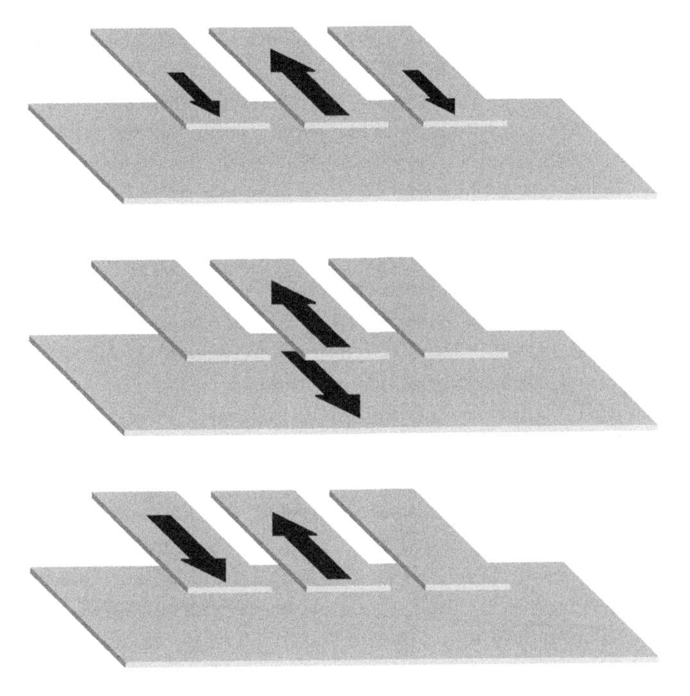

FIGURE 4.5

Grounded coplanar waveguide (CPW), with CPW mode (top) can easily have higher-order mode propagation when ground return current flows along the bottom ground plane (microstrip mode, middle) or the ground return current flows preferentially along one of the ground strips (slot line mode, bottom).

What might go wrong depends on the specific nature of the DUT connected to the other end of the CPW. For example, the DUT might force some of the ground return current to travel along the ground plane, see Figure 4.5 (middle). This is referred to as a microstrip mode. In addition, see Figure 4.5 (bottom), the DUT might cause one of the two ground strips to carry more return current than the other. This excites a slot line mode. Both of these modes have different characteristic impedances and different propagation constants as compared to the desired CPW mode.

For physical measurement, the CPW wafer probe usually has the ground strips shorted together right at the probe tips. This suppresses the slot line modes. The wafer probe does not have an underlying ground plane, so there is no microstrip mode.

To suppress these modes in our DUT, we must place frequent wire bonds or air bridges shorting the two ground strips together, especially at transmission line junctions. In addition, we must place frequent vias from the ground strips to the underlying ground plane. The transmission line now actually starts to look very much like a coaxial cable, and, provided all cross-sectional dimensions are small compared to the wavelength, we should have a good single-mode transmission line.

For EM analysis we should set up the ports so only the desired CPW mode is generated. For example, when placing port 1 on the center conductor, we should also place ports – 1 on the end of both of the ground strips, see Figure 4.6 (top). This port configuration also forces the underlying ground plane to have zero current, so there are neither slot line modes nor microstrip modes generated. Only the desired CPW mode is generated by the port.

The above port configuration keeps the ports from generating undesired modes. However, the desired CPW mode incident on an improperly designed DUT can be converted into reflected slot line and microstrip modes. To achieve a good EM analysis, and to build a DUT that actually works as expected, it must be designed to suppress undesired modes. If this is not possible, then, instead of one CPW port, the designer should treat the circuit as having three separate microstrip ports, see Figure 4.6 (bottom). Calibration of multiple closely spaced ports (whose fringing fields interact with each other) is described in a later section.

When properly designed, all higher-order modes are evanescent. They can be viewed as the fringing fields surrounding the port discontinuity. However, under certain circumstances, the fringing fields can still cause problems—for example, if the *L*-length through is so short that the fringing fields from one port interact with the other port. In this case, our non-over-moded transmission line assumption fails once more. This failure mode is detected by plotting the circuit theory equivalent Z_0 as a function of the through length, see Figure 4.7.

When there is fringing field interaction between ports on either end of the *L*-length through line, Z_0 becomes a function of the through length and the calibration fails. For this geometry (described in the figure caption), we should not use an *L*-length standard less that 1.2 mm, or about twice the substrate thickness. Keep in

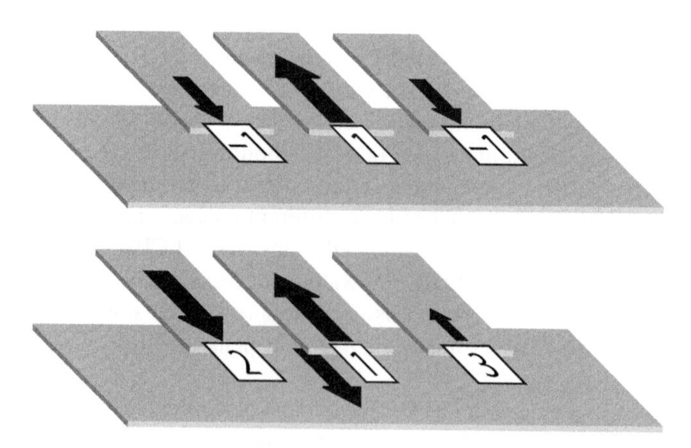

FIGURE 4.6

Proper EM port configuration (top) for prevention of excitation of undesired modes on CPW. This still does not prevent an improperly designed DUT from converting the CPW mode into undesired modes, in which case, separate ports should be used for each conductor (bottom).

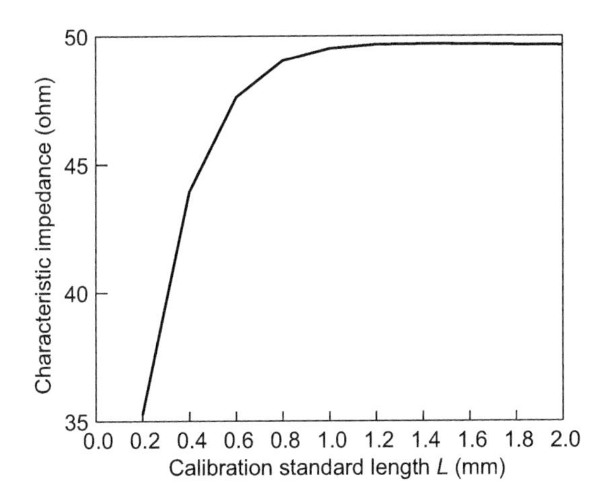

FIGURE 4.7

When the L-length standard is too short, port calibration fails as can be seen by Z_0 incorrectly becoming a function of L for L small. Test circuit is a 0.635 mm (0.025 in)-wide line on 0.635 mm-thick alumina, relative dielectric constant 9.8 at 10 GHz.

mind that the easily inferred rule of thumb (twice the substrate thickness) does not apply for all situations.

As long as we de-embed whatever length of port connecting line we connect to a DUT, we would hope to always get the same result no matter what that length is. And that will be the case, as long as the calibration has not failed. As in the previous example, when the L-length through is too short, fringing fields from one end interact with the other end, and Z_0 incorrectly appears to be a function of L. Likewise, in this case, the DUT result incorrectly changes when we change the length of the port connecting line that we de-embed. This is a sure sign of failure.

A necessary condition for any calibration correctness is that the determined Z_0 is independent of calibration standard length. Equivalently, when we change the length of the port connecting line but we shift the reference plane to the same point on the DUT, thus removing that length of port connecting line, we should obtain the same DUT S parameters.

In fact, this is a necessary condition for port calibration correctness, for both physical and EM analysis calibration. Change the length of the port connecting lines and de-embed that length. The DUT result should remain unchanged. The amount of change indicates the amount of error that the calibration procedure is actually inserting into the measurement.

4.2.4 Exact evaluation of EM analysis and calibration error

Unlike physical measurements, thermal noise is not of concern. Dynamic range is limited by numerical precision and can be quantified, for example, by evaluation of a very deep stop band filter [3]. A good EM analysis has a dynamic range of at least 100 dB and can approach 200 dB when using double precision. Other error sources generally dominate and are the topic of this section.

Exact evaluation of error is not possible unless we have a problem for which an exact answer is known. One such case is lossless, infinitely thin stripline [4–6]. Stripline is a planar transmission line sandwiched between two identical dielectric substrates with a full ground plane on both the top and the bottom.

For example, we can select the dimensions of a stripline so it is exactly 50 ohm. Here, we use the word *exactly* as shorthand for "to within numerical precision." We must also take care to make sure the stripline is not over-moded. Thus, we should keep the ground plane-to-ground plane spacing well under 0.1 wavelength at the highest frequency of interest.

The characteristic impedance of stripline is (from [4,5]):

$$Z_0\sqrt{\varepsilon_r} = \frac{\eta_0}{4}\frac{K(k')}{K(k)}, \tag{4.9}$$

where

$$k = \tanh\left(\frac{\pi}{4}\frac{w}{b}\right), \quad k' = \sqrt{1-k^2}, \quad \eta_0 = 376.7303136, \tag{4.10}$$

and $K(k)$ is a complete elliptic integral of the first kind. Numerical approximations are available in [7]. Be sure to note the errata, $m_1 = 1 - m^2$, not $1 - m$. Also note the use of the η_0 term. Expressions for characteristic impedance that use the more commonly quoted 30 π term, are actually approximations. The approximation is in error by only 0.07%, but we wish to validate to far higher precision.

The velocity of propagation for a lossless stripline exactly corresponds to the dielectric constant of the substrate, which is uniform everywhere in the volume of the stripline. For our example, we use free space, relative dielectric constant exactly equal to unity. For a 50.000000 ohm characteristic impedance, we use a line width of 1.0000000 mm, and a ground plane-to-ground plane spacing of 0.69329396 mm. A line 3.74740573 mm long is 90.000000° long. All digits are significant.

The exactly correct reflection coefficient (S_{11} magnitude) for a 50 ohm line is zero. When the line is 90° long, the difference from zero is to first order the fractional error in Z_0. For example, an S_{11} magnitude of 0.01 corresponds to 1% error in Z_0. For small subsection size, error in Z_0 is normally determined almost entirely by subsection width. Error in velocity of propagation (corresponding to error in S_{21} phase) is normally determined almost entirely by subsection length.

Figure 4.8 shows convergence in Z_0 error as subsection width decreases. The fact that the error converges uniformly indicates that error due to subsection width is the only significant error source. If the error does not converge uniformly, additional error sources are indicated. Error in Z_0 due to subsection length can be a factor when subsection length is large compared to subsection width. In that case, Z_0 error can actually drop to and then past zero as subsection width error cancels subsection length error. As convergence is taken even further, the analysis does converge, but to the wrong answer unless subsection length is also reduced.

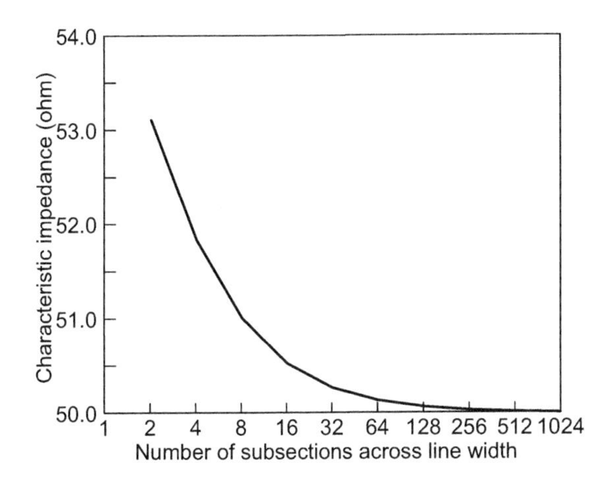

FIGURE 4.8

As subsection width is reduced, error in Z_0 decreases uniformly indicating we have only EM analysis meshing error. If the error does not decrease uniformly, multiple error sources are indicated, possibly including calibration error.

Figure 4.8 shows a well-behaved convergence plot. If there is any form of port calibration error, the convergence will fail, sometimes dramatically. Analysis times for the data of Figure 4.8 range from 0.28 to 33 s.

An area that provides extreme stress for EM analysis is deep submicron Si RFIC (silicon radio frequency integrated circuit) design. These designs require performance up to and beyond 100 GHz and feature dielectric stack-ups of 40–50 layers or more. The layers can be relatively thick, or extremely thin and of widely varying dielectric constants. We do not have an exact solution for this kind of extreme structure complexity. However, an EM analysis does not care what the dielectric constants are. So, we make a stripline using a dielectric stack-up with comparable layer complexity but with all dielectric layers having the same dielectric constant. We then analyze the full complexity of the stack-up, without simplifying to a single dielectric. Results of a convergence analysis for such a stripline in SiO_2 (relative dielectric constant = 3.9) at 100 GHz with 50 dielectric layers ranging in thickness from 0.01 to 4 μm are shown in Figure 4.9. One-quarter wavelength is 379.51453 μm, the line width is 10.000000 μm, and the ground plane-to-ground plane spacing is 19.390404 μm. Analysis times range from 10 to 447 s.

The stripline standard evaluates the error for an EM analysis to an extremely low level, much lower than we typically need in practice. However, the low error validation is important as it can reveal additional error sources that might not become important until we are working with large circuits at high frequency. Such error sources are revealed by failure to converge to the correct answer, by lack of a uniform convergence, and, in some cases, by outright divergence. With a complete deep

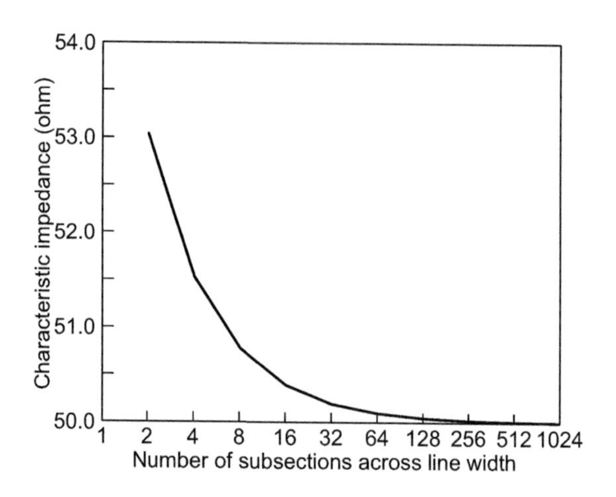

FIGURE 4.9

Convergence analysis for the stripline standard at 100 GHz in SiO_2 composed of 50 dielectric layers ranging in thickness from 0.01 to 4 μm. In spite of being a radically different problem, convergence is nearly identical to that seen in Figure 4.8. Very few EM analyses can even attempt analysis of a dielectric stack-up with this complexity.

submicron RFIC design cycle costing millions of dollars, leaving error identification and evaluation to chance is not advised.

The data plotted here confirm that the subject EM analysis is free of any and all non-meshing related errors to well under the 0.01% error level.

4.3 Multiple coupled port calibration and de-embedding

For the above, we have treated de-embedding and calibration only of single, isolated ports. In practice, multiple closely spaced ports are common. If the ports are close enough that the fringing fields from one port interact with those of another nearby or adjacent port, the calibration algorithms must be appropriately modified. The interacting fringing fields can be viewed as a fringing capacitance connected between the two ports. This mutual fringing capacitance must be removed when high accuracy is desired.

For both physical measurement and for EM analysis, extension of calibration/de-embedding algorithms to multiple coupled ports can be a major research-and-development effort. However, if the original single port problem is solved carefully, the multi-port problem is trivial. And, of course, that is the case here.

For the single port problem, which we solve above, the important variables, like A, B, C, D, Y_{ij}, etc., are complex scalars. Note that all of the previous equations are equally valid if these variables are themselves matrices. For example, for two coupled ports, A, B, C, D, Y_{ij}, all become 2×2 matrices. For three coupled ports, they all become 3×3 matrices. Equations (4.3), (4.6) and (4.8) remain fully valid.

Equation (4.4), used for evaluation of characteristic impedance and velocity of propagation (derived from wave number, β) is an exception. A characteristic impedance matrix and a wave number matrix benefit from special treatment. The matrices become diagonal when transformed to the underlying modes. For a symmetric two-coupled line, the underlying modes are known as the "even" and "odd" modes. We do not describe this process here.

Thus, perhaps without even realizing it, we already have a fully developed theory for calibration and de-embedding of multiple coupled ports. All that remains is to test the theory to identify and quantify error sources. A competent designer has a complete and detailed understanding of what can go wrong.

As for characterizing error, there are exact equations for infinitely thin lossless coupled stripline [8] and nearly exact equations for thick lossless coupled [8] and single [4] stripline. These may be applied as in a previous section for single ports.

We can also test multiple coupled port calibration by repeatedly analyzing the same DUT, only with different lengths of port connecting (coupled) transmission lines. If the length of port connecting line (whatever length it is) is de-embedded from the analysis, the same answer should be obtained each time for the same DUT. Differences can be attributed to calibration and de-embedding error.

A third benchmark is to take a long length of coupled line, open ended on both ends. Lightly couple into the resonator at one end, and observe the resonances. The resonances of a coupled line come in pairs, one corresponding to the even mode, and the other corresponding to the odd mode.

It would seem difficult to use a coupled line resonator as a benchmark because we do not know the exact answer. The open end discontinuities (we have two coupled open ends on each end of the resonator), while well understood, lack an exact solution. However, since we are primarily interested in evaluating calibration and de-embedding error, rather than the analysis error of the underlying EM analysis, we can still use this for calibration error diagnostics.

For precise analysis of large filters, it is often desirable to cut a filter into pieces. Each piece is EM analyzed and then the overall result connected back together for the complete result. Of course, care must be taken to make sure that, for example, we do not cut the filter into pieces that place one side of a coupled line in one sub-circuit and the other side in different sub-circuit. Sub-circuit cut lines must go perpendicular to coupled lines.

In order for this technique to work, we must have essentially perfect port calibration. Even the tiniest unremoved port discontinuity can result in failure at and near certain frequencies. Thus, we can give port calibration an acid test by taking a long resonator, dividing it into small pieces and analyzing each piece and then use circuit theory to connect the pieces together. If the port calibration is good, then the analysis of the entire resonator (in a single EM analysis run) should be identical to the piece-by-piece analysis. If there is a problem, then spikes (perhaps even gain) appear at frequencies for which the pieces of the piece-by-piece analysis are a multiple of a half wavelength long.

Figure 4.10 shows how we divided a 25.4 cm-long coupled line resonator into small sections [9]. The resonator is coupled to input and output ports at one end (bottom left, 1.27 cm long) using the tabs of an SMA coaxial bulkhead connector. The coupled line section (middle, 2.54 cm long) is EM analyzed once and then repeatedly connected (nine times) to form most of the resonator. The coupled line open end (top right, 1.27 cm long) is then connected on the end of the cascade to complete the resonator.

At the upper end of its frequency range (16 GHz) the resonator is about 25 wavelengths long. In order to have high-order resonances with high Q, radiation cannot be allowed. Thus the resonator is placed in a completely shielding enclosure. In fact, more than just radiation cannot be allowed. Loss-inducing propagating rectangular waveguide modes cannot be allowed either. The dimensions of the enclosure form a rectangular waveguide with a lowest cutoff frequency of 17 GHz. Thus, we can obtain useful data up to around 16 GHz. A complete physical description is provided in [9].

To perform an acid test of the port calibration, we EM analyze and double-delay de-embed 11 separate pieces from Figure 4.10 (the middle piece is analyzed once and connected nine times to build the entire length). We compare this to an analysis of the entire resonator, all done in one EM analysis. So that we can compare results, it is important to use the same meshing in both cases. The cell size is 2.54 μm (0.001 in). Each line in the coupled line resonator is 60 cells wide and 10,000 cells long, for a total of 1.2 million cells. So that we use exactly the same meshing, the resonator polygons are not a single 10,000 cell long polygon but rather a series of

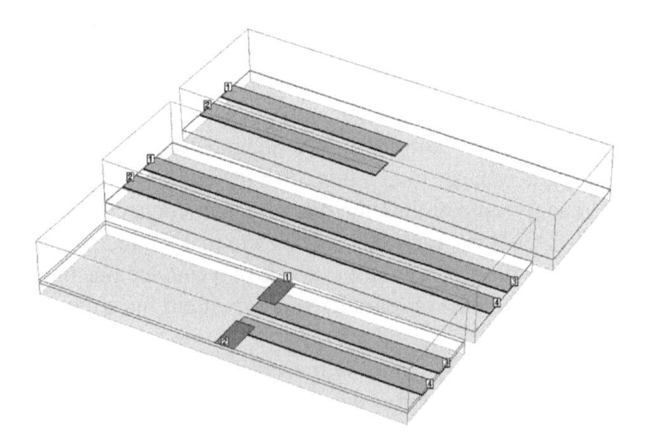

FIGURE 4.10

Analysis of a long 25.4 cm coupled line resonator is composed of EM analysis of the input coupling section (bottom), multiple instances of the coupled line section (middle), and the terminal coupled line open end section (top). All pieces are connected together with circuit theory to obtain the result for the complete resonator.

From [9]

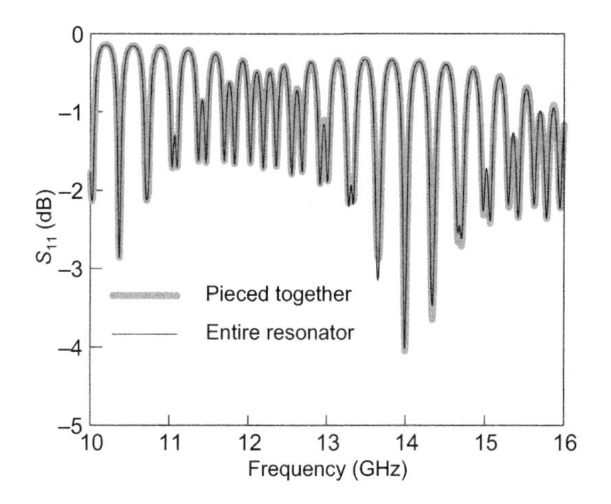

FIGURE 4.11

Comparison of EM analysis of the entire 25.4 cm-long resonator to the response calculated from using circuit theory to connect together 11 pieces 2.54 cm long, each separately EM analyzed and de-embedded using double-delay. If there were any problem with port calibration, the two results would be different.

11 polygons, with the end polygons 500 cells long and the nine interior polygons each 1000 cells long.

The piece-by-piece analysis is much faster than the analysis of the entire resonator. The important question is, do the two approaches give the same answer?

The result, Figure 4.11, shows essentially identical answers for both the piece-by-piece analysis and the analysis of the entire resonator. The comparison at all lower frequencies is even better. We are showing from 10 GHz and up as this is the most extreme test. It should be noted that analysis of a very long resonator (in this case, about 25 wavelengths long at 16 GHz) is also an excellent acid test for the validity of an EM analysis. If improper approximations are made in Green's function, especially in regard to coupling between far-spaced subsections, noisy results, sometimes even indicating substantial nonphysical gain, can be seen. That would be an EM analysis problem, not a de-embedding problem.

4.4 Short-open calibration

The double-delay calibration described previously requires the port discontinuity to have no series impedance. The port discontinuity must be pure shunt admittance. Such a port is easily realized in EM analysis of shielded planar circuits as shown in Figure 4.4. However, unshielded planar EM tools are also commonly used. It would be nice to be able to use modern calibration techniques for that type of EM analysis as well. The short-open calibration (SOC) fulfills this need.

The fundamental difficulty in calibrating ports in unshielded EM analysis is that the ports lack an immediate ground reference. Sometimes we forget that a port is two terminals, a signal terminal and a ground terminal. Sometimes we think only of the signal terminal and ignore the ground terminal. However, the two terminals are equally important. It is easy to demonstrate this with a battery, two wires, and a light bulb. If the ground return wire is not connected to the negative terminal of the battery, the light will not light. The ground return path is of importance equal to the signal path.

The lack of an immediate ground means we must make a special effort to provide a ground to the voltage source that excites our port. Using the previous light bulb analogy, we must connect the ground return wire or nothing happens. This can be done by placing a gap port at the end of the port connecting line and then connecting the negative terminal end to ground through a via, as shown in Figure 4.12. This assumes a microstrip-like ground plane below the circuit. We address the situation where there is no ground plane in a later section.

The reason the double-delay requires no series impedance in the port discontinuity is so the matrix square root of the double port discontinuity is unique, see Eqns (4.1) and (4.2). In order to resolve the nonuniqueness of the matrix square root in the general case, which includes a series impedance in the port discontinuity, we need more information. The SOC obtains this additional information by monitoring the current in the center of the $2L$-length standard. This is represented by the current meter in the center of the through line in Figure 4.12.

In its original form, SOC does not determine the single port discontinuity. Rather it determines the port discontinuity cascaded with the L-length port connecting transmission line. Thus, with SOC, the reference plane must be shifted to the interior of the DUT by length L from the port. The reference plane is not left at the port itself.

SOC involves several trade-offs as compared with double-delay. First, the double-delay allows a necessary check on calibration validity by requiring $A = D = 1$ and

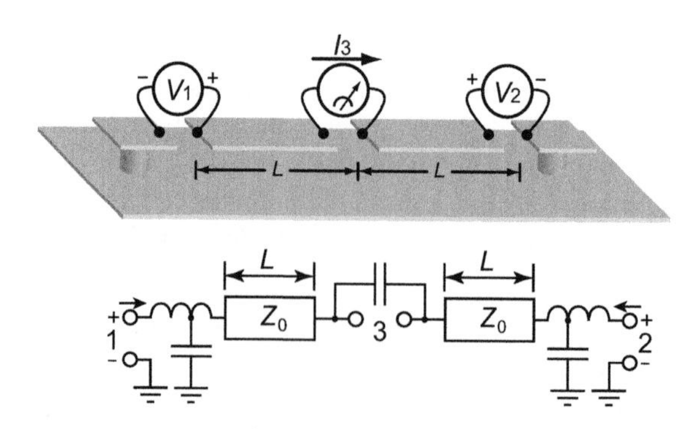

FIGURE 4.12

(Top) The short-open calibration standard is used to calibrate unshielded EM analysis results that require removal of series inductance from the port discontinuities shown in the equivalent schematic (below, *from* [10]). The central port (Port 3) is used to measure current.

$B = 0$ (see Eqn (4.2)). Since the SOC port discontinuity includes a series inductance and it also includes the L-length of port connecting transmission line, its ABCD de-embedding matrix is fully populated. This means that double-delay's independent check for calibration validity is not available when using SOC.

A second trade-off is that the port discontinuity, when viewed as either a Pi-network (Figure 4.12, bottom), or as a Tee-network, looks like an LC ladder network, which in turn looks just like a transmission line. In other words, the length L is fuzzy. It is not precisely defined. For example, in Figure 4.12, it is not clear exactly where the end of the transmission line is. In double-delay, as applied to a shielded EM analysis, the location of the end of the through line is exactly at the grounding sidewall. There is no fuzziness or ambiguity.

A third trade-off is related to the length uncertainty, and that is characteristic impedance uncertainty. Just like with double-delay, an SOC calibration can be used to de-embed a through line. The ABCD parameters of that through line can then be used to infer the equivalent circuit theory characteristic impedance. However, as described below, if the through line is near a multiple of a half wavelength long, the errors inserted by the length uncertainty translate into error in calculation of characteristic impedance.

Nevertheless, SOC is a valuable technique for calibrating and de-embedding unshielded EM analysis results.

4.4.1 SOC theory

SOC theory for EM analysis calibration was first proposed in [11]. It was extended to include multiple coupled ports in [12] and [13]. Later, SOC was unified with double-delay and extended further in [10]. The theory presented here is adapted from [10].

The SOC standard of Figure 4.12 is first analyzed as a three-port, with the calculated Y parameters,

$$\begin{bmatrix} I_1 \\ I_2 \\ I_3 \end{bmatrix} = \begin{bmatrix} Y_{11}^{\mathrm{SOC}} & Y_{12}^{\mathrm{SOC}} & Y_{13}^{\mathrm{SOC}} \\ Y_{21}^{\mathrm{SOC}} & Y_{22}^{\mathrm{SOC}} & Y_{23}^{\mathrm{SOC}} \\ Y_{31}^{\mathrm{SOC}} & Y_{32}^{\mathrm{SOC}} & Y_{33}^{\mathrm{SOC}} \end{bmatrix} \begin{bmatrix} V_1 \\ V_2 \\ V_3 \end{bmatrix}. \tag{4.11}$$

As mentioned before, we are interested only in the current flowing through port 3 with port 3 short-circuited across its terminals. Thus, we set V_3 to zero and discard the third column of the Y parameter matrix. Note that setting V_3 to zero does not short port 3 to the ground plane. Rather, it shorts one side of port 3 to the other side, completing the through line.

We need to determine the input impedance of port 1 with a perfect short circuit to the ground plane at the location of port 3. We also need to know the short circuit current at the port 3 location. Port 3 shorted to the ground plane represents the short-circuited stub, or the "S" of the SOC.

To short port 3 to the ground plane, we need a perfectly conducting vertical plane at that location. This is realized by setting $V_2 = -V_1$, effectively placing a perfect electric conducting wall at the location of port 3 and providing a perfect ground reference connection to the ground plane at the location of port 3. Thus, at least for this port, our port is now equivalent to the sidewall port of a shielded EM analysis. Given the three-port Y parameters of Eqn (4.11), the short circuit, or perfect electric conductor (PEC) Y parameters are

$$\begin{bmatrix} I_1 \\ I_3 \end{bmatrix} = \begin{bmatrix} Y_{11}^{\mathrm{SOC}} - Y_{12}^{\mathrm{SOC}} \\ Y_{31}^{\mathrm{SOC}} - Y_{32}^{\mathrm{SOC}} \end{bmatrix} [V_1] \triangleq \begin{bmatrix} Y_{11}^{\mathrm{PEC}} \\ Y_{31}^{\mathrm{PEC}} \end{bmatrix} [V_1]. \tag{4.12}$$

We also need a perfect open circuit. This condition is simulated by setting $I_2 = I_1$. This places a perfect magnetic conductor (PMC) at the location of port 3. For this, we invert the EM calculated SOC standard Y parameter matrix to give Z parameters,

$$\begin{bmatrix} V_1 \\ V_2 \\ V_3 \end{bmatrix} = \begin{bmatrix} Z_{11}^{\mathrm{SOC}} & Z_{12}^{\mathrm{SOC}} & Z_{13}^{\mathrm{SOC}} \\ Z_{21}^{\mathrm{SOC}} & Z_{22}^{\mathrm{SOC}} & Z_{23}^{\mathrm{SOC}} \\ Z_{31}^{\mathrm{SOC}} & Z_{32}^{\mathrm{SOC}} & Z_{33}^{\mathrm{SOC}} \end{bmatrix} \begin{bmatrix} I_1 \\ I_2 \\ I_3 \end{bmatrix}. \tag{4.13}$$

The electric current, I_3, is zero into the perfect open circuit, which means we discard the third column. Since there is no need for information about the open circuit port 3 voltage, we also discard the third row. Reducing, we have

$$[V_1] = \left[Z_{11}^{\mathrm{SOC}} + Z_{12}^{\mathrm{SOC}}\right][I_1] \triangleq \left[Z_{11}^{\mathrm{PMC}}\right][I_1], \tag{4.14}$$

$$[I_1] = \left[Z_{11}^{\mathrm{PMC}}\right]^{-1}[V_1] \triangleq \left[Y_{11}^{\mathrm{PMC}}\right][V_1]. \tag{4.15}$$

With these quantities in hand, and under the assumption of reciprocity ($AD - BC = 1$), we can now determine the de-embedding matrix for the port discontinuity plus L-length port connecting line as

$$\begin{bmatrix} A & B \\ C & D \end{bmatrix}_{\mathrm{SOC}} = \begin{bmatrix} Y_{31}^{\mathrm{PEC}}\left(Y_{11}^{\mathrm{PMC}} - Y_{11}^{\mathrm{PEC}}\right)^{-1} & \left(-Y_{31}^{\mathrm{PEC}}\right)^{-1} \\ Y_{11}^{\mathrm{PMC}} Y_{31}^{\mathrm{PEC}}\left(Y_{11}^{\mathrm{PMC}} - Y_{11}^{\mathrm{PEC}}\right)^{-1} & -Y_{11}^{\mathrm{PEC}}\left(Y_{31}^{\mathrm{PEC}}\right)^{-1} \end{bmatrix}^{-1}. \tag{4.16}$$

DUT de-embedding is realized in a manner similar to Eqn (4.3) by first premultiplying the EM analysis DUT result by Eqn (4.16), but then post-multiplying by the transpose of Eqn (4.16). The transpose is required because port 1 and port 2 of the SOC de-embedding matrix are different (unlike double-delay) and must be swapped as in

$$\begin{bmatrix} A & B \\ C & D \end{bmatrix}_{\mathrm{DebedDUT}} = \begin{bmatrix} A & B \\ C & D \end{bmatrix}_{\mathrm{SOCPort1}}^{-1} \begin{bmatrix} A & B \\ C & D \end{bmatrix}_{\mathrm{DUT}} \left(\begin{bmatrix} A & B \\ C & D \end{bmatrix}_{\mathrm{SOCPort2}}^{-1}\right)^{T}. \tag{4.17}$$

Multiple coupled ports, as with double-delay calibration, are handled by changing all (complex) scalars in the above equations into appropriate matrices.

4.4.2 Relationship between SOC and double-delay

As one might expect from the previous discussion, double-delay and SOC have significant similarities. In fact, double-delay and the original SOC (as described herein) are both special cases of an extended SOC, as described in [10].

First, for a shielded environment, note that both of the double-delay standards, see Figure 4.1, can be derived from the single SOC standard provided it is modified for a shielded environment. The SOC standard in a shielded environment has perfectly conducting box walls backing the ports, in contrast to the vias of Figure 4.12. The double-delay $2L$-length through is derived from this shielded version of the SOC standard by setting V_3 to zero. This is equivalent to crossing out the third row and column of Eqn (4.11).

The first column of the double-delay L-length through Y parameters represents the input current and the output current with the output port shorted to ground. This has already been derived above; see Eqn (4.12). Since the through line is symmetrical, the Y parameter matrix must be symmetric, yielding

$$Y_{L\ \mathrm{Length}} = \begin{bmatrix} Y_{11}^{\mathrm{SOC}} - Y_{12}^{\mathrm{SOC}} & Y_{31}^{\mathrm{SOC}} - Y_{32}^{\mathrm{SOC}} \\ Y_{31}^{\mathrm{SOC}} - Y_{32}^{\mathrm{SOC}} & Y_{11}^{\mathrm{SOC}} - Y_{12}^{\mathrm{SOC}} \end{bmatrix}, \tag{4.18}$$

Again, this assumes double-delay is appropriate i.e., we are working in a shielded environment. Thus, from a single $2L$-length shielded environment SOC standard, we can derive data for both double-delay standards. Unfortunately, this does not provide a significant speed improvement. Assuming our EM analysis is limited by an order N^3 matrix solve, and given that the L-length double-delay standard requires T analysis time, then the $2L$-length double-delay standard requires $8T$. The $2L$-length SOC standard also requires $8T$. Thus the fractional reduction in analysis time realized by deriving the two double-delay standards from the single SOC standard is 8/9, about 10%.

However, the shielded EM analysis does not need to infer the short and open circuit standards from analysis of the SOC standard. Rather, the short and open circuit standards can be analyzed directly. The PEC wall for the short-circuited L-length stub is natively available in a shielded analysis. The PMC wall for the open-circuited L-length stub is created across the middle of the shielding box by modifying the shielded Green's function.

The shielded Green's function treats the perfectly conducting sidewalls of the shielding box as a rectangular waveguide with a weighted sum of waveguide modes propagating vertically, perpendicular to the planar substrate surface. To place a PMC wall down the center of the box, all we need to do is exclude all waveguide modes that have tangential H-field along that wall. Thus, we zero out half of the terms in the Green's function summation and we have a PMC wall down the center of the shielding box. We EM analyze a stub whose far end terminates in that PMC wall, and we have the open circuit stub data.

The directly EM-analyzed short- and open-circuited stubs are used in Eqn (4.16) to perform DUT de-embedding. They may also be used to create both double-delay standards. The simplest way to obtain the Y parameters of the L-length through is to add a second port to the short-circuited stub to make it an L-length through and analyze it directly. The Y parameters of the $2L$-length through are obtained from the short- and open-circuited stub results as

$$Y_{2L\ \text{Length}} = \begin{bmatrix} \left(Y_{11}^{\text{PMC}} + Y_{11}^{\text{PEC}}\right)/2 & \left(Y_{11}^{\text{PMC}} - Y_{11}^{\text{PEC}}\right)/2 \\ \left(Y_{11}^{\text{PMC}} - Y_{11}^{\text{PEC}}\right)/2 & \left(Y_{11}^{\text{PMC}} + Y_{11}^{\text{PEC}}\right)/2 \end{bmatrix}. \tag{4.19}$$

Note that $Y_{11}^{\text{PEC}} = Y_{11}$ of the L-length through. With this approach, we still require analysis of two standards; however, they are both L-length, requiring a total time of $2T$. Comparing with the previous time requirement of $9T$, we realize a speed increase of 4.5 times.

As mentioned before, the original SOC calculates the port discontinuity plus the port connecting transmission line. It does not determine the single port discontinuity by itself. Fortunately, it is possible to extend the unshielded SOC so it too can calculate the single port discontinuity.

This is done by adding a second standard to the SOC calibration, specifically, an L-length through. Then proceed by calculating the SOC ABCD de-embedding matrix, namely Eqn (4.16). Use this to pre-multiply the EM analysis of the L-length through. The result is the ABCD matrix of a single port discontinuity as in

$$\begin{bmatrix} A & B \\ C & D \end{bmatrix}_{\text{SinglePort}} = \begin{bmatrix} A & B \\ C & D \end{bmatrix}_{\text{SOCPort1}}^{-1} \begin{bmatrix} A & B \\ C & D \end{bmatrix}_{L\ \text{Length}}. \tag{4.20}$$

The single port discontinuity ABCD matrix can be then used to de-embed the L-length through, and we may proceed as with double-delay in determining characteristic impedance, namely Eqn (4.4).

With SOC typically called upon to remove the effect of larger port discontinuities (i.e., ports with series discontinuity impedance), there is a price to be paid. Using Eqn (4.4) to infer circuit theory equivalent characteristic impedance assumes that the electrical length of the line, βL, is not equal to a multiple of one half wavelength. When this is the case, we have a line that is electrically zero length, and characteristic impedance simply cannot be inferred from ABCD data for a zero-length line.

In practice, when βL is nearly equal to a multiple of one half wavelength, there can still be problems. Figure 4.13 shows characteristic impedance inferred from a length of transmission line as a function of frequency. Notice that the data from SOC for a line with ports that include a series port discontinuity impedance fails over a much broader range and more dramatically than for the pure shunt admittance port discontinuity. This is due to the ambiguity in the length of the line being emphasized when it is close to a multiple of a half wavelength. The electrically nearly zero-length line introduces large error compared to the uncertainty in the line length.

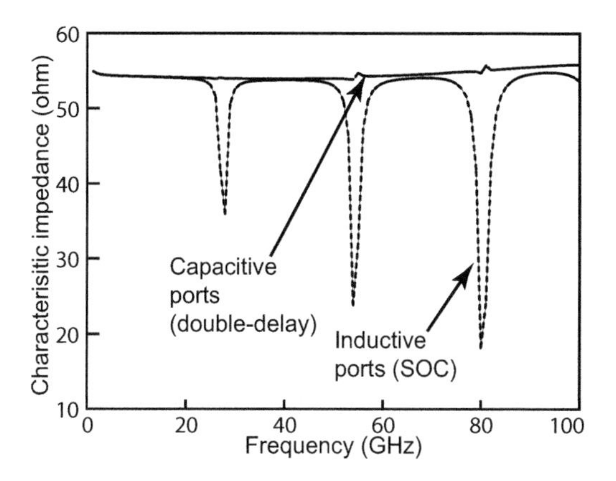

FIGURE 4.13

Presence of the series port discontinuity impedance, which, when combined with the shunt port discontinuity admittance, introduces uncertainty in line length. This in turn causes the unshielded SOC calculation of characteristic impedance to fail when the line length is close to a multiple of a half wavelength. The problem is significantly reduced when the port discontinuity has only a shunt capacitance, allowing double-delay calibration.

From [10]

It would be good if we could perform high-accuracy EM port calibration in locations, perhaps buried deep within a circuit, where a good ground reference is not conveniently placed, or perhaps even missing altogether. We discuss this problem in the next section.

4.5 Local ground and internal port de-embedding

Ground is a human concept. As Maxwell postulated when he introduced displacement current, electric current must always flow in complete loops. That we arbitrarily decide to call one portion of that loop "ground" leaves the underlying physics unchanged.

There are many kinds of ground. There is earth ground, chassis ground, power ground, logic ground, and of course, ground planes. They all form a necessary piece of a complete loop of current.

The concept of ground originated in the fact that a voltage of a single point cannot be measured. Our voltmeter has two leads, and we must connect both leads somewhere before we can measure voltage. A voltage must be measured between two points. To make things simple, we can select the one point for the negative lead of our voltmeter and all voltage measurements are taken with the negative lead attached to that same point for every measurement. If we like, we can call that point "ground."

This is a concept used in circuit theory nodal analysis. A given point, arbitrarily selected, is designated ground. All voltages are defined with respect to that ground. This works well for purely lumped, or low frequency, circuits. Problems arise when we consider higher frequencies.

Take, for example, a 100-m-long coil of coaxial cable. We have two ends that we can hold side by side (not touching). Our circuit theory model for this transmission line, see Figure 4.14, includes the effect of propagation delay, characteristic impedance, and loss (resistance). However, even if we select all of these parameters correctly, this model has problems. The problems have to do with ground.

For the circuit theory model of Figure 4.14, we have zero resistance between the ground terminals of port 1 and port 2. In the coax cable, we can connect an ohmmeter between the ground shields of the two ends and measure, let's say, 1.0 ohm of resistance. Then we measure the resistance between the two ends of the center conductor. Our model gives, let's say, 3.0 ohms of resistance. But we only measure 2.0 ohms! Our model takes all the resistance of both the center conductor and the ground shield and lumps it together into the path from port 1 to port 2, with nothing in the ground path. How can the model possibly work if it fails this simple experiment?

It turns out that this model works well if we realize that the ground reference for port 1 is different from the ground reference for port 2 and, as a result, we then realize that we should never connect circuitry between port 1 and port 2. If we do connect circuitry between port 1 and port 2 using the (incorrect) model of Figure 4.14, we might (depending on the circuitry) obtain unexpected and incorrect results. If all we do is connect one end to an antenna and the other to a transmitter, with no other connection between the antenna and transmitter, all will be well.

Thus, there is absolutely no problem with multiple ports in a circuit where different ports are referenced to different grounds provided we make no connections between ports with different grounds. Engineers working at electric power plants know this rule well.

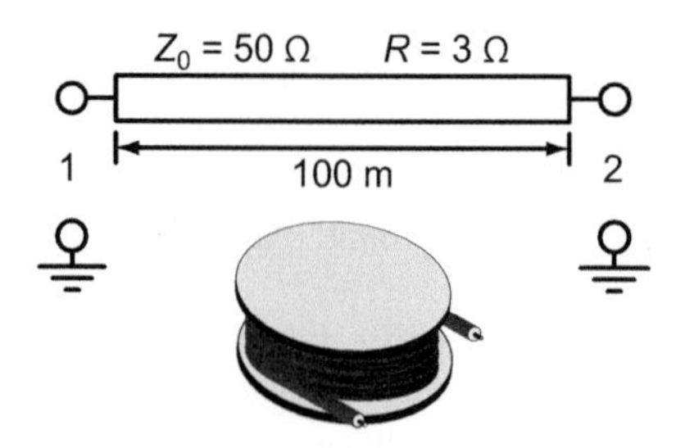

FIGURE 4.14

A simple model for a length of coax seems unlikely to be a problem. However, measuring the DC resistance between the two ends of the ground shield of the actual coax yields a value greater than the perfect short circuit suggested by the model.

We call these different grounds "local" grounds. In some cases, we might be moved to refer to one of the larger or more significant grounds as a "global" ground. But that is an arbitrary, human decision.

In the next section, we explore the theory of local ground calibration and de-embedding. This theory is useful not only for ports that lack anything resembling a nearby ground but also for ports that are deep in the interior of a microwave circuit. This includes, for example, power RF transistors embedded deeply inside the input and output matching circuitry. In fact, even with a nearby microstrip ground plane, a lossy ground plane can result in a situation similar to that of the two ends of a 100-m-long coaxial cable. With the techniques described in the following section, modern EM analysis calibration techniques can be applied in all of these situations.

4.5.1 Theory of local ground and internal port de-embedding

This section provides an overview of local ground and internal port de-embedding, summarizing the full description of the theory presented in [14]. For a complete and detailed description of this theory, Ref. [14] should be consulted.

Figure 4.15 shows the calibration standard for internal ports. Ports 1 and 2 are external box wall ports. Box wall ports (only a portion of the box sidewall is shown) allow double-delay calibration. This is beneficial because the location of the internal port can be critical in certain design situations. The position of external non-box wall ports with a ground reference provided by a via to ground is not precisely defined. That can translate into problems with the precise location of de-embedded internal ports.

The small patch in Figure 4.15 with ports 3 and 4 attached is the local ground for the internal ports. The via in the center attaches the local ground to global ground. The via has inductance. There are capacitive fringing fields coupling between all parts of the circuit, including the local ground. Figure 4.16 shows all these

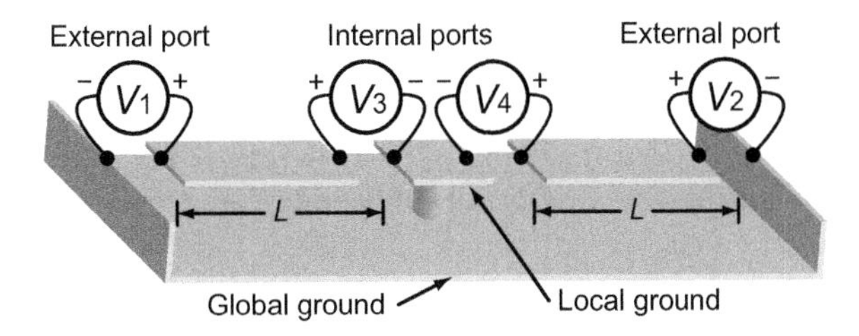

FIGURE 4.15

The calibration standard for internal ports includes external ports 1 and 2. Using double-delay (or SOC if box wall ports are not available), the *L*-length port connecting lines and the associated port 1 and 2 port discontinuities are removed leaving the local ground acting as the DUT in this analysis. Note the polarity of each port. The problem is solved by removing the effect of the now known local ground from the desired circuit.

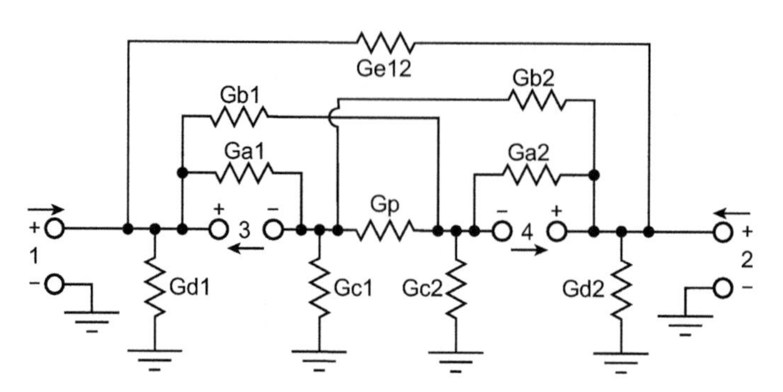

FIGURE 4.16

This is the equivalent lumped model of the local ground once it has been treated as though it is a DUT and the external ports and connecting lines are removed from the standard of Figure 4.15. For simplicity, lumped conductances are shown representing general lumped admittances.

Adapted from [14]

couplings. The schematic shows conductances (resistors) as a simplified representation of more general admittances.

The objective of internal/local ground port calibration is to determine the local ground ABCD parameters and then invert those ABCD parameters. This gives us the "de-embedding adapter." We connect the de-embedding adapter to the internal ports in our actual circuit to yield de-embedded internal ports. This process is identical to that described in previous sections for calibrating and de-embedding external ports.

It would seem that we are done. By EM analyzing and de-embedding the standard of Figure 4.15, we know what the local ground is. Once we know what it is, we remove its effects from the actual circuit. Unfortunately, there is considerably more complexity required to make this technique robust.

The first problem is what size matrix is required for the de-embedding adapter. Let's assume we have two internal ports, as in Figures 4.15 and 4.16. The instantaneous, without thinking, answer is simple. The matrix size is 2×2, just like we used for all previous two-port de-embeddings. Of course, that is not the right answer.

Look at the figures again. Our de-embedding adapter (i.e., the inverted ABCD parameters) is not a two-port, it is a four-port, a 4×4 matrix. Now let's imagine a situation where we might apply it: An amplifier with an input port, an input matching network, two internal ports for a transistor, an output matching network, and then an output port, as illustrated in Figure 4.17. This is also a four-port. We must organize the two four-port matrices (one for the DUT and one for the de-embedding adapter) so when we multiply the matrices, the amplifier's internal transistor ports (which include the effect of the local ground) are cascaded with the de-embedding adapter. When we cascade (i.e., multiply the ABCD parameters) of the amplifier with the de-embedding adapter, we have

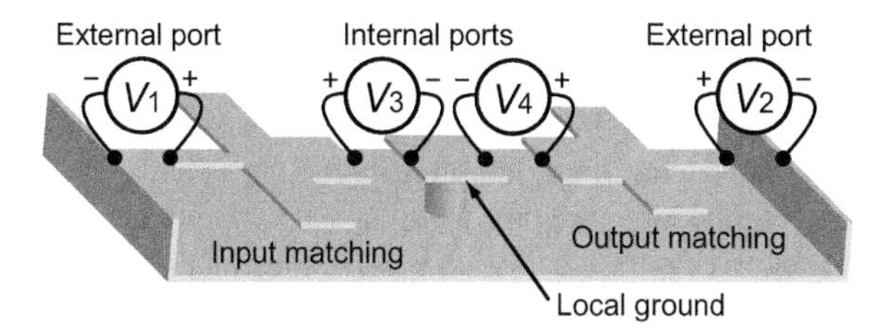

FIGURE 4.17

Once we have characterized the local ground, Figure 4.15, then we can remove the effect of the local ground from the circuit we are designing. Note that this amplifier circuit has the same local ground, and the internal port connecting transmission line widths are the same as in Figure 4.15.

de-embedded the internal ports, as illustrated in Figure 4.18. The internal ports are now de-embedded and referenced to global ground. The external ports must still be calibrated and de-embedded using double-delay, SOC, or equivalent.

For this problem, there are an equal number of external and internal ports. We organize the ports, Figure 4.18, so all the external ports are on one side and all the internal ports are on the other side. This allows us to solve the problem by inverting one 4×4 matrix and multiplying by another.

The general situation is more complicated. The number of external and internal ports is not necessarily going to be equal. We might be able to connect the de-embedding adapter using nodal analysis. But note that in Figure 4.18, we have not drawn all the ground terminals for the ports directly to a ground symbol as we did, for example, in Figures 4.1–4.3. There is a very good reason for this. The internal and the external ports have different ground references. We need to make connections to both signal and ground terminals of each port separately. Nodal analysis assumes all ground terminals are connected to global ground. That assumption is no longer valid.

Thus, it is most convenient to continue using a cascading matrix analysis. To do so, we must make the cascading matrices square. In other words, there must be an

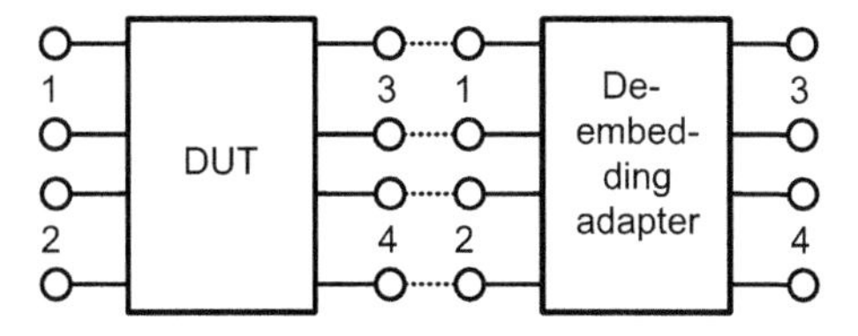

FIGURE 4.18

The internal ports, 3 and 4, of the DUT (Figure 4.17) are de-embedded by cascading with the de-embedding adapter, the inverted ABCD parameters of the calibration standard, Figure 4.15. The external ports, 1 and 2, must still be calibrated and de-embedded.

equal number of input and output (left and right side, Figure 4.18) ports. If there are fewer internal ports, then we add supplemental ports to the internal port side. Likewise for the external port side. These supplemental ports have no effect on the result; they are only there so we can invert and multiply square matrices.

Note that the de-embedding adapter is always a square matrix, with one output (calibrated) port for each input (uncalibrated) port. The embedded DUT must have at least the number of internal ports as the de-embedding adapter, and is likely to have quite a few more. The constraint we face is that both the de-embedding adapter and the embedded DUT must have the same number of ports, and they must have half of the ports considered to be input (on the left-hand side); and half of the ports must be output (on the right-hand side).

Figure 4.19 shows how we supplement both the embedded DUT and the de-embedding adapter to realize these constraints. The de-embedding adapter is supplemented by ports passing straight through. Since the numbers of input and output ports for the de-embedding adapter are always equal, these ports are just zero-length pass-through connections as illustrated.

The embedded DUT is more difficult. If it already has an even number of ports, we just make sure the internal ports (to be de-embedded) are all on the right-hand side. If convenient, we can place some of the external ports on the "output" side as well, as long as they are connected to the supplemental pass-through de-embedding adapter ports.

However, if we have an odd number of ports, or if there are fewer external ports than internal ports, we must add supplemental input ports to the left-side embedded DUT. The ports have to connect to something. If we leave the ports open circuited, then Z parameters do not exist. If we leave them short circuited, then Y parameters

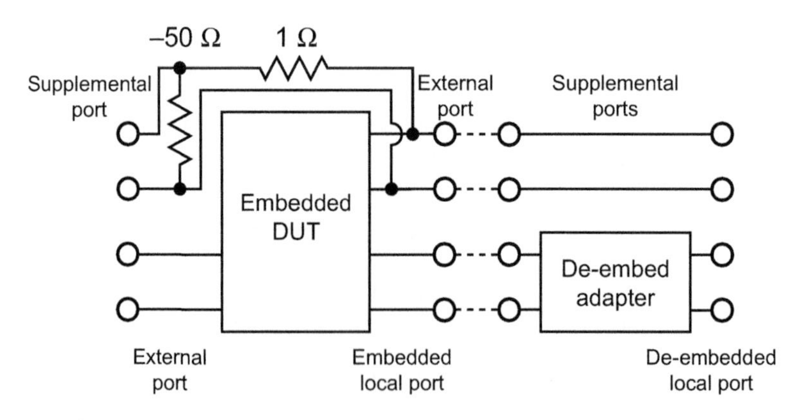

FIGURE 4.19

Supplemental ports are added so we can multiply and invert square matrices for the de-embedding operation.

From [10]

do not exist. The solution is to connect a −50 ohm resistor across the terminals of the supplemental port. *S*-parameters are determined with a +50 ohm resistor across each pair of port terminals. The two resistors in parallel yield an open-circuited port, which in turn has no effect on the *S*-parameters of the rest of the circuit. To make sure *Y*- and *Z*-parameters are still defined, connect the signal terminal of the supplemental port to any other point in the de-embedding adapter with anything other than a short circuit.

After cascading the embedded DUT with the de-embedding adapter, we have de-embedded the internal ports. Discard all rows and columns dealing with the supplemental ports and we have our solution. Perhaps now, we are done. Not quite.

An increasingly common situation is design of circuits that have no ground plane. This is typical, for example, in Si RFIC. Now, the via of Figure 4.15 has no place to go. The solution is to just leave it out. If there is no connection to global ground (in this case, because there is no global ground), then everything should work exactly as just discussed, only the resulting de-embedded internal ports are referenced to a floating ground.

Everything does not work as expected. As formulated herein, there are situations where the cascading matrix becomes singular and the de-embedding fails. These situations are caused directly by there being no connection at all between the grounds of the external and internal ports. There are two measures that must be taken to resolve this problem.

First, we must use a "double port" connection, as shown in Figure 4.20. Each of the 12 ports illustrated represents a group of ports. Note that after cascading the two illustrated circuits, we have exactly the same result as Figure 4.18—provided we treat port R_3 (or L_3) as the input port group (i.e., the external ports) and port R_2 (or L_2) as the output port group (i.e., the calibrated internal ports), and provided

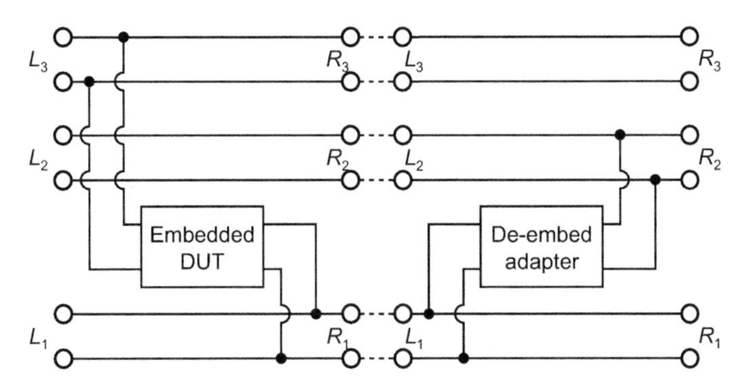

FIGURE 4.20

Cascading the embedded DUT with the de-embedding adapter after connecting each into the "double port" configuration eliminates matrix singularities that arise when there is no connection at all between the ground references of the internal and external port groups. *From* [14]

we leave all other ports open circuited. Note that these ports play the role of the supplemental ports of Figure 4.19 in that they allow us to multiply and invert square matrices no matter how many ports are in each group.

Second, we switch from ABCD parameters, which are based on current and voltage, to R parameters [15], which are based on incident and reflected waves. The necessary conversion equations, analogous to Eqns (4.5)–(4.8), are

$$R = \begin{bmatrix} S_{12} - S_{11}S_{21}^{-1}S_{22} & S_{11}S_{21}^{-1} \\ -S_{11}^{-1}S_{22} & S_{21}^{-1} \end{bmatrix} \tag{4.21}$$

$$S = \begin{bmatrix} -R_{11}^{-1}R_{12} & R_{11}^{-1} \\ R_{22} - R_{21}R_{11}^{-1}R_{12} & R_{21}R_{11}^{-1} \end{bmatrix} \tag{4.22}$$

In all work above, simply replace use of ABCD parameters with R parameters. However, even with R parameters, singularities can occur during inversion of the calibration standard matrix when there is no connection from the internal ports to global ground. Thus, we obtain the S parameters of the de-embedding matrix by first converting to Y parameters (the internal results of an EM analysis are typically Y parameters to begin with). Then we change the sign on all the Y parameters and convert back to S parameters using Eqn (4.6).

With the S parameters of both the embedded DUT and the de-embedding adapter in hand, we can place them in the double port connection of Figure 4.20 with

$$\begin{bmatrix} b_{L_1} \\ b_{L_2} \\ b_{L_3} \\ b_{R_1} \\ b_{R_2} \\ b_{R_3} \end{bmatrix} = \begin{bmatrix} S'_{11} & S'_{12} & 0 & 1+S'_{11} & S'_{12} & 0 \\ S'_{21} & S'_{22} & 0 & S'_{21} & 1+S'_{22} & 0 \\ 0 & 0 & 0 & 0 & 0 & 1 \\ 1+S'_{11} & S'_{12} & 0 & S'_{11} & S'_{12} & 0 \\ S'_{21} & 1+S'_{22} & 0 & S'_{21} & S'_{22} & 0 \\ 0 & 0 & 1 & 0 & 0 & 0 \end{bmatrix} \begin{bmatrix} a_{L_1} \\ a_{L_2} \\ a_{L_3} \\ a_{R_1} \\ a_{R_2} \\ a_{R_3} \end{bmatrix}. \tag{4.23}$$

The a are incident wave amplitudes and the b are reflected wave amplitudes. The primed S parameters are the S parameters of the de-embedding adapter or the embedded DUT with an extra 50 ohm resistor connected across every one of its ports. This can be determined either with a cascading analysis or a nodal analysis. The elements of Eqn (4.23) need not be square, but the overall matrix is square, making it appropriate for cascading analysis.

Our problem is solved by placing both the de-embedding adapter and the embedded DUT into the double port configuration. Then convert both to R parameters, and multiply the matrices (in either order). Next, convert the result to Z parameters and discard the rows/columns corresponding to all but the R_3 and R_2 port groups. This corresponds to leaving the unneeded ports open circuited. Convert back to S parameters and the internal port group is de-embedded.

4.5.2 Failure mechanisms of local ground and internal port de-embedding

Internal port calibration is based on external port calibration and de-embedding using (preferably) double-delay or (in unshielded situations) SOC. Thus, any failure of the external port calibration will likewise result in failure of the local ground de-embedding. These failure modes can include over-moded port connecting transmission lines due to, for example, overly thick substrates for the frequency of operation, radiation, de-embedding calibration standard length, *L*, too short, ground return path too far from the signal path, etc. These issues are described in detail in Section 4.2.2.

Above, we describe de-embedding a single group of internal ports. In general, there are multiple groups. For example, there could be multiple transistors, each with its own group of internal ports. Some circuits use numerous surface mount devices (SMD). Each SMD can have its own group of internal ports. To calibrate multiple groups of internal ports, we repeat the port calibration procedure for each group. The trade-off is that while all port fringing field coupling between ports in the same group is completely and exactly removed, any fringing field coupling between port groups is not removed.

The problem that can arise in this case is if groups of ports are so close together that their fringing fields interact. This coupling between groups of ports is not removed. Situations of concern include placement of ports closer than the substrate thickness. The vias from local ground to global (microstrip) ground are now close enough that they might inductively couple to each other. Another situation of concern is if there is other circuitry that has significant fringing field coupling to the local ground. For example, a transmission line that passes beneath the local ground might have some capacitive coupling to the local ground via. That capacitance is not removed by the calibration.

Another grounding via situation can arise in Si RFIC, which normally has no bottom side ground plane. Thus, a floating ground should always be used for Si RFIC. If a ground via is accidentally used, its length is equal to the substrate thickness, typically a few hundred microns. Place another group of internal ports within a few hundred microns and the grounding via will have strong mutual coupling. We emphasize that if there is no nearby, convenient global ground, then a floating ground should always be used.

Many of these failure mechanisms, as described herein, include coupling to the grounding via. In some of these cases, the solution is simple. If a global ground reference is not needed, don't bother including the grounding via. This results in a floating ground. Let's say we are connecting an SMD resistor model between two ports in the same group. Let's say also, as is typical, the resistor model does not include the effect of a nearby ground plane. In this case, the resistor model could not care less what ground reference the ports use, so use a floating ground. There are no grounding vias for a floating ground, and thus no problem with stray coupling to or from the grounding vias.

In contrast, let's say we are going to connect a circuit theory microstrip transmission line between two calibrated ports in the same group. A transmission line is critically dependent on current flowing in the ground return—recall the light bulb experiment described in a previous section. In this case, the internal ports must be calibrated using a global ground reference. If the ports are calibrated with a floating ground reference, the ground return for the transmission line model is left open and the transmission line does not work.

As also mentioned before, do not make connections between ports of different groups unless those ports all have the same ground reference. If this rule is violated, unexpected and incorrect results are the likely outcome.

Calibrated internal ports with different or even floating ground references do require some extra thought and care. In return, the opportunity for achieving fast, accurate analysis of complicated designs rewards our efforts.

4.6 Circuit subdivision and port tuning: application of calibrated ports

In the old days of tubes, radios and televisions were built from discrete inductors, capacitors, and resistors. Of course, the critical components could not be reliably or inexpensively manufactured to the extreme tolerances sometimes required. For example, the LC circuit in a narrow band intermediate frequency filter must have exactly the correct resonant frequency in order to work properly.

To solve this problem, tuning elements were added. There might be a small screwdriver adjust trimming capacitor connected in parallel with a large capacitor. The trimming capacitor provides enough range to tune the LC circuit to the desired resonance in spite of manufacturing tolerances.

For most of today's microwave circuits, physical tuning is a very expensive option to be avoided, if it is even available. However, by using strategically placed calibrated internal ports, we can tune our EM analyses quickly and effectively. Once we have, say, a filter design somewhere in the vicinity of our design goals, we can make small "trimming capacitor" adjustments to realize the final design quickly. Only a few EM analyses are required, not hundreds.

Perfectly calibrated ports also allow us to aggressively divide circuits, analyze the pieces, and connect the pieces back together. These techniques are introduced in this section.

An early form of port tuning is described in [16]. An initial description of both port tuning and circuit subdivision is included in [17]. Subsequently, there has been substantial research in the area, much of it summarized in [18]. To illustrate, we explore a hairpin filter example cited in [17] and shown in Figure 4.21 (top).

This filter is already easily analyzed as it is. Typical analysis times are just several seconds. However, it makes a nice test bed for experimenting with various techniques that have promise of significant speed increases for large microwave circuits.

One way to achieve faster analysis is to note that the filter is perfectly symmetrical about a horizontal axis of symmetry. Dividing the filter into two identical

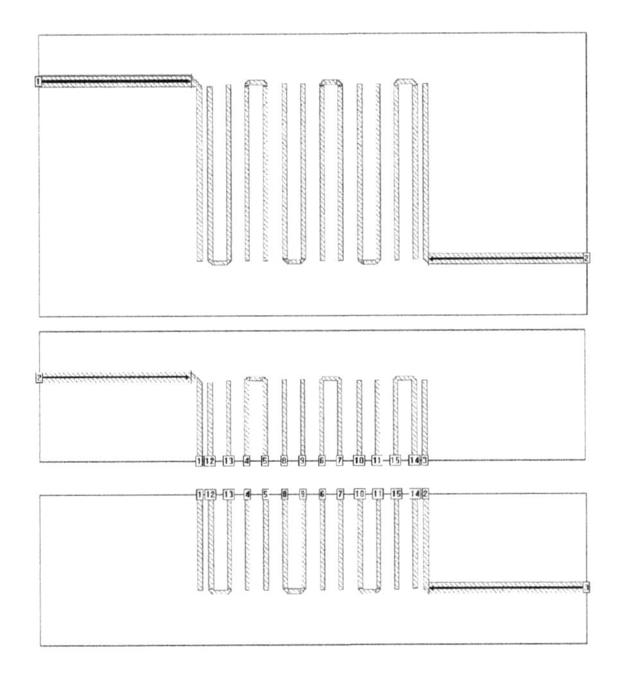

FIGURE 4.21

This band pass filter is used to illustrate port tuning and circuit subdivision (top). One way to speed EM analysis is to divide the filter in two identical halves (bottom), analyze, and then connect together.

From [17]

(except for rotation) pieces is shown in Figure 4.21 (bottom). Noting that they are identical, we need only to EM analyze one of them. After it is analyzed, we connect it with itself, while being very careful to keep each port connected correctly. The frequency response of the entire filter (not shown here) is exactly identical to the same filter analyzed in two pieces. Note that this is actually another case of what we did in Figures 4.10 and 4.11 for validating the coupled port double-delay calibration. In that case, we had a long coupled line divided into 11 pieces and connected together. In this example we have divided the circuit into only two sections, but the sections each have 14 coupled ports.

For problems dominated by large a large matrix, for which solve times are long, the speedup is substantial. If it takes T time to analyze one half of the filter, then it would take about $8T$ time to analyze the entire filter. The speed increases by a factor of eight. When possible, this is an important technique to master.

A common reaction to the circuit subdivision technique is that coupling between the two halves of the filter in Figure 4.21 is not included in the EM analysis, so it could not possibly work. The first part of this reaction is true; coupling between the two halves is not included. However, we lose only coupling between sections of the filter in which current flows parallel to the cut line. If we take a transmission line and cut it up into 100 pieces, or 1000 pieces, and reconnect, there is no problem. As long as the little piece we analyze is done correctly, we get the correct results for the entire transmission line. The same is true for a coupled transmission line, as in Figure 4.10, and for a 14-line coupled transmission line, as in Figure 4.21.

Notice that there is a second line of symmetry in the filter of Figure 4.21. This line is vertical and passes between the two central coupled lines. With a vertical cut line, the coupling between the two central resonators is not included and the filter analysis fails. So, when placing cut lines for analyzing a circuit by subdivision, make sure that any transmission lines parallel to the cut line are a reasonable distance from the cut line. Transmission lines that are perpendicular to the cut line, and even cross the cut line, are fine.

Analysis of circuits much faster by using circuit subdivision is nice, but we would like to fine-tune circuits in real time, but with nearly full EM accuracy. This is achieved by placing calibrated internal ports in key locations and adding tuning elements.

Figure 4.22 shows the same hairpin filter only now with a short section removed from the input and the output coupled line section. In place of each missing section is a group of four calibrated internal ports. Connected into those four ports is a short length of coupled line. All of this is detailed in the insets of the figure.

The short, inserted coupled line is known as a "tuning element." In this case, the tuning element is from separate EM analysis. However, it could just as easily be from a circuit theory program. In fact many circuit theory programs include the ability to quickly and interactively tune element values (like transmission line length) in

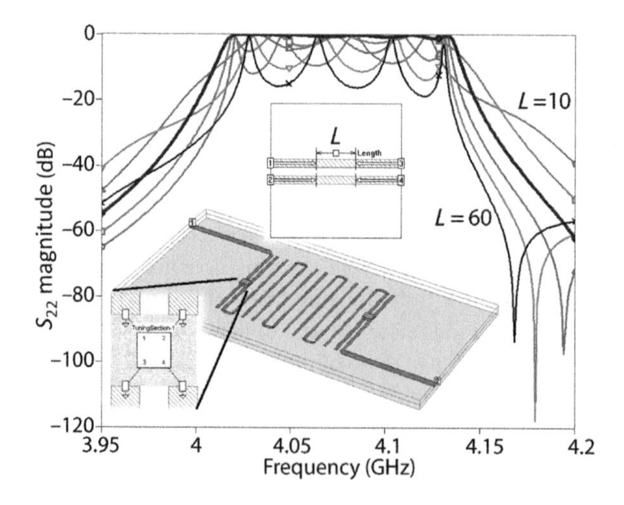

FIGURE 4.22

For a simple case of port tuning we insert calibrated internal ports into the input and output coupled line sections of the filter. Then we connect a short length of coupled transmission line in its place. Filter tuning is quickly realized by adjusting the length of the inserted transmission line without repeating the entire EM analysis.

From [17]

real time, perhaps by sliding a slider bar. Similar tuning elements can be inserted in all the coupled line sections of the filter. Tuning can be done either manually or under the control of an optimization algorithm. The key point is that we can tune all the filter resonators in real time without repeating the EM analysis of the entire filter. Figure 4.22 also shows a few representative filter responses for different lengths of the inserted coupled line.

Keep in mind that for this application the four ports in each of the internal calibrated port groups must have a global ground reference. That is because we are inserting a transmission line (actually a coupled transmission line) and it needs access to ground for the ground return current or it simply will not work. Internal ports referenced to a floating ground reference are inappropriate in this example.

The range of values explored for the inserted length of line in Figure 4.22 is large. For highest accuracy, it is best if the filter design is already somewhat close to the desired response so the tuning element range can be kept small. The reason for this is that the coupling between the tuning element and the rest of the circuit is not included. As long as the tuning element is kept small, this has only a small effect.

Not obvious is how to tune the coupling between resonators. Since this is difficult to see, let's ease into this concept with a simple thought experiment. To start with, we ignore inductive coupling that occurs across the gap between the coupled lines. Let's just think about capacitance. If we decrease the distance between the coupled lines, the capacitance between the lines increases. We can model this by connecting a small lumped capacitor across the gap using our internal calibrated ports, as shown in Figure 4.23. (The length tuning coupled line is left out of the figure.) If we increase the capacitance, we increase the coupling.

If we set the tuning capacitor to zero, then we simulate zero change in the coupled line gap. Now, suppose we want to simulate increasing the coupled line gap. This decreases the coupling between the lines. To simulate decreased coupling we need to add a negative capacitor between the lines.

A wider line has more capacitance to ground. So, if we want to simulate a small increase in the width of a line, we just add a small positive capacitance to ground. A small decrease in width is simulated by adding a small negative capacitance.

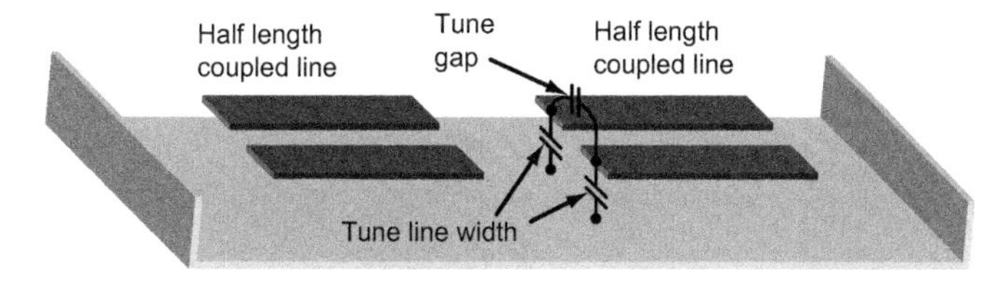

FIGURE 4.23

To illustrate tuning the gap between coupled lines, we add a capacitance across the gap. Positive capacitance increases the coupling, which decreases the gap. To simulate less coupling, or increased gap, we connect a negative capacitance across the gap. Capacitance to ground can simulate small changes in the width of the lines.

Now, let's return to the real world. The coupled line also experiences inductive coupling. So instead of the three capacitors shown in Figure 4.23, we connect a second, circuit theory coupled line across the two internal ports, as illustrated in Figure 4.24. A coupled line can be specified in terms of even and odd mode characteristic impedances, Z_{0e} and Z_{0o}. We won't go into coupled line theory here, but it is important to understand that Z_{0e} is roughly analogous to the capacitance to ground, and Z_{0o} is roughly analogous to the capacitance between the coupled lines. A large Z_{0e} or Z_{0o} means a low capacitance and a small value means high capacitance. Infinite Z_{0e} or Z_{0o} means zero capacitance.

If we connect a coupled line with infinite Z_{0e} and Z_{0o}, there is no change in the filter response. If we start decreasing Z_{0o}, we start increasing the capacitance between the coupled lines of the tuning coupled line. Since this is connected in parallel with the actual coupled line, we are increasing the capacitance between the actual coupled lines and thus simulating a decrease in the actual coupled line gap. To simulate a small increase in the actual coupled line gap, we need to specify a large negative Z_{0o}. Not all circuit theory programs allow negative characteristic impedances, so we might have to revert to the simpler capacitor model described previously.

Line width is tuned in a similar manner by adjusting Z_{0e}. As a practical matter, in most coupled line filters, the line width is not a critical parameter. We should be able to get good results even if we leave Z_{0e} at infinity. Advanced designers will realize that a simple open stub could be used in this case.

In the illustrated hairpin filter, the coupled line sections are about one-quarter wavelength long. A common technique is to introduce a group of four internal ports, as in Figure 4.23, at the halfway, or one-eighth wavelength point. One circuit theory tuning coupled line one-eighth wavelength long is connected across the gap using the internal tuning ports. To make the coupled line section whole again, a short length of circuit theory coupled line is connected so as to replace the short section of coupled line that was removed.

One might think we need two one-eighth wavelength–long circuit theory coupled line sections to tune the actual coupled line gap when connected to the center of a one-quarter wavelength–long coupled line section. However, both of

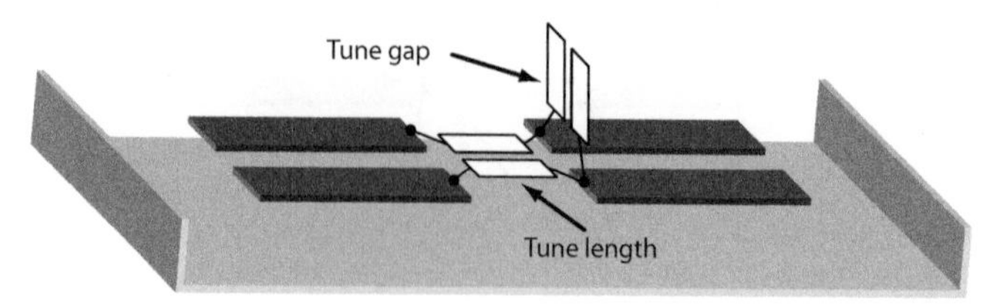

FIGURE 4.24

Replacing the tuning capacitors of Figure 4.23 with a coupled line, we can tune the coupling between the lines by tuning the odd mode characteristic impedance. We can tune the line width by tuning the even mode characteristic impedance.

these one-eighth wavelength-long tuning coupled line sections would be connected in parallel, so they would be exactly equivalent to a single one-eighth wavelength-long coupled line section with a modified Z_{0e} and Z_{0o}.

Once we have found the desired values for the even and odd mode impedances of the gap tuning coupled line sections, and the optimal length for the length tuning coupled line section, we need to translate those values into a new, slightly modified, filter design. This is easy for the new coupled line length. Just add the length (which might be negative, if the optimal length is shorter than the new length) of the length tuning coupled line section to the original length.

For the coupled line gap width, we need to know how much to change the gap width for a given value of Z_{0o} of the gap tuning coupled line. This is done by running a secondary optimization where we optimize the gap of a coupled line model so it matches the original coupled line plus tuning coupled line. From this, we can determine the sensitivity of the gap to the added coupled line Z_{0o}. Since infinite Z_{0o} corresponds to zero change, we actually determine the sensitivity to $1/Z_{0o}$. Once the sensitivity is determined, we can quickly calculate a new value for the gap based on the value of $1/Z_{0o}$ of the gap tuning coupled line.

When selecting optimization variables for a filter, it is important to restrict our choices to as few variables as possible. Even with the speed of circuit theory analysis, an excessive number of optimization variables can cause difficulty. For the hairpin filter discussed here, with seven coupled line sections, symmetry allows us to use only four unique variable lengths for tuning coupled line lengths. Three of those lengths are used twice. The same is true for tuning coupled line gap widths. Tuning line width is not needed for this filter.

When adding a large number of tuning elements, it is helpful if the circuit theory program can display "layout look-alike" components. Then we don't need to know or remember what port number corresponds to which resonator. If instead we must deal with a black box component, i.e., one with just port numbers, it is very easy to make a mistake when hooking up all the tuning elements.

There is a wealth of additional techniques possible using the concepts of circuit subdivision and port tuning. We hope that this introduction provides inspiration to advance the art even further.

References

[1] Rautio JC. A de-embedding algorithm for electromagnetics. Int J Microwave Millim Wave Comput Aided Eng July 1991;1(3):282–7.

[2] Rautio JC. A new definition of characteristic impedance. IEEE MTT-S Int Microwave Symp Dig, vol. 2 (Boston, MA, USA) July 1991. pp. 761–4.

[3] Rautio JC. Testing limits of algorithms associated with high frequency planar electromagnetic analysis. Eur Microwave Conf Dig (Munich, Germany), October 2003. pp. 463–6.

[4] Cohn SB. Problems in strip transmission lines. IRE Trans Microwave Theory Tech March 1955;MTT-3(2):119–26.

[5] Gupta KC, Garg R, Chadha R. Computer Aided Design of Microwave Circuits (Dedham, MA, USA): Artech House; December 1981. p. 57.
[6] Rautio JC. An ultra-high precision benchmark for validation of planar electromagnetic analyses. IEEE Trans Microwave Theory Tech November 1994;42(11):2046–50.
[7] Abramowitz M, Stegun I. Handbook of mathematical functions with formulas, graphs, and mathematical tables. (New York, NY, USA): Dover; December 1972. pp. 590–2.
[8] Gupta KC, Garg R, Chadha R. Computer Aided Design of Microwave Circuits (Dedham, MA, USA): Artech House; December 1981. p. 72.
[9] Rautio JC, Carlson RL, Rautio BJ, Arvas S. Shielded dual-mode microstrip resonator measurement of uniaxial anisotropy. IEEE Trans Microwave Theory Tech March 2011;59(3):748–54.
[10] Rautio JC, Okhmatovski VI. Unification of double-delay and SOC electromagnetic deembedding. IEEE Trans Microwave Theory Tech September 2005;53(9):2892–8.
[11] Zhu L, Wu K. Unified equivalent-circuit model of planar discontinuities suitable for field theory-based CAD and optimization of M(H)MIC's. IEEE Trans Microwave Theory Tech September 1999;47:1589–602.
[12] Farina M, Rozzi T. A short-open deembedding technique for method of moments based electromagnetic analyses. IEEE Trans Microwave Theory Tech April 2001;49(4):624–8.
[13] Okhmatovski VI, Morsey J, Cangellaris AC. On deembedding of port discontinuities in full-wave CAD models of multiport circuits. IEEE Trans Microwave Theory Tech December 2003;51:2355–65.
[14] Rautio JC. Deembedding the effect of a local ground plane in electromagnetic analysis. IEEE Trans Microwave Theory Tech February 2005;53(2):770–6.
[15] Kerns DM, Beatty RW. Basic theory of waveguide junctions and introductory microwave network analysis. New York: Pergamon; January 1967.
[16] Swanson DG. Narrow-band microwave filter design. IEEE Microwave Mag October 2007;8(6):105–14.
[17] Rautio JC. Perfectly calibrated internal ports in EM analysis of planar circuits. IEEE MTT-S Int Microwave Symp Dig (Atlanta, GA, USA) June 2008. pp. 1373–6.
[18] Cheng QS, Rautio JC, Bandler JW, Koziel S. Progress in simulator-based tuning—the art of tuning space mapping. IEEE Microwave Mag June 2010;11(4):96–110.

CHAPTER

5 Large-Signal Time-Domain Waveform-Based Transistor Modeling

Iltcho Angelov[1], Gustavo Avolio[2], Dominique M.M.-P. Schreurs[2]
[1] *Chalmers University of Technology, Göteborg, Sweden,* [2] *ESAT-TELEMIC, KU Leuven, Leuven, Belgium*

5.1 Introduction

Nonlinear transistor models are essential for the accurate design of high-frequency (HF) nonlinear electronic circuits, such as power amplifiers and mixers. Among the existing techniques exploited for the extraction of nonlinear models, measurement-based approaches are widely employed. In particular, methods relying on the use of nonlinear measurements have been widely exploited in the last few decades. The aim of this chapter consists of outlining a methodology to generate empirical models of microwave transistors starting from nonlinear measurements. In many situations, the transistors that are used in the design of power amplifiers or mixers experience excitations that drive them to operate nonlinearly. Therefore, one can benefit from nonlinear measurement systems [1] in order to extract the transistor models in conditions as close as possible to one of the final applications. Among the existing instruments, the large-signal network analyzer (LSNA) has become very popular in recent decades. The LSNA enables vector calibrated measurements, and therefore it provides the current and the voltage in the time and frequency domain at the terminals of the device under test. Different types of LSNA exist [2–5]. The measurement of high-frequency time-domain waveforms is performed by exploiting down-conversion to intermediate bandwidth followed by acquisition and digitalization (Figure 5.1(a)). The down-conversion step is not necessary when measuring directly in the low-frequency (LF) range, i.e., kHz–MHz (Figure 5.1(b)). A solution that combines both low- and high-frequency capabilities is presented in [5]. Although the LSNA is originally intended for the characterization of high-frequency electronic components, in this chapter it will be shown how LSNA operating in different frequency ranges can provide a powerful tool for the extraction of nonlinear transistor models.

Commonly, microwave transistors are described in terms of nonlinear current and charge sources. By sweeping the measurement frequency from low (kHz–MHz) to high (GHz) values, one can isolate the resistive current contribution from the displacement current originated by charge variations. When dealing with transistors based on III–V compounds, the resistive current can be strongly affected by low-frequency dispersion [6]. In this perspective, low-frequency measurements are

Microwave De-embedding. http://dx.doi.org/10.1016/B978-0-12-401700-9.00005-7

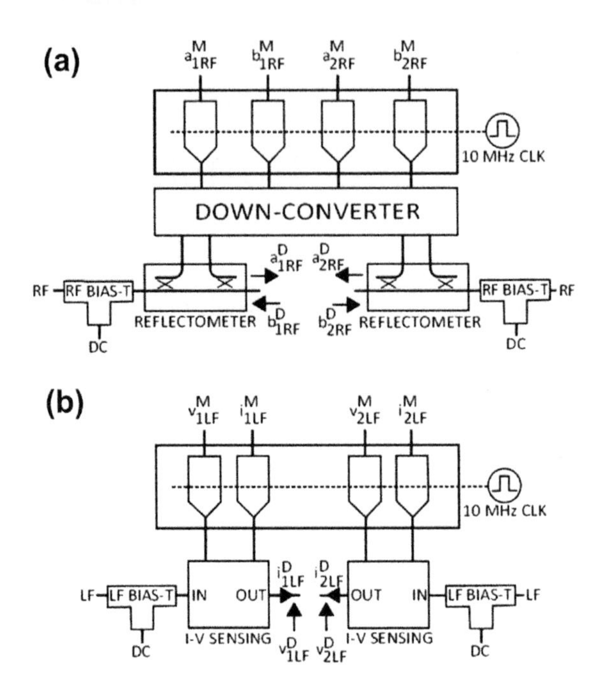

FIGURE 5.1

High-frequency (a) and low-frequency (b) LSNA architectures.

very useful as the effect of the charge sources is minimized, i.e., "de-embedded," already within the experiment itself. Therefore one can directly investigate and subsequently model the impact of dispersive phenomena on the transistor low-frequency current and voltage time-domain waveforms. On the other hand, high-frequency measurements are essential to boost the contribution of the displacement current.

This chapter begins with an overview of the most commonly adopted modeling technique. Next, the equations of the adopted empirical model are reported and a general extraction procedure of the model parameters is described. In the chapter's last section, large-signal models for microwave transistors are generated starting from time-domain voltages and currents measured at different frequencies. The extracted models are principally oriented for the design of power amplifiers or mixers, which, in many situations, operate under nonlinear conditions.

5.2 Large-signal transistor modeling: overview

A standard topology of large-signal (LS) equivalent circuit of microwave field-effect transistors is shown in Figure 5.2. I_{ds}, gate-source, gate-drain diodes ($I-V$ functions) and capacitances C_{gs} and C_{gd} ($C-V$ or $Q-V$ functions) are bias dependent. All the

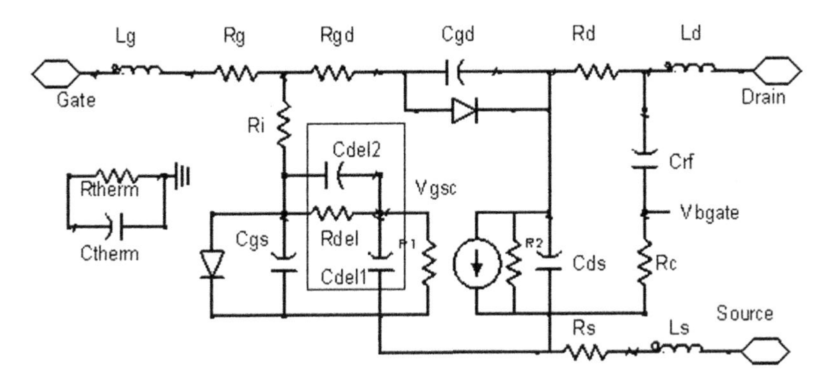

FIGURE 5.2

Large-signal equivalent circuit of a transistor.

other elements of the equivalent circuit are assumed to be linear. Moreover, the LS equivalent circuit includes thermal and delay sub-circuits. The bias-dependent capacitances can be replaced by bias-dependent charge sources. For convenience, in the circuit in Figure 5.2 one can define an "intrinsic" section that encompasses transistor capacitances, diodes, and current source representing the active channel and an "extrinsic" section that also includes the parasitic network. Similarly, the intrinsic voltages (and currents) are the ones across the terminals of the elements encompassed in the intrinsic section.

Nonlinear models can be constructed either starting from small-signal measurements or directly from large-signal measurements. In the former case, the transistor nonlinearities, described by the current ($I-V$) and capacitance/charge ($C-V/Q-V$) functions, are derived from the bias-dependent small-signal transistor parameters that are extracted from S parameter measurements. These small-signal parameters are then used to build the large-signal $I-V$ and $Q-V$ models. In the course of time, significant experience has been developed in FET transistor modeling, which has resulted in usable CAD models such as those discussed in [7–10].

When starting from small-signal measurements, large-signal models are generated by applying mathematical integration to the bias-dependent small-signal parameters [11–13]. The values of the $I-V$ and the $Q-V$ functions, which depend on the instantaneous intrinsic voltages, are then stored in look-up tables (LUT). It is noteworthy that the result of the mathematical integration may not be unique as the values of the $I-V$ and $Q-V$ functions may be dependent on the selected integration starting point and path, which must be properly chosen selected to correctly convert the small-signal model into the LS one. Alternatively, nonlinear models can be directly extracted from large-signal measurements [14–23]. The $I-V$ and $Q-V$ relationships can be either represented by a table [16] or by analytical expressions whose parameters are estimated directly from measurements or by numerical optimization against the measurement data [17]. The use of large-signal experiments is very convenient for two reasons: first and importantly because the models are extracted from data obtained with the device operating close to real life; in this case one can benefit from a more accurate identification step as especially the *I*–*V* and *Q*–*V* functions give an approximate description of phenomena that are inherently nonlinear. Secondly, the number of experiments needed for model extraction can be significantly reduced, though this aspect has to be counterbalanced with the cost of the available measurement equipment. Finally, it is worth mentioning that the use of large-signal excitations, as opposed to DC and small-signals, allows one to instantaneously excite the transistor in an extreme operating region, such as breakdown, while minimizing degradation of the device's electrical characteristics.

Several approaches aimed at nonlinear model extraction have been reported in the literature. These approaches either make use directly of large-signal measurements or rely on the combination of large- and small-signal measurements. The works proposed by Schreurs et al. in [17,18], and by Curras-Francos et al. [19,20], belong to the first group, and nonlinear models are extracted only by using high-frequency vector calibrated nonlinear measurements. The values of the nonlinear currents and charges in the model of Figure 5.2 are calculated as a function of the intrinsic voltages $v_{GS}(t)$ and $v_{DS}(t)$ starting from the experimental time-domain waveforms. More precisely, in [18] high-frequency continuous wave (CW) one-tone signals are injected at both input and output ports with a slight shift in frequency. In this way an optimal coverage in the $v_{GS}(t) - v_{DS}(t)$ plane, over which the values of the *I*–*V* and *Q*–*V* functions are calculated, is achieved. In [19] and [20], instead, the amplitude and the phase of the signals injected at the input and output ports is varied in order to obtain a non-looping $v_{GS}(t) - v_{DS}(t)$ characteristic at the intrinsic plane. Therefore, for a fixed pair of values ($v_{GS}(t)$, $v_{DS}(t)$), the values of both the *I*–*V* and *Q*–*V* functions can be computed. Alternatively, in [17] the *I*–*V* and *Q*–*V* functions are approximated by artificial neural networks (ANNs) whose parameters are determined by numerical optimization. An ANN-based nonlinear model is also proposed in [21], and the model parameters are extracted by numerical optimization against DC, small-, and large-signal measurements. All these approaches are very robust, although they are limited in many cases as they do not account for low-frequency dispersion and their extrapolation capability might be inaccurate outside the range of the data set used to extract the model.

In order to account for dispersive phenomena, an ANN-based modified formulation is proposed in [22] and the *I*–*V* and *Q*–*V* model parameters are extracted starting from a large number of large-signal measurements performed under several bias, power, and load conditions. The modeling technique proposed by Raffo et al. in [23] relies, instead, on the use of low-frequency large-signal measurements to accurately capture dispersive effects. The LF experimental time-domain waveforms can be directly exploited for the identification of the *I*–*V* functions only whereas the *Q*–*V* relationships are replaced by LUTs where values derived from multi-bias small-signal measurements are stored. In this chapter, empirical models are generated starting from nonlinear measurements, performed in both low- and high-frequency ranges. Specifically, the equations of the model in [10], which is available in commercial CAD tools, are used to describe the I–V and Q–V functions. The parameters of the model are then obtained from numerical optimization against the experimental data. It is noteworthy that nonlinear measurements can be combined with other type of measurements, such as DC and S parameter measurements, in order to obtain an initial estimate of the model parameters to be used within the optimization routine. The basic idea in the model is to directly connect some model parameters to measurement data [24]. In this way the model extraction is simplified.

5.3 Modeling currents (I–V) and charges (Q–V): procedure

In this section the equations of the adopted model are described in detail, and a general extraction procedure for the parameters of these equations is outlined. Also, a subsection focuses on the proper selection of the modeling functions.

5.3.1 Model equations

In the model described in [10] the nonlinear currents and charges (or capacitances) are expressed as analytical functions of the intrinsic voltages. The model equations are with continuous derivatives, without poles from $-\infty$ to $+\infty$, without switching or conditioning.

The current associated with the current source between drain and source is expressed as follows:

$$I_{ds} = I_{pk0}{}^{*}(1+\tanh(x_{1p}))^{*}\tanh(\alpha_p V_{ds})^{*}(1+\lambda^{*}V_{ds}+L_{SB0}{}^{*}\exp(V_{dg}-V_{tr})) \quad (5.1)$$

where

$$x_{1p} = P_{1m}{}^{*}(V_{gs}-V_{pkm})+P_2{}^{*}(V_{gs}-V_{pkm})^2+P_3{}^{*}(V_{gs}-V_{pkm})^3$$
$$P_{1m} = P_1{}^{*}(1+\mathrm{B}_1/\cosh(\mathrm{B}_2{}^{*}V_{ds}))$$
$$\alpha_p = \alpha_R+\alpha_S{}^{*}(1+\tanh(x_{1p}))$$
$$V_{pkm} = V_{pks}-\mathrm{D}V_{pks}+\mathrm{D}V_{pks}{}^{*}\tanh(\alpha_S)-V_{SB2}{}^{*}(V_{dg}-V_{tr})^2$$

The model is optimized to work in the saturation region for $V_{ds} > V_{knee}$ and V_{gs} at the peak of the transconductance (V_{pkm}). For saturated V_{ds} and $V_{gs} = V_{pkm}$ the function $\tanh(\alpha V_{ds})(1+\lambda V_{ds}) \sim 1$ if $\lambda << 1$, and the drain current is $I_{ds} = I_{pk0}$ by definition. The parameter $P_1 = G_m/I_{pk0}$ will automatically define the transconductance g_m at this point. Parameters V_{pkm}, I_{pk0}, and $P_1 = G_m/I_{pk0}$ are taken directly from the measurements, and as result, the extraction is very simple with only five parameters I_{pk0}, V_{pkm}, P_1 at saturated V_{ds}, λ, and α. The model and derivatives are strictly defined at V_{pkm} and in the vicinity of V_{pkm} where the maximum of the transconductance occurs. With five parameters the transconductance is automatically generated with a bell shape with typical global accuracy better than 10%.

The parameter α together with R_d (and all DC line resistances in the measurement setup) define the slope of I_{ds} vs. V_{ds} at small drain voltages $V_{ds} < V_{knee}$. The parameter λ will define the slope of I_{ds} vs. V_{ds} at high $V_{ds} > V_{knee}$ and is extracted at small currents to avoid the influence of self-heating. These two parameters α and λ are commonly used in many FET models.

For devices with complicated doping profiles a more sophisticated model structure can be used. The I_{ds} dependence on the gate voltage is described as a power series using more terms, such as P_2, P_3, to track a variety of I_{ds} cases. The parameter P_2 will introduce asymmetry of the I_{ds} vs. V_{gs} and will influence the second harmonic, and parameter P_3 will trim drain current at gate voltages close to pinch-off. These parameters directly influence the respective second and third harmonics. Typically three terms are enough to provide accuracy better than 5%.

Some of the model parameters, like V_{pkm}, P_1, are bias and temperature dependent [25]. Parameter V_{pkm} describes the change of V_{pks} due to the drain voltage, and parameters α_r and α_s change the slope of I_{ds} at small drain voltages. A good fit in this area of small or negative drain voltages can be important for circuits working at low V_{ds} like resistive mixers or switches. The parameters are rather independent in adjusting I_{ds}. For example α_r will influence the drain current at small V_{ds} and small currents, and α_s will influence the drain current at small V_{ds} and high currents, close to the knee (see Figure 5.3). Above the knee the slope of I_{ds} vs. V_{ds} is adjusted with parameter λ. Breakdown modeling, if required, can be treated with parameters V_{tr}, L_{sb0}, and V_{sb2} [25].

Many of these parameters are typical for all FETs. For example, transconductance parameter P_1 is typically $P_1 = 1.2-1.5$ for MESFET, $P_1 = 2-4$ for HEMT,

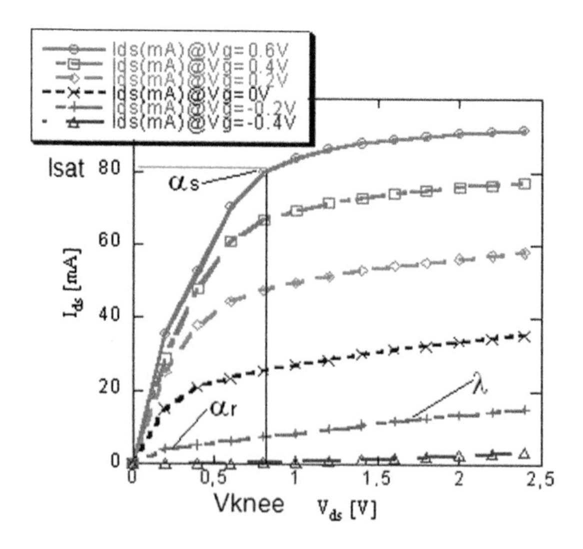

FIGURE 5.3

Typical I_{ds} vs. V_{ds} characteristic for a FET; the parameter is V_{gs}.

Table 5.1 Basic model parameter values for FETs

Parameter	MESFET	HEMT	HighGain HEMT	Linear HEMT	SiC	GaN	LDMOS
I_{ch} (A/mm)	0.3–0.6	0.3–0.6	0.17–0.25	0.3–0.6	0.35	1.5	0.75
P_1 (V^{-1})	1.1–1.5	2–3	4.5–5.5	1.5	0.1	0.3	2
V_{pks} (V)	−0.5	−0.2	+0.05	−1.4	−9	−3	3.5
V_{knee} (V)	0.75	0.75	0.75	0.75	9	4	3
α_s	1.3–1.5	2–2.5	3.7	1.5	0.14	0.4	1.5
Cap (pF/mm)	1	1	1.0	1.3	0.8	0.7	0.6

$P_1 = 0.3$ for GaN, and $P_1 = 2$ for LDMOS. A high value of P_1 will produce higher gain for the same current, which is good for low-noise high-gain applications. But if P_1 is very large, the gate voltage swing (input power) can be limited, and this will influence the linearity and intermodulation characteristics. Transistors with low P_1, like MESFETs, GaAs HEMTs are specially designed for linear applications, and SiC and GaN FETs will have better intermodulation properties, but lower gain. This means that some compromise should be made if we want to have a high-efficiency, high-power, and linear amplifier. Depending on the application we can select the best P_1 for our application. Nowadays the physical simulators are fast enough and can help to optimize the device structure for specific application. In Table 5.1 some basic data for different FET devices are given.

Normally, the devices operate at positive drain voltages, and it seems obvious that there is no need to look at negative V_{ds}. When drive level is small this is correct, but when the device is used as power amplifier, switch, or mixer, the instantaneous drain voltage is swinging into the negative V_{ds} region. Therefore, the drain current model should describe properly the I_{ds} at negative V_{ds} even if the device is biased with positive V_{ds}. Usually, in circuit simulators the model switching at negative V_{ds} is arranged in a simple way. When the drain voltage V_{ds} is positive, the gate voltage V_{gs} controls the drain current. When V_{ds} is negative, the control voltage is switched to V_{gd} and I_{ds} current is calculated from the same equation with reversed sign (I_{ds} is negative). If the device is symmetrical, this is correct. But at the switching point $V_{ds} = 0$ V, there will be a singularity and the derivative of I_{ds} is not defined. As a consequence, it will be more difficult for the simulator, e.g., harmonic balance (HB) analysis, to converge and the results of the simulations can be wrong in the vicinity of $V_{ds} = 0$ V. A solution to this is a continuous, single-model equation for I_{ds} valid for all control voltages from $-\infty$ to $+\infty$.

For cases like switches and resistive mixers applications operating at low and negative V_{ds} the drain current equation is composed from two sources, I_{dsp} and I_{dsn}, which are controlled respectively by V_{gs} and V_{gd} [26]:

$$I_{ds} = 0.5 * \left(I_{dsp} - I_{dsn}\right) \tag{5.2}$$

with

$I_{dsp} = I_{pk0}*(1 + \tanh(x_{1p}))*(1 + \tanh(\alpha_p V_{ds})*(1 + \lambda_p * V_{ds} + \lambda_{p1} * \exp(V_{ds}/V_{kn}))$
$I_{dsn} = I_{pk0}*(1 + \tanh(x_{1n}))*(1 - \tanh(\alpha_n V_{ds})*(1 + \lambda_n * V_{ds} + \lambda_{n1} * \exp(V_{ds}/V_{kn}))$
$x_{1n} = P_{1m}*(V_{gd} - V_{pkm}) + P_2*(V_{gd} - V_{pkm})^2 + P_3*(V_{gd} - V_{pkm})^3$
$\alpha_n = \alpha_R + \alpha_S*(1 + \tanh(x_{1n}))$
$\lambda_n = \lambda + L_{VG}*(1 + \tanh(x_{1n}))$
$\lambda_p = \lambda + L_{VG}*(1 + \tanh(x_{1p}))$
$\lambda_{n1} = \lambda_1 + L_{VG}*(1 + \tanh(x_{1n}))$
$\lambda_{p1} = \lambda_1 + L_{VG}*(1 + \tanh(x_{1p}))$

There are cases with very complicated I_{ds} vs. V_{gs}, V_{ds} dependencies, and it is very difficult to obtain a good correspondence between the model and measurements. In this case the power series can be replaced with a data set calculated directly from measured data. The empirical-equivalent circuit models in this way are combined with table-based models [10,24]. By using a mixed empirical-table approach it is possible to combine and extract the best from both types of model. The empirical model is serving as an envelope for the table-based model, and the problems with spline function selection, extension out of the measurement region, and convergence are solved. This is because FET model equations are used as a spline. The derivatives are continuous, correct, and the model will converge well. The linear extrapolation out of the measured data range will be adequate because the empirical model will limit the solution. The model will be limited and valid out of the measured range because the data set is naturally limited by using the measured data for the extraction.

Quite often there is a spread of parameters and it is important to give the users some flexibility to tune basic model parameters in the empirical or mixed empirical-table-based model. For example there are always some tolerances in G_m, pinch-off voltage, thermal resistance etc., and the model can be arranged in such a way that the user, without making complete measurement and extraction set, can change only the required parameter. This can be done with a proper arrangement of the mixed empirical-table-based model. The mixed empirical-table-based model can be arranged to access the basic parameters combining the benefits of the empirical and the table-based models. The LS model is extracted for a typical device, but later it should be possible to trace process tolerances.

So far, equations describing the drain-source current have been reported. The equations for the current associated with the gate-source and gate-drain diodes are given by [26]:

$$I_{gs} = I_j^*\left(\exp\left(P_G^*\tanh\left(V_{gs} - V_{jg}\right)\right) - \exp\left(-P_G^* V_{jg}\right)\right) \quad (5.3a)$$

$$I_{gd} = I_j^*\left(\exp\left(P_G^*\tanh\left(V_{gd} - V_{jg}\right)\right) - \exp\left(-P_G^* V_{jg}\right)\right) \quad (5.3b)$$

wherein P_G is the gate current parameter, which is linked to the ideality factor (N_e) of the Schottky diode, I_j the gate-forward saturation current, and V_{jg} the turn-on voltage of the diode.

The model is completed with the thermal network, including the thermal resistance and thermal capacitance. The dependence of the model equations on the temperature (T) is accounted for as in Eqn (5.4):

$$I_{pk0} = \text{IPK0}^{*}\left(1 + T_{CIPK0}{}^{*}(T - T_{nom})\right) \tag{5.4a}$$

$$P_1 = P_1{}^{*}\left(1 + T_{CP1}{}^{*}(T - T_{nom})\right) \tag{5.4b}$$

$$L_{SB0} = \text{Lsb0}^{*}\left(1 + T_{CLSB0}{}^{*}(T - T_{nom})\right) \tag{5.4c}$$

where I_{pk0}, P_1, and L_{SB0} are the same as in (5.1) and IPK0, P1, and Lsb0 are the values of these parameters at $T = T_{nom}$.

5.3.2 Extraction of basic parameters

The model extraction procedures can be summarized as follows:

1. Extraction of on-resistance
2. Parameter extraction and fit of the gate current
3. Extraction of the drain current parameters below the knee
4. Extraction of I_{pk0}, P_1, and V_{pks}
5. Extraction of λ
6. Extraction of thermal resistance R_{TH} and thermal parameters
7. Preliminary extraction of P_2 (second derivative) and P_3 (third derivative) parameters.

Repeat the procedure 4–7, because parameters are interdependent.
At this point one should get typically <10% fit.

8. Global IV optimization (<2–5% fit)
9. Breakdown parameter extraction
10. S parameter extraction
11. S parameter fit including dispersion part
12. Fitting nonlinear RF measurements.

Steps 1–12 represent a general procedure that, obviously, has to be adapted depending on the device under test and the application. In the following, techniques for the direct extraction of some model parameters are outlined:

5.3.2.1 On resistance (R_{on})

Nonlinear models for currents and capacitances are controlled by intrinsic voltages. Therefore, one needs to extract R_s and R_d to account for the voltage drop. R_s and R_d can be extracted from S parameter measurements [27], but DC extraction provides a very good starting point and allows limiting the R_s and R_d parameter ranges for optimizations. It should be considered that extraction from S parameter "cold" measurements, i.e., $V_{ds} = 0$, is not good for GaN. In fact for GaN devices, the value of R_s and R_d can be dependent on bias and temperature and can be larger than the values extracted under the "cold" condition.

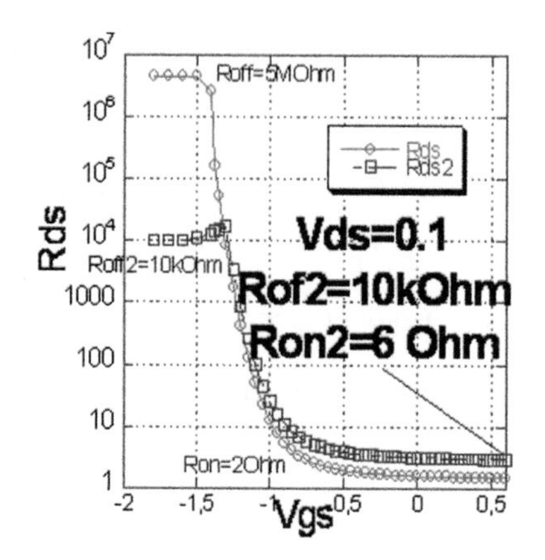

FIGURE 5.4

R_{ds} as a function of V_{ds} in the linear region.

The $R_{ds} = V_{ds}/I_{ds} = R_d + R_s + R_{ch}$ extraction is done using the data I_{ds} vs. V_{gs} for low V_{ds} in the linear part of the IV, below the knee.

For GaAs, $V_{ds} = 0.1-0.2$ V is a good choice, while for GaN, V_{ds} is 1–2 V.

R_{dsOn} vs. V_{gs} (Figure 5.4) will give info:

1. OFF-resistance—when device is pinched off. The R_{dsOff} gives preliminary info on how the device will pinch at high V_{ds}. A device with a poor R_{dsOff} will not behave properly at high V_{ds} but might be usable at low V_{ds} and low RF power. A good quality device will show $R_{dsOff} > 1$ MΩ. A device that is not pinching properly will have lower breakdown and not good characteristics in circuits such as switches, power amplifiers, resistive mixers, etc.
2. R_{dsOn} is equal to $R_d + R_s + R_{ch}$. This is a very important parameter that should be recorded and monitored on process control monitoring (PCM) samples on the wafer. The on-resistance gives info on the quality of the ohmic contacts, recess etching, channel doping, gate voltage control, pinch-off characteristics, etc. When R_{dsOn} resistance degrades (i.e., R_d, R_s increase), the output power of the amplifier and power added efficiency (PAE) will drop. Typically $R_{dsOn} < 0.3$ Ω/per mm device size.

Normally, it is considered that R_{dsOff}/R_{dsOn} should be >1000, but for some applications, like resistive mixers, 100 can be sufficient. However, we should not expect to get very impressive, low conversion losses from the resistive mixer or performance from the switch.

For a gate placed in the middle of source and drain, one can consider $R_d = R_s = R_{ch} = R_{ds}/3$. These are very good starting values for optimization. Roughly, R_d and R_s can be estimated as follows:

- HEMT GaAs
  $R_{dw} = R_{sw} = 0.065$ Ω mm

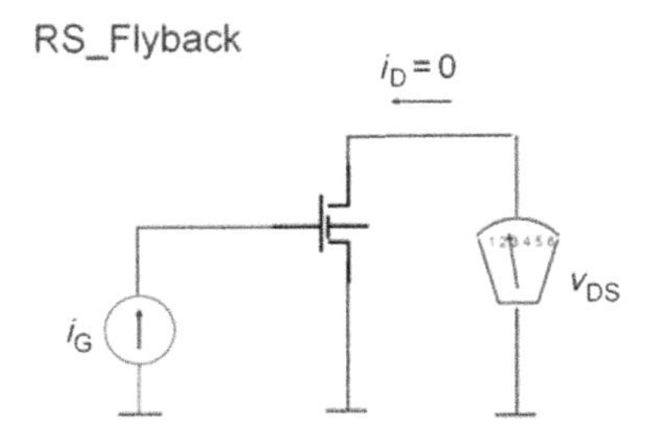

FIGURE 5.5

Equivalent circuit for simulations of R_{on}.

$R_s = 0.1 + R_{sw}/W_{tot}$
$R_d = 0.1 + R_{dw}/W_{tot}$
$W_{tot} = N_{fing}{*}W_{fing}/1000$

- GaN

$R_{dw} = R_{sw} = 0.35\ \Omega\ \text{mm}$
$R_s = 0.05 + R_{sw}{*}(0.2 + L_{sg}/3)/W_{tot}$
$R_d = 0.05 + R_{dw}{*}(0.2 + L_{sd}/3)/W_{tot}$

Another method to extract R_s, and also R_d, is to use the fly-back method (Figure 5.5). The method is based on injecting high current I_{gs} via the gate and measuring the voltage drop across R_s on the drain side, when the drain current is 0 A or very small:

$R_{meas} = R_s + R_{ch}$
$R_s = V_{dmeas}/I_{gs}$

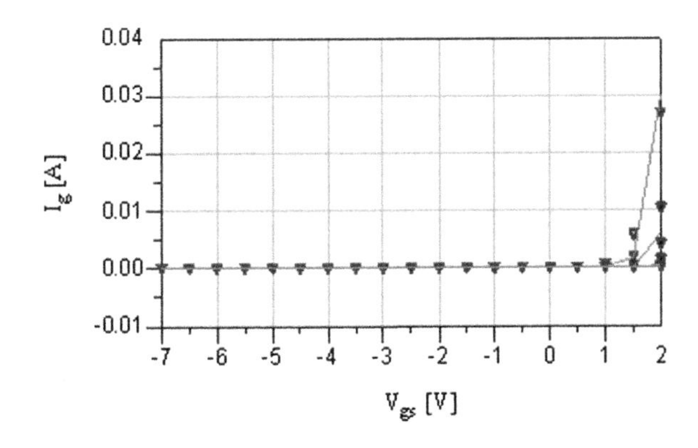

FIGURE 5.6

I_g versus V_{gs}.

Flipping drain and source, we can also get $R_d = V_{smeas}/I_{gs}$.

The fly-back method will not work with devices with oxide isolation under the gate. Attempts to push current via the gate can damage the device.

5.3.2.2 I_{gs} extraction

After extracting access resistances, before continuing with I_{ds} extraction and fit, it is recommended to continue with I_{gs} extraction. This is because for forward bias device and enhancement type devices, I_{gs} will influence the intrinsic voltage. In addition, I_{gs} will influence significantly the S_{11} parameter. Therefore, it is important to have this part of the FET model working before starting to do global fit for S parameters.

The extraction of the parameters in Eqn (5.3) is rather straightforward:

1. Read the measured current I_J at forward bias device V_{jg}. For GaAs V_{jg} is typically equal to 0.8 V; for GaN HEMT it is in the range 0.9–2 V. This voltage is the model parameter that is used directly. The current parameter I_J is the measured current for the built-in voltage V_{jg}. The I_{gs} model is exact at this point.
2. Adjust the slope using the parameter P_G. Typically, $P_G = dI_{gs}/dV_{gs}$ is in the range of 12–15 (decades of current per volt). P_G is the model parameter internally connected with the Schottky diode ideality parameter N_e. Start with $P_G = 15$. By default, P_G is used. From P_G, the intrinsic N_e is calculated.

It is important to consider that R_s, R_g, and R_d will influence the slope at high I_{gs} and I_{gd} currents and will not influence the slope at low currents. When the I_{gs} current is small, i.e., <0.1 mA, this influence from R_s and R_g is minimal, and therefore the measured value for I_{gs} can be used directly.

5.3.2.3 Drain current extraction and fit

Extracting the current part of the model is a very important part of creating the FET large-signal model. For the extraction of the drain current model, one can use DC measurements, small-signal measurements, or large-signal measurements. When DC or small-signal measurements are used, the drain current is measured in a wide range of biases sweeping both V_{gs} and V_{ds}. Typically we will need at least 10 gate voltages and 5 to 10 drain voltages depending on the voltage and power range of the transistor. When measurements and extraction are done properly, we can expect that at low frequency, where the contribution from reactive components (capacitance and inductances) is small, the model will be correct. The situation is different when the device is affected by dispersive phenomena [23,28–36], which cause the deviation between the static and dynamic characteristics measured at low frequency. In this case the use of large-signal measurements under CW or pulsed conditions is mandatory to correctly catch this deviation.

5.3.2.4 Extraction of λ, α_S, and α_R

Parameter λ should be extracted from the slope at small currents and high V_{ds}, above the knee as illustrated in Figure 5.7. In particular, $\lambda \sim (I_{ds2} - I_{ds1})/(\text{Delta\_VDS})$. Assume Delta_VDS is equal to 1 when extracting λ.

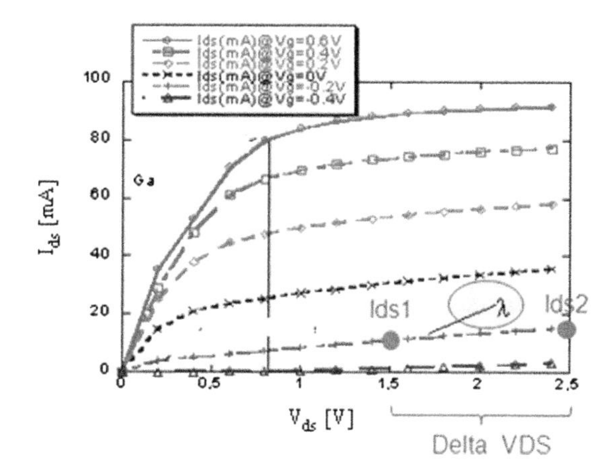

FIGURE 5.7

Typical $I_{ds}-V_{ds}$ characteristics of a FET.

For GaAs devices, Delta_VDS is in the range 1–3 V and for GaN Delta_VDS is 10–15 V. This is due to the fact that at high currents the self-heating will decrease the current and lower the slope. At very high power the slope will become negative. Unfortunately, some researchers try to fit this negative slope arranging or adjusting some of the nonlinear current source. The negative slope at higher dissipated power (above 0.15–0.2 W) should be handled by the self-heating model.

The slope I_{ds} vs. V_{ds} for V_{ds} below the knee-voltage is dependent on the gate voltage. The modeling is done with the α parameters α_R and α_S.

For devices biased at high V_{gs} (all carriers are collected), the slope is max and equal to α_S. Selection $\alpha_S = P_1$, is a good starting point.

Parameter α_S is the slope of I_{ds} vs. V_{ds} at high currents and low V_{ds}. At high currents in the knee region several parameters have combined influence on I_{ds}. When R_d has been extracted accurately, α_S can be fitted adjusting the slope in the linear region, carefully fitting the saturated current at the knee (I_{sat}). A very good starting point for α_S is the value of P1. This is because the sensitivity of I_{ds} versus controlling voltages V_{gs}, V_{ds} should be similar.

For GaAs devices, α_S is in the range 1–4; for GaN devices it is in the range 0.2–1.0.

α_R is the slope at small currents and small V_{ds}. A typical value is in the range 0.01–0.2.

Once the thermal model resistance R_{TH} and the other thermal parameters have been extracted from the high current fit, the value of λ can be refined. The easiest is to define λ at V_{gs} voltage close to G_{mpeak} with V_{ds} above saturation. In this case the equation for the current simplifies to $I_{ds} = I_{pk0}*(1 + \lambda V_{ds})$ and λ is found directly.

For GaAs devices λ can be between 0.05 for high voltage devices and 0.2 to 0.5 for millimeter wave FET devices with high-In content in the channel. For GaN FET, λ is very low, typically 0.01–0.04.

5.3.2.5 Extraction of I_{pkO}, V_{pks}, P_1, DV_{pks}

I_{pkO} can be taken as $I_{ds}/2$ drain current for V_{ds} at the knee, channel fully opened, i.e., $V_{gs} = 0.8$ V, and $I_{pkO} = I_{sat}/2$. The explanation is that usually the maximum of the transconductance occurs in the middle of I_{ds} vs. V_{gs} dependence. In the example in Figure 5.8, $I_{sat} = 0.4$ A, so $I_{pkO} = 0.2$ A is a very good starting point. The gate voltage for this current is $V_{gs} = -0.5$ V. At higher V_{ds}, the self-heating (see Figure 5.9) will start to influence the $I-V$ characteristic, so $I_{pkO} = I_{dsknee}/2$ selected at the knee is a very good starting point. A typical value for I_{sat} for GaAs device is I_{sat} GaAs $= 0.3-0.8$ A/mm; for GaN it can be higher, I_{sat} GaN $= 0.7-1.2$ A/mm. V_{pks} is the V_{gs} of I_{pkO}.

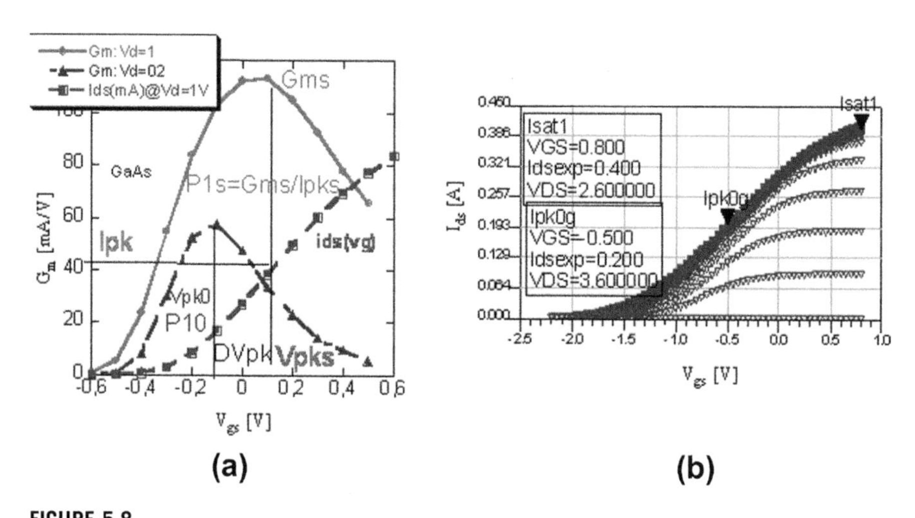

FIGURE 5.8

I_{ds} vs. V_{gs}: I_{pkO}, V_{pks}, and P_1 (a); I_{ds} vs. V_{gs}: $I_{pkO} = I_{sat}/2 = 0.2$ A, $V_{pks} = -0.5$ V (b).

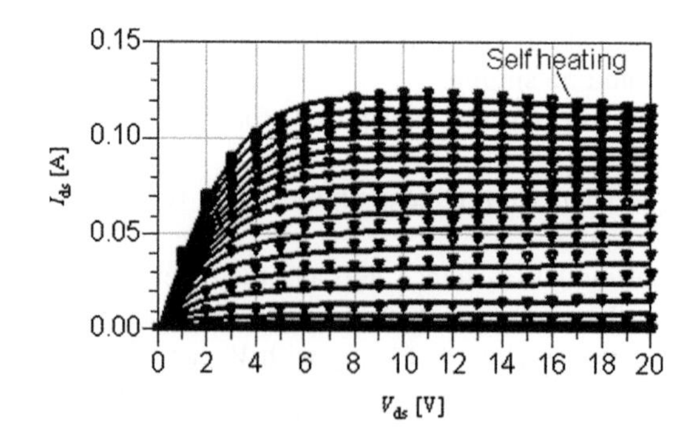

FIGURE 5.9

Measured (symbols) and simulated (line) I_{ds} vs. V_{ds}, V_{gs} as a parameter.

A common mistake is to forget that this refers to the intrinsic voltage. Due to this, it is possible to select a wrong value of the current for peak transconductance I_{pk0}. In a similar way, if the transconductance is with complex shape and it is difficult to decide where the peak of G_m is, the current should be selected as $I_{sat}/2$ and fitted with the derivative parameters P_2 and P_3. The complex shape can be due to various reasons, such as R_s, temperature, and bias dependence [37–40], early breakdown influence, and gate leakage. So first it is important to find the reason for the shape of transconductance and then to model.

The derivative (quality) factor of the transistor P_1 is found from either $P_1 = G_m/I_{pk0}$ with I_{pk0} as shown in the above plot.

Parameters DV_{pks}, B_1, B_2 (see Eqn (5.1)) will determine the change of P_1 when V_{ds} is increased from low voltages to V_{ds}, well above the knee (P_1 is usually decreasing and saturating).

DV_{pks} refers to intrinsic change of V_{pks} and it is much smaller than extrinsic change DV_{pk}, due to the voltage drop on the source resistance R_s.

Figures 5.5–5.8 shows typical dependencies of I_{ds} and G_m for GaAs FET. The gate voltage V_{pks}, G_m and the drain current I_{pk0} at which the maximum transconductance occurs can be used to link measured and modeled I_{ds}. Typically, this inflection point occurs at the gate voltage for which we have half of the channel current.

When the drain voltage is changed, there is a change of the gate voltage for which we have the maximum of the transconductance V_{pks}, as can be seen in Figure 5.8. At low drain voltage $V_{ds} = 0.2$ V, the peak of G_m is at $V_{gs} = -0.1$ V and at high $V_{ds} > V_{knee}$ the $V_{pks} = 0.1$ V. Above V_{knee} there is some increase of the drain current, due to the channel opening from the drain voltage influence. If the drain voltage is further increased, breakdown can occur. Typically, high power devices are biased for high-efficiency operation, i.e., at high voltages and low currents. A properly constructed load line will keep the devices away from the breakdown area and they will be switched from high voltage/low current to high currents/low voltages (close to the V_{knee}). If this is the case, there is no sense in spending much time making a very detailed and accurate breakdown model. Only if the device will be operated in the breakdown area is it worth doing this.

Transconductance and the ratio $P_1 = G_m/I_{ds}$ also change when the drain voltage is changed. This means that the models should have a functional dependence for the peak voltage $V_{pkm} = f(V_{ds})$ and $P_1 = f(V_{ds})$ to describe the changes of V_{pks} and G_m due to drain voltage influence. Figure 5.10 shows the I_{ds} vs. V_{gs} dependence when stepping drain voltage V_{ds} from negative to positive. As can be seen, the device is not completely symmetrical; this is in part due to the shift of V_{pks} when V_{ds} is negative.

Quite often due to large device size, high dissipated power, and dispersive effects [23,28–36], the modeled I–V characteristics are far from ideal. The self-heating will decrease the drain current at high dissipated power (see Figure 5.9). The decrease of I_{ds} at high dissipated power will critically depend on the thermal resistance R_{TH} and for high power devices it is important to select a proper material with a high thermal conductivity, to make a good thermal design of the transistor, i.e.,

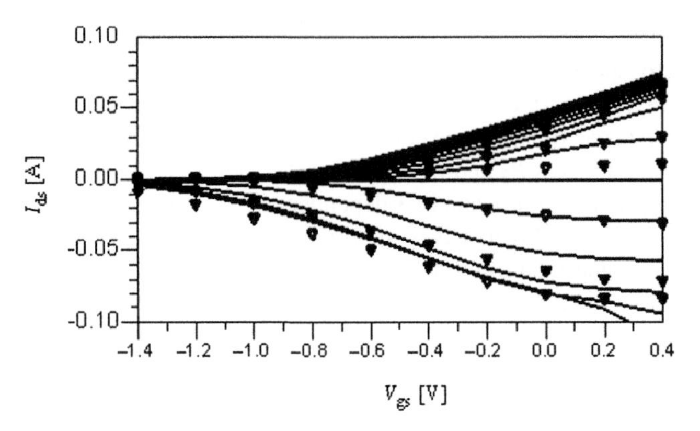

FIGURE 5.10

Measured (symbols) and simulated (line) $I_{ds} - V_{gs}$ characteristics at negative V_{ds}, V_{gs} as a parameter.

using properly placed via holes, thermal shunts, and thin substrate. The technology for the new GaN and SiC devices is very promising but still not settled, and there is substantial activity to improve these devices. We can expect that the I–V curves and all parameters for these new, high-power devices will gradually improve, to a greater extent than for devices with established technology like GaAs.

5.3.3 Modeling functions

The bases for the FET operation are two dependencies: the carrier velocity and carrier concentration. Their bias and temperature dependencies will be the main factors that will determine the transistor behavior. The I_{ds} vs. V_{gs} dependence is similar to the carrier concentration dependence versus gate voltage, and the corresponding modeling function should be selected. Generally, the solution of the Schrödinger and Poisson equations are *erf* types of functions, but an *erf* function is usually not available in circuit simulators. That is why it can be replaced with another suitable function, like *tanh*, which is accurate enough for this application.

In GaAs FET devices associated with some electric field (V_{ds}, V_{gs}) we observe a maximum of the carrier velocity and transconductance. With Si, we have gradual increase of the carrier velocity, which will produce a quite different shape of I_{ds}, G_m, G_{ds} vs. V_{gs}, V_{ds} in comparison with GaAs; see Figure 5.8. The G_m for the Si CMOS device increases with increasing drain voltage and will change the shape of I_{ds} vs. V_{gs} as well. The different shape of G_m for Si CMOS (Figure 5.11) will produce different harmonic content in comparison with the GaAs FET. This means that in FET models we should have respective parameters describing these dependences.

There are some general requirements for the selection of the modeling functions in empirical models. In FET and HBT the device parameters can be considered dependent on two voltages $I = f_1(V_{gs}) * f_2(V_{ds})$, or respective V_{be}, V_{ce}. The best

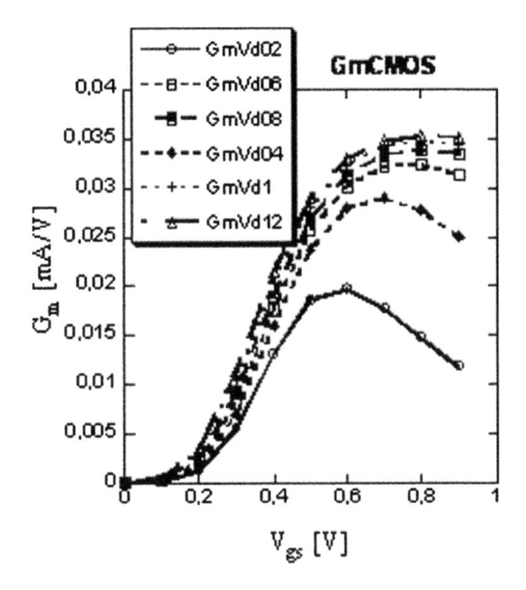

FIGURE 5.11

G_m vs. V_{gs} for CMOS.

solution from the extraction and user understanding point of view is to make both parts, f_1 and f_2, completely independent. This will greatly simplify extraction. However, what follows from device physics is that we have inter-coupling between the $f_1(V_{gs})\cdot f_2(V_{ds})$ parts, which should be implemented in a proper way. Then, with a very small number of additional parameters, the model will describe the device behavior accurately. When the proposed modeling function is correct and the device is ideal, from the measured data we should obtain a linear function for the extracted argument of f_1 or f_2. If from the reverse extraction we can get two values of the argument, as is shown for the example function P_{si2} in Figure 5.12 (i.e., we have $\partial\psi^2/\partial V_{gs}^2 = 0$), this is an indication that our choice for modeling function is not very good. This is because the selected function P_{si2} will work in the simulations but will create problems in the extraction. This is valid also for the sub-functions responsible for the inter-coupling between f_1 and f_2. For example, if the function we guess is $y = Ax^2$, this will work well in simulation. But obviously there is a problem in the reverse extraction because the same value of y can be produced by two values of the argument $x = \pm\sqrt{y/A}$.

Quite often the device is not ideal and we need some flexibility to tune the model. It seems logical that a complex model is more likely to be accurate. This is correct, within limits, because we should always keep in mind that there are processing tolerances, and there is no sense making a model 1% accurate when process tolerances are 10%. The representation of the argument as a power series (APS) will offer the possibility of fitting a variety of devices. Fitting a polynomial function is a rather

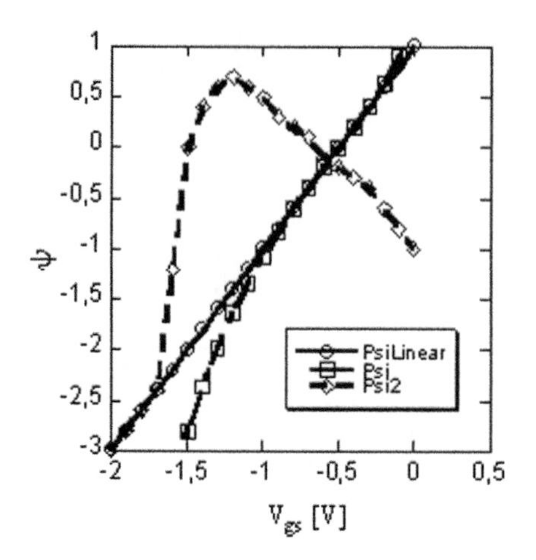

FIGURE 5.12

Extracted argument ψ vs. V_{gs}.

simple task, but even in this case, parameters of the APS should be selected properly. For example, when we have a negative second term in APS we should always add a positive third term and so on. This will exclude the possibility of a local maximum and dual argument reading and provide required trimming at pinch-off.

5.3.4 Capacitances model and implementation in simulators

In multiple extraction and physical simulations on different FET structures it was evaluated that the main device capacitances (C_{gs} and C_{gd}) are bias dependent on both voltages. This is normally expected, but the problem is how to implement this in circuit simulators. The charge implementation and conservation problem is very old, and several good works are devoted to the topic that propose solutions [41–43]. Traditionally FET total gate charge has been modeled by two nonlinear charges: gate-source Q_{gs} and gate-drain charge $Q_{gd.}$ A consequence of the dependence of the capacitances on the remote voltage is that we need an additional charge control element, which D. Root called transcapacitances [42,43].

There are several ways to implement the gate charges in two individual components: division by capacitances or division by charge [41].

As FET devices have both gate-to-source capacitance C_{gs}, and gate-to-drain capacitance C_{gd}, it seems natural to use them directly. In this case:

$$I_g = \frac{\partial Q_g}{\partial t} = \frac{\partial Q_g}{\partial V_{gs}}\frac{dV_{gs}}{dt} + \frac{\partial Q_g}{\partial V_{gd}}\frac{dV_{gd}}{dt} \tag{5.5a}$$

and

$$I_s = \frac{\partial Q_g}{\partial V_{gs}}\frac{dV_{gs}}{dt} \tag{5.5b}$$

$$I_d = \frac{\partial Q_g}{\partial V_{gd}} \frac{dV_{gd}}{dt} \tag{5.5c}$$

where I_s and I_d are reactive parts of the source and drain current, and

$$C_{gs} = \frac{\partial Q_g}{\partial V_{ds}} \tag{5.6a}$$

$$C_{gd} = \frac{\partial Q_g}{\partial V_{gd}} \tag{5.6b}$$

In the case we use capacitances in the implementation, the currents I_s and I_d depend only on the time derivative of their own terminal voltage and not on the changes in any remote voltage. The resulting small-signal equivalent circuit is completely consistent with the large signal equivalent circuit and requires no transcapacitances.

Another option is to divide the gate charge Q_g into two independent charges. Then

$$Q_g = Q_{gs} + Q_{gd} \tag{5.7}$$

where both Q_{gs} and Q_{gd} are functions of V_{gs} and V_{gd}. Differentiating Q_g with respect to time gives:

$$I_g = I_s + I_d \tag{5.8a}$$

$$I_s = \frac{\partial Q_{gs}}{\partial t} = \frac{\partial Q_{gs}}{\partial V_{gs}} \frac{dV_{gs}}{dt} + \frac{\partial Q_{gs}}{\partial V_{gd}} \frac{dV_{gd}}{dt} \tag{5.8b}$$

$$I_d = \frac{\partial Q_{gd}}{\partial t} = \frac{\partial Q_{gd}}{\partial V_{gs}} \frac{dV_{gs}}{dt} + \frac{\partial Q_{gd}}{\partial V_{gd}} \frac{dV_{gd}}{dt} \tag{5.8c}$$

In this case the reactive source and drain currents result from both capacitances and transcapacitances, and both definitions, charge and capacitance, are not equivalent.

A common approach to implement the charge part of a transistor model is to directly use the charge approach. In this case the current of the capacitance is easy to calculate by taking the time derivative of the charge, i.e., multiplying by $j\omega$. This operation is very reliable, because making the derivative will always produce only one solution. This works very well with capacitances that depend only on their own terminal voltage. The problem with all FET transistors is that the gate capacitance depends on two controlling voltages. When we multiply by $j\omega$ we are in fact making the full derivative of the charge, and the end result is not correct if the charge is obtained as integrating the capacitance equation by the terminal voltage. It is obvious that partial (considering the remote part constant) and full derivatives are different.

This can be shown with the case of the capacitance model using Eqns (5.9)–(5.14). Integrating the C_{gs} capacitance by the terminal voltage V_{gs}, we obtain

Eqn (5.15). It is assumed that V_{ds} part is constant. If the ordinary charge approach is used, multiplying by $j\omega$ will bring obviously different results. In fact we need to compensate for the difference due to the partial derivative, and therefore we need an extra term, the transcapacitance [41–43].

$$C_{gs} = C_{gsp} + C_{gs0}{}^{*}(1+\tanh(\psi_1))^{*}(1+\tanh(\psi_2)) \tag{5.9}$$

$$C_{gd} = C_{gdp} + C_{gd0}{}^{*}(1-P_{111}+\tanh(\psi_3))^{*}(1+\tanh(\psi_4)+2^{*}P_{111}) \tag{5.10}$$

$$\psi_1 = P_{10} + P_{11}{}^{*}V_{gs} + P_{111}{}^{*}V_{ds} \tag{5.11}$$

$$\psi_2 = P_{20} + P_{21}{}^{*}V_{ds} \tag{5.12}$$

$$\psi_3 = P_{30} - P_{31}{}^{*}V_{ds} \tag{5.13}$$

$$\psi_4 = P_{40} + P_{41} * V_{gd} - P_{111}{}^{*}V_{ds} \tag{5.14}$$

$$Q_{gs} = \int C_{gs}{}^{*}\partial V_{gs} = C_{gsp}{}^{*}V_{gs} + C_{gs0}{}^{*}\left(\psi_1 + L_{c1} - Q_{gs0}\right)^{*}(1+\tanh(\psi_2))/P_{11} \tag{5.15}$$

where

$$L_{c1} = \ln(\cosh(\psi_1))$$
$$Q_{gs0} = P_{10} + P_{111}{}^{*}V_{ds} + L_{c10}$$
$$L_{c10} = \ln(\cosh(P_{10} + P_{111}{}^{*}V_{ds}))$$

In some advanced circuit simulators and for compiled models, the derivatives of the charges are calculated analytically using the selected terminal voltage. Then the problem is solved in a better way. The CAD tool is making the derivative versus respective terminal voltage, considering the remote voltage constant. In this case we will have the capacitance described as a derivative of the charge at the terminal voltage, and the capacitances calculated by both methods should be similar.

In the first case we need a correct description of the charge that will compensate for the difference between the partial and full derivatives, otherwise the model will not be charge conservative. The consequence that the model is not charge conservative is that this difference will create additional current, implying that the solution will become path dependent and the harmonic balance of the circuit simulator will have difficulties to converge [42,43].

Figure 5.13 shows the typical shape of simulated C_{gs} and C_{gd} capacitances. When the device is symmetrical, for $V_{ds} = 0$ V, capacitances C_{gs} and C_{gd} are equal. For gate voltage values close to pinch-off, capacitances C_{gs} and C_{gd} have their minimum values C_{gspi} and C_{gdpi}, and this should be used in the capacitance models to define the capacitance at pinch-off. Cgs_infl and Cgd_infl are the capacitances at the inflection point from which the parameter P_{11} (Eqn (5.11)) can be derived. Increasing V_{gs} will increase C_{gs} and C_{gd}. Generally, when V_{ds} increases, C_{gs} will increase and saturate at voltages around $V_{ds} = 2$ V. Large FET devices can show linear change of C_{gs} vs. V_{ds} at high V_{ds}, and these problems should be addressed.

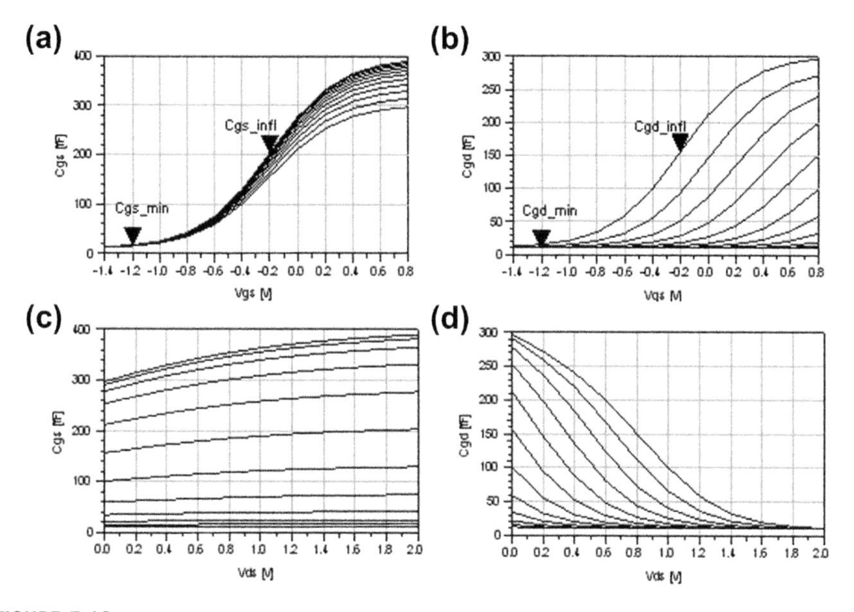

FIGURE 5.13

Simulated capacitance (a) C_{gs} vs. V_{gs}, V_{ds} as parameter, (b) C_{gs} vs. V_{ds}, V_{gs} as parameter, (c) C_{gd} vs. V_{gd}, V_{ds} as parameter, and (d) C_{gd} vs. V_{ds}, V_{gs} as parameter.

The capacitance C_{gd} decreases when V_{ds} increases, and for drain voltages around V_{knee}, the capacitance C_{gd} settles to a small value. The shape of capacitance dependencies will depend on the doping profile and material, and in some specific cases a special capacitance model can be developed.

A reasonably good description of the capacitance shape for FET can be obtained using Eqns (5.9)–(5.14) [26].

Independently of the implementation (capacitance or charge) and the type of model, in order to have the capacitance model charge conservative, it is mandatory to fulfill the following basic requirement:

$$\frac{\partial C_{gs}}{\partial V_{gd}} = \frac{\partial C_{gd}}{\partial V_{gs}} \tag{5.16}$$

This means that the equations for the capacitances C_{gs} and C_{gd} should be symmetrical, and the model coefficients should be selected properly. In the case of Eqns (5.9)–(5.14), this means that $P_{11} = P_{41}$ and $P_{21} = P_{31}$. The consequences can be non-convergence in HB analysis. A good test for the consistency of the capacitance models is to simulate the S parameters in the small-signal case, and on the other hand simulate S parameters in the LS case with HB, but with very small input power. If this difference is small, this means that the capacitance model is correct and has been implemented properly.

For capacitances described with Eqns (5.9)–(5.14), the charges are: Q_{gs} (see Eqn (5.15)) and Q_{gd}:

$$\begin{aligned} Q_{gd} &= \int C_{gd}{}^{*}\partial V_{gd} \\ &= C_{gdp}{}^{*}V_{gd} + C_{gd0}{}^{*}(\psi_4 + L_{c4} - Q_{gd0})^{*}(1 - P_{111} + \tanh(\psi_2))/P_{41} \end{aligned} \quad (5.17)$$

where

$$\begin{aligned} L_{c4} &= \ln(\cosh(\psi_4)) \\ Q_{gd0} &= P_{40} + P_{111}{}^{*}V_{ds} + L_{c40} \\ L_{c40} &= \ln(\cosh(P_{40} - P_{111}{}^{*}V_{ds})) \end{aligned}$$

Figure 5.14 shows the typical shape of the simulated Q_{gs} and Q_{gd} charges.

The functions for capacitances, charges, and their derivatives are symmetrical and defined from $-\infty < V_{gs}, V_{gd}, V_{ds} < +\infty$.

A problem that should be accounted for is the boundary condition problem, i.e., what will be the capacitances (charges) when the capacitance terminal is shorted and there is a voltage on the remote terminal. For example, when the gate source junction is shorted ($V_{gs} = 0$ V), the capacitance C_{gs} will continue to exist and the charge Q_{gs} should be equal to 0 C independent from the remote voltage V_{ds}. This puts additional

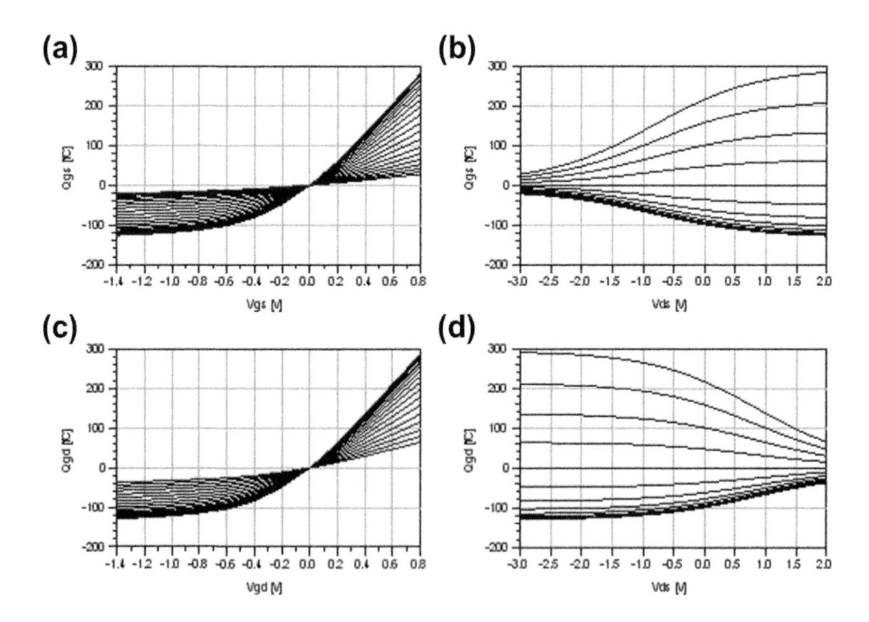

FIGURE 5.14

Simulated charge (a) Q_{gs} vs. V_{gs}, V_{ds} as parameter, (b) Q_{gs} vs. V_{ds}, V_{gs} as parameter, (c) Q_{gd} vs. V_{gd}, V_{ds} as parameter, and (d) Q_{gd} vs. V_{ds}, V_{gs} as parameter.

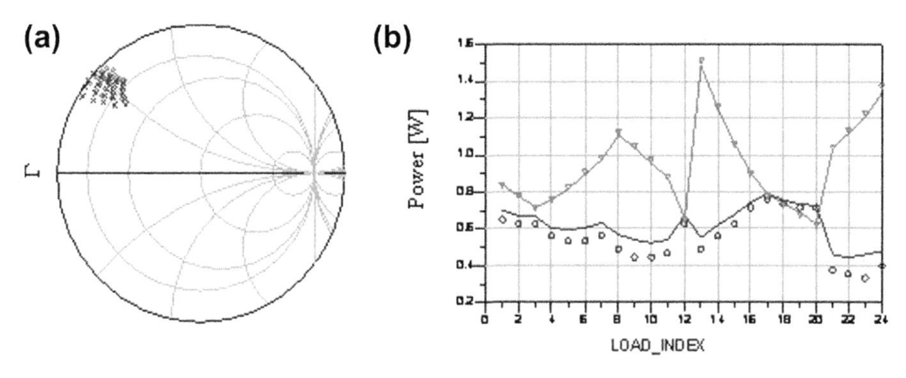

FIGURE 5.15

Load pull evaluation of GaN HEMT at 7 GHz, $V_{ds0} = 20$ V, and $P_{in} = 18$ dBm. (a) Measured (triangles) and simulated (crosses) load impedances, and (b) measured DC power (symbols) and modeled (line) DC (circle) and RF (triangle) power. *Load_index* refers to the load impedances in (a).

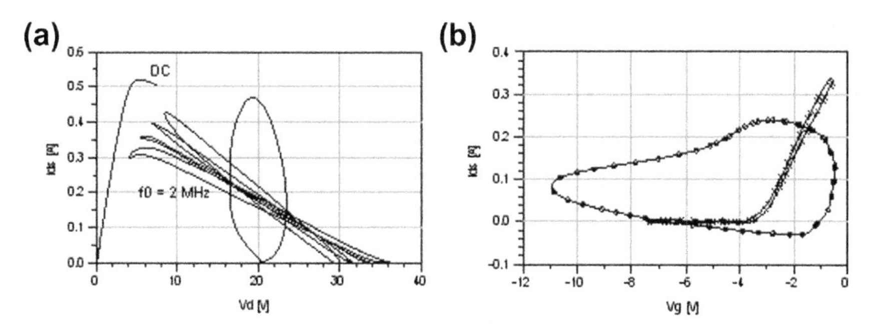

FIGURE 5.16

(a) Measured load lines at $f_0 = 2$ MHz and DC characteristic; (b) measured $i_d - v_g$ characteristic at the extrinsic plane (probe tips) of a GaN HEMT at $f_0 = 2$ MHz (crosses) and $f_0 = 4$ GHz (circles).

constraints on the boundary conditions for the charge definition. For these reasons some circuit simulators use separate Q_{gs}, Q_{gd}, but taking into account the boundary condition with charges Q_{gs0} and Q_{gd0}. As can be seen from Figures 5.19–5.21, when $V_{gs} = 0$ V, the charge $Q_{gs} = 0$ C and when $V_{gd} = 0$ V, the charge $Q_{gd} = 0$ C, independently of the remote voltage V_{ds}.

Generally, most circuit simulators use either the standard charge approach or direct capacitance approach.

It is important to know that there always should be some small difference in the calculated $S(Y)$ parameters depending on the implementation type, capacitance or charge, even if the same model parameters for the capacitances are used. The origin of this difference in the calculated S parameters depending on the implementation is

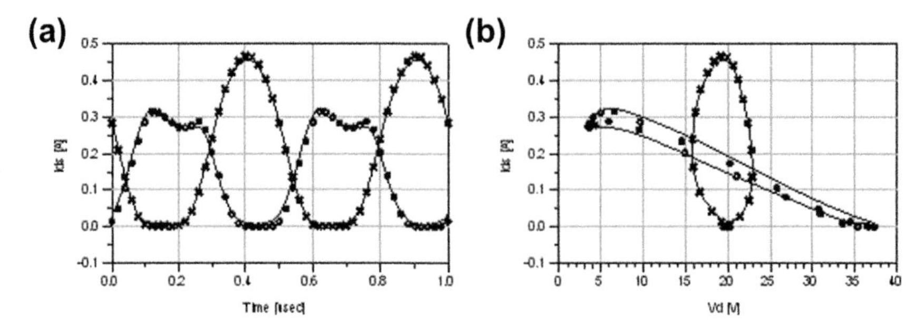

FIGURE 5.17

Time-domain waveform of drain-current of a GaN HEMT ($L = 0.7\ \mu m$, $W = 800\ \mu m$) (a) and corresponding load line (b) at: $V_{gsO} = -2$ V, $V_{dsO} = 20$ V, $f_O = 2$ MHz, and $P_{in,av} = 14.7$ dBm. $Z_L(f_O) = 1.6 + j11.6\ \Omega$ (crosses) and $Z_L(f_O) = 110.7 - j17.3\ \Omega$ (circles).

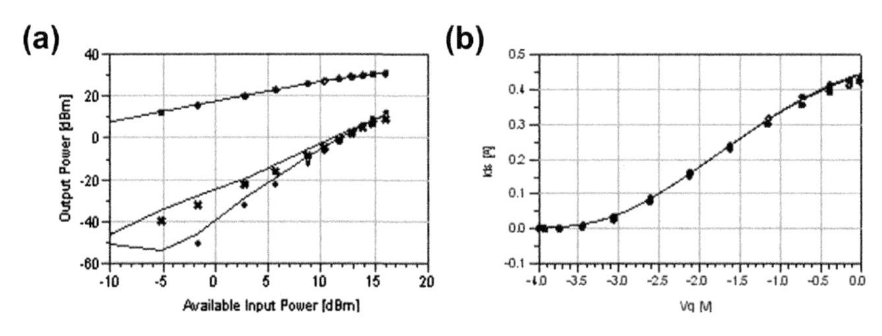

FIGURE 5.18

(a) Output power of a GaN HEMT ($L = 0.7\ \mu m$ and $W = 800\ \mu m$) at f_O (circles), $2f_O$ (crosses), and $3f_O$ (diamonds). (b) $i_d - v_g$ characteristic at the highest level of input power (14.7 dBm). operating conditions: $V_{gsO} = -2$ V, $V_{dsO} = 20$ V, and $f_O = 2$ MHz.

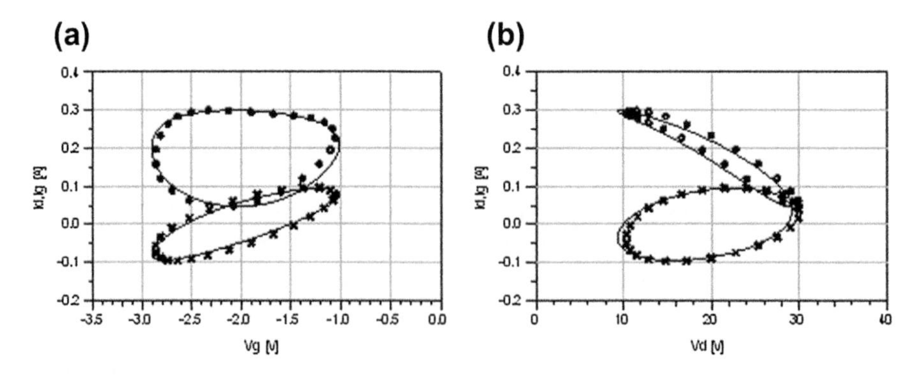

FIGURE 5.19

Input (a) and output (b) loci of a GaN HEMT ($L = 0.7\ \mu m$ and $W = 800\ \mu m$) at $V_{gsO} = -2$ V, $V_{dsO} = 20$ V, $f_O = 4$ MHz, $Z_L(4\text{ GHz}) = 77.1 + j16.2\ \Omega$, and $P_{in,av} = 21$ dBm.

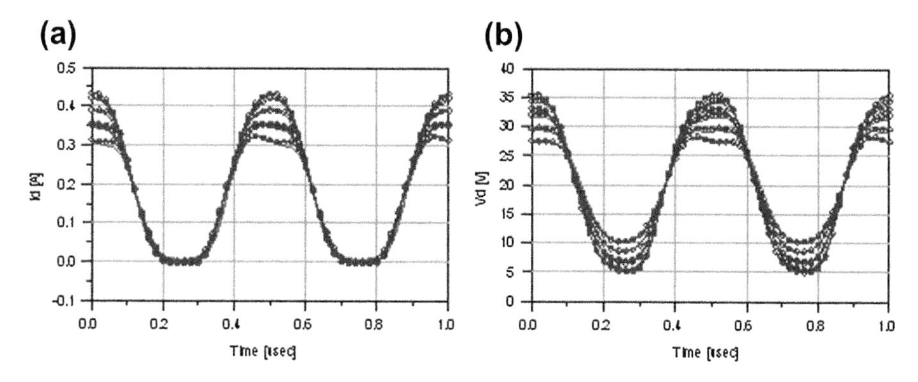

FIGURE 5.20

Current and voltage time-domain waveforms of GaN HEMT ($L = 0.7$ μm and $W = 800$ μm) after numerical optimization at $f_0 = 2$ MHz, $V_{gs0} = -2$ V, and $V_{ds0} = 20$ V for different load conditions. Measurements (symbols) and simulation (line).

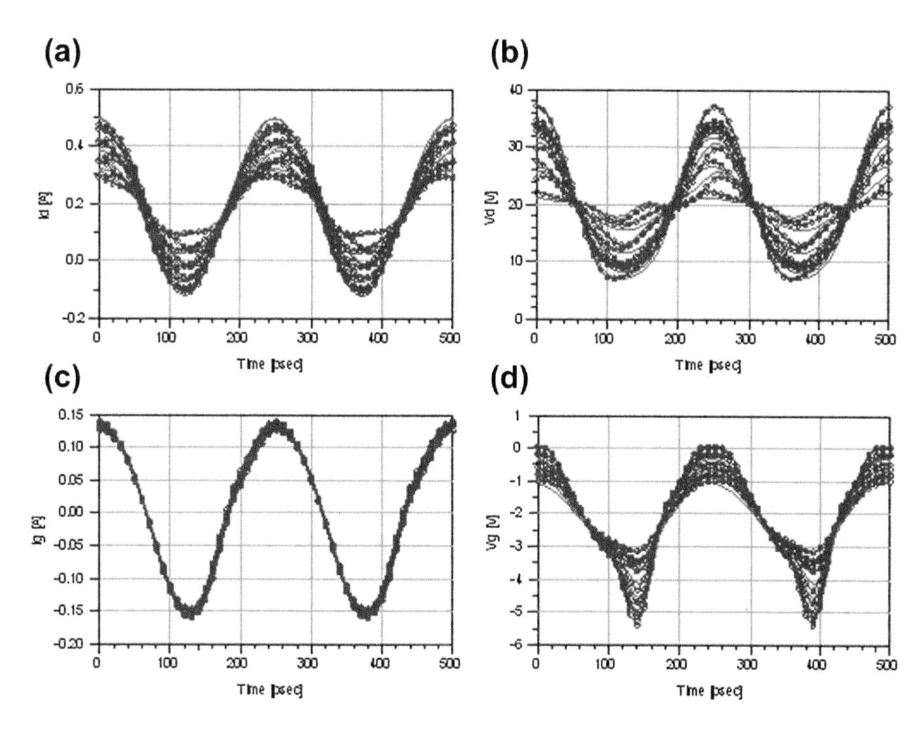

FIGURE 5.21

Current and voltage time-domain waveforms of GaN HEMT ($L = 0.7$ μm and $W = 800$ μm) after numerical optimization at $f_0 = 4$ GHz, $V_{gs0} = -2$ V, and $V_{ds0} = 20$ V for different loading conditions. Measurements (symbols) and simulation (lines).

very well described by S. Maas [41]. As a consequence, it is important to keep the same type of model in extraction and later in the circuit simulations, because this small difference can be accounted for by fitting the S parameters with the selected capacitance model and fulfilling the necessary condition Eqn (5.16).

5.4 Time-domain waveform-based models extraction

In the previous sections the general procedure for the extraction of the model parameters has been described. The adopted model is such that some of its parameters can be obtained directly from experimental data, e.g., DC measurements, for the nonlinear currents, and multibias S parameter measurements for the nonlinear capacitances and charges. Large-signal measurements can be also used for the parameter extraction. In this case the model equations are fitted against time-domain current and voltage waveforms at the DUT terminals. Nevertheless, one can still use a few DC measurements and S parameter measurements to determine the initial values of some of the model parameters. These initial values can be then used directly in the large-signal model and used as starting values for the fitting procedure against the experimental time-domain waveforms. In most of the examples reported in the following sections the large-signal model parameters are obtained as a result of numerical optimization against low- and high-frequency nonlinear measurements.

5.4.1 Modeling of currents and charges from high-frequency time-domain waveforms

The parameters nonlinear current and charge sources in the large-signal model of Figure 5.1 can be extracted from time-domain voltage and current waveforms measured at the transistor terminals [44]. In this way, the device under test is driven under operating conditions that are similar to the ones of the real-life application. In addition, the dynamic large-signal operation allows one to reach regions of the device operating area (e.g., breakdown, high-power dissipation) without damaging or degrading device characteristics.

Basically, the model parameters are estimated by numerical optimization, which runs in combination with a harmonic-balance solver [45]. It is worth observing that also the value of the parasitic elements can be included within the optimization procedure, provided that starting values are known. As mentioned in Section 5.2, well-established techniques are available in order to estimate starting values of the elements in the parasitic network. An example of a model extracted for a GaN HEMT is illustrated in Figure 5.15.

5.4.2 Low-frequency waveforms-based modeling of currents

Similarly to the approach in Section 5.4.1, time-domain waveforms measured in the low-frequency range (kHz to MHz) can be exploited to determine the parameter of only the current sources (i.e., $I-V$ functions). In fact, under low-frequency operation the displacement current generated by the intrinsic transistor capacitances can be neglected as compared with the one generated by the current sources. Moreover, all the reactive elements can be neglected in the parasitic network. In this way, the contribution of all the intrinsic and extrinsic reactive elements in the network shown in Figure 5.2 is minimized. In other words, the selection of the operating frequency in the kHz–MHz range enables one to "de-embed" linear and nonlinear reactive effects already within the experimental phase itself. Compared to the way it is classically intended, de-embedding here is not a post-measurement procedure, but it somehow happens within the experiment design. At low frequencies, the model in Figure 5.2 simplifies to the current sources, the resistive parasitic elements, and the thermal network only.

Therefore, the use of low frequency is advantageous as follows:

- First, as reported in Chapter 9, one can clearly highlight low-frequency dispersive phenomena related to charge trapping mechanisms. In fact, the latter significantly affect the behavior of the current sources (see Figure 5.16(a)) and, consequently, the measured low-frequency voltage and currents directly provide information about dispersion.
- Second, the identification of the parameters of the $I-V$ model can be split from the charge sources, thus enabling a simplified yet more consistent optimization procedure. At high frequency, instead, the actual contribution of the current sources can barely be observed in the measured voltages and current when the (nonlinear) effects of the transistor capacitances start to be significant, as illustrated in Figure 5.16(b).

Clearly, the dynamic characteristic measured at low frequency directly provides the response of the drain-source current generator. Therefore, the extraction of the basic parameters, as outlined in Section 5.3.2 can be straightforwardly performed. The same considerations apply for the gate current, which commonly is dominated at high frequency by the capacitive component, which masks the resistive contribution arising from the gate-source and gate-drain diodes.

Regarding the selection of the frequency of the excitations used for the nonlinear measurements, the following assumptions are made:

1. The frequency is above the cut-off of dispersive phenomena and thermal dynamics (Chapter 9). This is device dependent as trapping mechanisms as well as thermal properties depend on the semiconductor materials.
2. If (1) is verified, the behavior of the current source is considered to be independent on the operating frequency.

(1) and (2) are obviously approximations, which, nevertheless, hold in many practical cases involving III–V based devices.

In Figure 5.17 the modeled and measured low-frequency load lines of a GaN HEMT are reported. The fundamental frequency (f_0) of the CW excitations is 2 MHz. The load is swept by adopting an LF time-domain load-pull system

(Figure 5.1(b)) with CW signals applied at input and output ports of the device under test. For illustrative purposes in Figure 5.17 only two load lines are shown. For the extraction of the current source parameters, a larger number of load lines has been used. In Figure 5.18 the model is simulated at a fixed bias condition, $f_0 = 2$ MHz, and swept input power with the output port terminated with a 50 Ω load. In Figure 5.18(b) the $i_d - v_g$ dynamic characteristic is reported for the highest input power level (14.7 dBm).

Once the parameters of the current sources are determined, the high-frequency response is obtained by embedding the contribution of the nonlinear capacitances along with the parasitic network. These can be derived from multibias, multifrequency *S* parameter measurements. The values of the bias-dependent capacitances can be stored in a look-up table (LUT). The high-frequency response obtained by combining the LF *I*–*V* model and a LUT for the nonlinear capacitances is illustrated in Figure 5.19.

5.4.3 Combining low- and high-frequency time-domain waveforms

In this section the parameters of the equations representing the nonlinear current and charge functions are extracted by making use only of nonlinear measurements at different frequencies [46]. With respect to the previous section, the charge sources are now replaced with the charge model equations whose parameters are obtained by fitting high-frequency nonlinear measurements. As highlighted previously in the text, nonlinear measurements better represent the condition at which the device is going to operate in real life. Moreover, current and charge variation within semiconductor materials are caused by physics-related phenomena that are intrinsically nonlinear.

The extraction procedure consists of the following steps:

- Extraction of the parameters of the *I*–*V* model by starting from the measured LF voltage and current time-domain waveforms. As mentioned earlier, in the model of Figure 5.1 all the reactive contributions are neglected. Also, within this optimization step the resistive parasitic elements R_g, R_s, and R_d can be optimized;
- Extraction of the parameters of the nonlinear capacitances (C_{gs} and C_{gd}) or charge sources (Q_{gs} and Q_{gd}) starting from a set of high-frequency measurements. In this phase of the optimization procedure the complete model in Figure 5.1 is considered wherein the parameters of the *I*–*V* sources, R_g, R_s, and R_d are known as they are determined from LF measurements. Also within this step the values of the parasitic elements can be optimized.

Consequently, this procedure allows one to separately obtain the parameters of the *I*–*V* and *Q*–*V* functions, and in both cases starting from large-signal measurements. This procedure is rigorously valid when the assumptions made in Section 5.4.2 are valid and represent a reasonable approximation. In other words, the *I*–*V* functions are assumed to be independent on the operating frequency. If this condition were not satisfied, for instance due to the presence of fast trapping mechanisms

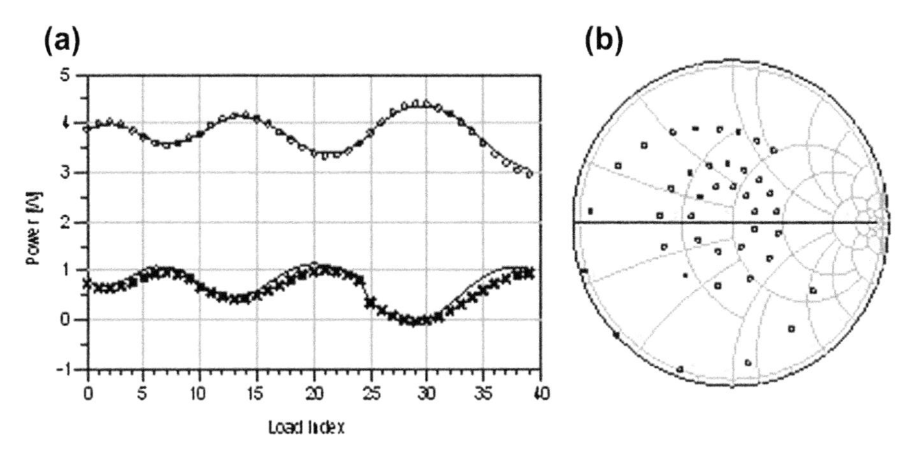

FIGURE 5.22

(a) Measured (symbols) and simulated (line) DC (circles) and RF power at $f_0 = 4$ GHz (crosses) of GaN HEMT ($L = 0.7$ μm and $W = 800$ μm) at various loading conditions as illustrated in (b).

with a time constant in the nano second range, the estimated parameters of the $I-V$ functions would have to be adjusted with further optimization in order to account for it.

In Figure 5.20 experiments and model predictions are compared. The $I-V$ model, R_s, and R_d are obtained from experimental load lines at 2 MHz. The parameters of the $Q-V$ functions are determined from load lines at $f_0 = 4$ GHz. Regarding the selection of the bias point in the case of dispersive devices [23], it is important to

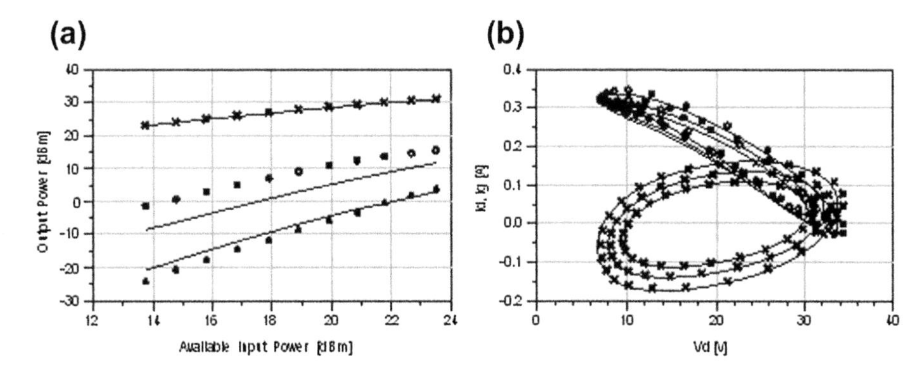

FIGURE 5.23

(a) Output power at f_0 (crosses), $2f_0$ (circles), $3f_0$ (triangles) and (b) $i_d - v_d$ and $i_g - v_d$ characteristics of a GaN HEMT ($L = 0.7$ μm and $W = 800$ μm) at $f_0 = 4$ GHz, $V_{gs0} = -2$ V, and $V_{ds0} = 20$ V. The load impedance is $77 + j^*16\ \Omega$. In (b) the characteristics at three levels of input power are shown.

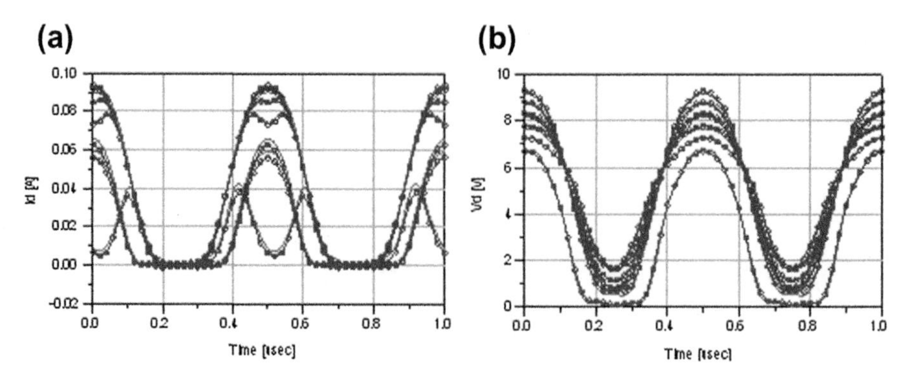

FIGURE 5.24

$i_d - v_g$ (a) and $i_d - v_d$ (b) characteristics of GaAs pHEMT ($L = 0.15$ μm and $W = 200$ μm) after numerical optimization at $f_0 = 1$ MHz. Measurements (symbols) and simulation (lines).

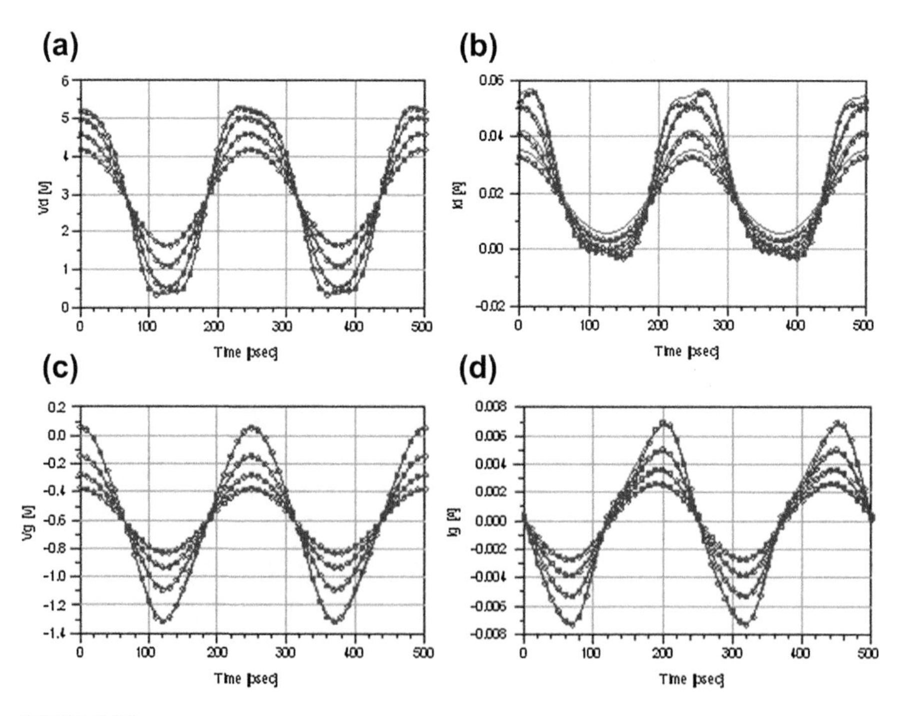

FIGURE 5.25

Measured (symbols) and simulated (line) voltage and current time-domain waveforms of GaAs pHEMT ($L = 0.15$ μm and $W = 200$ μm) at $f_0 = 4$ GHz, $V_{gs0} = -0.6$ V, and $V_{ds0} = 3$ V. Input power is swept. The load impedance at 4 GHz is equal to $83 + j{*}20\,\Omega$.

observe that in order to account for bias-dependent low-frequency dispersion either the model equations have to be modified [25] or the parameters of the $I-V$ functions, which are the most sensitive to dispersion, should be re-extracted when the bias point is changed.

Measured and simulated voltage and current time-domain waveforms after optimization against high-frequency load lines are illustrated in Figure 5.21.

The measured and simulated DC power and RF power at $f_0 = 4$ GHz corresponding to the time-domain waveforms in Figure 5.21 are shown in Figure 5.22. In Figure 5.23 measurements and simulation results are compared at different input power levels for a fixed load condition at the fundamental frequency. The $i_d - v_d$ and $i_g - v_d$ trajectories at three power levels are plotted as well.

Results obtained for a GaAs pHEMT are illustrated in Figures 5.24–5.27. For this device, the large-signal model is extracted from measurements at 1 MHz (see Figure 5.24) and 6 GHz. In Figures 5.25–5.27 model predictions are compared with experimental data. In Figure 5.25 the time-domain waveforms at $f_0 = 4$ GHz, $V_{gs0} = -0.6$ V, $V_{ds0} = 3$ V, and swept input power level are shown. In Figure 5.26

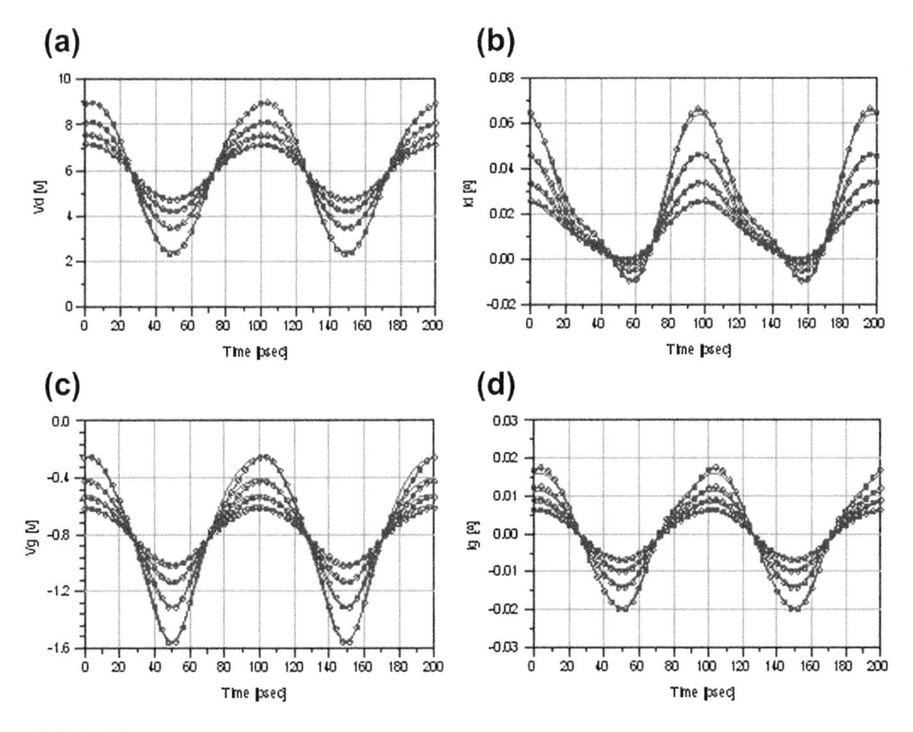

FIGURE 5.26

Measured (symbols) and simulated (line) voltage and current time-domain waveforms of GaAs pHEMT ($L = 0.15$ μm and $W = 200$ μm) at $f_0 = 10$ GHz, $V_{gs0} = -0.8$ V, and $V_{ds0} = 6$ V. Input power is swept. The load impedance at 10 GHz is equal to $90 + j^{*}34\ \Omega$.

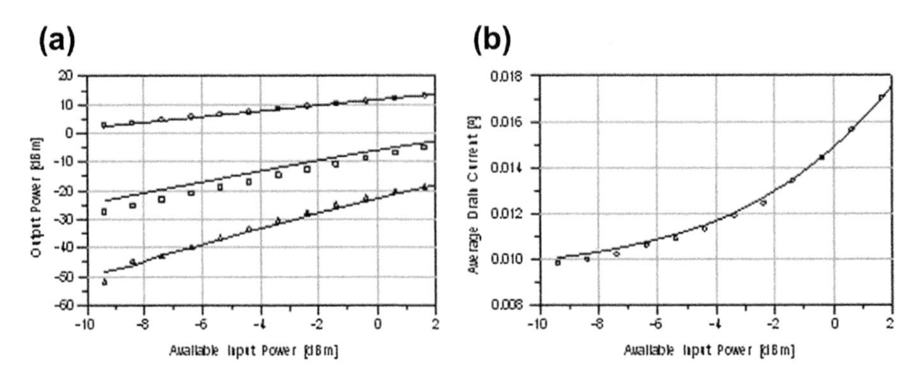

FIGURE 5.27

(a) Output power at $f_0 = 10$ GHz (circles), $2f_0$ (squares), $3f_0$ (triangles) and (b) average drain current a GaAs pHEMT ($L = 0.15$ μm, $W = 200$ μm). Measurements (symbols) and simulation (line).

and 5.27 the frequency is 10 GHz and the bias point $V_{gs0} = -0.8$ V and $V_{ds0} = 6$ V. For all the reported cases the model yields accurate predictions.

Acknowledgments

This work was supported by the GigaHertz Centre, Chalmers University of Technology, Sweden, FWO-Vlaanderen, Belgium, and KU Leuven GOA project.

References

[1] Rolain Y, Van Moer W, Vandersteen G, Schoukens J. Why are nonlinear microwave systems measurements so involved. IEEE Trans Instrum Meas June 2004;53(3):726–9.

[2] Van Moer W, Gomme L. NVNA versus LSNA: enemies or friends? IEEE Microwave Mag February 2010;11(1):97–103.

[3] Verspecht J. The return of the sampling frequency converter. In: 62nd Automatic RF Tech Group Conf (ARFTG) (Boulder, CO, USA), December 2003. pp. 155–64.

[4] Blockley P, Gunyan D, Scott JB. Mixer based, vector corrected, vector signal/network analyzer offering 300kHz–20GHz bandwidth and traceable phase response. In: IEEE MTT-S Int Microwave Symp Diges (Long Beach, CA, USA), June 2005. pp. 1497–500.

[5] Avolio G, Pailloncy G, Schreurs DMM-P, Vanden Bossche M, Nauwelaers B. On-wafer LSNA measurements including dynamic-bias. In: Eur Microwave Conf (Rome, Italy), September 2009. pp. 930–3.

[6] Raffo A, Di Falco S, Vadalà V, Vannini G. Characterization of GaN HEMT low-frequency dispersion through a multiharmonic measurement system. IEEE Trans Microw Theory Tech September 2010;58(9):2490–6.

[7] Curtice W. A MESFET model for use in the design of GaAs integrated circuit. IEEE Trans Microw Theory Tech May 1980;28(5):448–55.

[8] Materka A, Kacprzak T. Computer calculation of large-signal GaAs FET amplifiers characteristics. IEEE Trans Microw Theory Tech February 1985;33(2):129–35.

[9] Brazil T. A universal large-signal equivalent circuit model for the GaAs MESFET. In: Eur Microwave Conf (Stuttgart, Germany), September 1991. pp. 921–6.

[10] Angelov I, Rorsman N, Stenarson J, Garcia M, Zirath H. An empirical table-based FET model. IEEE Trans Microw Theory Tech December 1999;47(12):2350–7.

[11] Crupi G, Raffo A, Schreurs D, Avolio G, Vadalà V, Di Falco S, et al. Accurate GaN HEMT nonquasi-static large-signal model including dispersive effects. Microw Opt Technol Lett March 2011;53(3):710–18.

[12] Fernández-Barciela M, Tasker PJ, Campos-Roca Y, Demmler M, Massler H, Sánchez E, et al. A simplified broad-band large-signal nonquasi-static table-based FET model. IEEE Trans Microw Theory Tech March 2000;48(3):395–405.

[13] Crupi G, Schreurs DMM-P, Xiao D, Caddemi A, Parvais B, Mercha A, et al. Determination and validation of new nonlinear FinFET model based on lookup tables. IEEE Microw Wireless Compon Lett May 2007;17(5):361–3.

[14] Verspecht J, Horn J, Betts L, Gunyan D, Pollard R, Gillease C, et al. Extension of X-parameters to include long-term dynamic memory effects. In: IEEE MTT-S Int Microwave Symp (Boston, MA, USA), June 2009. pp. 741–4.

[15] Schreurs D, Verspecht J, Vandenberghe S, Carchon G, van der Zanden K, Nauwelaers B. Easy and accurate empirical transistor model parameter estimation from vectorial large-signal measurements. In: IEEE MTT-S Int Microwave Symp (Anaheim, CA, USA), June 1999. pp. 753–6.

[16] Carmen Currás-Francos M. Table-based nonlinear HEMT model extracted from time-domain large-signal measurements. IEEE Trans Microw Theory Tech May 2005;53(5):1593–600.

[17] Schreurs DMM-P, Verspecht J, Vanderberghe S, Vandamme E. Straightforward and accurate nonlinear device model parameter-estimation method based on vectorial large-signal measurements. IEEE Trans Microw Theory Tech October 2002;50(10):2315–19.

[18] Schreurs D, Verspecht J, Nauwelaers B, Van de Capelle A, Van Rossum M. Direct extraction of the non-linear model for two-port devices from vectorial non-linear network analyzer measurements. In: Eur Microwave Conf (Jerusalem, Israel), September 1997. pp. 921–6.

[19] Curras-Francos MC, Tasker PJ, Fernandez-Barciela M, Campos-Roca Y, Sanchez E. Direct extraction of nonlinear FET Q-V functions from time domain large signal measurements. IEEE Microwave Guided Wave Lett December 2000;10(12):531–3.

[20] Curras-Francos MC, Tasker PJ, Fernandez-Barciela M, O'Keefe SS, Campos-Roca Y, Sanchez E. Direct extraction of nonlinear FET I-V functions from time domain large signal measurements. Electron Lett October 1998;34(21):1993–4.

[21] Xu JJ, Gunyan D, Iwamoto M, Cognata A, Root DE. Measurement-based non-quasi-static large-signal FET model using artificial neural networks. In: IEEE MTT-S Int Microwave Symp (San Francisco, CA, USA), June 2006. pp. 469–72.

[22] Xu JJ, Horn J, Iwamoto M, Root DE. Large-signal FET model with multiple time scale dynamics from nonlinear vector network analyzer data. In: IEEE MTT-S Int Microwave Symp (Anaheim, CA, USA), May 2010. pp. 417–20.

[23] Raffo A, Vadalà V, Schreurs DMM-P, Crupi G, Avolio G, Caddemi A, et al. Nonlinear dispersive modeling of electron devices oriented to GaN power amplifier design. IEEE Trans Microw Theory Tech April 2010;58(4):710−18.

[24] Root D, Fan S, Meyer J. Technology-independent large-signal FET models: a measurement-based approach to active device modeling. In: 15th ARMMS Conf (Bath, UK), September 1991. pp. 1−21.

[25] Angelov I, Bengtsson L, Garcia M. Extensions of the Chalmers nonlinear HEMT and MESFET model. IEEE Trans Microw Theory Tech October 1996;46(11):1664−74.

[26] Advanced Design System (ADS) Agilent. Ansoft Designer User Manuals.

[27] Dambrine G, Cappy A. A new method for determining the FET small-signal equivalent circuit. IEEE Trans Microw Theory Tech July 1988;36:1151−9.

[28] Kunihiro K, Ohno Y. A large-signal equivalent circuit model for substrate-induced drain-lag phenomena in HJFET's. IEEE Trans Electron Dev June 1996;43(9):1336−42.

[29] Conger J, Peczalski A, Shur M. Modeling frequency dependence of GaAs MESFET characteristics. IEEE J Solid State Circuits January 1994;29(1):71−6.

[30] Scheinberg N, Bayruns N. A low-frequency GaAs MESFET circuit model. IEEE J Solid State Circuits April 1988;23(2):605−8.

[31] Canfield P. Modeling of frequency and temperature effects in GaAs MESFETs. IEEE J Solid State Circuits February 1990;25(1):299−306.

[32] Lee M, Forbes L. A self-back-gating GaAs MESFET model for low-frequency anomalies. IEEE Trans Electron Devices October 1990;37(10):2148−57.

[33] Camacho-Penalosa C, Aitchison C. Modeling frequency dependence of output impedance of a microwave MESFET at low frequencies. Electron Lett June 1985;21(12):528−9.

[34] Reynoso-Hernandez J, Graffeuil J. Output conductance frequency dispersion and low-frequency noise in HEMT's and MESFET's. IEEE Trans Microw Theory Tech September 1989;37(9):1478−81.

[35] Ladbrooke P, Blight S. Low-field low-frequency dispersion of transconductance in GaAs MESFETs with implication for other rate-dependent anomalies. IEEE Trans Electron Dev March 1988;35(3):257−67.

[36] Teyssier JP, Campovecchio M, Sommet C, Portilla J, Quere R. A pulsed S-parameter measurement set-up for the nonlinear characterization of FETs and bipolar transistors. In: Eur Microwave Conf (Madrid, Spain), September 1993. pp. 489−93.

[37] Manohar S, Pham A, Evers N. Direct determination of the bias-dependent series parasitic elements in SiC MESFETs. IEEE Trans Microw Theory Tech February 2003;51(2):597−600.

[38] Sommer V. A new method to determine the source resistance of FET from measured S-parameters under active-bias conditions. IEEE Trans Microw Theory Tech March 1995;43(3):504−10.

[39] Palacios T, Rajan S, Chakraborty A, Heikman S, Keller S, DenBaars PS, et al. Influence of the dynamic access resistance in the gm and fT linearity of AlGaN/GaN HEMTs. IEEE Trans Electron Dev October 2005;52(10):2117−23.

[40] Campbell CF, Brown SA. An analytic method to determine GaAs FET parasitic inductances and drain resistance under active bias conditions. IEEE Trans Microw Theory Tech July 2001;49(7):1241−7.

[41] Maas S. Nonlinear microwave and RF circuits (Dedham, MA, USA): Artech House; January 2003.

[42] Root D, Hughes B. Principles of nonlinear active device modeling for circuit simulation. In: 32nd Automatic Radio Frequency Tech Group (ARFTG) Conf (Tempe, AZ, USA), December 1988. pp. 3–26.

[43] Root D. Nonlinear charge modeling for FET large-signal simulation and its importance for IP3 and ACPR in communication circuits and systems. IEEE Midwest Symp, Vol. 2. (Dayton, OH, USA), August 2001. pp. 768–72.

[44] Angelov I, Andersson K, Schreurs D, Xiao D, Rorsman N, Desmaris V, et al. Large-Signal Modeling and Comparison of AlGaN/GaN HEMTs and SiC MESFETs. APMC (Yokohama, Japan), December 2006. pp. 279–82.

[45] Bandler J, Zhang QJ, Ye S, Chen SH, et al. Efficient large-signal FET parameter extraction using harmonics. IEEE Trans Microw Theory Tech December 1989;37(12):2099–108.

[46] Avolio G, Schreurs D, Raffo A, Crupi G, Angelov I, Vannini G, et al. Identification technique of FET model based on vector nonlinear measurements. Electronics Lett November 2011;47(24):1323–4.

[illegible]

[illegible]

[illegible]

[illegible]

[illegible]

CHAPTER

6 Measuring and Characterizing Nonlinear Radio-Frequency Systems

Wendy Van Moer, Lieve Lauwers, Kurt Barbé
Dept. ELEC, Vrije Universiteit Brussel, Brussels, Belgium

6.1 Introduction

During the last few decades, linear system theory dominated all engineering applications. Almost all engineers are raised in the well-established linear framework: a lot of techniques are available to measure and model linear devices. In this linear framework, everything that does not behave linearly or cannot be linearized is treated as a perturbation.

However, the continuous pressure of low power and high bandwidth applications pushes an increasing number of devices beyond the edges of their linear range into the nonlinear operation region. Consider the example of a demanding customer who wants an optimal working and cheap-priced mobile phone with a large autonomy. Most of these demands can be granted by increasing the efficiency of the transistors. The problem, however, is that increasing the efficiency of a transistor means pushing its operating point closer to the compression region. Since compression is synonymous with nonlinearities, it becomes essential to acquire insight into the nonlinear behavior of the component. By getting familiarized with the nonlinear world, it will hence be easier to model the nonlinear behavior of radio-frequency integrated circuits (RFICs). This example illustrates that it is important to first explore the nonlinear behavior in the measured data before starting the actual modeling step.

Pushed by the modeling world, which clearly needs good nonlinear measurements to build good nonlinear models, different nonlinear measurement instruments were developed to correctly measure the nonlinear time-domain waveforms. The two major measurement principles being pursued are the sampler-based and the mixer-based methodologies.

In this chapter, the working principle of both types of measurement instruments and their differences, as well as their advantages and disadvantages, will be highlighted. Next, the calibration procedure for nonlinear measurements is discussed. To obtain accurate high-frequency measurements, a calibration procedure is mandatory. However, for nonlinear measurement instruments the calibration procedure becomes even more involved. This is due to the fact that relative measurements are no longer sufficient, but the knowledge of the absolute time-domain waveforms is required. Hence, besides a relative calibration, two additional calibration steps need to be performed: a power and a phase calibration. The process of

Microwave De-embedding. http://dx.doi.org/10.1016/B978-0-12-401700-9.00006-9

de-embedding or mathematically shifting the reference calibration planes as close as possible to the ports of the device under test (DUT) requires an additional calibration step in the case of on-wafer measurements. This is due to the fact that until the present time no on-wafer power and phase calibration elements have been available [1].

The measurements of nonlinear behavior can then be used to gain more insight into the DUT by recycling the good ideas from the linear world. A linear transfer function is well known and has proven its efficiency for many decades. Therefore, we will reuse this established idea and introduce the so-called best linear approximation. The theory behind this concept will be elaborated on and some examples will be given. The classical in-band best linear approximation is extended to an out-of-band best linear approximation in order to reduce the adjacent co-channel power ratio of radio-frequency (RF) amplifiers.

In conclusion, two important questions will get a clear answer in this chapter: "How to measure the nonlinear behavior of RF devices?" and "How to obtain a simple and robust characterization of the nonlinear behavior of an RF system?"

6.2 Measuring the nonlinear behavior of an RF system

In this section, the differences between a sampler- and a mixer-based measurement instrument will be explained, as well as the pitfalls that may be encountered when measuring the nonlinear behavior of RF systems.

6.2.1 Sampling and calibration issues

The major problem when measuring high-frequency signals resides in the fact that it requires a very fast analog-to-digital conversion (ADC). Measuring signals up to 50 GHz means that, according to Nyquist [2], sampling frequencies of at least 100 GHz are required. These high sampling rates are simply not available or very expensive. To circumvent this sampling problem, the RF signals are first down converted to much lower intermediate frequency (IF) signals, which can then be digitized by much slower ADCs. Two major methodologies can be followed to down convert RF signals to IF signals: a sampler-based and a mixer-based methodology. A sampler-based measurement instrument uses samplers to perform the down conversion and is based on the harmonic sampling principle [2]. A mixer-based measurement instrument uses the heterodyne principle [3] to down convert the RF signals to the IF spectrum.

On top of the down conversion issue, an additional problem pops up when measuring the nonlinear behavior of an RF device. Indeed, contrary to the classical relative S-parameter measurements that are used to characterize the linear behavior of an RF device, the knowledge of the absolute time-domain waveforms is now required [4]. In other words, to measure the nonlinear behavior, one must be able to measure the absolute magnitude of the spectral components as well as the absolute phase relationships between the harmonics present in the measured spectra. As a result, a relative calibration, as used in the case of linear vector network analyzer measurements, is no longer sufficient. Two additional calibration steps are required: a power and a phase calibration [4].

In the following, a detailed answer will be given to a frequently posed question amongst RF engineers: "What exactly is the difference between a sampler-based and a mixer-based nonlinear measurement instrument?" Next, the extended calibration procedure for these two types of nonlinear measurement instruments will be presented. The focus in this chapter is put on the basic calibration steps needed for nonlinear measurement results. The reader is referred to Chapter 1 to gain more insight in the differences between a linear and nonlinear de-embedding technique.

6.2.2 Sampler-based measurement instruments

Eager to explore the nature of nonlinearities, the first prototype of a sampler-based measurement instrument was developed by the industry in 1993: the Large-Signal Network Analyzer (LSNA) [4]. The goal was to build an absolute wavemeter that allows capturing the whole wave spectrum in one single take. The LSNA was the first instrument able to measure the absolute magnitude of the waves as well as the absolute phase relations between the measured harmonics. Hence, the LSNA can be seen as an absolute fast Fourier transform (FFT) analyzer for microwaves. Figure 6.1 shows a simplified block schematic of a two-port LSNA to perform connectorized continuous wave measurements.

The DUT can be excited at one or both ports by an RF generator (input source). The incident and reflected waves at both ports of the DUT are then measured through couplers, which have a bandwidth from about 500 MHz to 50 GHz. The high-frequency content of the signals does not allow digitizing these signals immediately. Therefore, the measured RF spectrum is down converted to an IF spectrum by using harmonic sampling [2]. This part of the setup is referred to as the down convertor of the LSNA (Downc, gray box in Figure 6.1) and is in fact the key component of the instrument. The down convertor is based on a microwave transition analyzer (MTA) [5]; four fully synchronized RF data acquisition channels are available. Before the waves are down converted, attenuators (ATT) can be used to bring the signal level at the input of the down convertor below −10 dBm. This is necessary to prevent the down convertor from being pushed into its nonlinear operation region. As a result, the linearity of the LSNA is better than 60 dBc. After down conversion, the measured data can be amplified and digitized by four synchronized ADC cards. These ADC cards sample the data at a rate of 25 MHz and have a usable bandwidth of 10 MHz. The four ADC cards, the down convertor, and the RF generator are clocked by a common 10 MHz reference clock in order to obtain a fully synchronized phase coherent measurement instrument.

It took a few years before the market was convinced of the power of this type of nonlinear measurement instrument. But since then many other companies have developed their own sampler-based measurement instrument, such as VTD (Verspecht-Teyssier-Degroote [6]) or Mesuro [7]. The latter is based on a Tektronix

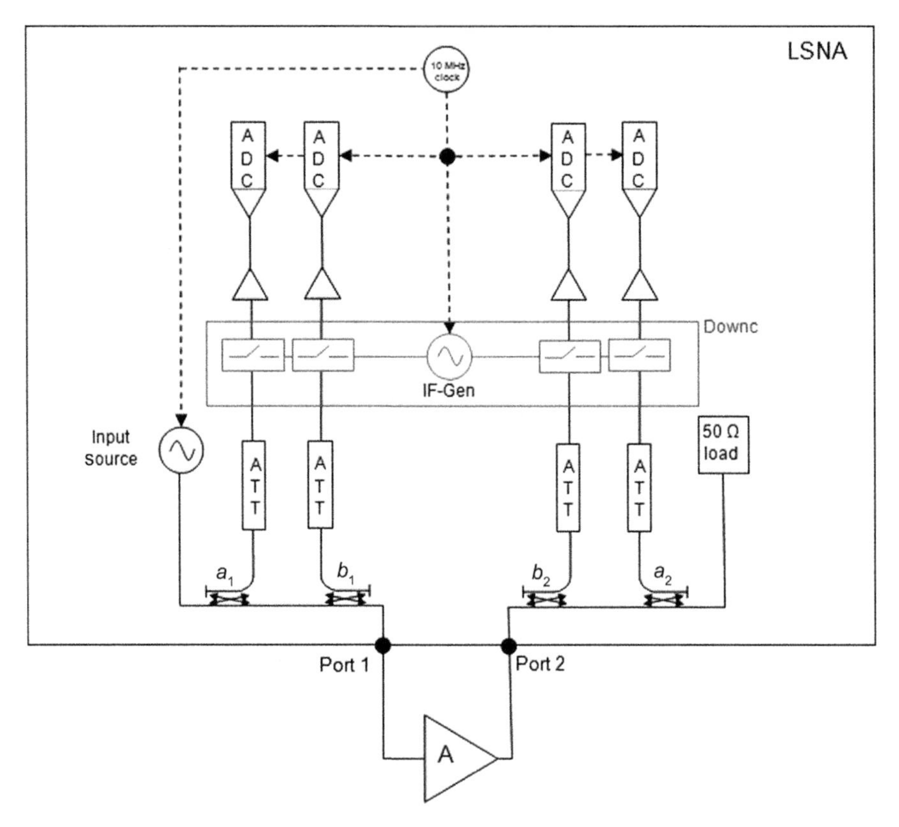

FIGURE 6.1

Simplified block schematic of a two-port LSNA.

digital serial analyzer sampling oscilloscope [8] that can measure frequencies up to 67 GHz with a dynamic range of 50 dB.

6.2.3 Mixer-based measurement instruments

A mixer-based measurement instrument does not use samplers to down convert the RF signals to an IF spectrum but instead uses mixers for the down conversion process. Hence, this type of nonlinear measurement instrument is based on the heterodyne principle [3]. While the sampler-based approach measures the whole wave spectrum in one single take, the mixer-based approach measures only one frequency component at a time. As a result, each measured spectral line needs then to be down converted separately, and all spectral lines must be correctly "stitched" together, especially with respect to the phase.

A mixer-based measurement approach can for instance be found in the nonlinear vector network analyzer (NVNA) [9]. Note that many other companies

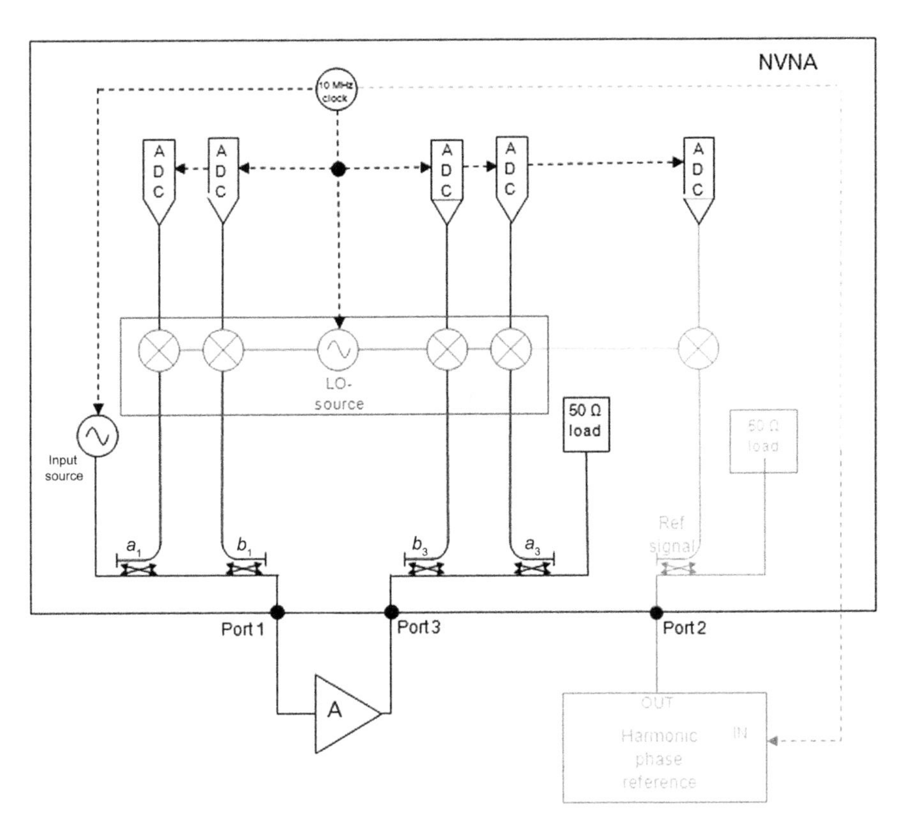

FIGURE 6.2

Simplified block schematic of a mixer-based measurement instrument (LO represents the local oscillator signal).

have built their own mixer-based measurement instrument, but they all have one thing in common: the design always starts from a classical vector network analyzer (such as that discussed in [10]).

In Figure 6.2, a simplified block schematic of a mixer-based measurement instrument is shown. The DUT can be excited at one or both ports. All four waves are measured through couplers and down converted by means of mixers that are all driven by the same local oscillator (LO) signal (see dark gray box in Figure 6.2). It should be noted that when sweeping its frequency, the phase of the LO signal will change by an unknown amount. As long as this phase change cannot be controlled, the phase relationships between the different harmonic components cannot directly be measured by using only undivided measured waves (a_1, b_1, a_3, b_3). To overcome this problem, relative measurements are required that compare the measured waves with a reference signal (a_1/ref, b_1/ref, a_3/ref, b_3/ref). This reference signal is generated by a harmonic phase reference (see light gray box in Figure 6.2). When the harmonic phase reference is excited with a fundamental tone, its output signal consists of an impulse signal. This means that in the frequency domain all harmonic components of the input fundamental tone of the phase reference are generated. Note that the cross-frequency phase relationship is static: the phase relationship between the different harmonic components of the phase reference signal remains constant over time. This static property allows restoring the phase relationships between the harmonic components present in the measured waves. The phase reference clock can be 10 MHz or can be driven by an external clock locked to a common 10 MHz reference.

As can be seen from Figure 6.2 (light gray part), a mixer-based instrument contains a fifth mixer that is used to select and to down convert the desired harmonic component generated by the harmonic phase reference. After down conversion by the mixers (dark gray box in Figure 6.2), the measured signals are digitized and the data can be represented in the time or frequency domain.

6.2.4 Calibration procedure for nonlinear measurement instruments

High-frequency measurement instruments are far from ideal and, hence, always introduce systematic errors due to imperfections. As a result, accurate high-frequency measurements not only require a good measurement instrument but also an accurate calibration procedure, which removes these systematic errors. A classical calibration procedure consists of connecting well-known and stable elements (standards) to the measurement instrument in order to determine the measurement error of the instrument.

Calibrating a classical vector network analyzer belongs to the common knowledge of each RF engineer. The most well-known procedure is the short-open-load-thru (SOLT) calibration [11,12]. Other types of relative *S*-parameter calibrations are the thru-reflection-line (TRL) and the load-reflect-match (LRM) calibration [11,12]. After applying such a relative calibration to a sampler- or a mixer-based measurement instrument, it can be used as a classical vector network analyzer to obtain *S*-parameter measurements. Unfortunately, *S* parameters do not fully describe a nonlinear system, such that the knowledge of the separate incident and reflected waves is required when measuring nonlinear devices. As a result, it will no longer be sufficient to calibrate wave ratios. Hence, the calibration procedure for nonlinear instruments needs to be extended by two additional steps: a power and a phase calibration [4].

In summary, the calibration procedure of a nonlinear (sampler- or mixer-based) measurement instrument consists of the following three steps, which need to be performed for all measured frequencies:

1. *Classical relative calibration*: SOLT, TRL, LRM, ...
2. *Power calibration*:
   During the power calibration, a calibrated power meter is connected to one of the ports of the instrument in order to measure the absolute power flowing into the

DUT. This allows setting the absolute amplitude of the measured time-domain waveforms [4].

3. *Phase calibration*:
   A third calibration step is required to obtain the phase relationships between the harmonics present in the measured spectra. The phase calibration is based on a known harmonic phase reference. The first harmonic phase reference available on the market was based on step-recovery diodes to produce the reference pulse. Nowadays, a monolithic microwave integrated circuit can be used to generate the required reference pulse. However, whatever technology is used, the phase relationships between the frequency components of this reference signal are assumed to be exactly known. Measuring the reference signal and comparing the measured phase relationships with the known phase relationships allows correcting the measured signals of the DUT in phase. The harmonic phase reference itself can be calibrated by two different approaches: a nose-to-nose procedure [8] or an electro-optical sampling (EOS)-based calibration [13,14]. Especially for frequencies above 20 GHz, the EOS approach gives the most accurate results.

6.2.5 On-wafer calibration

The two additional calibration steps for nonlinear measurement instruments are seemingly only valid for connectorized measurements, since the power meter and the harmonic phase references are connectorized devices. Unfortunately, on-wafer power meters and harmonic phase reference are not (yet) available. As a result, when calibrating on-wafer measurements, an alternative approach is required that allows using the connectorized standards for an on-wafer calibration [1]. Thereto, the setup of Figure 6.3 will be used. The idea consists of interconnecting port 1 and port 2 by means of an on-wafer THRU-element. The auxiliary port 3 will then be used to connect the power meter and the harmonic phase reference indirectly to port 1. In order to reconstruct the signals at port 1, a SOLT calibration is first performed between port 2 and port 3, which results in the transmission matrix between

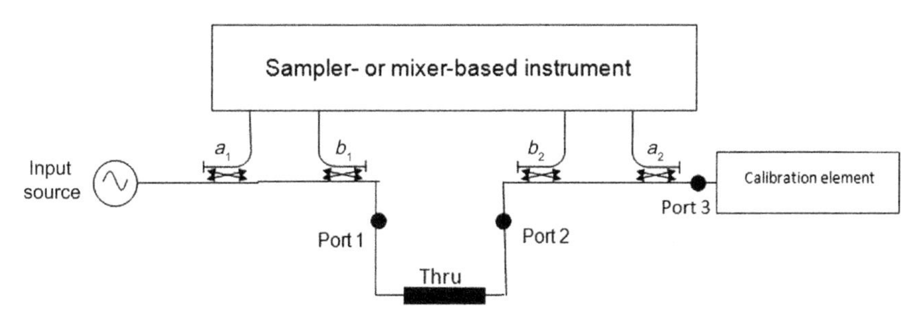

FIGURE 6.3

On-wafer calibration setup.

port 2 and port 3 [1]. This transmission matrix allows calibrating the waves at the reference plane of port 1 and port 2.

6.2.6 Advantages and disadvantages

The sampler- and mixer-based measurement principles are both indispensable if one wants to characterize the nonlinear behavior of RF components or systems. However, they both have advantages and disadvantages.

1. *Sampler-based instruments*

   Advantages:

- The phase relationship between the measured harmonics can be calibrated and is thus absolutely known. As a result, the time-domain waveforms can be perfectly reconstructed.
- Since the whole spectrum of the incident and reflected waves is measured in one single take, there simply are no phase synchronization problems between the spectral components, and the measurement time is very short.

   Disadvantage:

- The signal-to-noise ratio (SNR) of the sampler-based instruments (QUOTE $\sim$60 dB) is much smaller than the SNR of a classical vector network analyzer ($\sim$100 dB). However, one should be careful when comparing these two numbers since they do not say everything. The SNR of the vector network analyzer has been optimized for linear DUTs and cannot simply be copied for the mixer-based nonlinear measurement instruments, since the linearity of the receivers should be taken into account when measuring the nonlinear behavior of a DUT.

2. *Mixer-based instruments*

   Advantages:

- The phase relationship between the measured harmonics can be calibrated and is thus absolutely known. As a result, the time-domain waveforms can be perfectly reconstructed.
- The phase calibration can be performed on a grid spacing of 625 kHz.
- The minimal tone spacing between calibrated modulation tones is 625 kHz, without the need for interpolation algorithms.
- The noise floor of the mixer-based instruments is comparable to the noise floor of a classical vector network analyzer (approximately -120 dBm, depending on the instrument settings such as IF bandwidth). However, when measuring the nonlinear behavior of a DUT, one must take into account the linearity of the receivers of the mixer-based instruments.

   Disadvantages:

- Since each spectral line is measured separately, special precautions are needed to obtain phase synchronization between the spectral components. This has been solved by means of an additional harmonic phase reference.
- The measurement time is longer due to the fact that each spectral line is measured separately. The measurement time is also influenced by the chosen IF bandwidth, or in other words, by the SNR.

6.3 Best linear approximation and nonlinear in-band distortions

Measuring the nonlinear behavior of a DUT is one thing; characterizing the nonlinear behavior of the DUT is a completely different story. This section will introduce the best linear approximate model of a nonlinear system, which can be considered as its linearized transfer function.

6.3.1 Approximate modeling

Currently available "white-box" models that use physical insight and knowledge of the system's internal structure are accurate at the component level. However, they mostly fail when used as a quantitative description of the operation of the RF circuit as a whole. The major drawback of white-box models is that their building process has to be restarted for every new RF design. An alternative modeling approach that does not rely on prior system knowledge is the black-box approach. The mathematical model is then built from observed input—output measurements. For example, modeling a loudspeaker by a black-box approach can be achieved by proposing an input—output relation taking the form of a high-order transfer function.

When modeling nonlinear systems, engineers like to start from the well-known linear framework. Linear approximations are a powerful tool to first gain some insight into the system's behavior. For this reason, we introduce the concept of the best linear approximation (BLA) of a nonlinear system. This frequently used black-box model can be considered as the linearized transfer function of the nonlinear system.

6.3.2 Best linear approximation: the concept

It is generally known that a linear, time-invariant, two-port system satisfies the following equation:

$$B(f) = S(f)A(f) \tag{6.1}$$

where $A(f)$ and $B(f)$ represent, respectively, the incident and reflected waves at the ports of the DUT (stacked as a 2×1 vector of complex numbers and given as a function of the frequency f), and $S(f)$ denotes the two-by-two linear scattering matrix. For a nonlinear system, it is no longer possible to satisfy the relationship in (Eqn (6.1)) exactly. However, as long as the system behaves mainly linear, the relation between $A(f)$ and $B(f)$ can be approximated by a linear scattering matrix. To impose the existence of such a linear approximation, we need to narrow the considered class of nonlinear systems. Hence, we restrict ourselves to the class of so-called nonlinear PISPO systems, i.e., systems that, when excited with a periodic input signal, return a periodic output signal with the same period as the input signal. This means that subharmonics, chaotic behavior, and hysteresis phenomena are excluded from the framework.

When a nonlinear PISPO system is excited with a spectrally rich large-signal (such as for instance Gaussian noise, a multisine signal, ...), the system's response can always be approximated by a linear scattering matrix consisting of the following contributions:

$$S(f) = S_O(f) + S_B(f) + S_S(f) + N_S(f) \tag{6.2}$$

Herein,

- S_O is the scattering matrix of the underlying linear system
- S_B is the systematic deviation between the underlying linear behavior and the compression (or expansion) characteristic due to odd-order nonlinear behavior
- S_S is the stochastic nonlinear contribution
- N_S is the measurement noise.

The first two terms form the best linear approximation (BLA) [15]:

$$S_{BLA}(f) = S_O(f) + S_B(f) \tag{6.3}$$

of the nonlinear PISPO system. This linear time-invariant approximation is called "best" in the sense that it minimizes the mean square error between the true system's response and the modeled response for a particular class of inputs:

$$S_{BLA} = \arg\min_S E\left\{[B(t) - S(A(t))]^2\right\} \tag{6.4}$$

with $A(t)$ and $B(t)$ the measured input and output waves, respectively, and S the linear transfer function operator. Hence, there is no other linear time-invariant system that approximates the nonlinear system with a smaller mean square error.

The BLA can be obtained in a nonparametric way by performing classical frequency response function (FRF) measurements [16]:

$$S_{BLA}(f_k) = \frac{P_{BA}(f_k)}{P_{AA}(f_k)} \tag{6.5}$$

where $P_{BA}(f_k)$ is the cross-power spectrum between the output B and the input A at frequency line f_k, and $P_{AA}(f_k)$ is the auto-power spectrum of the input. Note that since the BLA is only determined at the excited frequencies, S_{BLA} remains unknown outside the excited frequency band.

In Figure 6.4, the alternative representation of a nonlinear single-input, single-output (SISO) system is shown.

The linearized transfer function S_{BLA} models (1) the underlying linear behavior of the nonlinear system and (2) the systematic contributions of the nonlinear behavior to the system's response. Furthermore, two additional output noise sources are present: one to describe the stochastic contributions of the nonlinear behavior of the system $S_S(f)$ and the second to describe the measurement noise $N_S(f)$.

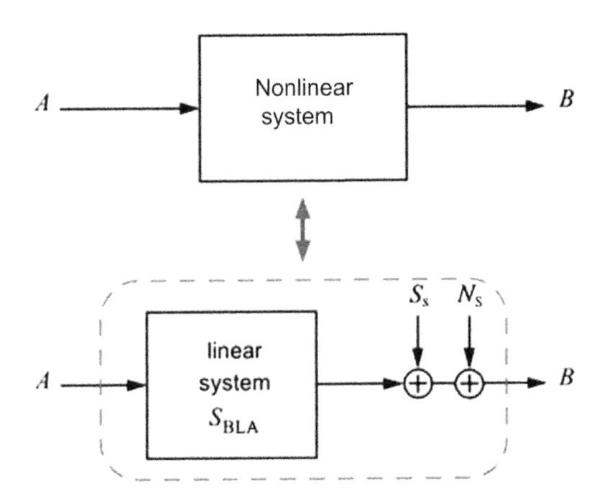

FIGURE 6.4

A nonlinear system with its alternative BLA representation.

6.3.3 Excitation signals

Since the BLA depends on the probability density function (pdf) of the input signal [17], we need to specify the considered class of excitation signals. We limit ourselves to the class of Gaussian-like excitation signals with a fixed power spectrum. Two signals belonging to this class are Gaussian noise and the random phase multisine.

In order to determine the BLA of a nonlinear device, the type of excitation signal is crucial. Since one needs to separate the stochastic nonlinear contributions $S_S(f)$ from the best linear approximation $S_{BLA}(f)$ (see Eqn (6.2)), the excitation signal should be stochastic. Indeed, when the DUT is excited by a stochastic excitation signal, the stochastic nonlinear contributions $S_S(f)$ act as a stochastic noise source with zero mean and a nonzero variance. However, Gaussian noise (whose noise contribution changes over time) does not lend itself easily as an input signal to microwave measurements due to its non periodicity. Hence, another stochastic excitation signal is needed. Since the theory of the BLA does not require the excitation signal to be stochastic over time, it can be stochastic over "another variable," for instance, over the different realizations of the excitation signal. An example of such an excitation signal is a random-phase multisine:

$$x(t,r) = \frac{1}{\sqrt{N}} \sum_{k=1}^{N} X_k \cos\left((k+k_0)2\Pi f_0 t + \phi_k^r\right) \tag{6.6}$$

in which X_k are the deterministic amplitudes of the different sine waves; ϕ_k^r are the random phases that are uniformly distributed in the interval $[0, 2\pi]$ and independent over different realizations r; f_0 is the fundamental frequency of the multisine. This multisine signal is periodic and consists of a sum of N sine waves that are commensurate in frequency. Each line has a fixed power, but its phase is a random variable, whose value changes for each realization of the signal [15].

Note that these random-phase multisine signals have a Gaussian probability density function. The use of Gaussian excitation signals when determining the BLA has a major advantage: the statistical properties of the stochastic nonlinear contributions are well known. For instance, the stochastic nonlinear contributions are independent of the frequencies [15]. This is very important, since the influence of the different nonlinear contributions on each other vanishes over the frequency and, hence, they can be treated as separate noise sources.

When determining the BLA of a system it is also important that the input signal resembles the actual communication signal applied to the DUT as much as possible. For most modern communication systems, such signals have a Gaussian-like pdf [18–20]. One of the major advantages of multisines over real communication signals is that they are periodic and hence easily measurable. Random phase multisines can therefore be seen as the most appropriate excitation signals to determine the BLA.

6.3.4 Estimating the best linear approximation

To obtain the best linear approximation with random-phase multisine signals, the experiment design shown in Figure 6.5 can be followed [21]. On top of determining the BLA, this scheme also allows making a distinction between the effect of the nonlinear distortions and the measurement noise on the BLA due to the periodicity of the excitation signal.

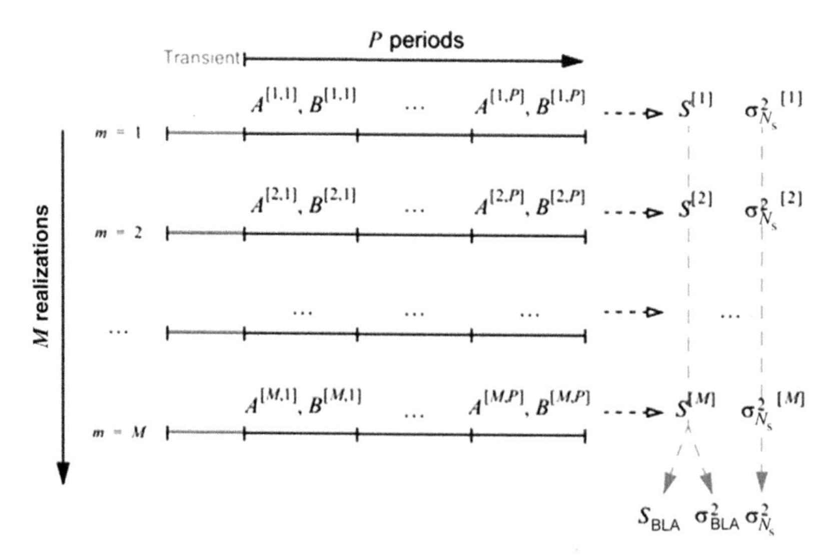

FIGURE 6.5

Experiment design to determine the stochastic nonlinear contributions and the measurement noise.

Different realizations (i.e. different phases ϕ_k^r in successive realizations) of a random phase multisine need to be applied to the DUT. After the transients have faded out, the linearized transfer function is obtained by taking the mean over the different phase realizations and time periods for each frequency f_k.

$$S_{BLA}(f_k) = \frac{1}{M}\sum_{m=1}^{M} S(f_k)^{[m]} \tag{6.7}$$

Note that variations over the periods are caused by measurement noise, and variations over the realizations are due to the combined effect of the measurement noise and the stochastic nonlinear behavior. Hence, the variance on the BLA denotes the combined effect of the stochastic nonlinear behavior and the measurement noise:

$$\sigma_{BLA}^2(f_k) = \sigma_{S_S}^2(f_k) + \sigma_{N_S}^2(f_k) \tag{6.8}$$

Calculating the variance over the different time periods returns a value for the measurement noise. A value for the variance $\sigma_{S_S}^2$ due to the stochastic nonlinear contributions can hence be obtained by observing the difference between σ_{BLA}^2 and $\sigma_{N_S}^2$.

6.3.5 Measurement example

As an illustration, the in-band BLA of a microwave amplifier is shown in Figure 6.6.

The amplifier is excited by a large-multisine signal generated by an arbitrary waveform generator. The multisine signal consists of 500 spectral components

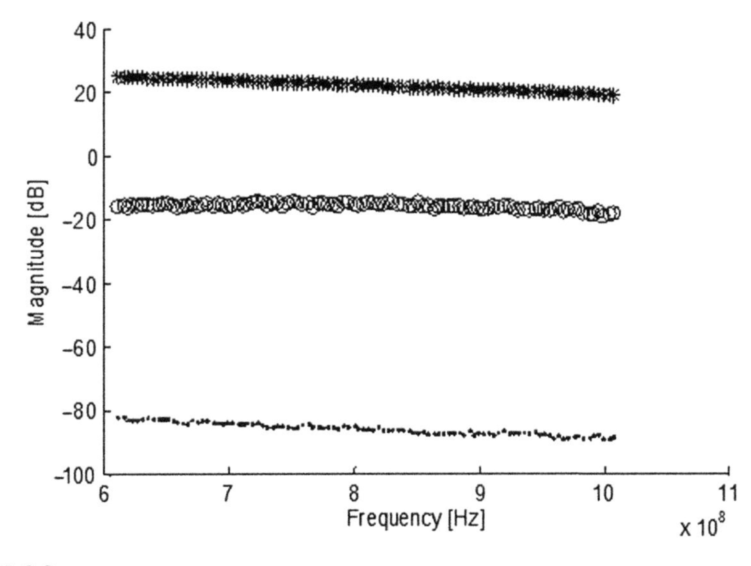

FIGURE 6.6

In-band BLA of a microwave amplifier: linearized transfer function (stars), stochastic nonlinear contributions (circles) and the measurement noise (dots).

between 700 MHz and 1 GHz. The peak voltage amplitude of the multisine in the time domain is 2 V.

The incident and reflected waves at both ports of the amplifier were measured by means of a four-port network analyzer (PNA-X N5242A, 10 MHz–26.5 GHz) which had been calibrated by an 8-error terms load-open-short-thru calibration. This measurement was repeated for 100 phase realizations of the multisine in order to obtain a measure for the stochastic nonlinear contributions. For each phase realization, 10 periods were measured. This allowed determining the measurement noise on the measurements by taking the variance of the spectral lines as obtained from the spectrum of each period separately. The stars represent the linearized transfer function $\widehat{S}_{\mathrm{BLA}}(f_{\mathrm{k}})$, while the circles denote the standard deviation of the stochastic nonlinear contributions $\sigma^2_{S_{\mathrm{S}}}(f_{\mathrm{k}})$ and the dots denote the measurement noise $\sigma^2_{N_{\mathrm{S}}}(f_{\mathrm{k}})$. It can be concluded that the amplifier was driven in compression, since the level of the stochastic nonlinear contributions is 60 dB higher than that of the measurement noise.

6.4 Out-of-band best linear approximation

The classical best linear approximation gives a lot of insight into the linear as well as the nonlinear behavior of a device inside the excitation band. However, no out-of-band information can be obtained. In this section, an extended measurement technique is proposed to extract the in-band as well as the out-of-band BLA of an RF device. The proposed measurement technique is based on a vector network analyzer measurement combined with a broadband multisine excitation of the DUT.

6.4.1 Out-of-band best linear approximation: the concept

Recall that the BLA of a nonlinear DUT is only valid inside the excited frequency band, since the linearized transfer function can only be determined in a nonparametric way at the excited frequencies. Hence, the BLA remains unknown outside the excited frequency band.

From a design point of view, this bandwidth limitation can hamper the use of the in-band BLA as a complete high-level design model for nonlinear systems. Indeed, the behavior of the system outside the excited frequency range is often required up to, for example, three times the excited frequency band in order to assess the importance of out-of-band interferers and blockers or to predict the influence of harmonic spectral content in the signal chain. On the other hand, the use of a linear time-invariant high-level design model significantly reduces the numerical evaluation cost and increases the ease of interpretation of the design properties. Hence, a similar concept to the in-band BLA is desired in order to gain information about the system's behavior outside the excited frequency band: the out-of-band BLA.

6.4.2 Determining the out-of-band best linear approximation

The most straightforward approach to extract the behavior of the system outside the excited frequency band is to extrapolate the in-band best linear approximation. However, it is generally known that extrapolation can lead to large errors and reduce the predictive capability of the model. In this case, the extrapolation approach will indeed be erroneous since the excited frequency band can be—and often is—very narrow compared to the bandwidth of the DUT.

An alternative but sound approach to extract out-of-band information is to perform an experiment that consists of exciting the system jointly by a large input signal and an additional excitation signal that lies outside the frequency band where the DUT is used. The additional excitation should be "small enough" to ensure that its contributions to the nonlinear behavior can be neglected. In other words, the system's nonlinear operation point may not be changed significantly by the additional excitation. The only purpose of this "small" input signal is to extract the BLA outside the excited frequency band.

In practice, for the in-band input signal a random phase multisine is chosen that ensures the nonlinear "bias" setting of the DUT. For the additional excitation, a small sinusoidal signal is used that is swept along the frequency band of interest to allow calculating the response of the circuit to spurious out-of-band signals.

The verification of whether or not the amplitude of the sinusoidal signal is "small enough" in order not to influence the nonlinear behavior can be done by calculating the BLA starting from in-band BLA measurements (multisine excitation) and by comparing it to the in-band BLA obtained from the out-of-band experiment (multisine + small sinusoidal signal). A stochastic-based comparison then detects whether the uncertainty bounds on both BLAs overlap. As a rule of thumb, one can allow a deviation between both in-band BLAs of 0.001% in a noise-free experiment.

In order to obtain the out-of-band BLA, one needs to perform repeated measurements and calculate the mean over the different time periods as was explained in Section 6.3.4 for the in-band BLA.

Although the measurements of the out-of-band BLA result in a quantification of the nonlinear behavior of the DUT, one only needs to perform linear, relative measurements. There is no need for "absolute" wave measurements; hence, no specialized measurement equipment (LSNA or NVNA) is required. The measurements that are required to obtain the BLA are performed by a classical vector network analyzer (VNA). Note that this is in contrast to the *X* parameters [22,23], which are an extension of the classical *S* parameters to large-signal conditions and which require the use of an NVNA. The measurement principle to determine the *X* parameters uses a large single sine wave in combination with a small probing sine wave that is swept over the different harmonics. Also in this case, the large sine wave fixes the nonlinear operation point of the DUT, while the small sine wave allows scanning the harmonics of interest.

6.4.3 Measurement example

In this section, the out-of-band BLA of an RF amplifier operating with and without compression is measured.

1. *Device under test*
   The DUT is a two-port microwave amplifier (Motorola, MRFIC-2006) with a usable bandwidth between 800 MHz and 1 GHz. The gain factor of the amplifier is about 23 dB at 900 MHz.
2. *Measurement setup*
   The incident and reflected waves at both ports of the DUT are measured by the measurement principle presented in Figure 6.7. The measurement setup consists of a four-port network analyzer (PNA-X N5242A, 10 MHz–26.5 GHz) and an arbitrary waveform generator. Note that the network analyzer is used in such a configuration that it is able to measure time-domain waveforms instead of *S* parameters. In order to make a distinction between the effect of the stochastic nonlinear distortions and the measurement noise on the best linear approximation, the measurement setup must allow measuring different phase realizations of the large excitation signal as well as a number of synchronized time periods of the applied signals. To measure the correct time-domain waveforms of the multisine, the frequency grid of both the network analyzer and the multisine signal must fall on top of each other.
3. *Measurement experiments*
   The DUT is excited by a small-signal and a large-multisine signal, called the "pump" signal. The small-signal consists of a single sine wave generated by the

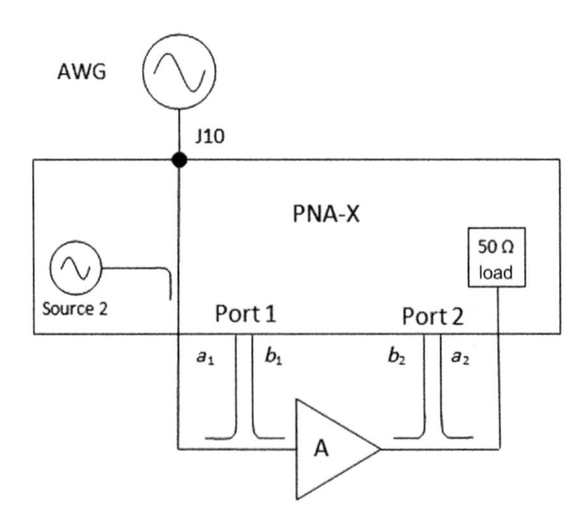

FIGURE 6.7

Measurement setup.

internal source of the network analyzer and is swept from 10 MHz to 2.41 GHz in 2401 frequency steps of 1 MHz. The power of the sine wave is kept constant at −50 dBm. The low power of this signal is chosen in order not to influence the nonlinear behavior of the DUT in the frequency band of interest. The "pump" signal is a multisine signal generated by the arbitrary waveform generator. The multisine signal consists of 500 spectral components between 700 MHz and 1 GHz. The peak voltage amplitude of the multisine in time domain is 0.5 V.

During the measurement experiments, the input port of the amplifier is excited simultaneously by the small-signal and the multisine signal. The output port of the amplifier is terminated in a 50 Ω load impedance. The incident (a_1 and a_2) and reflected (b_1 and b_2) waves are measured by the network analyzer. This measurement is repeated for 100 phase realizations of the multisine in order to obtain a measure for the stochastic nonlinear contributions. For each phase realization, 10 periods are measured. This allows determining the measurement noise on the measurements by taking the variance of the spectral lines as obtained from the spectrum of each period separately. The vector network analyzer is calibrated by an eight-error terms load-open-short-thru calibration. The IF bandwidth of the VNA is set to 100 Hz.

4. *Measurement results*
   a. *Linear transfer function*
      A first experiment consists of only applying the small sine wave signal to the input of the amplifier, without the multisine pump signal. This allows measuring the small-signal transfer function of the amplifier as can be seen in Figure 6.8.

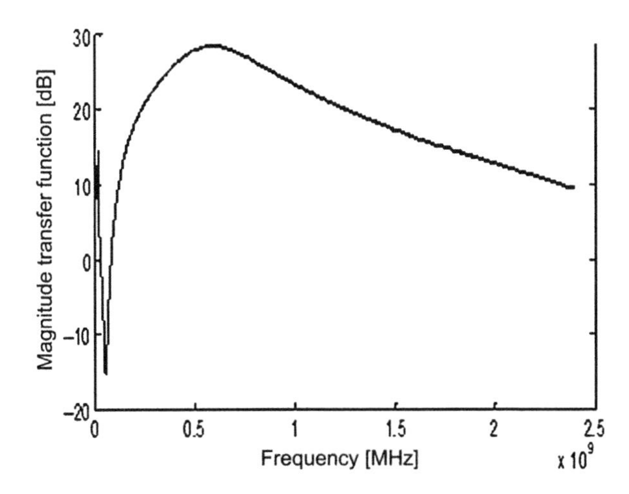

FIGURE 6.8

Linear transfer function of the amplifier.

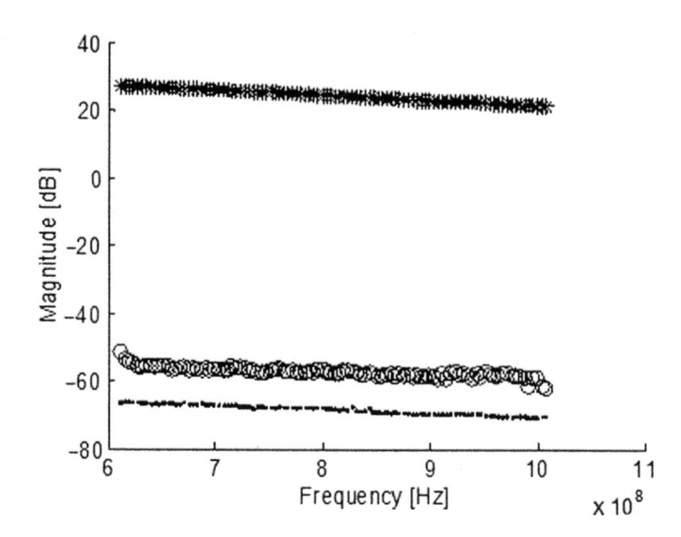

FIGURE 6.9

In-band BLA of the amplifier in linear regime: linearized transfer function (stars), stochastic nonlinear contributions (circles), and the measurement noise (dots).

b. In-band BLA measurement
In a second experiment, the amplifier is excited by the multisine signal only. Figure 6.9 shows the results for the amplifier measurements in a linear regime (the peak voltage amplitude of the multisine in the time domain is 0.5 V). The results of the amplifier in compression are shown in Figure 6.10 (the peak voltage amplitude of the multisine in the time domain is 5 V).

In both cases, the stars represent the linearized transfer function, while the circles denote the standard deviation of the stochastic nonlinear contributions, and the dots denote the measurement noise.

One can clearly see that when the amplifier goes into strong compression, the level of nonlinear contributions becomes much larger than the measurement noise.

Note that the measurement noise is lower for the experiment with compression than for the experiment without compression. This is due to the very small-signal amplitudes that were measured in the experiment without compression.

c. Extracting the in- and out-of-band BLA
In a third experiment, the combined multisine—sine wave excitation is applied to the amplifier. The measurement is repeated for 100 phase realizations of the multisine and for each phase realization, 10 periods are measured.

The upper black curve in Figure 6.11 represents the out-of-band BLA of the amplifier in linear regime. The lower black curve represents the out-of-

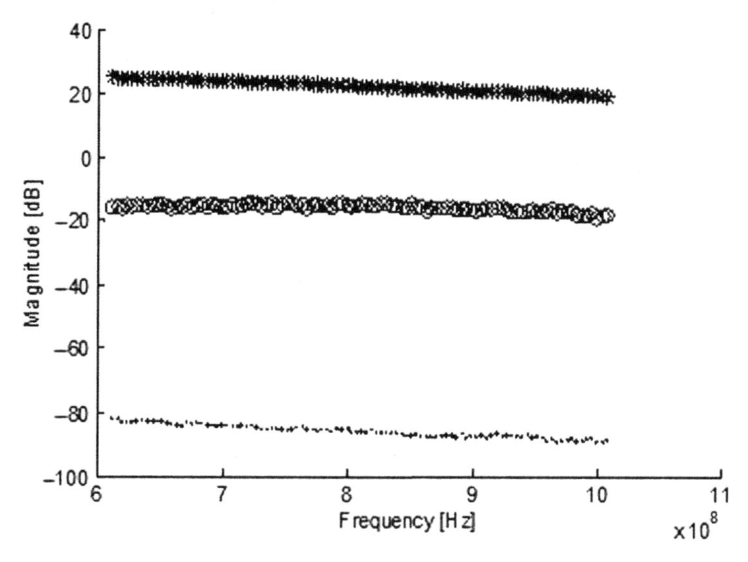

FIGURE 6.10

In-band BLA of the amplifier under compression: linearized transfer function (stars), stochastic nonlinear contributions (circles), and the measurement noise (dots).

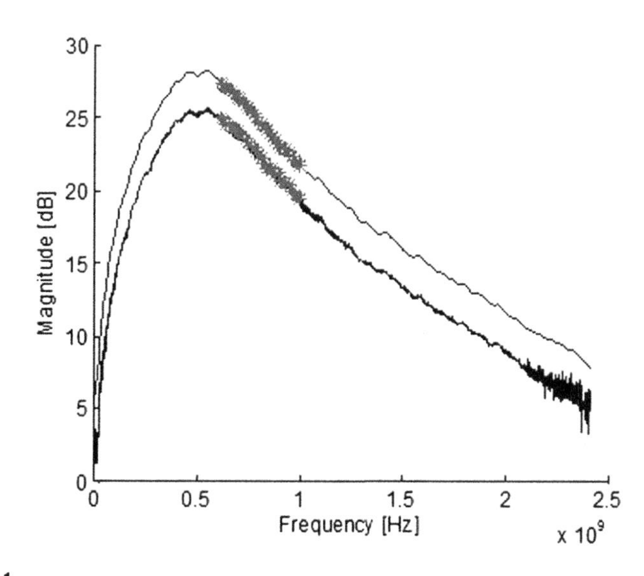

FIGURE 6.11

Out-of-band BLA of the amplifier in linear regime (upper black curve) and compression (lower black curve). The gray stars on each curve represent the in-band BLA out of the multisine experiment.

band BLA of the amplifier in compression. The gray stars on the upper black curve represent the in-band BLA of the amplifier in linear regime, while the gray stars on the lower black curve represent the in-band BLA of the amplifier in compression, obtained by only applying the multisine excitation.

The amplifier is pushed into compression by augmenting the power of the multisine. The power of the sine wave excitation remains the same when performing the measurements of the amplifier in compression and in linear regime.

From Figure 6.11 one can clearly see the influence of the level of nonlinearities on the out-of-band BLA. In case of compression, the BLA (lower black curve in Figure 6.11) shows a much "noisier" behavior than for the linear operation (upper black curve on Figure 6.11) mode. This "noisy" behavior is due to the nonlinearities that arise. For instance, between 2.1 GHz and 2.5 GHz the behavior is very noisy due to the presence of the third harmonic component, which becomes larger when the compression becomes stronger.

Note that these out-of-band BLAs could never have been measured without the presence of the multisine excitation. The multisine excitation is responsible for setting the nonlinear operating point of the amplifier.

Figure 6.12 shows a zoom of Figure 6.11 for the first 56 components of the multi-sine. One can clearly see that the respective linearized transfer functions from the two experiments, with (full black curve) and without sine wave excitation (gray stars), fall on top of each other, and this is valid not only for the amplifier in linear regime but also for the amplifier in compression.

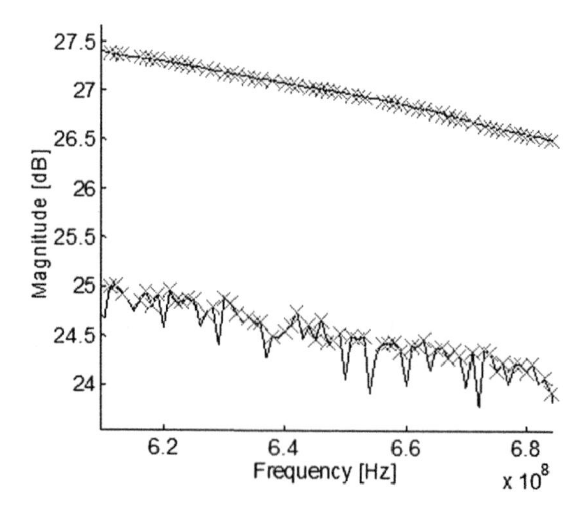

FIGURE 6.12

Zoom of Figure 6.11 for the first 56 components of the multisine.

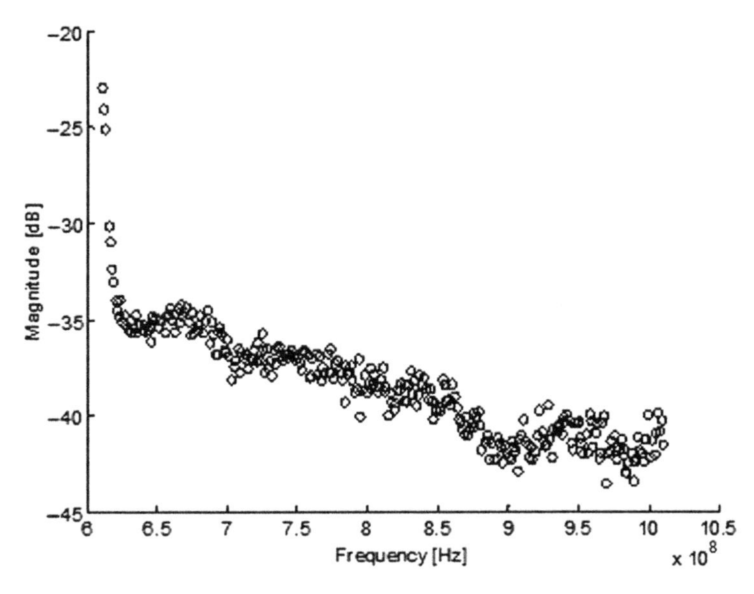

FIGURE 6.13

Difference (circles) between the in-band BLA of the multisine experiment and the in-band BLA of the combined multisine–sine wave experiment.

Figure 6.13 represents the comparison between both in-band BLAs of Figure 6.12. The circles represent the magnitude of the complex difference between both in-band BLAs, which is approximately −40 dB. This can be considered small enough in order to conclude that the sinusoidal signal does not influence the nonlinear behavior of the amplifier.

6.5 Compensating nonlinear out-of-band distortions

Based on the out-of-band BLA, it becomes possible to compensate the nonlinear output distortions of an RF device for the nonlinear distortions present in the input signal. As a result, a source-pull free level of the nonlinear output distortions is obtained, which is mandatory for determining an accurate adjacent co-channel power ratio.

6.5.1 Adjacent co-channel power ratio

Frequency bandwidth is very scarce and expensive nowadays. Thousands of telecommunication applications are used daily, and they all require their own piece of bandwidth. Sharing these limited frequencies requires that each application does not "leak" into the frequency band of the other. In practice, this "frequency leaking" is caused by the nonlinear behavior of components and systems. Hence, when designing RF systems it is very important to know the amount of in- and out-of-band nonlinear distortions that will be generated by the devices. The amount of interference, or power, in the adjacent frequency channel is represented by the Adjacent co-Channel Power Ratio (ACPR), which is defined as the ratio of the average power in the adjacent frequency channel to the average power in the transmitted frequency channel [24,25]. The ACPR describes the level of distortions generated by the nonlinear behavior of RF components and is often used to characterize the linearity of a device. A high ACPR corresponds to a strong nonlinear device.

When the input signal of the device under test contains nonlinear distortions due to for instance source-pull, the ACPR will result in an over- or underestimation of the level of nonlinear distortions. In order to compensate for these nonlinear input distortions, a linearized transfer function is required that is able to describe the behavior of the DUT in-band as well as out of the frequency band where the device operates.

In the previous section, we have shown that the BLA can not only be determined inside but also outside the frequency band where the DUT operates. To obtain this out-of-band BLA, the system is jointly excited by a large-signal and an additional small-signal that lies outside the BLA band. Now, we will show that this additional small excitation signal allows compensating for the nonlinear distortions present in the input signal and, hence, results in a correct source-pull free level of nonlinear output distortions.

6.5.2 Compensating source-pull

1. *Experimental setup*

   Consider a two-port microwave amplifier (Motorola, MRFIC-2006) with a usable bandwidth between 600 MHz and 1 GHz. The gain factor of the amplifier is about 23 dB at 900 MHz.

   This DUT is excited by a multisine signal generated by an arbitrary waveform generator. The multisine signal consists of 500 spectral components uniformly distributed between 611 MHz and 1.01 GHz. The peak voltage amplitude of the multisine in the time domain is respectively 0.5 and 5 V when measuring the amplifier in linear operation mode and compression mode. The normalized multisine signal has a root-mean-square value of 0.2 V.

   The measurements are performed by the setup of Figure 6.7. Ten repeated measurements over time are performed. The IF bandwidth of the network analyzer is set to 100 Hz.

2. *Spectrum measurements*

   Figures 6.14 and 6.15 represent, respectively, the measured in- and output spectra (black stars) of the microwave amplifier in linear operation mode and in compression. The gray stars represent the measurement noise on the measured spectra obtained by calculating the standard deviation of the spectral lines over the repeated time periods.

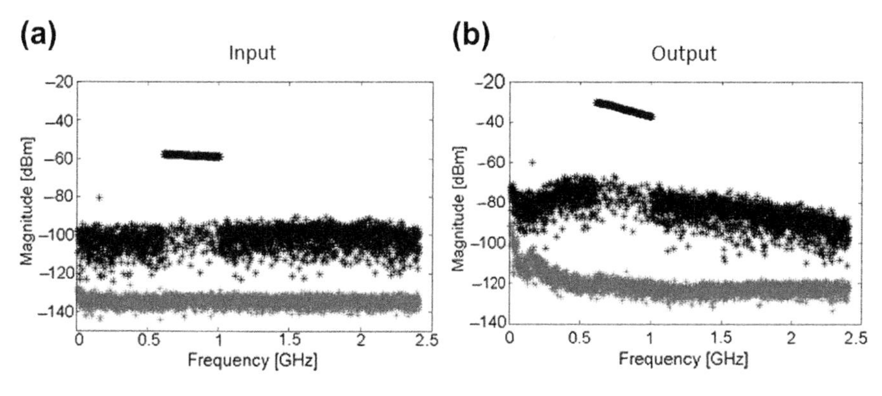

FIGURE 6.14

Input (a) and output (b) spectrum (black stars) of the amplifier in linear operation mode, and the measurement noise (gray stars).

From Figure 6.14, we observe that the out-of-band spectral components lie above the measurement noise. As a result, some nonlinear distortions are present even if the amplifier is working in linear operation mode [15].

From the measured spectra of Figure 6.15, we can see that spectral regrowth is present in the input and the output signal. Since the amplifier is working in compression, nonlinearities appear also in the frequency band of the second and third harmonic.

Using the output spectra of Figures 6.14 and 6.15 to calculate the ACPR would result in an overestimation, since nonlinear distortions are present in the input signal, which influence the amount of nonlinear output distortions. So, in order to obtain a correct ACPR value, the output spectrum must be compensated for the nonlinear distortions present in the input signal.

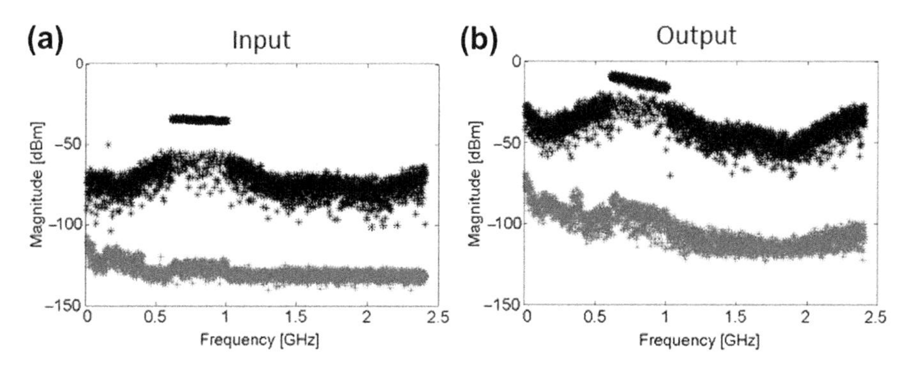

FIGURE 6.15

Input (a) and output (b) spectrum (black stars) of the amplifier in compression, and the measurement noise (gray stars).

3. *Measuring the out-of-band BLA*
   In order to calculate the out-of-band BLA, we have to excite the system out of the frequency band where the DUT is used. This means that besides the in-band multisine input excitation the DUT will also be excited by a small sinusoidal signal. Here, the results of Section 6.4.3c will be used to experimentally show the efficiency of the source-pull compensation technique.

   Calculating the mean over the different phase realizations and different time periods of these measurements results in the out-of-band BLA. This linearized transfer function describes the behavior of the amplifier in- as well as out-of-band and, hence, allows compensating the output signal for the nonlinear distortions present in the input signal. Figure 6.16 represents the in- and out-of-band BLA for the amplifier in linear regime (black curve) and the amplifier in compression (gray curve). As expected, the BLA of the amplifier in compression lies below the BLA of the amplifier in linear operation. Note that the BLA of the amplifier in compression looks much noisier due to the presence of the nonlinear distortions.
4. *Compensating the nonlinear distortions*
   The out-of-band output spectrum of the amplifier when excited by a multisine signal only is denoted by $Y^{\text{multi}}(f)$ and can be written as follows:

$$Y^{\text{multi}}(f) = G_{\text{BLA}}^{\text{m+s}}(f)U^{\text{multi}}(f) + Y_{\text{S}}(f) \tag{6.9}$$

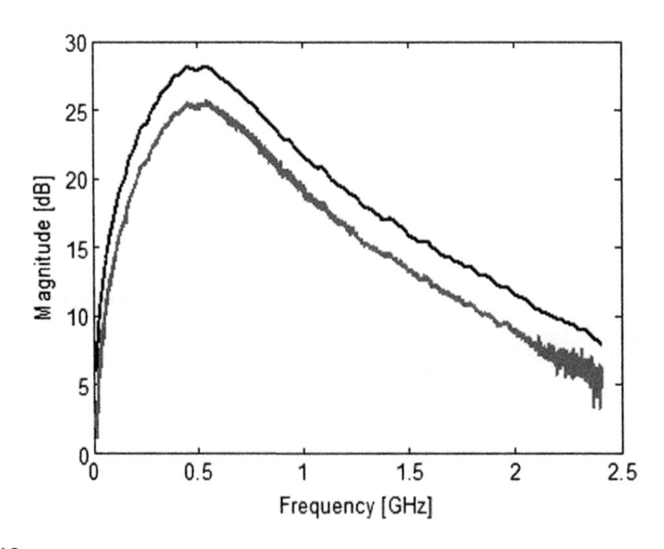

FIGURE 6.16

Out-of-band BLA for the amplifier in linear operation (black curve) and compression (gray curve).

Herein,

- $U^{\mathrm{multi}}(f)$ represents the multisine input signal of the amplifier at the out-of-band frequencies,
- $G_{\mathrm{BLA}}^{\mathrm{m+s}}(f)$ is the out-of-band BLA determined by the combination of a multisine and sine wave excitation,
- $Y_{\mathrm{S}}(f)$ are the compensated stochastic nonlinear distortions of the output signal.

  Note that the out-of-band output spectrum $Y^{\mathrm{multi}}(f)$ is interpreted in Figures 6.14 and 6.15 as nonlinear output distortions.

  As a result, the corrected nonlinear output distortions can be calculated as follows:

$$Y_{\mathrm{S}}(f) = Y^{\mathrm{multi}}(f) - G_{\mathrm{BLA}}^{\mathrm{m+s}}(f)U^{\mathrm{multi}}(f) \tag{6.10}$$

Applying Eqn (6.10) on the multisine measurements of Figures 6.14 and 6.15 results in the compensated nonlinear distortions shown in Figure 6.17. Figure 6.17(a) and (b) represent the output spectrum (black stars), the compensated nonlinear distortions (light gray stars), and the measurement noise (gray stars), respectively, for the amplifier in linear operation and compression. Note that for the amplifier in compression the measurement noise is not plotted in order not to overload the plot type shown in Figure 6.17(b).

Without compensating the nonlinear output distortions, one would conclude from Figure 6.14(b) that nonlinear distortions are present in the output spectrum. However, when looking at the compensated nonlinear distortions in Figure 6.17(a) one can clearly see that after compensation the level of nonlinear distortions lies much lower (20 dB). For the amplifier in compression, the compensation also results in a decrease of the level of nonlinear distortions

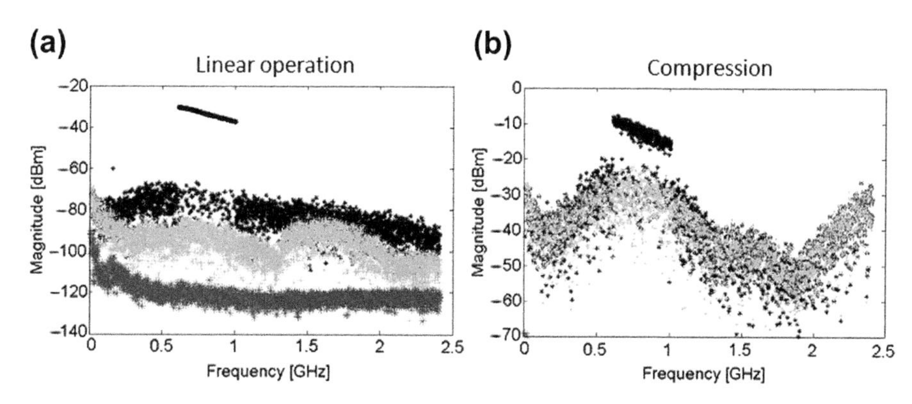

FIGURE 6.17

Output spectrum (black stars), compensated nonlinear distortions (light gray stars), and measurement noise (gray stars) for the amplifier in linear operation (a) and in compression (b). In (b) no measurement noise has been plotted.

(Figure 6.17(b) light gray stars). After compensation, the spectral regrowth is smaller, i.e., the "shoulders" around the multisine excitation are lower (3 dB) after correction (gray stars) than before (black stars). This is due to the fact that the nonlinear contributions of the input signal have been taken into account. As an additional verification, the ACPR is calculated for the output spectrum with and without compensation. Recall that the ACPR is defined as the ratio of the average power in the adjacent frequency channel to the average power in the transmitted frequency channel. In this work, the power in the transmitted frequency channel is defined as the total power of the applied multisine. For the adjacent frequency channel, the frequency band from 1.02 GHz to 1.3 GHz has been considered. For the non-compensated measurements this results in an ACPR of −18.5 dB. For the compensated measurements, an ACPR of −20.7 dB is obtained. Hence, by compensating the data for nonlinear distortions present in the input, a gain in ACPR of 2 dB is obtained.
To conclude, applying a small single sine wave on top of the multisine excitation allows acquiring a correct interpretation of the nonlinear distortions in the output spectrum.

5. *Validation of the compensation algorithm*
To validate whether or not the developed compensation method is able to compensate for all nonlinear distortions present in the input signal, a simulation experiment is set up. The compensation algorithm will be used in two cases: with and without nonlinear input distortions. If the algorithm is able to compensate for all nonlinear input distortions, the compensated spectra for both cases should be identical.
As device under test, we used a Wiener–Hammerstein system [26,27] which is a block structure consisting of two linear time-invariant blocks G_1 and G_1 with a static nonlinearity in between f (see Figure 6.18).

The Wiener–Hammerstein system was excited by a multisine excitation similar to the multisine used in the measurement experiment. The multisine consists of 270 components between 700 and 1100 MHz. On top of the multisine excitation, a small sine wave excitation was applied to the nonlinear DUT and was swept over the whole frequency band. The input and output signals are disturbed by white Gaussian noise.

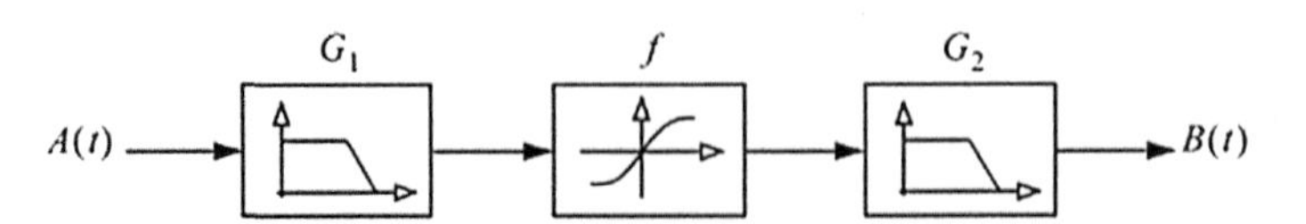

FIGURE 6.18

Wiener–hammerstein system.

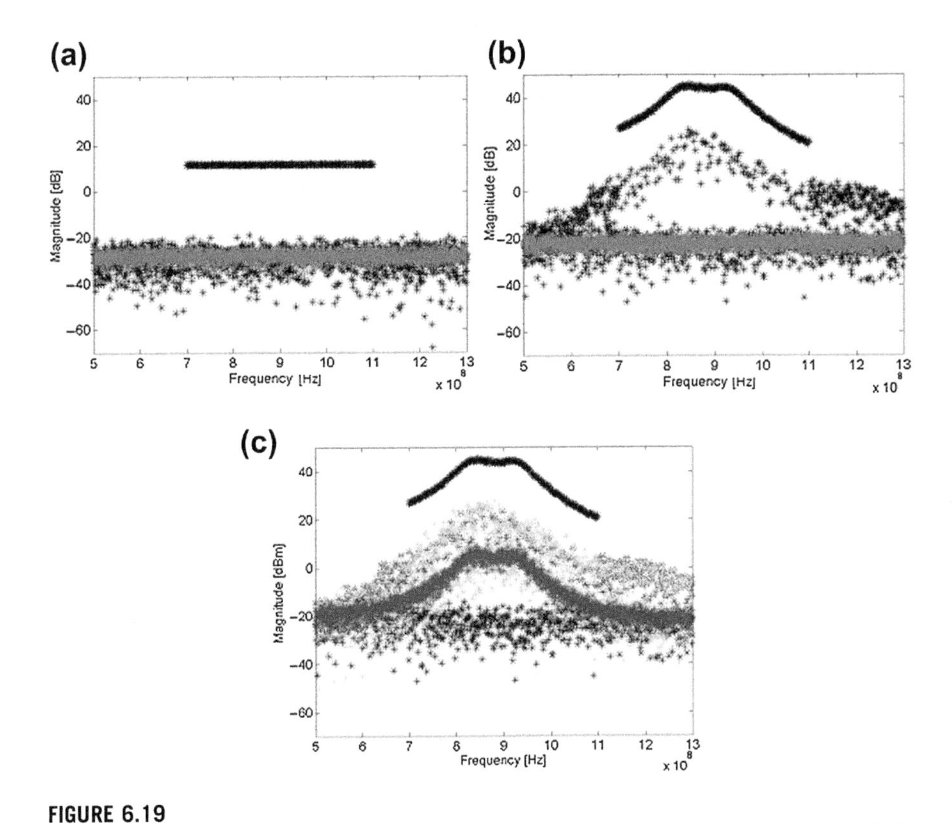

FIGURE 6.19

Input (a) and output (b) spectrum (black stars) of the nonlinear DUT, when no nonlinearities are present in the input signal. The measurement noise is represented by the gray stars. In (c) the compensated nonlinear distortions are represented by the light gray stars and the compensated measurement noise is represented by the gray stars.

In a first simulation, the input signal is free of nonlinear distortions as can be seen from Figure 6.19(a), since the out-of-band spectral components do not lie above the measurement noise floor. The output spectrum, however, contains nonlinear distortions as can be seen from Figure 6.19(b), due to the nonlinear DUT.

Since no nonlinear distortions were present in the input, the level of nonlinear output distortions should remain the same after applying the compensation algorithm as before compensation (Figure 6.19(c)). If this were not the case, the compensation algorithm would not be correct. The dark gray stars in Figure 6.19(c) represent the measurement noise of the compensated output spectrum at the non-excited frequency components. Note that the level of the in-band measurement noise is higher after compensation than before compensation. This is due to the fact that the measurement noise present in the input signal

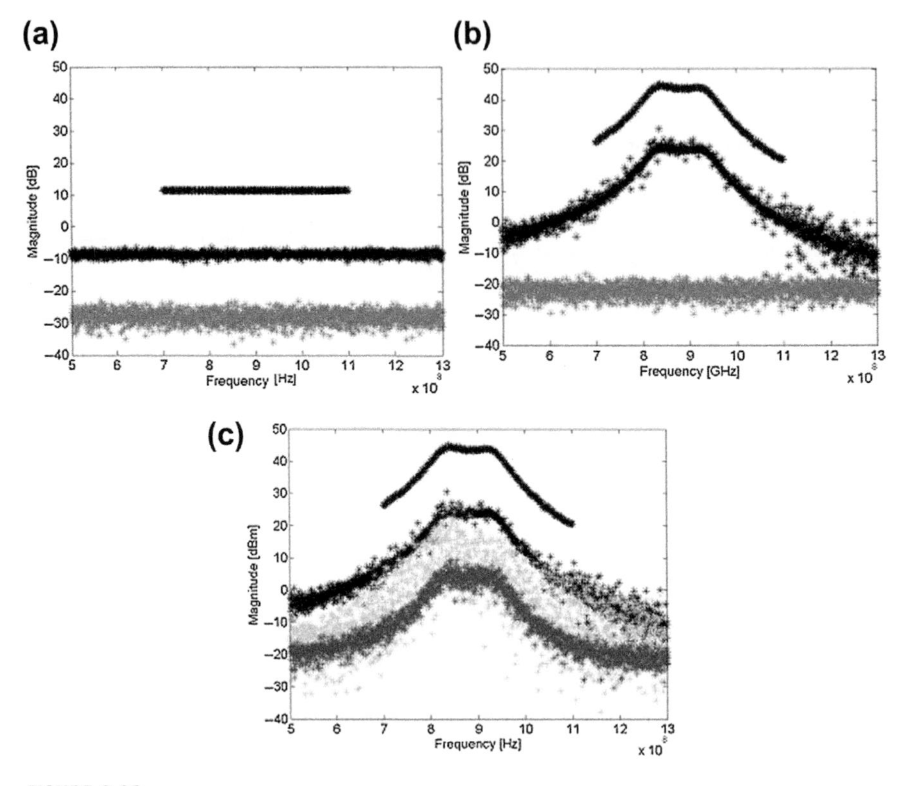

FIGURE 6.20

Input (a) and output (b) spectrum (black stars) of the nonlinear DUT, when nonlinearities are present in the input signal. The measurement noise is represented by the gray stars. In (c) the compensated nonlinear distortions are represented by the light gray stars and the compensated measurement noise is represented by the gray stars.

$U^{\text{multi}}(f)$ is taken into account during the compensation ($Y_{\text{S}}(f) = Y^{\text{multi}}(f) - G_{\text{BLA}}^{\text{m+s}}(f)U^{\text{multi}}(f)$).

In a second simulation, the input signal contains nonlinear distortions that lie about 15 dB above the measurement noise floor (see Figure 6.20(a)). As a result extra nonlinear distortions are present in the output spectrum (see Figure 6.20(b)). After compensation, the level of nonlinear distortions clearly decreased (see Figure 6.20(c)).

To prove that the compensation method indeed compensated all nonlinearities present in the input signal, the level of the compensated nonlinear distortions for both cases, with and without input nonlinearities, should be equal. Figure 6.21 clearly shows that this is the case. Hence, this simulation test validates the correctness of the compensation method.

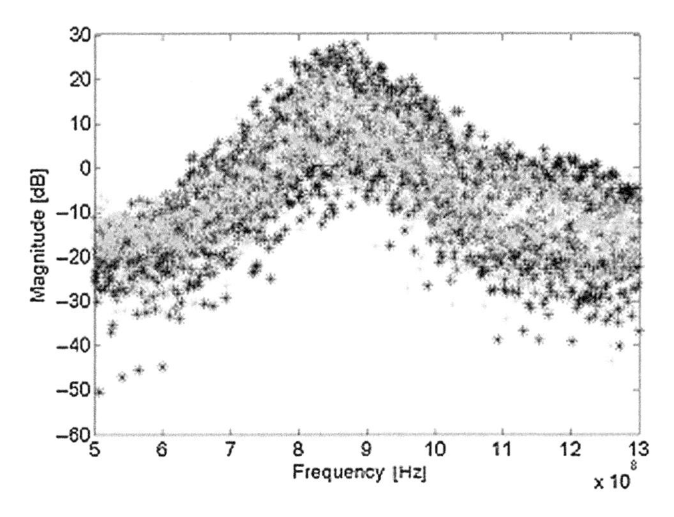

FIGURE 6.21

Compensated nonlinear distortions of the output spectrum for an input signal with (light gray stars) and without (black stars) nonlinear distortions.

Acknowledgments

This research was funded by a postdoctoral fellowship of the Research Foundation-Flanders (FWO).

References

[1] Roblin P. Nonlinear RF circuits and nonlinear vector network analyzers. Cambridge (New York, NY, USA): Cambridge University Press; 2011.

[2] Kahrs M. 50 years of RF and microwave sampling. IEEE Trans Microwave Theory Tech June 2003;51(6):1787−805.

[3] Blockley PS, Gunyan D, Scott JB. Mixer-based, vector-corrected, vector signal/network analyzer offering 300 kHz−20 GHz bandwidth and traceable phase response. IEEE MTT-S Int Microwave Symp Dig (Long Beach, CA, USA), June 2005;4: 1497−500.

[4] Van Moer W, Rolain Y. A large-signal network analyzer: why is it needed? IEEE Microwave Mag December 2006;7(6):46−62.

[5] HP71500A Microwave transition analyzer, Product Note 70820-11, http://www.home.agilent.com/upload/cmc_upload/All/Product_Note_70820_11.pdf.

[6] http://www.vtd-rf.com/.

[7] http://www.mesuro.com/technology.htm.

[8] Verspecht J, Rush K. Individual characterization of broadband sampling oscilloscopes with the nose-to-nose calibration procedure. IEEE Trans Microwave Theory Tech April 1994;43(2):347−54.

[9] http://cp.literature.agilent.com/litweb/pdf/5989-8575EN.pdf.
[10] Agilent PNA-X Series, 5989-8041EN, May 2009.
[11] Kruppa W, Sosomsky KF. An explicit solution for the scattering parameters of a linear two-port measured with an imperfect test set. IEEE Trans Microwave Theory Tech January 1971;19(1):122−3.
[12] Rehnmark S. On the calibration process of automatic network analyzer systems. IEEE Trans Microwave Theory Tech April 1974;22(4):457−8.
[13] Williams D, Hale P, Clement T. Calibrating electro-optic sampling systems. IEEE MTT-S Int Microwave Symp, (Phoenix, AZ, USA), May 2001;3:1527−30.
[14] Williams D, Hale P, Clement T. Electrical-phase traceability to NIST's EOS system. Research update presented at the 4th ARFTG NVNA User's Forum, http://www.arftg.org/LSNA/4th/UsersForum_June2004_Minutes2.pdf; 2004.
[15] Schoukens J, Pintelon R, Dobrowiecki T. Linear modeling in the presence of nonlinear distortions. IEEE Trans Instrum Meas August 2002;16(5):786−92.
[16] Bendat JS, Piersol AG. Engineering applications of correlation and spectral analysis (New York, NY, USA): Wiley; 1980.
[17] Pintelon R, Vandersteen G, De Locht L, Rolain Y, Schoukens J. Experimental characterization of operational amplifiers: a system identification approach − part I: theory and simulations. IEEE Trans Instrum Meas June 2004;53(3):854−62.
[18] Rolain Y, Van Moer W, Pintelon R, Schoukens J. Experimental characterization of RF amplifiers: a system identification approach. IEEE Trans Microwave Theory Tech August 2006;54(8):3209−18.
[19] Dinis R, Palhau A. A class of signal-processing schemes for reducing the envelope fluctuations of CDMA signals. IEEE Trans Commun May 2005;53:882−9.
[20] Banelli P. Theoretical analysis and performance of OFDM signals in nonlinear fading channels. IEEE Trans Wireless Commun March 2003;2:284−93.
[21] D'haene T, Pintelon R, Schoukens J, Van Gheem E. Variance analysis of frequency response function measurements using periodic excitations. IEEE Trans Instrum Meas August 2005;54(4):1452−6.
[22] Verspecht J. Large-signal network analysis. IEEE Microwave Mag December 2005; 6(4):82−92.
[23] Verspecht J, Root DE. Polyharmonic distortion modeling. IEEE Microwave Mag June 2006;7(3):44−57.
[24] Adjacent Channel Power Ratio (ACPR), Anritsu Application Note (11410-00264).
[25] Kenney JS, Avis SE. The relationship between IMD and ACP. Wireless Des Dev 2000:7−8.
[26] Billings SA, Fakhouri SY. Identification of systems containing linear dynamic and static non-linear elements. Automatica January 1982;18(1):15−26.
[27] Brillinger DR. Identification of a particular nonlinear time-series system. Biometrika December 1977;64(3):509−15.

CHAPTER

7 Behavioral Models for Microwave Circuit Design

José C. Pedro, Telmo R. Cunha

Instituto de Telecomunicações, Universidade de Aveiro, Aveiro, Portugal

7.1 Introduction

Microwave device modeling—and, in particular, nonlinear active device modeling—has been an insatiable field of research due to the substantial advances faced on microwave transistor technologies. Some of the technologies that have made a great impact on circuit design include the gallium arsenide (GaAs) metal-semiconductor field-effect transistor (MESFET), the GaAs and indium phosphide (InP) high electron-mobility transistor (HEMT), the GaAs and silicon germanium (SiGe) heterojunction bipolar transistor (HBT), the silicon light-diffusion metal-oxide (LDMOS) field-effect transistor (FET), and, more recently, the gallium nitride (GaN) HEMT. In fact, physics-based models (directly derived from the device geometry and semiconductor physics) and equivalent-circuit models are both driven by the transistor technology, and so have to follow any technology changes.

On the contrary, linear modeling appears to be much more resilient to these technology shifts. The same model structures, such as the frequency-domain admittance matrix or scattering matrix formulations, have been successfully used for many transistors and even passive devices. This naturally raises one question regarding the apparent dissimilarity existing between linear and nonlinear characteristics—or, at least, on the generality of linear and nonlinear modeling tools. As a matter of fact, the essential difference resides in that these technology-independent linear models are not based on device physics; rather, they are behaviorally based models. That is, they are nothing but measurement-based mathematical representations of the observable input—output characteristics of the devices, regardless of their technology or internal composition. In addition, contrary to linear models, which have already more than four decades of existence and usage, nonlinear behavioral modeling is still taking its first steps. Nevertheless, because behavioral models protect the intellectual property of the device's manufacturer and they are only limited by the available laboratory measurement-based capabilities, they are rapidly receiving increasing attention from researchers in both academia and industry. Unfortunately, they also require a great deal of ingenuity in the model formulation. Their predictive capability is generally worse than that of physics-based models and, because they are measurement-based, they require expensive and dedicated broadband microwave instrumentation.

Microwave De-embedding. http://dx.doi.org/10.1016/B978-0-12-401700-9.00007-0

Because behavioral models are elaborate ways of representing interpolated input–output observation data, they completely rely on device characterization. Therefore, and similarly to what happens in the traditional physics-based or equivalent-circuit models, parameter extraction procedures of behavioral models also make use of specific instrument calibration methods and embedding and de-embedding techniques.

Accordingly, the aim of this chapter is two-fold. First, it will provide the reader with a (necessarily brief) overview of microwave device behavioral models, addressing both their mathematical formulation and their laboratory parameter extraction. Second, it will discuss the embedding and de-embedding procedures that appear to be associated with the behavioral model extraction process.

The chapter is divided into three sections, including this introduction. Section 7.2 is focused on the identification of behavioral models and their distinction from the most commonly seen physics-based and equivalent-circuit models. Then, it presents an overview of the three most important types of behavioral model forms (polynomial, artificial neural networks, and table-based models) and their instantiation in the context of microwave transistors. Finally, Section 7.2 addresses the behavioral model parameter extraction procedures, reviewing their associated specific microwave instrumentation and correspondent calibration and de-embedding of measurement data. Section 7.3 discusses some actual examples of behavioral model embedding and de-embedding procedures, with a particular emphasis on the necessary behavioral model inversion.

7.2 Behavioral modeling tools

The definition of a model structure for characterizing the behavior of a physical system can follow the system's inner processes handling the signals, or it can be simply based on a canonical mathematical representation whose aim is to fit a set of realizations of the system's input–output signals. A mixture of these two approaches can also be used. After defining the model structure, the second step is to determine the values of the model parameters so that the model mimics the system behavior up to a desired level. This modeling process is described in the following subsections, where the distinction between the referred modeling approaches (physics-based and behavior-based modeling) is highlighted. The main canonical mathematical structures used to represent behavior-based models are described and analyzed. Finally, model parameter extraction techniques are addressed, with particular emphasis on microwave applications, focusing on the use of microwave instrumentation for such a task, and its inherent processes of calibration and de-embedding.

7.2.1 Physics-based and behavior-based models

Finding models to analytically simulate, design, understand, and control physical systems has been one of the predominant objectives in science, not only in recent history but also from centuries ago. In fact, most of the science work developed in the eighteenth and nineteenth centuries was on the development of models that could be useful in predicting the behavior of physical systems. By analyzing the strategies used in model development, one can clearly identify two distinct approaches: physics-based and behavioral modeling.

Physics-based modeling consists of applying a priori known physics laws in the establishment of the relationships between the signals or variables of interest that are processed by the system. Because this strategy requires that the system under analysis is completely known, it was also named "transparent-box modeling." As will be detailed in Section 7.2.1.1, given the physical description of a system and the definition of the input and output signals, a model is then obtained by following the signals' flow throughout the system components and, step by step, specifying the physical relationships that describe the signal transformations when passing through such components.

The behavioral modeling approach, described in Section 7.2.1.2, follows a completely different strategy. The underlying idea is that only the system input—output mapping is relevant, regardless of its internal constitution. Therefore, this approach (also called "black-box modeling") requires only the observation of input—output signal sets and the specification of a model topology that is to be tuned to mimic the observed input—output relationship. The model structure may not reflect any of the inner physical laws of the system; it is simply a mathematical descriptor of the input—output mapping. No information is considered from the internal constitution of the system to be modeled (this is a black box where only the inputs and outputs are accessible).

As will be described in the following sections, both physics-based and behavioral modeling approaches have advantages and disadvantages. Therefore, a mixed modeling strategy is often used, in which some of the information is mostly behavioral and other information is obtained from the physical relationships of its subsystems. This mixed approach is generally called "gray-box modeling," for obvious reasons, and it is also detailed in Section 7.2.1.3.

It could be argued that, in the limit, most of what we recognize as physics laws is, in fact, a set of behavioral models. For instance, Newton's motion laws are mathematical models that were motivated by experimental observations and testing. Also, Coulomb's law of electrostatic forces (one of the pillars of electromagnetism) can be considered as a behavioral model that was designed from the observation of the interaction of charged particles submitted to specific laboratory tests. Nevertheless, whenever a system model is obtained from the mathematical description of its constitution (even if the mathematical models that describe each of its components could be interpreted as behavioral) then it is a physics-based model. As an example, consider the physics-based input—output model of a linear electronic circuit, with interconnected lumped elements, that is retrieved from the application of Kirchhoff's laws (these are not behavioral) and the laws that describe the voltage—current behavior of the lumped elements. A behavioral model of such a linear circuit could be defined by, for example, a table of amplitude and phase of its response to a continuous wave (CW) signal, for a set of distinct frequencies.

7.2.1.1 Physics-based or transparent-box models

An evident case of application of physics-based modeling in electronics engineering is the concept, design, and understanding of semiconductor-based devices, such as diodes and transistors. In fact, if we consider a forward-biased silicon p–n junction of two differently doped semiconductors, as shown in Figure 7.1, we can derive the mathematical relationship between the average current crossing such junction and the voltage across it by applying the physical laws that relate uncovered charge concentration, charge carrier concentration, mobility, and diffusivity. Such analysis yields the well-known current–voltage relationship of a p–n junction:

$$I = Aqn_{\mathrm{i}}^{2}\left(\frac{D_{\mathrm{p}}}{L_{\mathrm{p}}N_{\mathrm{D}}}+\frac{D_{\mathrm{n}}}{L_{\mathrm{n}}N_{\mathrm{A}}}\right)\left(e^{V/V_{\mathrm{T}}}-1\right) = I_{\mathrm{S}}\left(e^{V/V_{\mathrm{T}}}-1\right) \qquad (7.1)$$

where A is the junction cross-sectional area, q is the charge of the electron, n_{i} is the concentration of free electrons or holes in the intrinsic semiconductor, and D_{p} and D_{n} are the diffusion constants (or diffusivity) for electrons in a p-type semiconductor and for holes in an n-type semiconductor, respectively. L_{n} and L_{p} are the diffusion lengths for holes in an n-type semiconductor and for electrons in a p-type semiconductor, and N_{D} and N_{A} are the concentrations of donor and acceptor atoms. V_{T} is the junction's thermal voltage, which, according to the Einstein relationship, is equal to the carrier diffusivity divided by its mobility. I_{S} is the junction's reverse saturation current (or drift current).

The model in Eqn (7.1) describes only the static current–voltage relationship of the intrinsic diode junction. Intrinsic dynamic behavior can also be modeled by deriving, again from physical rules, the junction's depletion and diffusion capacitances. However, it would not be simple to get an empirical nonlinear dynamic model of the p–n junction because this would require testing the junction in a laboratory and, as is known, the junction is available only in the packaged format of a diode component. It is not possible, in a standard electronics laboratory, to access the junction terminals directly in order to test its behavioral response to a set of excitations; only the diode package terminals or the die pads are accessible. Therefore, the elements that are interfacing the intrinsic junction to the outside world (i.e., the extrinsics) will also cause an impact on the observed response of the diode when tested in the laboratory. For instance, the bond wires interfacing the die pads, the pads and

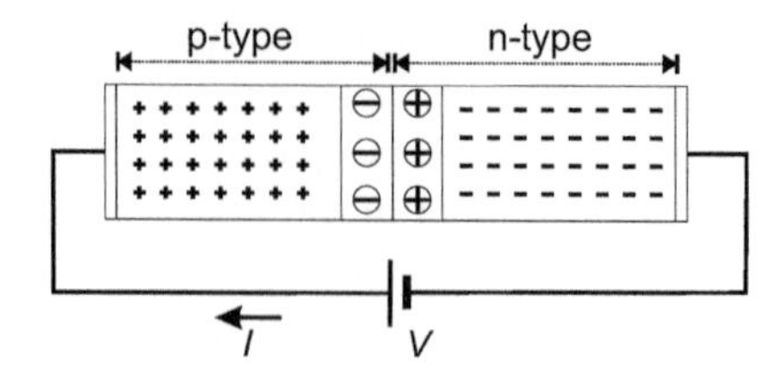

FIGURE 7.1

Forward-biased silicon p–n junction.

traces in the board holding the die, and the die pads themselves show inductive and capacitive effects (in general, they present complex electromagnetic behavior) that, for certain frequency ranges of operation of the diode, will produce a nonnegligible effect on the diode's current–voltage relationship.

This example evidences one of the advantages of physics-based modeling: the model is obtained without the necessity of setting a dedicated (and often elaborate) laboratory setup to test the system under modeling. In this case, testing the junction would require first the knowledge of the behavior of the diode's extrinsics (i.e., given a physical description of the extrinsics structure and components, a model for it would have to be obtained, which is usually achieved by resorting to electromagnetic simulations) and then such a model would have to be processed in order to calculate an estimate of the virtual measurements at the intrinsic connections, based on the accessible extrinsics measurements. This is the process of de-embedding the diode extrinsic behavior.

In past decades, an obvious example of intrinsic device modeling is that of transistor modeling, with particular emphasis on the microwave FET. In the published literature, examples of physics-based and behavior-based (with de-embedding) approaches can be found. Following a strategy similar to that of the p–n junction analysis, a physics-based mathematical model can be derived for the intrinsic FET, whose physical description is shown in Figure 7.2. Considering the effects in the channel caused by the gate-generated electric field, the charge that is accumulated on the channel, and the physics-based equations derived from solid-state physics and electromagnetism, the relationships between the electric potential in the semiconductor, $\psi(x,y)$, and the physical characteristics of the intrinsic device can be determined from the differential equation (which has support on Gauss's law relating charge and electric potential):

$$\nabla^2\psi(x,y) = -\frac{q}{\varepsilon}\left[p(x,y) + N_D^+(x,y) - n(x,y) - N_A^-(x,y)\right] \tag{7.2}$$

where ε is the permittivity of the semiconductor, $p(\cdot)$ and $n(\cdot)$ are the hole and electron concentrations, and $N_D^+(\cdot)$ and $N_A^-(\cdot)$ are the donor and acceptor doping concentrations, respectively. By applying the physics law that implies continuity of the electron and hole concentrations in time, we obtain the following:

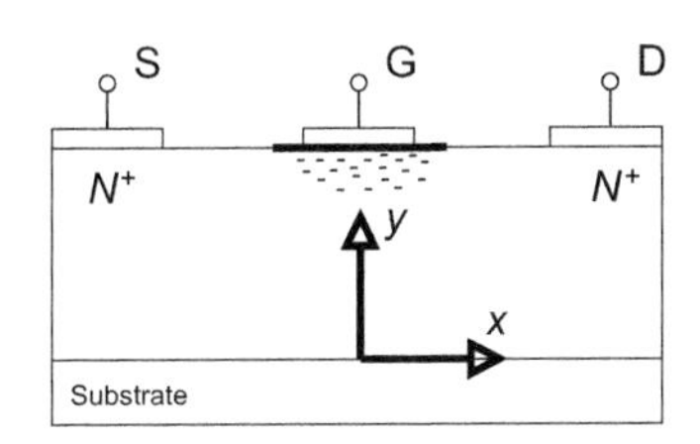

FIGURE 7.2

Representation of an intrinsic FET structure (transversal view).

$$\begin{cases} \dfrac{dn(x,y)}{dt} = -\mu_n \nabla^2 \psi(x,y) n(x,y) - \mu_n \nabla \psi(x,y) \nabla n(x,y) + D_n \nabla^2 n(x,y) - G_n + R_n \\ \dfrac{dp(x,y)}{dt} = -\mu_p \nabla^2 \psi(x,y) p(x,y) - \mu_p \nabla \psi(x,y) \nabla p(x,y) + D_p \nabla^2 p(x,y) - G_p + R_p \end{cases} \tag{7.3}$$

for the electron and hole mobility μ_n and μ_p, and electron and hole generation (G_n and G_p) and recombination (R_n and R_p) rates. Finally, imposing the boundary conditions for the potential at the terminals of the intrinsic FET, we obtain the remaining expressions of the FET physics-based model:

$$\begin{cases} \psi(x,y)|_{\text{source}} = v_S = 0 \\ \psi(x,y)|_{\text{gate}} = v_G = v_{GS} \\ \psi(x,y)|_{\text{drain}} = v_D = v_{DS} \end{cases} . \tag{7.4}$$

Equations (7.2)–(7.4) constitute a physics-based intrinsic FET model that, for a set of properties of the materials composing the FET and for the specified boundary conditions, predicts the potential distribution and the electron/hole concentrations through the semiconductor, from which other variables can be calculated. This aspect highlights another advantage of physics-based models: they predict the signals of interest with high accuracy and such accuracy is (in general) independent of the excitation signal.

However, this example also indicates some difficulties inherent to physics-based modeling. A useful model will only be achieved if the properties of the physical system are known in detail. For instance, for the model of Eqns (7.2)–(7.4) to be evaluated, it would be necessary to know the concentration of donor and acceptor atoms in the FET semiconductor—values that cannot be accurately predicted from the doping process. Thus, measurements would still be necessary to tune the parameters of the model. Nevertheless, physical knowledge defines the structure of the model, which is guaranteed to be adequate for the system under analysis.

Like the diode, the FET is also commonly available in a packaged format, where the package parasitics (the extrinsic FET) mask the intrinsic FET (when in the die form, the FET intrinsics are also masked by the die parasitics). Therefore, measuring the intrinsic FET for tuning its physics-based model parameters, or for validating it, would require the package parasitics to be known and modeled. Such a model would have to be de-embedded so that the measurement reference plane could be moved from the extrinsic FET terminals to the intrinsic FET ports. As mentioned before, electromagnetic simulation is commonly used to determine the model for the package behavior, given its physical description (trace and pad dimensions and material, dielectric material, bond wire geometry, and so on).

Finally, we must stress the Achilles heel of physics-based models, which is their high complexity. As can be observed by the FET example above, the simple calculation of the static current–voltage characteristics of an intrinsic FET would require, at least, solving a set of space-time differential equations. If you imagine the model of a chip holding several intrinsic FETs, where each FET is modeled by Eqns (7.2)–(7.4), it is obvious that the computational burden required to estimate the chip's output signals (given certain excitation signals) would be tremendous and the simulation time would make such a model impractical. For cases like this one, where the modeling interest is in the overall input–output relationship of a physically complex system (and not in the inner details of the system), it is usually more appropriate to find simpler mathematical equations relating the observed input–output correlation, which do not correspond to the physical description of how the system internally processes the signals. This is behavioral modeling.

7.2.1.2 Behavior-based or black-box models

Instead of looking into every component of the system and the way they are internally connected (as was performed in physics-based modeling), the behavioral modeling approach limits the observation of the system to its input and output signals. All the inner details of the system are not taken into consideration in the model formulation or extraction, which is why this approach is also referred to as black-box modeling. Therefore, based only on the observation of the system's response to a certain set of input excitations, a mathematical formulation (the model) is designed to reproduce the same input–output mapping (up to a certain modeling accuracy).

Let us illustrate this approach with the diode example considered in Section 7.2.1.1. Figure 7.3 shows the current–voltage characteristic of a tested packaged

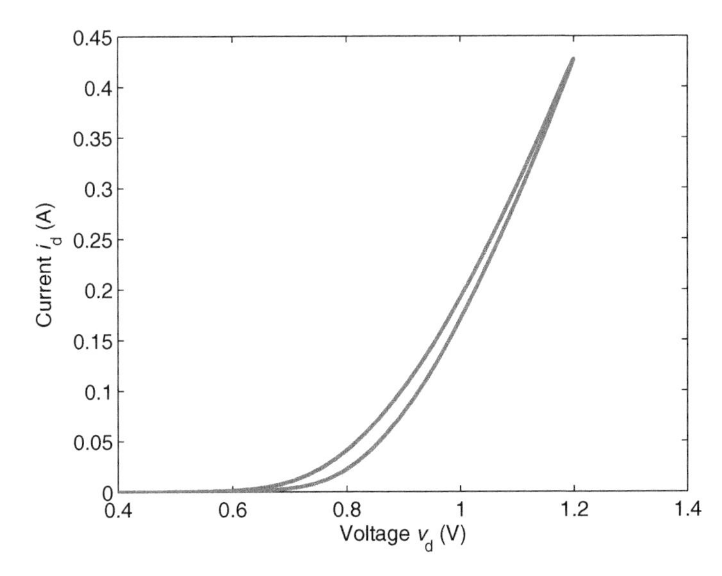

FIGURE 7.3

Diode current–voltage characteristic when excited with a high-frequency sinusoidal voltage waveform.

diode to which was directly applied a sinusoidal voltage waveform of amplitude 0.4 V, with a direct current (dc) offset of 0.8 V. The frequency of the sinusoidal voltage was set high enough so that the diode's internal resistance and dynamic effects were already noticeable (this is observed by the loop shown in Figure 7.3, which indicates that the current–voltage trajectory follows different paths when the voltage is rising or falling).

Assuming no other knowledge other than that from the observation of Figure 7.3, we could try to recreate such an input–output characteristic by the following mathematical expression:

$$i_{\mathrm{d}}(n) = \sum_{p=1}^{5} \left[a_{p,0} v_{\mathrm{d}}(n)^p + a_{p,1} v_{\mathrm{d}}(n-1)^p \right]. \tag{7.5}$$

The parameters $a_{p,0}$ and $a_{p,1}$ of this selected model were determined by linear least-squares fitting and the resulting modeled current–voltage characteristic is depicted in Figure 7.4, superimposed on to the measured one. It should be mentioned that such model parameters do not represent any physical characteristic of the diode. They are just the parameters of an arbitrarily selected model topology, which are determined so that the model would best fit the measured input–output data, according to a predefined error minimization criterion.

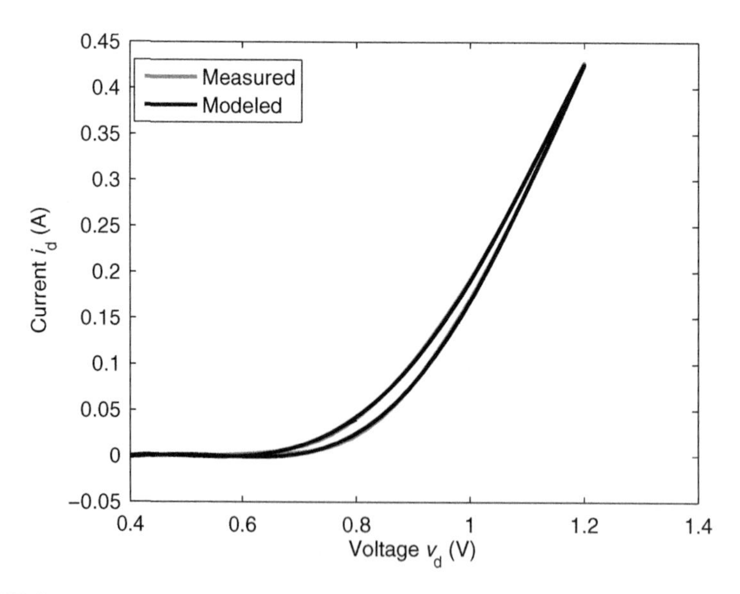

FIGURE 7.4

Measured and modeled diode current–voltage characteristic when excited with a high-frequency sinusoidal voltage waveform.

As the model parameters are extracted for particular input signals, its achieved accuracy is always inherently related to the characteristics of such excitation inputs. Therefore, the behavior-based model prediction ability is always limited to a set of signals whose characteristics are not too distinct from those of the signals used to extract the model parameters. To exemplify this issue, let us test the diode with a sinusoid of slightly higher voltage and use Eqn (7.5), with the previously determined parameters, to predict the new current–voltage characteristic. The obtained result, shown in Figure 7.5, indicates a noticeable degradation of the model's ability to predict the diode's behavior (at the edges of the current–voltage characteristic, the zones not excited in the parameter determination process), a modeling characteristic not shared by physics-based models.

Considering a similar approach, a behavioral model for the microwave FET device described in the previous section can be developed assuming that it is a two-port voltage-dependent system, as illustrated in Figure 7.6. Port 1 can be attributed to the gate-source terminal, whereas port 2 is set to the drain-source; in fact, this same model could also be used for a bipolar junction transistor device, with the base-emitter attributed to port 1 and collector-emitter to port 2. Notice that choosing the port voltages as independent variables and the currents as dependent variables is arbitrary. The case shown in Figure 7.6 is an admittance model; an impedance model could be obtained if currents were set as independent variables and voltages as dependent ones.

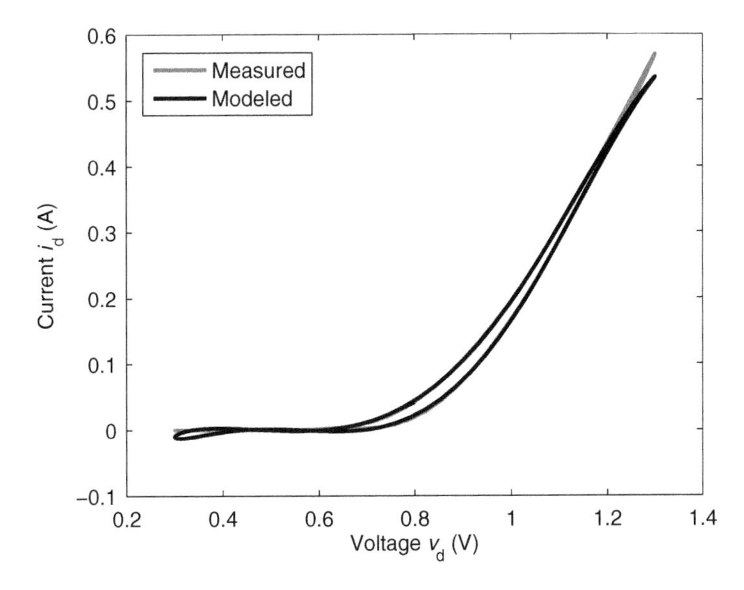

FIGURE 7.5

Diode model prediction degradation as the input signal characteristics differ from those of the signal used for extracting the model parameters.

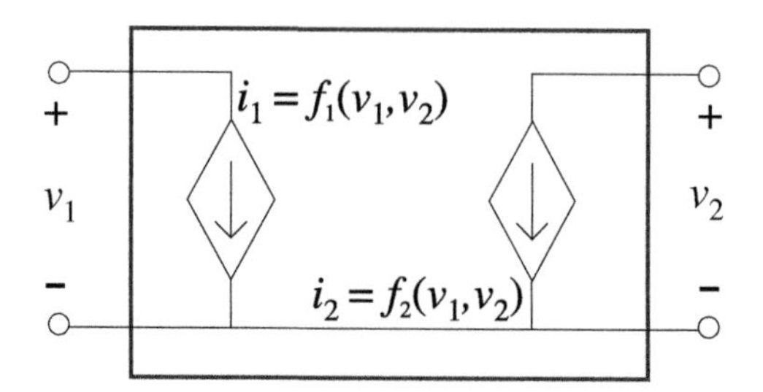

FIGURE 7.6

Behavioral model structure considered for a microwave transistor.

Given a set of current–voltage measurements of the transistor and defining a set of mathematical equations to implement the dual-input single-output mappings f_1 and f_2, a behavioral model for this particular transistor can be obtained.

Even though this illustrative example has considered a nonrecursive formulation for the model operators f_1 and f_2, these could have perfectly been chosen to be recursive:

$$i_k = f_k(v_1, i_1, v_2, i_2), \quad k = 1, 2. \tag{7.6}$$

The transistor model complexity is dependent only on the definition of the operands f_1 and f_2. In practice, it is expected for these to be much less complex than the physics-based equations shown in Section 7.2.1.1. Therefore, the execution of such a model in a simulator would be much more efficient and simple than the physics-based model, which could make the difference between a feasible simulation of a more complex circuit with several transistors (with the cost of predictive capability reduction) and an impractical simulation (although the physics-based model would provide the highest predictive ability and accuracy) due to the excessive required simulation time.

7.2.1.3 Gray-box models

The previous sections highlighted the tradeoffs between physics-based and behavior-based models. Because the benefits of one approach seem to compensate for the deficiencies of the other, it is often the case where system models are considered that some information from the physical constitution of the system and other information is retrieved from the observation of the system's behavior. The modeling technique that mixes physics-based and behavior-based information is called gray-box modeling.

This strategy is commonly used, for instance, in the description of electronic devices by means of their equivalent circuit models, used in most circuit simulators. The usual approach for generating such equivalent circuit models is to use physics-based knowledge to define the model topology and the type of lumped elements to use in each branch of such topology. Then, the values of such components are determined by means of input–output observations of the system (or, when possible, of some of its subsystems). That is, the model parameters are tuned to optimize the ability of the model to mimic the behavior of the system.

Gray-box models usually have high predictive ability. Even when the characteristics of the input signals differ from those considered in the model parameter extraction procedure, the estimated output signals still show a good accuracy when compared to their measured counterparts in the physical system. Therefore, the inner complexity of the model is kept reasonably low (leading to acceptable simulation efficiency).

Figure 7.7 presents a possible gray-box model of the diode device under consideration. It contains a nonlinear voltage-dependent current source (which describes the exponential static current–voltage characteristic of the p–n junction), a series resistor that models the energy dissipation within the diode, and a capacitor that represents the charge accumulation (i.e., the dynamic behavior) due to diffusion and depletion.

An equivalent circuit model can be also determined for the packaged transistor device by modeling the intrinsic charge accumulation effects by means of nonlinear capacitors, the resistive paths by resistors, the semiconductor junctions by ideal diodes, and the package parasitics by inductors (for the bond wires) and capacitors (for the pads). Figure 7.8 shows a possible gray-box model for the transistor.

7.2.1.4 System-level and circuit-level behavioral models

In electronic circuit modeling, two behavioral modeling approaches can be distinguished—one that considers only the global behavior of the circuit as seen from its accessible ports (the input and output ports) where the signals (currents and voltages) are measured; and another one that describes the circuit by a network of interconnected basic electronic components (e.g., inductors, capacitors, resistances) using behavioral models to describe the input–output relations of each electronic component. The former is usually denominated as system-level behavioral modeling (it is a purely behavioral approach whose objective is to replace the system by a single global model) and the latter is referred to as circuit-level behavioral modeling.

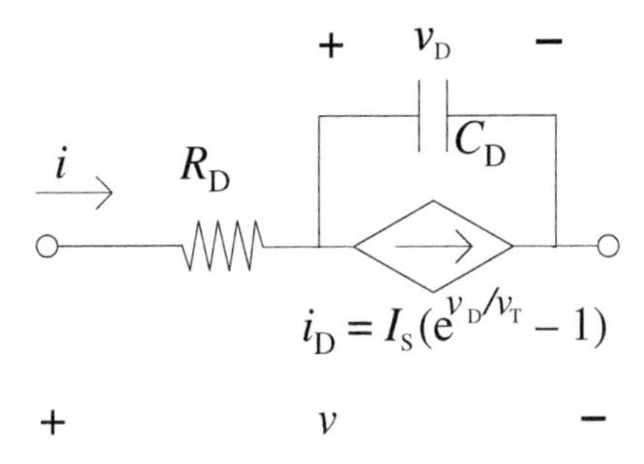

FIGURE 7.7

Possible gray-box model for the diode device.

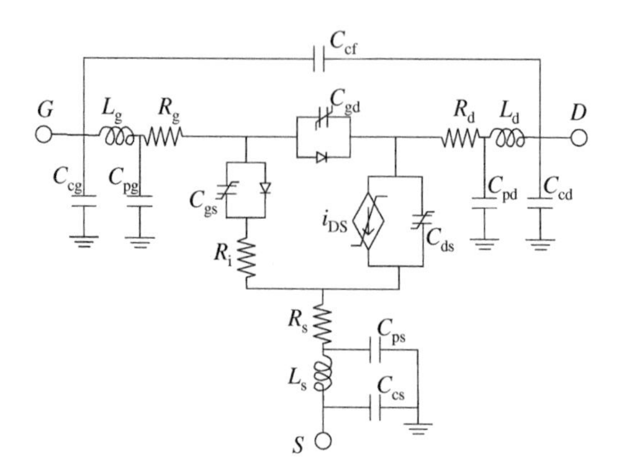

FIGURE 7.8

Possible gray-box model for a packaged transistor device.

For instance, a system-level model of a linear electronic filter circuit (with input x and output y) can be obtained from the finite impulse response (FIR) representation of a filter:

$$y(n) = \sum_{m=0}^{M} h(m)x(n-m) \tag{7.7}$$

where M is the memory depth of the filter and parameters $h(\cdot)$ constitute the sampled impulse response of the filter circuit. Alternatively, its infinite impulse response (IIR) form could be also used:

$$y(n) = \sum_{m_D=0}^{M_D} h_D(m_D)x(n-m_D) + \sum_{m_R=1}^{M_R} h_R(m_R)y(n-m_R). \tag{7.8}$$

Even if no access is given to the filter circuit schematics, an equivalent circuit model can be designed in which the values of the lumped elements (constituting the equivalent circuit) are optimized for the model to fit the global behavior of the filter circuit. For this purpose, in modeling linear dynamic systems, the π- and T-networks of lumped admittances and impedances are commonly encountered, as depicted in Figure 7.9. It is usual to construct series or parallel connections of these basic networks when developing an equivalent circuit model of a linear electronic system, conferring more degrees of freedom to the model which, after proper parameter optimization, contributes to a higher degree of modeling accuracy.

Equivalent circuit modeling of linear systems is commonly considered when simulating electronic systems with distributed elements (e.g., transmission lines) that are characterized by means of S parameters (which are table-based models expressed in the frequency domain, as will be shown later in this chapter). Because time-domain circuit simulators cannot directly incorporate S-parameter submodels

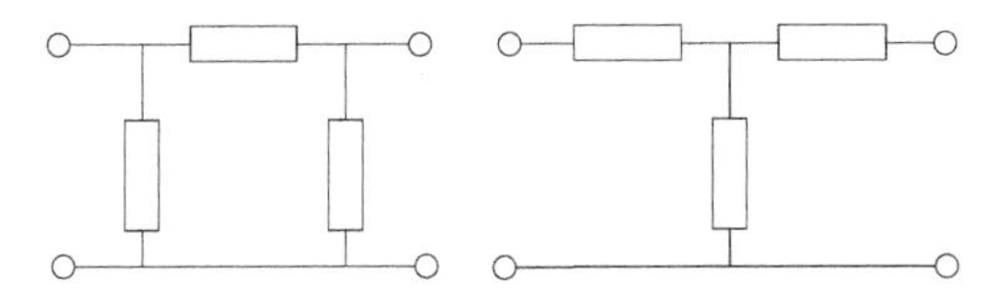

FIGURE 7.9

Basic admittance π-network (left) and impedance T-network (right) topologies.

(because these are frequency-domain entities), it is necessary to first convert such submodels into time-domain equivalent blocks that share the same behavior. This is commonly achieved by designing a network of lumped elements whose port relationships are similar to those of the *S*-parameter submodels. This network can now be directly incorporated into the circuit simulator because the lumped elements are time-domain models. Therefore, they can be used in the time-domain differential equations to be solved by the simulator's algorithms.

Although the description here focuses on typical three-terminal systems (actually a two-port network where a terminal is shared by both ports), this discussion can be directly extended to multiple-port systems. In such case, instead of single-input single-output (SISO) models, one would have to implement multiple-input multiple-output (MIMO) models for such systems.

7.2.2 Behavioral modeling technology

In behavior-based modeling, it is required to specify the mathematical operators that process the input signals and generate the estimated outputs, as highlighted in the previous sections. The selection of such mathematical operators is far from being a trivial task. In fact, the success of a behavioral model is a direct reflex of the adequate choice of such mathematical formulations. Fortunately, extensive theory has been developed on this topic, which can be used to support or conduct such selections.

This section presents the most widely used mathematical formulations that, under determined (but sufficiently general) conditions, were demonstrated to lead to universal approximator models of nonlinear dynamic systems (i.e., models whose prediction error can be theoretically reduced to an arbitrarily small value). This is, indeed, a very important property. These formulations are (theoretically) general models which, in principle, accommodate all the required signal processing combinations for capturing all the effects that are produced by the system under study (as long as the required assumptions hold). This property will be presented in detail in the following sections.

7.2.2.1 The two basic behavioral modeling questions

Section 7.2.1.2 presented a behavioral model for a diode device (Eqn (7.5)) that was designed in an ad-hoc way, simply based on the observation of the current–voltage characteristic of the diode (responding to a sine wave). This mathematical model considered that the diode current could be approximated by powers of the diode's instantaneous voltage added to powers of that same voltage delayed by one sample (it assumes that the current and voltage signals are sampled versions of the continuous real signals). Can it be guaranteed that these selected mathematical combinations are sufficient to describe (i.e., to approximate within the desired accuracy) the input–output behavior of the diode? Or, would it be possible that the diode would produce some internal effect that would require mixed terms in the model of the form $v_d(n)^k v_d(n-1)^{p-k}$, $k = 1, \ldots, p$? What about any other combinations that would result from any arbitrary operation on some or all of the samples of the input signal? In a generalized way, this question could be stated as: Is there a definition of a set of algebraic combinations of the input samples that is guaranteed to be able to successfully mimic the behavior of any nonlinear dynamic system (i.e., to form a complete basis of the model description)? Because the model is to be used with input excitations different from those considered in the model parameters extraction procedure (in fact, the properties of the input signals to be processed by the model are, in principle, unknown), another important question would be: Is it possible to produce a behavioral model with predictive capabilities, not limited to reproducing the measured data used for its extraction, from a finite set of observations?

A positive answer to the first question is of utmost importance to behavioral modeling because it avoids the ad-hoc model formulation approach that, when unsuccessful in representing the observed model behavior, would leave the model designer wondering what could be missing from the model. Such a positive answer would guarantee that the model could have the possibility to "capture" the system's entire behavior. The answer to the second question would guarantee that the model would be useful (i.e., would predict accurate outputs) under any excitation signal (not tested previously), thus conferring on the model a general scope of application.

Let us analyze these important issues by considering, first, the simpler case of a static SISO nonlinear system. By definition, a static system is one whose current output value depends only on the current (i.e., instantaneous) input value. Therefore, it can be characterized graphically by a static injective function, such as the one exemplified in Figure 7.10(a), which has the typical shape of an input–output voltage characteristic of an amplifier entering into compression. The problem of function approximation is widely covered in the mathematical literature and thus will not be thoroughly covered here. However, we would like to call the reader's attention to some demonstrated theorems that are very important in the context of behavioral modeling. For instance, the Weierstrass theorem (1885) states that any continuous function defined in a closed interval can be uniformly approximated by a polynomial function as closely as desired, within that interval. Figure 7.10(b) shows the successive approximations of the static function of Figure 7.10(a) for polynomials of increasing order. Therefore, the Weierstrass theorem demonstrates that the set of monomials (which is an infinite set) constitutes the basis of input signal combinations, which allows any static nonlinear system (continuous in the interval of interest) to be modeled to any desired accuracy level.

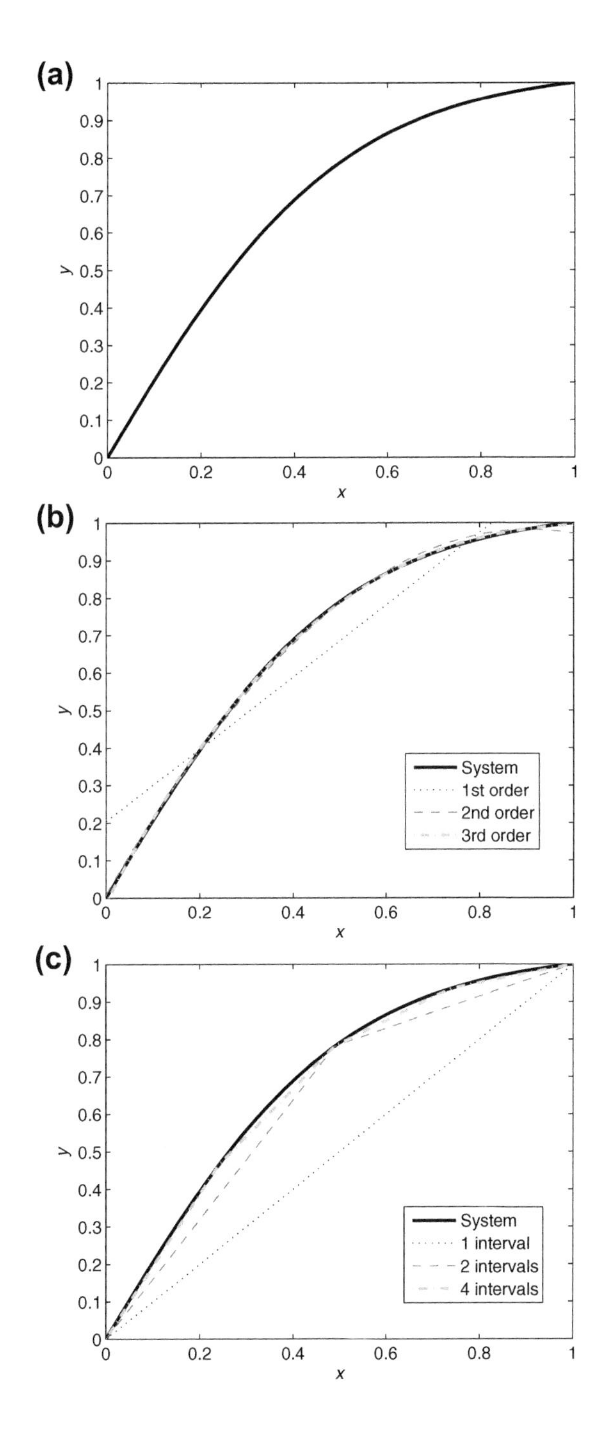

FIGURE 7.10

(a) Example of a static nonlinear system characteristic. (b) Approximation by polynomial functions of increasing order. (c) Approximation by the LUT-based approach using piecewise-linear interpolation functions.

Another possible approach for modeling continuous static nonlinear systems is to discretize the observed input–output relationship, storing a set of N input–output pairs in memory (this stack of values is usually called a lookup table [LUT]) and then use an interpolation function to determine the output corresponding to input values that fall between two points (e.g., n and $n+1$, $1 \le n \le N-1$). It can be demonstrated that this LUT-based approach can approximate the static function to any desired accuracy level, either by increasing the number of N sampling points or, for instance, by using polynomials as interpolation functions and applying the Weierstrass theorem to each of the $N-1$ intervals produced by the sampling. The LUT approach permits one to reduce the complexity (the polynomial order in the case of polynomials) of the interpolation function (notice that the polynomial-based approach mentioned previously can be thought of as a particular case of the LUT approach where only one interval is considered, defined by the minimum and maximum values of the input range). With an increase of the polynomial order, the least-squares polynomial approximation is degraded near the edges of the input range—an effect known as Runge's phenomenon. In this sense, the interpolation process used in the LUTs differs from the standard least-squares polynomial approximation because it usually imposes restrictions on the interpolation in the edges of each interval. For instance, the use of cubic splines as interpolating functions compels the splines to pass through the edge points and to maintain the first derivative at both sides of each edge point. Figure 7.10(c) shows the results of LUT-based approximations of the nonlinear characteristic of Figure 7.10(a), using linear interpolation functions (in this case, the approximation is denominated as piecewise linear) for different sampling schemes.

Although modeling of continuous static systems is quite straightforward, the complexity level of the problem is significantly increased when the system also presents dynamic effects. The generation of the current output value of the system depends not only on the instantaneous input values but also on past action (i.e., the system has energy storage processes). It is commonly mentioned that dynamic systems possess memory (and the corresponding effects are denominated memory effects). To properly characterize systems with memory, it is required to use a dynamic model. Depending on the topology selected for such model, the level of complexity can range from relatively simple models (although with limited generality) to highly complex models (these can reach a high level of generality). For instance, two well-known dynamic models of reduced complexity are the Wiener and the Hammerstein topologies, shown in Figure 7.11. These are constituted by a cascade of a nonlinear static block and a linear dynamic filter, and their usual advantage is exactly their simplicity (which results from the assumption that nonlinearity and dynamics are completely separated).

Even though the Wiener and the Hammerstein models have been extensively used, real systems are rarely composed of a cascade of a static nonlinear block and a linear dynamic one. In other words, these two-box models are not general representations of nonlinear dynamic systems. Nonlinearity and dynamics are usually entangled in complex forms in physical systems.

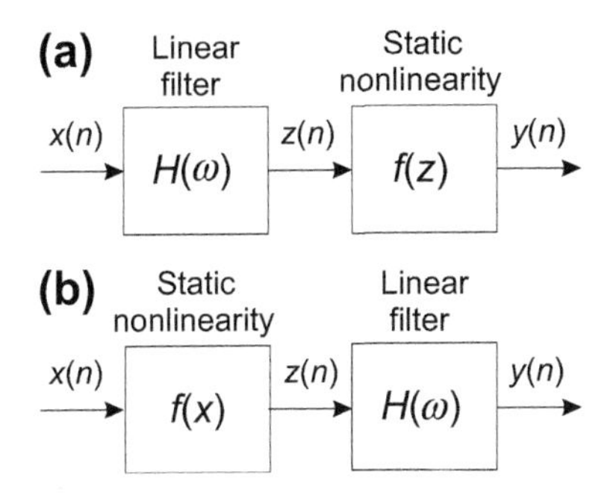

FIGURE 7.11

(a) The Wiener two-box model. (b) The Hammerstein two-box model.

In fact, we can write a general nonlinear equation for a SISO dynamic system (this analysis is directly extendable for MIMO systems) by defining a set of state variables, assembled in a vector $\mathbf{s}(t)$, in the equivalent forms:

$$\frac{\mathrm{d}\mathbf{s}(t)}{\mathrm{d}t} = \mathbf{f}_1[\mathbf{s}(t), x(t)] \quad \text{or} \quad \mathbf{f}_2\left[\mathbf{s}(t), \frac{\mathrm{d}\mathbf{s}(t)}{\mathrm{d}t}, x(t)\right] = 0 \tag{7.9}$$

where $\mathbf{f}_1$ and $\mathbf{f}_2$ are two vector of adequate nonlinear functions, and $x(t)$ is the input signal.

Because the output signal $y(t)$ is a composition of the state variables, $y(t) = g[\mathbf{s}(t)]$, then this general model can be written in the implicit recursive form:

$$f_{\mathrm{R}}\left[\frac{\mathrm{d}^p y(t)}{\mathrm{d}t^p}, \cdots, \frac{\mathrm{d}y(t)}{\mathrm{d}t}, y(t), \frac{\mathrm{d}^r x(t)}{\mathrm{d}t^r}, \cdots, \frac{\mathrm{d}y(t)}{\mathrm{d}t}, x(t)\right] = 0. \tag{7.10}$$

Nonlinear recursive (or feedback) models (even though being more compact models than nonrecursive ones because they require, in principle, fewer parameters to perform the same representation) are known to be susceptible to unstable conditions, and extreme care must be taken to guarantee that such models are stable for all their possible states of operation. On the other hand, direct (or feed-forward) models—that is, models that require only the current and past samples of the input signals to compute the current sample of the output signal—are inherently stable. Therefore, it would be very interesting to have a general formulation such as that in Eqn (7.10) but with a direct form. Fortunately, system identification theory has demonstrated that a continuous and stable system with fading memory can be modeled (up to any desired accuracy) by a direct model of the following form:

$$y(t) = f_{\mathrm{D}}\left[\frac{\mathrm{d}^q x(t)}{\mathrm{d}t^q}, \cdots, \frac{\mathrm{d}x(t)}{\mathrm{d}t}, x(t)\right]. \tag{7.11}$$

A very important theoretical result from system identification theory is that the model of Eqn (7.11) can be used to accurately predict the system's response to an input signal that was never considered before, if the parameters of such model are determined from testing the system with a set of sufficiently rich excitations (in the sense of covering a wide range of signal characteristics to be representative of all possible signals that are to be applied to the system) [1,2].

Nowadays, models and systems are handled in the discrete domain and not in continuous time, usually assuming a constant sampling frequency, f_s, in the continuous–discrete transformation. Therefore, it is important to represent the general formulations of Eqns (7.10) and (7.11) in the discrete domain (assuming that the time between two consecutive samples has a duration of $T_s = 1/f_s$). These can be expressed as follows.

For the explicit recursive case:

$$y(n) = f_{\text{Re}}\left[y(n-1), y(n-2), \cdots, y\left(n - M_{\text{y}}\right), x(n), x(n-1), \cdots, x(n - M_{\text{x}})\right] \tag{7.12}$$

For the implicit recursive case:

$$y(n) = f_{\text{Ri}}\left[y(n), y(n-1), y(n-2), \cdots, y\left(n - M_{\text{y}}\right), x(n), x(n-1), \cdots, x(n - M_{\text{x}})\right] \tag{7.13}$$

For the nonrecursive (direct) case:

$$y(n) = f_{\text{D}}[x(n), x(n-1), \cdots, x(n - M_{\text{x}})] \tag{7.14}$$

M_{x} and M_{y} are denominated the memory depth (or memory span) of the system in terms of its input $x(n)$ and output $y(n)$, respectively.

Not only due to the stability precautions required in the implementation of recursive model formulations, but also due to the significant advantages that direct models present in terms of parameter extraction (recursivity increases the complexity of this process), the most used modeling approaches adopt the nonrecursive formulation (which, like the recursive one, has theoretically guaranteed predictive ability if the system is stable and of fading memory). In general terms, any direct model formulation can be framed into the topological description of the general Wiener model, depicted in Figure 7.12, where the system's dynamics and static nonlinearity are split into two distinct blocks [1]. This topological separation is very important because the system dynamics are all concentrated in a linear block and all the nonlinear behavior is considered as static (or memoryless) and is condensed in a distinct block. Notice that this structure is completely different from the Wiener two-box model of Figure 7.11(a). In this case, the linear dynamic block is a single-input multiple-output structure, and the nonlinear static block is a multiple-input single-output structure.

In the general model formulations presented here, the indicated functions are not explicitly defined. Different modeling variations can be obtained by selecting different descriptions for those functions. However, it must be guaranteed that such descriptions provide the general approximation capabilities to the model. The following sections present the general model formulations that are most

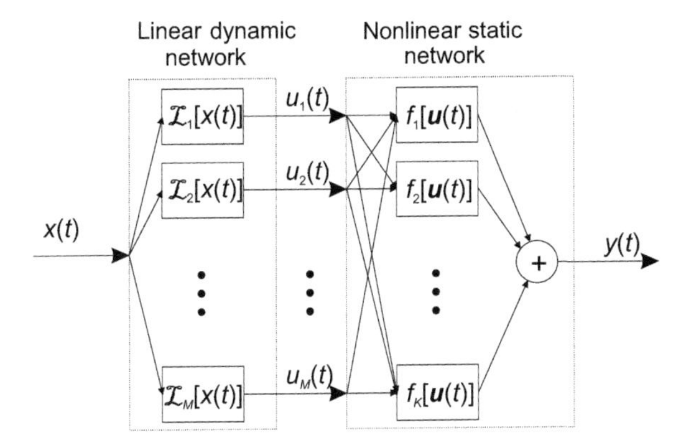

FIGURE 7.12

The general Wiener model.

commonly used. These formulations directly result from distinct implementations of the two blocks of the general Wiener model of Figure 7.12. The first case considers the nonlinear static block implemented by a multidimensional polynomial, giving rise to the polynomial filter model (or Volterra filter). If specific amplitude limited functions, such as radial basis functions or sigmoids, are used instead, it can be observed that the resulting model is the time-delay artificial neural network (TD-ANN). It was theoretically demonstrated that both polynomial filters and TD-ANNs have a global approximation capability for a stable system with fading memory. Finally, we will illustrate how interpolated LUTs can also be used in the nonlinear static block, producing the topology known as the LUT model.

7.2.2.2 Polynomial-based models

The first general model topology is based on polynomial approximations and is generally known as the polynomial filter. As an extension (in the polynomial sense) to the linear convolution, it consists of a multidimensional convolution integral (or sum, in the discrete form) associated with a polynomial-based approximation of the system's behavior. This extension is highlighted in the polynomial filter formulation:

$$y(n) = \sum_{p=1}^{P} y_p(n), \quad y_p(n) = \sum_{m_1=0}^{M-1} \cdots \sum_{m_p=0}^{M-1} h_p(m_1, \cdots, m_p) \prod_{q=1}^{p} x(n - m_q) \tag{7.15}$$

where the output signal sample $y(n)$ is computed from a sum of the kernels' outputs (one kernel per polynomial order), and the parameters weighting product combinations are the elements of the p-dimensional matrices $h_p(m_1, \cdots, m_p)$.

To illustrate how this formulation fits into the general Wiener model of Figure 7.12, we present, in Figure 7.13, the diagram representation of the first- and second-order kernels of the polynomial filter. The linear dynamic block consists

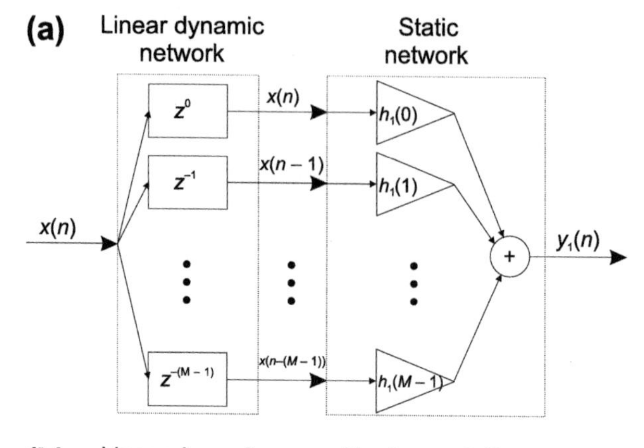

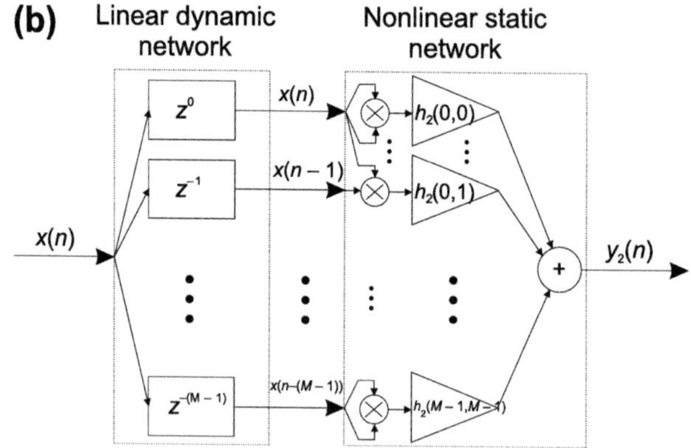

FIGURE 7.13

Diagram representations of the polynomial filter kernels: (a) first order; (b) second order.

of a cascade of time delays, and the nonlinear static structure is that of polynomial terms processing different combinations of delayed input signal samples.

The polynomial filter formulation of Eqn (7.15) shows one interesting feature of this model: The model is linear in the parameters, meaning that the parameter estimation process can be reduced to the linear form of $\mathbf{X\Theta} = \mathbf{Y} \Leftrightarrow$

$$\begin{bmatrix} x(M-1) & \cdots & x(0) & x(M-1)x(M-1) & \cdots & x(0)x(0) & \cdots \\ \vdots & \ddots & \vdots & \vdots & \ddots & \vdots & \ddots \\ x(N) & \cdots & x(N-M+1) & x(N)x(N) & \cdots & x(N-M+1)x(N-M+1) & \cdots \end{bmatrix}$$

$$\begin{bmatrix} h_1(0) \\ \vdots \\ h_1(M-1) \\ h_2(0,0) \\ \vdots \\ h_2(M-1,M-1) \\ \vdots \end{bmatrix} = \begin{bmatrix} y(M-1) \\ \vdots \\ y(N) \end{bmatrix} \tag{7.16}$$

where $\mathbf{\Theta}$ is the vector containing parameters of the kernels, $\mathbf{X}$ is the matrix containing the mixing products of the input signal samples, and $\mathbf{Y}$ is the vector of the measured output signal samples. This property allows the calculation of the parameters by means of the linear least-squares approach that guarantees that the minimization of the squared errors (where the error is defined by the difference between the measured output samples and those predicted by the model, for a set of excitation signals) has only one minimum: the global minimum.

One disadvantage of the use of polynomials as basis blocks for modeling the system's nonlinear characteristic is that polynomials (as has already been seen in the static case) suffer from catastrophic error degradation when used to extrapolate the system's behavior outside the range considered during the parameter extraction procedure. Because polynomials (and monomials) are not bounded in amplitude, the level of predictability of polynomial-based models is highly reduced outside the amplitude range of the input signals considered for parameter determination. At high-input amplitude values, the model output is built through the joint compensation of the effects of all monomials or polynomials of different order, and, because these are unbounded, a slight deviation over that range will surely break such joint compensation and some terms will dominate over the others. The model prediction rapidly deviates from the measurements. This characteristic means that polynomial-based models (including polynomial filters) are unsuitable for extrapolation.

7.2.2.3 Artificial neural network models

The implementation of TD-ANNs assumes that, in its direct form, the multidimensional nonlinear static block of the general Wiener model is built from basic structures of the form [3]:

$$y(n) = b_o + \sum_{k=1}^{K} w_k f_k \left[b_k + \sum_{m=0}^{M-1} w_{m,k}\, x(n-m) \right] \tag{7.17}$$

where $f_k[.]$ is a SISO nonlinear function (commonly denominated activation function), b_k is an input bias for the function $f_k[.]$, $w_{m,k}$ is the input weighting factor applied to the delayed input sample $x(n-m)$, w_k is an output weight that is applied to the output of the function $f_k[.]$, and b_o is the output bias. Figure 7.14 illustrates, in a diagram format, the TD-ANN basic structure.

The activation functions can be selected from a wide range of possible functions. However, the most commonly selected functions (especially in the microwave electronics field) are squashing functions, such as the hyperbolic tangent or other sigmoidal functions, or localized functions, such as the bell-shaped Gaussian (or radial basis function) [3–5]. Figure 7.15 depicts some of these commonly selected activation functions.

It is common to use the same type of activation function for all $f_k[.]$ functions of the TD-ANN model (although different types could also be used within the same network). The weights and biases associated to a function $f_k[.]$ are used to build scaled and shifted (in both input and output domains) versions of the basic activation

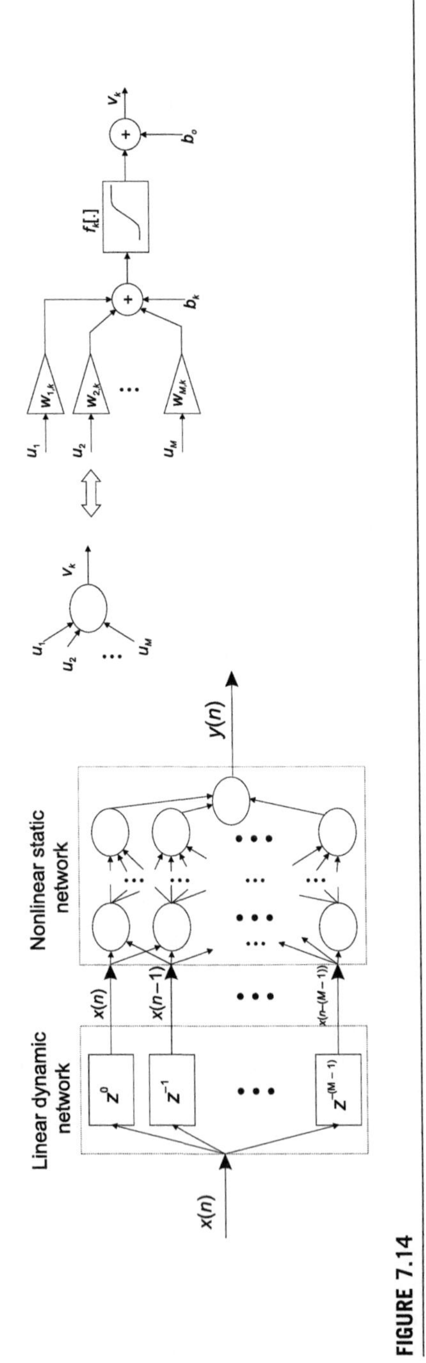

FIGURE 7.14

Basic structure of a TD-ANN model.

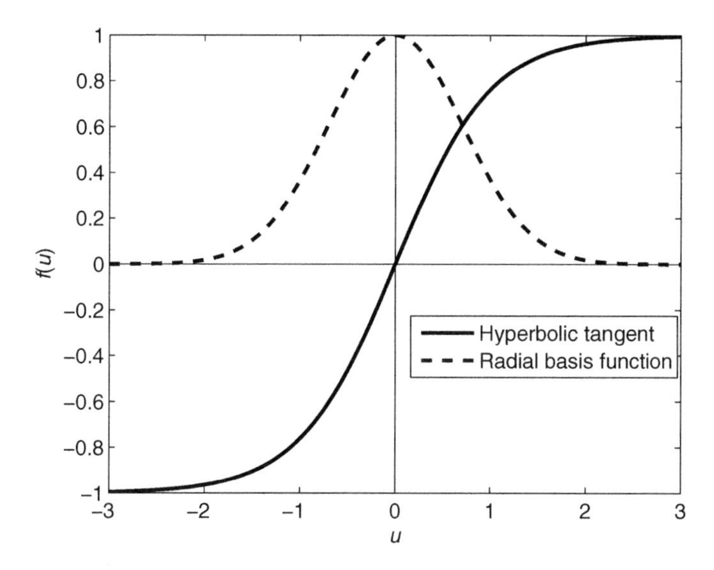

FIGURE 7.15

Two common activation functions: hyperbolic tangent $\left(f(u) = \frac{e^u - e^{-u}}{e^u + e^{-u}}\right)$ and radial basis function $(f(u) = e^{-u^2})$.

function type, thus creating the diversity required by the TD-ANN structure to model nonlinear dynamic systems.

One advantage that TD-ANN models have over Polynomial Filter models is that the TD-ANN can have a better extrapolation performance because the activation functions can be selected to be bounded in amplitude. Therefore, TD-ANN models do not suffer from the catastrophic extrapolation typical of polynomial-based models. However, TD-ANNs have the significant disadvantage of not being linear in the parameters and therefore not extractable through linear least-squares techniques. Even in a single layer TD-ANN (i.e., one with the exact form of Eqn (7.17) and not with cascaded layers in the form of Eqn (7.17), which is a strategy usually implemented to augment the modeling generality of the TD-ANN model), the input weight and bias coefficients cannot be linearly estimated from input–output measurements because these coefficients are arguments of nonlinear functions (the activation functions). This means that, for extracting the TD-ANN parameters, it is required to solve a nonlinear optimization problem whose error function easily has several minimum points (one global minimum and several local minima) and there is no guarantee that the obtained solution is that of the global minimum error. In fact, nonlinear optimization algorithms start from a set of initial parameter values which, as is known, strongly influence the minimum point that is reached by the algorithm. Most currently available nonlinear optimization algorithms are based on the error back propagation algorithm, a very interesting and computationally efficient approach to estimate the TD-ANN parameters [3].

7.2.2.4 Table-based models

Section 7.2.2.1 presents an example of LUT-based approximation of a static one-dimensional nonlinear function. This approach can be extended to multidimensional $\mathbb{R}^m \rightarrow \mathbb{R}$ nonlinear static functions

$$y = f(x_1, x_2, \cdots, x_m) \tag{7.18}$$

where a set of m-dimensional points, selected from a set of performed measurements, and the output value of y corresponding to each of such points are stored in a tabular form. This table of m-dimensional input entries to one-dimensional output values is then used to compute the estimated output signal by mapping the new values of the m input variables into the estimated output values. As in the one-dimensional case, only a subset of the possible input values is considered (not only due to limited measurement capabilities but also to limited storage space for the LUT), and every time the m-dimensional input point does not match any of the stored input points of the table, the corresponding output value estimate can be approximated by interpolation, based on the value of the table points in the vicinity of each considered input point. Different interpolating functions can be used to perform this task, the most commonly used being the linear and cubic spline interpolating functions.

One good example of this LUT-based approximation strategy is the canonical piecewise-linear representation (also known as the canonical piecewise-linear model) published in Refs [6,7], where a general function in the form of Eqn (7.18) is approximated by piecewise-linear sections. This constitutes a universal model in the sense that its structure remains unchanged for modeling different nonlinear functions; only the values of the model coefficients vary from function to function.

We can apply this modeling approach to the nonlinear static block of the general Wiener model of Figure 7.12, which is a multidimensional $\mathbb{R}^m \rightarrow \mathbb{R}$ nonlinear static function where the m input variables are now the delayed samples of the system's input signal:

$$y(n) = f[x(n), x(n-1), \cdots, x(n-(M-1))]. \tag{7.19}$$

Even though we have assumed that the linear dynamic block of the general Wiener model simply produces the successively delayed samples of the input signal, other more complex linear dynamic mappings, such as general filter banks, could be considered.

Besides the evident benefits that LUT-based models can have in terms of computational efficiency (for each output value estimate, they require only a search within the table domain for the nearest points and the execution of the interpolating function, which is usually of very low complexity), LUT models have the advantage of consisting of a set of local approximations to the system's behavior, instead of trying to approximate (with just one function) the global behavior of the system (like polynomial filters).

In other words, instead of adjusting the coefficients of only one function to fit the input—output behavior of a system in the whole considered domain of the input signal, LUT-based models divide the input signal domain into small sections, within which a dedicated function will be used for the fitting. This guarantees that the level of accuracy provided by a LUT-based model is high for every point of its domain, requiring for this purpose a set of very simple fitting functions. This is the opposite of the case for polynomial-based models, which use the same coefficients to fit the input—output data within the whole of the input range, and so need a very high number of parameters to meet high accuracy in the complete excitation domain. Consequently, extraction of polynomial coefficients easily leads to ill-conditioning in the linear regression process for extracting such coefficients, being also prune to oscillatory fitting behavior near the input domain edges and catastrophic approximation beyond such domain.

7.2.3 Behavioral models for RF and microwave devices

The previous section presented the general behavior model topologies that are known to be theoretically able to model the input—output behavior of a wide variety of nonlinear dynamic systems. Now we restrict this wide field to the case of radiofrequency (RF) and microwave devices. The core of most active circuits used in telecommunications is based on transistors. Transistors are nonlinear devices. The drain current of an FET depends, in a nonlinear way, on the gate-source and drain-source voltages, which also present dynamic effects (e.g., due to the capacitive effects between its intrinsic terminals). Therefore, accurate models for transistor behavior representation are required for multiple RF and microwave design and analysis applications.

Even though the general models of Section 7.2.2 could be directly applied to model transistor behavior, reduced (or pruned) versions of these have been, in fact, implemented. This pruning can have distinct support. In some cases, it is supported by the physical characteristics of the device (a transistor is not a general nonlinear dynamic system and, therefore, a more compact version of a general model should be sufficient to describe it). In other cases, it is supported in more empirical assumptions. This pruning process, which is essential to avoid models with a high number of coefficients (unmanageable, most of the time), is still not yet completely solved for the case of microwave device modeling.

This section describes different approaches used to model transistor devices, divided into two domains: time-domain models and frequency-domain models.

7.2.3.1 Time-domain behavioral models

We start by analyzing the microwave device models that implement time-domain formulations. The output signal is specified in the time-domain (in continuous or discrete form) and is calculated by the model as a mathematical operation on the input (and, possibly, the output) time-domain signals. This section enumerates the most common approaches used to model microwave transistors using time-domain formulations.

The modeling approach first described is based on nonlinear time series analysis [8] and consists of considering all the involved signals, including the input signals, as internal state variables of an equivalent autonomous dynamical system (i.e. one that responds only to its internal states, and to which no external signal is applied). Let a system, with input and output time series $\mathbf{x_n} = [x(0), x(1), \cdots, x(n)]$ and $\mathbf{y_n} = [y(0), y(1), \cdots, y(n)]$, respectively, be modeled by the nonlinear state equation (in the discrete-time form):

$$\begin{cases} \mathbf{s}(n+1) = \mathbf{f}[\mathbf{s_n}, \mathbf{x_n}] \\ y(n+1) = \mathrm{h}[\mathbf{s_{n+1}}] \end{cases} \tag{7.20}$$

where $\mathbf{s_n} = [\mathbf{s}(0), \mathbf{s}(1), \cdots, \mathbf{s}(n)]$ is the set of the state vector values for each time instant (from the first sample to the nth sample), and $f[\cdot]$ and $h[\cdot]$ are a vector of functions and a function, respectively. A consequence of Takens' delay embedding theorem [9] is that if a significant past memory is retained for the output signal, forming the delay vector

$$\mathbf{v_n} = [y(n), y(n-1), \cdots, y(n-L+1)] \tag{7.21}$$

then there exists a smooth function $g : \mathbb{R}^L \rightarrow \mathbb{R}$, which can be estimated from time series data, which verifies $y(n+1) = g[\mathbf{v_n}]$ for all $\mathbf{v_n}$ [10].

Even though it can be inferred from this theory that, like any other nonlinear dynamic system, a microwave device, excited with a particular input signal, can also be considered as an autonomous system modeled only through the observation of the evolution of its output signal, the time series analysis is not usually the chosen approach to implement device models. (It is usually preferred to take into account the fact that also the device input signal can be measured and, thus, it is not required to consider it as an additional internal state variable.) However, some applications can be found in the literature, such as those described in Ref. [11].

The second modeling approach considers polynomial filters. Polynomial-based models have been used to model microwave devices and transistor-based circuits, such as power amplifiers. One example of their application to device modeling is the Volterra input–output map (VIOMAP) model, which was developed from Volterra series analysis (although in the frequency domain) [12]. Device models are usually assumed to have the two-port admittance form of Figure 7.6 (herein replicated in Figure 7.16 for convenience), where the currents in ports 1 and 2 are

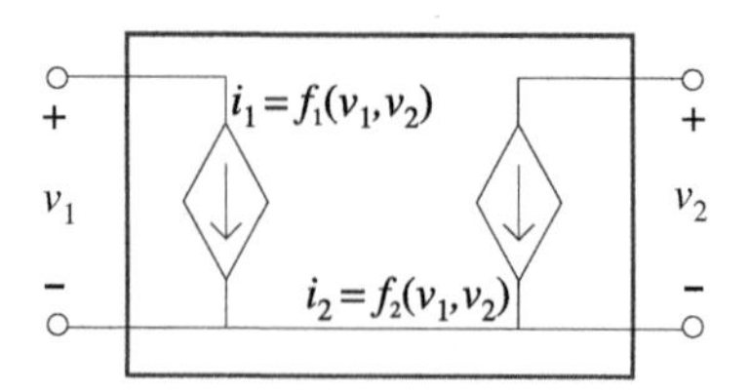

FIGURE 7.16

Conceptual microwave device two-port admittance model.

nonlinear dynamic functions of the ports' voltages. Even though these functions are represented as direct mappings, these could also be conceived in the recursive form, where both f_1 and f_2 would depend on i_1 and i_2.

In microwave device modeling, it is usual to assume that the nonlinear dynamic functions of this model can be simplified into separable-function forms. For instance, the i_{DS} current in an FET is usually assumed to be the product of a nonlinear dynamic function dependent only on v_{GS}, by another function only dependent on v_{DS} (i.e., $i_2 \approx f_{2,1}(v_1)f_{2,2}(v_2)$). This simplification replaces a bidimensional function into two single-dimensional ones.

The functions f_1 and f_2 can be implemented as general polynomial filters; however, the number of coefficients rapidly achieves unmanageable values as the polynomial order and memory depth are increased. Adequately pruning these general polynomial filters to significantly reduce the number of parameters without jeopardizing the modeling accuracy and prediction capability is a fundamental problem that is not yet completely solved, even though several attempts have been published. Most of the pruning approaches applied to polynomial filter models for microwave device modeling consist only of empirical reductions of the possible product combinations between delayed input samples, followed by accuracy evaluation tests to observe if such reduction leads to acceptable modeling results. A simple and widely used restriction consists in considering only the products between samples delayed by the same amount. Taking the form of Eqn (7.22) for the SISO case (for input signal $x(\cdot)$ and output $y(\cdot)$), it is commonly called the memory polynomial (MP) model, following the model proposed in Ref. [13] for modeling the inverse of the envelope behavior of power amplifiers.

$$y(n) = \sum_{p=1}^{P} \sum_{m=0}^{M-1} h_p(m)x(n-m)^p \tag{7.22}$$

Other variants of the MP model have been proposed, which consider the inclusion of additional product terms between distinct delayed input samples, such as the envelope MP [14] and the generalized MP [15].

Polynomial-based models are simple to deal with and are linear in the parameters, which eases the polynomial coefficients' extraction. However, as mentioned in Section 7.2.2, their use is limited when the model is applied for extrapolation.

The nonlinear dynamic admittance functions $f_1(\cdot)$ and $f_2(\cdot)$ of Figure 7.16 (or their recursive counterparts) can also be implemented through TD-ANN models, avoiding the catastrophic modeling degradation of polynomial models when under extrapolation conditions. As mentioned previously, this advantage is gained at the cost of losing the linearity in the parameters during model extraction, and a nonlinear optimization problem needs to be solved to tune (or train) the TD-ANN parameters. Similarly to the polynomial-based approach, it is most suitable, in practice, to assume the approximation of the bidimensional functions $f_1(\cdot)$ and $f_2(\cdot)$ by two pairs of single-dimensional separable functions $f_{k,1}(\cdot)$ and $f_{k,2}(\cdot)$, with $k = 1,2$. Then, a TD-ANN structure is used to model each of these single-dimensional functions. (Their parameters are estimated by fitting preconceived test data of the device under modeling.)

Again, the problem of extreme model complexity arises if general TD-ANN formulations are used for each function. For that case, the number of model parameters reaches impractical values as their extraction process becomes cumbersome and hard to control. For example, it may happen that a slightly different extraction signal set could lead to a completely different set of model values. It also becomes necessary to prune the general TD-ANN models, resulting in a model of good computational efficiency (both from the simulation and extraction points of view) and also guaranteeing the desirable prediction accuracy.

Neural network models are usually pruned on an empirical basis, where the model complexity is iteratively increased (by the addition of more neurons and more layers) and, for each step, checked for accuracy (with the data set selected for training/extracting the model). However, a wise selection of the activation functions of TD-ANN models can actually be considered as a form of pruning because intuition would lead us to believe that a TD-ANN model whose activation functions are similar in shape to some of the characteristics observed in the device being modeled (such as the DC behavior) would require less complexity than other TD-ANN models (with different activation functions) for the same goal accuracy. This is probably the reason why hyperbolic tangent and sigmoid functions (the latter is a mere transformation of the former) are so often used in microwave device modeling [16,17]. An interesting approach to evaluate the sensitivity (or, in other words, the robustness) of FET ANN models was proposed in [18].

Device models based on TD-ANNs are recognized to be computationally efficient and to present very smooth characteristics, which is a key benefit over other approaches that are not infinitely differentiable [19]. For example, LUT-based device models are limited to the flexibility of their interpolation functions, which is usually limited to cubic splines. Therefore, derivatives over the third one are always null within each interval, and discontinuous at the LUT data points. Nowadays, products can be found in the industry that consider ANN-based modeling of microwave devices (e.g., see Ref. [20]).

The next device modeling approach is based on the use of LUTs. Most of the device models encountered in literature follow a table-based implementation of the nonlinear dynamic admittance functions of the model of Figure 7.16. From simple implementations that consider only the static current–voltage (I–V) characteristics to multidimensional LUTs for nonlinear dynamic characterization, table-based device models present the advantage of being determined directly from measured data; the LUTs are the measured input–output points. However, because these measured data do not cover every possible state of the device under modeling, interpolation between LUT entries is required. Cubic splines are probably the most common interpolating functions. They show a significant advantage over piecewise-linear interpolation because the latter have constant first derivative in the interpolating intervals (and null higher derivatives) and derivative discontinuity on the LUT points. This discontinuity also appears in cubic spline interpolation but for derivatives of third or higher order; the splines are settled to maintain the continuity of the function and of its first and second derivatives, at the data points. As mentioned previously, this discontinuous behavior is recognized as a drawback of table-based device models because, as is known, the successive derivatives of $I-V$ curves are important in transistor simulation. In some applications, this supports the use of higher-order polynomial and TD-ANN structures to model the transistor behavior.

In the previously exposed approaches, the $f_1(\cdot)$ and $f_2(\cdot)$ functions of the transistor model of Figure 7.16 are implemented through reduced versions of general model formulations that are established until the desired modeling accuracy is met for a particular measured data set. Therefore, these approaches are generally considered to be of empirical nature. However, knowledge from the physical intrinsic behavior of the device can be used to prune the microwave device model. One of the most usual approaches consists of separating the nonlinear DC characteristic of the current functions $f_1(\cdot)$ and $f_2(\cdot)$ from the nonlinear dynamic behavior, which is assumed to originate from the nonlinear capacitive effects that are observed across the terminals of the intrinsic device. Using the physical knowledge that explains both the static nonlinear and capacitive-driven dynamic behavior of transistors, the model shown in Figure 7.17 is one of the most common approaches to simplify (i.e., prune) the device model, known as the current/charge, in-phase/quadrature or simply the I/Q model.

The I/Q model can be seen as the first-order dynamic approximation of the expansion of the two-dimensional functions $f_1(\cdot)$ and $f_2(\cdot)$ in terms of its successive dynamic terms, as given in

$$i_k\left(v_1, v_2, \frac{dv_1}{dt}, \frac{dv_2}{dt}, \frac{d^2v_1}{dt^2}, \frac{d^2v_2}{dt^2}, \cdots\right) \approx i_{Ik}(v_1, v_2) + i_{Qk}\left(\frac{dv_1}{dt}, \frac{dv_2}{dt}\right) + \cdots, \quad k = 1, 2 \tag{7.23}$$

where $i_{Ik}(\cdot)$ is a static function and

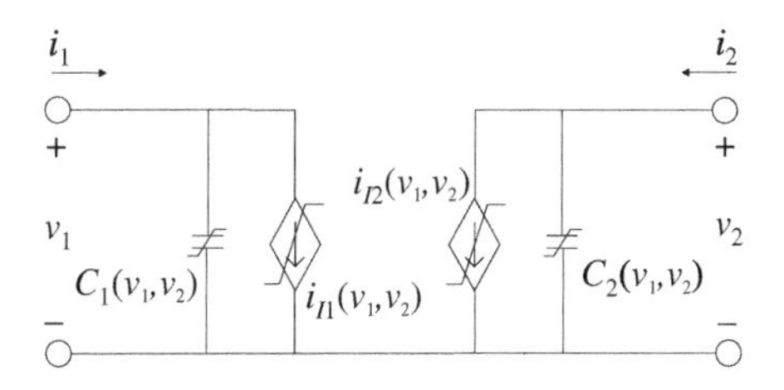

FIGURE 7.17

The I/Q intrinsic device model.

$$i_{Qk}\left(\frac{dv_1}{dt},\frac{dv_2}{dt}\right)=\frac{d}{dt}\left[q_k\left(\frac{dv_1}{dt},\frac{dv_2}{dt}\right)\right]=\left.\frac{d\,q_k(v_1,v_2)}{dv_1}\right|_{[v_1,v_2]}\cdot\frac{dv_1}{dt}+\left.\frac{d\,q_k(v_1,v_2)}{dv_2}\right|_{[v_1,v_2]}\cdot\frac{dv_2}{dt}$$
$$=C_{k1}(v_1,v_2)\frac{dv_1}{dt}+C_{k2}(v_1,v_2)\frac{dv_2}{dt}. \tag{7.24}$$

Several published transistor models make use of the structure of the I/Q model, considering different implementations for the nonlinear static functions $i_{Ik}(\cdot)$ and $C_1(\cdot)$, through LUTs, ANN, or polynomial functions. Some examples are found in Refs [21–25]. Another interesting characteristic of the I/Q model that makes it very appealing for transistor modeling is the fact that it can be extracted through bias-dependent S-parameter measurements, which is an easy procedure to implement in most RF and microwave laboratories.

7.2.3.2 Frequency-domain behavioral models

The previous time-domain formulation, namely that condensed in Figure 7.16, can be converted into the frequency-domain. Consider an initial simple case, in which the two current functions $f_1(\cdot)$ and $f_2(\cdot)$ are linear dynamic. In this case, the frequency-domain representation is

$$I_k(\omega)=Y_{k,1}(\omega)V_1(\omega)+Y_{k,2}(\omega)V_2(\omega),\quad k=1,2 \tag{7.25}$$

where $Y_{k,m}(\omega), k, m\in\{1,2\}$ are linear admittance frequency-domain functions. This model is only applicable to devices under small-signal excitation because it is a linear one. One obvious first step into a nonlinear frequency-domain formulation is to simply evaluate the device behavior under sinusoidal excitation plus a DC component and to consider that these admittances are also a function of the sinusoidal voltage and DC amplitudes, resulting in the so-called large-signal admittances matrices. This is a commonly used frequency-domain formulation because it is directly extractable from relatively simple laboratory measurements. This also has the benefit of being a much simpler model than general frequency-domain formulations, such as the equivalent frequency-domain Volterra model (or polynomial filter). However, as expected, the loss of generality is usually accompanied by global accuracy reduction.

In fact, for periodic input excitations, the generic time-domain polynomial filter formulation can be analytically converted into the frequency-domain by application of the multidimensional Fourier transform [2] (this analytical transformation is not possible with TD-ANN of table-based formulations). This has been used to formulate, for instance, the VIOMAP device model of Ref. [12]. However, the high number of required parameters of this formulation usually leads to reduced implementations, such as in the time-domain case.

Here, we present the most commonly used frequency-domain strategies for modeling microwave transistors. As will be perceived, these formulations are intimately related to the available laboratory procedures to measure device behavior in the frequency-domain.

S parameters representing the relationship (in amplitude and phase shift) over frequency between incident and reflected waves in multiport systems, under sinusoidal excitation, form the most common representation of linear behavior of RF and microwave systems. S parameters can be directly measured through a vector network analyzer (VNA), which returns the multiport transmission and reflection matrices for a range of predefined frequencies. This can be easily transformed into the admittance formulation of Eqn (7.25).

As mentioned earlier, the linear model is limited to the small-signal excitation regime. However, S-parameter measurements can be taken for different DC bias points of the port terminals. Therefore, the variation with bias point of the linear dynamic device behavior can be measured with a VNA, which is already useful to determine a reduced version of the $f_1(\cdot)$ and $f_2(\cdot)$ functions of Figure 7.16 or the device I/Q model. Moreover, different magnitudes of the sine waves exciting the device can be considered, and for each one the corresponding S parameters can be measured and registered; this results in the large-signal S parameters model.

For the device model shown in Figure 7.16, let us consider the following voltage excitation signal on port k (sinusoidal waves at frequency ω_0, with a DC component):

$$v_k(t) = V_{k,1}\sin(\omega_0 t) + V_{k,0}, \quad k = 1,2 \tag{7.26}$$

where the subscript m of $V_{k,m}$ represents the frequency to which such parameter is referred (in multiples of ω_0; $m=0$ is the DC component, and $m=1$ is the fundamental frequency component).

By measuring the S parameters for different $V_{k,0}$ and $V_{k,1}$ values, observing also the harmonic components that appear at multiples of ω_0, the large-signal frequency-domain admittance model can then be constructed, yielding, for the case of the two-port model of Figure 7.16:

$$\begin{aligned} I_{k,m}\left(\omega_0, V_{1,0}, V_{2,0}, V_{1,1}, V_{2,1}\right) &= Y_{1k,m}\left(\omega_0, V_{1,0}, V_{2,0}, V_{1,1}, V_{2,1}\right)V_1(\omega_0) \\ &\quad + Y_{2k,m}\left(\omega_0, V_{1,0}, V_{2,0}, V_{1,1}, V_{2,1}\right)V_2(\omega_0). \end{aligned} \tag{7.27}$$

In Eqn (7.27) the $Y_{1k,m}$ coefficient is interpreted as the admittance (or transadmittance) from the voltage applied (at fundamental frequency ω_0) at port 1 to the current observed at port k, at the frequency $m\omega_0$. The $Y_{2k,m}$ coefficient has identical interpretation but regarding the voltage applied at port 2 (also at the fundamental frequency ω_0). This approach is generally known, in control theory, as the sinusoidal input describing function (SIDF).

A drawback of this formulation is that it assumes that the voltages at the two ports of the device are sinusoidal (plus DC). Naturally, because the device is nonlinear, harmonic voltage components will appear at both ports of the device, and further relationships (admittances and transadmittances) need to be determined for increasing the accuracy and generality of this formulation. The relationship between harmonics of the port voltages to all the frequency components of the port currents needs to be measured and registered too. This will considerably increase the complexity of this model (imagine the admittances and transadmittances depending also on the magnitudes of the voltage harmonics), jeopardizing its practical usefulness.

An interesting approach to reduce the number of possible combinations between input and output tones (fundamental and harmonics) is to consider that the magnitude of the harmonic tones, at both ports, is much smaller than the magnitude of the signals at the fundamental frequency ω_0. This approximation, which is commonly verified in practice, permits treating the harmonic input terms as if they were linearly processed by the device (i.e., the harmonics' magnitudes are not sufficiently high, by themselves, to produce nonlinearly generated terms at the output whose impact would be noticeable in the observations). Therefore, the combinations that need to be accounted for in the frequency-domain formulation are those resulting from nonlinear dynamic processing of the fundamental input voltage terms and their mixing with linearly processed input harmonic terms, and also the terms produced by linear dynamic processing of the input harmonics. This approximation is generally known as the harmonic superposition principle, and it is the basic principle that supports the polyharmonic distortion (PHD) model [26–28]. The PHD formulation is the supporting model of the X parameters, an extension of the S parameters for the nonlinear case proposed by Agilent Technologies [29]. This formulation has been gaining increased attention in recent years from the academic and industrial communities. One of the reasons for this is the deployment of laboratory equipment that is able to directly measure the X parameters (i.e., the multiple relationships between input–output fundamental and harmonic sinusoidal terms), resulting in a direct extraction of PHD device models. Moreover, circuit-level and system-level simulation applications are also presenting the flexibility to incorporate directly the PHD model formulation, thus providing an easy and efficient process from device characterization by measurements to device (or device-based circuits) simulation.

7.2.4 De-embedding measurement data

As seen in the previous subsection, behavioral modeling can be understood as a set of systematic techniques to represent and interpolate measurement data in an organized and compact way. Therefore, it is determined by one's ability to observe the actual device's input–output characteristics. Indeed, it can even be said that pure behavioral modeling exclusively depends on measurement data, which means that its success is a direct consequence of the accuracy with which observed behavior is captured.

Therefore, if there are branches of knowledge in which calibration and de-embedding procedures are important, behavioral modeling is certainly among them. Therefore, the next two subsections are dedicated to these key aspects of measurement.

7.2.4.1 Measurement error mitigation techniques

According to conventional measurement theory, measurement error is any captured data that cannot be attributed to the device under test (DUT) but instead to the measurement instrumentation or setup. In other words, error is any measurement result that cannot be classified as desired signal or DUT behavior. So, assuming we are testing a system whose input and output are, say, $x(t)$ and $y(t)$, respectively, we will denote the captured input and output as

$$\hat{x}(n) = x(n) + F_X[x(n)] + p_X(n) \tag{7.28}$$

and

$$\hat{y}(n) = y(n) + F_Y[y(n)] + p_Y(n) \tag{7.29}$$

where the sampled time is such that $t = nT_S$ (in which T_S is the sampling period), $F_X[x(n)]$ and $F_Y[y(n)]$ are deterministic linear or nonlinear and static or dynamic operations of $x(n)$ and $y(n)$, and $p_X(n)$ and $p_Y(n)$ stand for measurement errors associated to the captured input and output, respectively. Within this definition, errors can be divided into distortions and additive perturbations, being further divided into random and systematic.

Distortions can be distinguished from additive perturbations (or simply perturbations) because the former are signal dependent, whereas the latter are not. This means that, although perturbations are always present, distortions vanish when the desired signal is turned off. Examples of perturbations are systematic DC offsets or interferences and random flicker or white additive noise. Examples of distortions can be unadjusted gain, bandwidth limitations due to linear filtering, or even nonlinear distortion such as clipping due to full-scale saturation. These distortions are modeled by $F_X[x(n)]$ and $F_Y[y(n)]$, whereas the additive perturbations are modeled by $p_X(n)$ and $p_Y(n)$ in the error models of Eqns (7.28) and (7.29).

In line with these definitions, systematic perturbations can be corrected by zeroing and subtraction, a technique that consists of performing a control measurement of a zero signal to capture the perturbation, followed by the real measurement and subsequent subtraction of the previously measured perturbation. In cases where these systematic perturbations drift in time, as are the ones related to aging or thermal phenomena, repeated zeroing and subtraction procedures may be needed.

Unfortunately, this procedure is useless for random perturbations as they may have already varied when the correction is to be applied. If this variation in time is such that the frequency content of the perturbation is significantly different from the desired signal, then it still can be eliminated by appropriate filtering. An example of filtering used when the random perturbation is much faster than the desired signal, or whenever the test can be repeated in exactly the same conditions, is averaging—a particular form of low-pass filtering. The true challenge is faced when the perturbation and signal components occupy the same spectrum and linear filtering cannot be applied. In that case, we assume that the only feature that distinguishes the useful signal from the undesired perturbation is the inherent deterministic relationship that exists between the observed system's input and output, and we rely on cross-correlation.

Cross-correlation between two measured entities, say $\hat{x}(n)$ and $\hat{y}(n)$, is defined as the averaged product between them:

$$R_{\hat{x}\hat{y}}(\sigma) \equiv \langle \hat{x}(n)\hat{y}(n+\sigma)\rangle = \frac{1}{N}\sum_{n=1}^{N} \hat{x}(n)\hat{y}(n+\sigma) \tag{7.30}$$

Substituting Eqns (7.28) and (7.29) into Eqn (7.30) and realizing that $p_x(n)$ and $p_y(n)$ are random measurement errors uncorrelated with both $x(n)$ and $y(n)$ and between each other, this cross-correlation becomes

$$\begin{aligned} R_{\hat{x}\hat{y}}(\sigma) \equiv \langle \hat{x}(n)\hat{y}(n+\sigma)\rangle &= \langle x(n)y(n+\sigma)\rangle + \langle F_x[x(n)]F_y[y(n+\sigma)]\rangle \\ &+ \langle F_x[x(n)]y(n+\sigma)\rangle + \langle x(n)F_y[y(n+\sigma)]\rangle \end{aligned} \tag{7.31}$$

which is only dependent on the desired input, output, and their distortions.

This insensitivity of cross-correlation to random measurement errors stands as the main reason why most system identification, or model extraction, methods rely on it. However, it should be noted that cross-correlation is a linear metric, in the sense that Eqn (7.31) only gives a non-null result when $x(n)$, $y(n)$, $F_x[x(n)]$, and $F_y[y(n)]$ are linearly related. For example, this means that if one wants to measure a component of the form of $x^2(n)$ in the output $y(n)$, then the appropriate cross-correlation to be used would be $\langle x^2(n)y(n+\sigma)\rangle$ as $\langle x(n)y(n+\sigma)\rangle$ would also be zero.

As seen from Eqn (7.31), after random perturbations are eliminated, only distortions remain. These are corrected by a set of methods known as calibration and de-embedding, which are so important in metrology that they deserve special treatment in a separate subsection.

7.2.4.2 Instrument calibration and de-embedding

After all additive errors have been corrected, only the measurement instrument's distortions remain. These can be either linear or, nonlinear, or even static or dynamic. Similarly to what we have done to correct offsets, we will first have to identify the distortions to then be able to correct them from the measured raw data.

Contrary to additive perturbations, distortions can only be manifested in the presence of signal, and thus have to be identified when measuring a signal. The question that arises is: When measuring a signal, how can one distinguish the "true" signal from the distortion? For example, if you suspect that a voltmeter may have a gain error and you have a reading of, say, 1.1 V, how can you know that the gain error is 1.1 because you were measuring a correct voltage of 1 V, or is there no gain error because it is the correct voltage that values 1.1 V? Obviously, the way to identify the gain error is to use a very particular and previously known signal. In fact, this signal has to be a standard, as its quality will determine the accuracy with which one can identify the instrument distortions and so, ultimately, the accuracy of the forthcoming measurements. This process of identifying the distortion and then correcting it is called calibration.

If the distortion is simply a gain error, its identification and correction is straightforward. If, however, this gain error should be assumed to possibly vary with frequency (i.e., it behaves as a filter, $G(\omega)$), then its identification must be done in a more elaborate way. To begin with, the standard signal used for calibration must have frequency components within the whole of the measurement, and thus calibration, bandwidth. One possibility is to use a standard noisy signal as white noise or simply to use a set of closely spaced tones. However, if the filter to be identified can be assumed linear, as is usually the practical case (e.g., because it is due to imperfections of any passive devices as cables, directional couplers, attenuators), superposition applies and we can substitute this single test with a K-tones multisine by K much easier tests, each one with a CW signal of different frequency. In this case, the filter can be identified using frequency-domain cross-correlation as follows:

$$G(\omega_k) = \frac{\langle Y(\omega_k)X^*(\omega_k)\rangle}{\langle X(\omega_k)X^*(\omega_k)\rangle} = \frac{\langle Y(\omega_k)X^*(\omega_k)\rangle}{\left\langle |X(\omega_k)|^2 \right\rangle} \tag{7.32}$$

where the input–output cross-correlation $\langle Y(\omega_k)X^*(\omega_k)\rangle$ can be computed via synchronous detection in a mixer-based signal analyzer (multiplication of $y(t)$ by a cosine and a sine of $\omega_k t$, followed by integration or low-pass filtering) and $\langle |X(\omega_k)|^2 \rangle$ is simply the input signal power at the tested frequency ω_k. Then, correction can be applied by simply dividing the measured response by the corresponding gain:

$$Y_c(\omega_k) = \frac{Y_r(\omega_k)}{G(\omega_k)} \tag{7.33}$$

in which $Y_c(\omega_k)$ and $Y_r(\omega_k)$ stand for the measured raw and corrected data, respectively.

In the field of network characterization, this conceptual calibration process needs to be elaborated. In fact, beyond this simple forward gain calibration, a reverse gain error correction and forward and reverse mismatch corrections are also needed. This corresponds to correcting the effects of undesired, but often present, parasitic networks from the actual DUT, or, as is usually accepted in microwave engineering, changing the reference plane from the one associated with the measured raw data to the one actually corresponding to the DUT—a procedure known as de-embedding.

These are the basic principles used in the workhorse of RF device characterization, the linear VNA. This instrument relies on the architecture shown in Figure 7.18, although a simpler version of only two receivers could also be used. It comprises a control/display unit, a receiver set, and a test set.

Because the VNA is intended to extract the most popular RF behavioral model, the S-parameter linear model, its test set is composed of a reflectometer of four directional couplers intended to capture the DUT's incident and reflected waves at each of the ports a_1, b_1 and a_2, b_2. Indeed, the S-parameter model assumes that the two-port

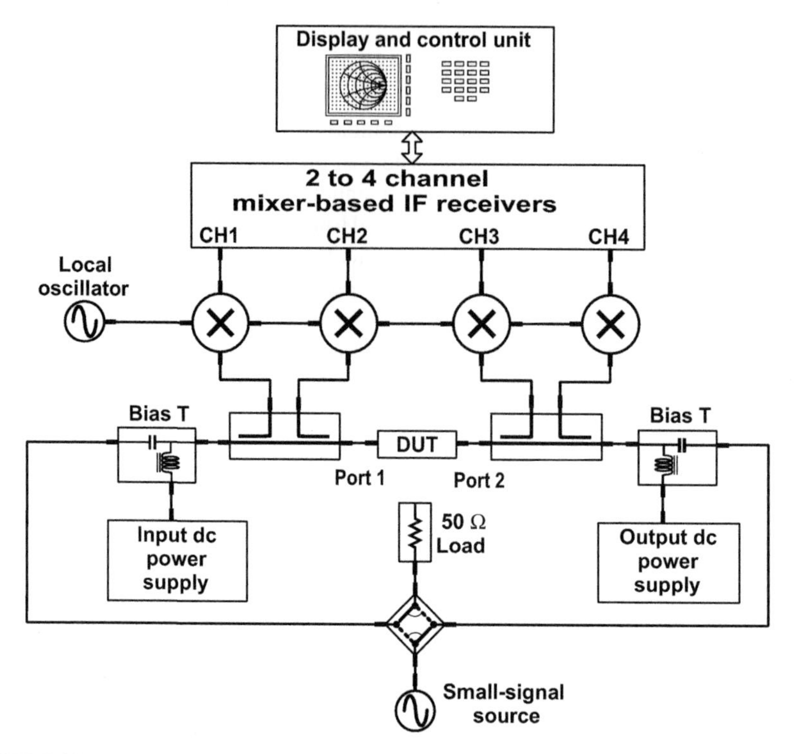

FIGURE 7.18

The linear vector network analyzer architecture.

DUT is linear. Therefore, its frequency-domain reflected or transmitted wave responses, $B_1(\omega)$ and $B_2(\omega)$, can be given as linear combinations of its frequency-domain incident wave excitations $A_1(\omega)$ and $A_2(\omega)$:

$$B_1(\omega) = S_{11}(\omega)A_1(\omega) + S_{12}(\omega)A_2(\omega) \tag{7.34a}$$

$$B_2(\omega) = S_{21}(\omega)A_1(\omega) + S_{22}(\omega)A_2(\omega) \tag{7.34b}$$

where these incident and reflected/transmitted waves can be related to the ports' currents and voltages by:

$$A_{1,2}(\omega) = \frac{V_{1,2} + Z_0 I_{1,2}}{2} \tag{7.35}$$

$$B_{1,2}(\omega) = \frac{V_{1,2} - Z_0 I_{1,2}}{2} \tag{7.36}$$

where Z_0 is the ports' characteristic impedance, usually assumed to be 50 Ω. According to Eqns (7.35) and (7.36), it is possible to convert the S parameters' matrix relationship of Eqn. Eqn (7.34), $\mathbf{B} = \mathbf{SA}$, into any other linear network matrix description such as the impedance matrix, $\mathbf{V} = \mathbf{ZI}$, the admittance matrix, $\mathbf{I} = \mathbf{YV}$, etc. [30].

Unfortunately, because of nonidealities of the receiver and test sets, the readings of the control/display unit are not the actual DUT's incident and reflected/transmitted waves, $\mathbf{a_c}$ and $\mathbf{b_c}$, but raw (uncorrected) versions of them, $\mathbf{a_r}$ and $\mathbf{b_r}$. Their correction requires a calibration model that is capable of describing the referred VNA nonidealities. Usually, these are assumed to be linear (because appropriate controlled attenuators and variable gain amplifiers are used to prevent any possible VNA receiver saturation) and associated with the test set's directional couplers in the following way.

A dual directional coupler (e.g., the one used to capture a_1 and b_1 in a VNA forward measurement, shown in Figure 7.18 and 7.19 (a)) is a device in which, ideally, the reading $\mathbf{a_{1r}}$ is transparent to the incident wave a_1 and insensitive to the reflected wave b_1, while the reading $\mathbf{b_{1r}}$ is transparent to b_1 and insensitive to a_1. As such, it has no loss and has infinite directivity. But it is also supposed to be completely matched: it should present the desired Z_0 to the DUT's input port, if its source and receiver ports are terminated with that same Z_0. However, not only does it distort the transmitted a_1 (forward incident error, E_{IF}); its reading is also dependent on b_1 (forward directivity error, E_{DF}) as it is not perfectly matched (source mismatch forward error, E_{SF}). Furthermore, the forward readings of b_2 and a_2 will also be affected by the forward transmit error, E_{TF}, and by the forward load mismatch error, E_{LF}. Adding the port-to-port crosstalk error, E_{XF}, we obtain the complete forward error model. This model is shown in Figure 7.19(b).

A similar model of seven error terms could be derived for the VNA when it is performing reverse measurements [see Figure 7.1(c)]. In the case of relative measurements, as with the linear S parameter model extraction, there is no need to measure the absolute values of $a_{1,2}$ and $b_{1,2}$ but only their ratios, which allows us to normalize all errors to E_{IF} and E_{IR}, leading to the well-known 12-error correction model of linear VNAs depicted in Figure 7.19(b) and (c).

The flow-graph analysis of this 12-error term model leads to the following relationships between the DUT's S parameters and the measured raw incident, reflected, and transmitted waves [31]:

$$R_F \equiv \frac{B_{r1}}{A_{r1}} = E_{DF} + \frac{S_{11}E_{RF}(1-S_{22}E_{LF}) + S_{12}S_{21}E_{LF}E_{RF}}{(1-S_{11}E_{SF})(1-S_{22}E_{LF}) - S_{12}S_{21}E_{SF}E_{LF}} \tag{7.37}$$

$$T_F \equiv \frac{B_{r2}}{A_{r1}} = E_{XF} + \frac{S_{21}E_{TF}}{(1-S_{11}E_{SF})(1-S_{22}E_{LF}) - S_{12}S_{21}E_{SF}E_{LF}} \tag{7.38}$$

$$R_R \equiv \frac{B_{r2}}{A_{r2}} = E_{DR} + \frac{S_{22}E_{RR}(1-S_{11}E_{LR}) + S_{12}S_{21}E_{LR}E_{RR}}{(1-S_{22}E_{SR})(1-S_{11}E_{LR}) - S_{12}S_{21}E_{SR}E_{LR}} \tag{7.39}$$

$$T_R \equiv \frac{B_{r1}}{A_{r2}} = E_{XR} + \frac{S_{12}E_{TR}}{(1-S_{22}E_{SR})(1-S_{11}E_{LR}) - S_{12}S_{21}E_{SR}E_{LR}} \tag{7.40}$$

The complete identification of this 12-error term model requires a total of 12 independent measurements, which can be gathered from the various calibration methods. From these, the most popular are the short-open-load-thru (SOLT) and the thru-reflect-line (TRL) [32,33], whose names refer to the used standards.

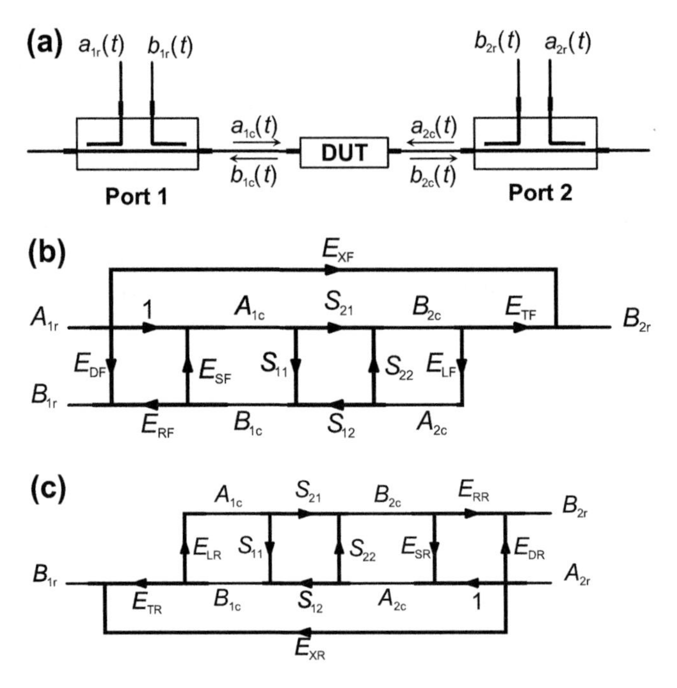

FIGURE 7.19

Twelve-term error model of the VNA reflectometer. (a) Reflectometer diagram. (b) Forward error model. (c) Reverse error model.

For example, the SOLT method starts by performing the isolation test. This is composed by one forward and one reverse transmission measurement, when the ports are isolated from each other and terminated by a matched load. From these results, the E_{XF} and E_{XR} crosstalk errors are determined.

Then, it passes to the S-O-L test phase where the S_{11} and S_{22} are measured for the Short, the Open, and the Load standards in a total of six measurements. As these six one-port measurements are made with a device in which $S_{12}S_{21} = 0$, they allow the determination of E_{RF}, E_{SF}, and E_{DF} and E_{RR}, E_{SR}, and E_{DR}.

Finally, with the Thru standard, four more measurements are made (forward and reverse reflection and transmission), which allow the determination of the remaining E_{LF}, E_{TF} and E_{LR}, E_{TR}.

The corresponding error correction is then conducted for each actual measurement, using [31]:

$$S_{11} = \frac{A_{11}(1 + A_{22}E_{SR}) - A_{12}A_{21}E_{LF}}{D} \tag{7.41}$$

$$S_{12} = \frac{A_{12}[1 + A_{11}(E_{SF} - E_{LR})]}{D} \tag{7.42}$$

$$S_{21} = \frac{A_{21}[1 + A_{22}(E_{SR} - E_{LF})]}{D} \tag{7.43}$$

$$S_{22} = \frac{A_{22}(1 + A_{11}E_{SF}) - A_{12}A_{21}E_{LR}}{D} \tag{7.44}$$

where

$$D = (1 + A_{11}E_{SF})(1 + A_{22}E_{SR}) - A_{12}A_{21}E_{LF}E_{LR} \tag{7.45}$$

the A_{ij} are defined by:

$$A_{11} = \frac{R_F - E_{DF}}{E_{RF}} \tag{7.46a}$$

$$A_{12} = \frac{T_R - E_{XR}}{E_{TR}} \tag{7.46b}$$

$$A_{21} = \frac{T_F - E_{XF}}{E_{TF}} \tag{7.46c}$$

$$A_{22} = \frac{R_R - E_{DR}}{E_{RR}} \tag{7.46d}$$

and R_F, R_R, T_F, and T_R are the raw measured ratios between the ports' reflected/transmitted waves B_{rj} and their corresponding incident waves A_{ri}, as defined in Eqns (7.37)–(7.40).

Unfortunately, this linear VNA is only capable of extracting linear behavioral models, such as the *S*-parameter model. If nonlinear model extraction is desired, then large-signal tests must be performed and a set of new instruments conceived. These can be divided into time-domain and frequency-domain instruments, as their genesis can be traced to the oscilloscope and the spectrum analyzer, respectively. For reasons that will be clear when the extraction of nonlinear behavioral models is discussed, both of these instruments bias the DUT with the conventional DC plus an input large-signal tone, and then perturb this desired cyclostationary quiescent point by another small-signal sinusoid.

The large-signal network analyzer (LSNA), whose architecture is shown in Figure 7.20 [34,35], is based on a four-channel subsampling oscilloscope that captures the raw $a_{ri}(t)$ and $b_{rj}(t)$. Then, these time-domain signals are used to obtain the desired frequency-domain data by Fourier transformation.

Unlike the LSNA, the nonlinear-vector network analyzer (NVNA), whose architecture is depicted in Figure 7.21 [36], directly measures the A_{ri} and B_{rj} in the frequency domain. Although it shares the same reflectometer with the LSNA and with its predecessor, the linear VNA, it no longer uses subsampling receiver technology because it is mixer-based, as are all spectrum analyzers. However, its receivers use synchronous detection, or cross-correlation, to be able to capture both amplitude and phase information.

Because both the LSNA and the NVNA are intended to measure actual amplitude values and harmonic phases (with respect to the fundamental phase), not only their

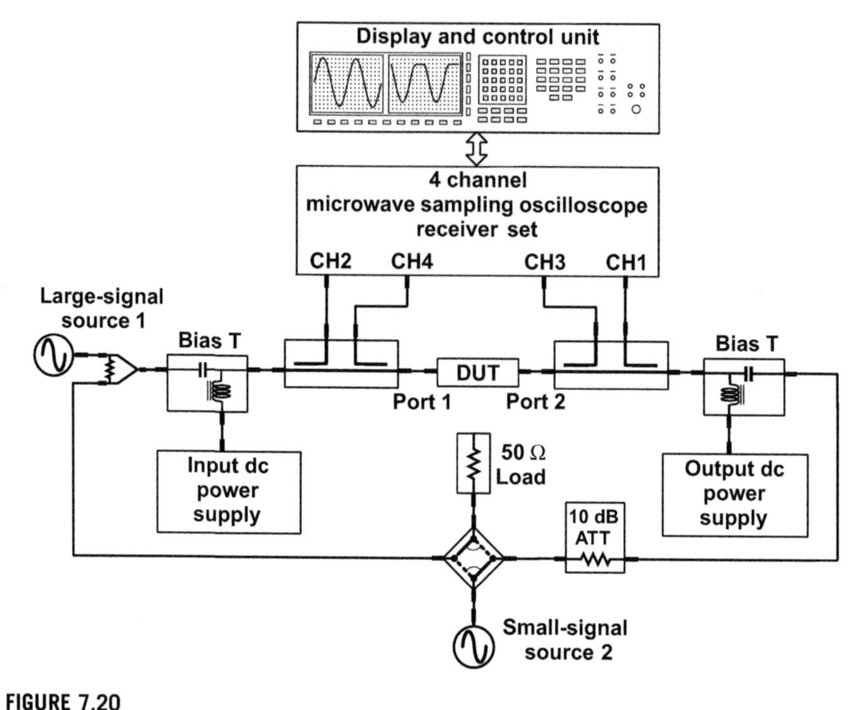

FIGURE 7.20

Simplified architecture of the large-signal network analyzer.

ratios (as in the linear VNA), their required calibration must be slightly more elaborated. Indeed, beyond the described 12-term error correction of their test set reflectometer, the calibration of these large-signal instruments must also be complemented by an absolute amplitude and relative phase measurement of the ports' incident and transmitted/reflected waves. The amplitude calibration is made with a power meter, whereas the phase calibration uses a standard comb-generated signal whose generated harmonics have a stable and traceable phase relationship [35].

7.3 Embedding and de-embedding behavioral models

The process of de-embedding usually requires the inversion of the model of the subsystem to be de-embedded. If a behavioral model is considered for modeling such a subsystem, then a proper procedure must be executed to adequately invert the model structure. In this section, behavioral model inversion techniques are analyzed for the purpose of system embedding and de-embedding. This analysis is accompanied by a presentation and detailed discussion of some illustrative examples. In addition, the analysis of embedding and de-embedding submodels connected in different arrangements—parallel, series, and cascade—is rigorously discussed and exemplified.

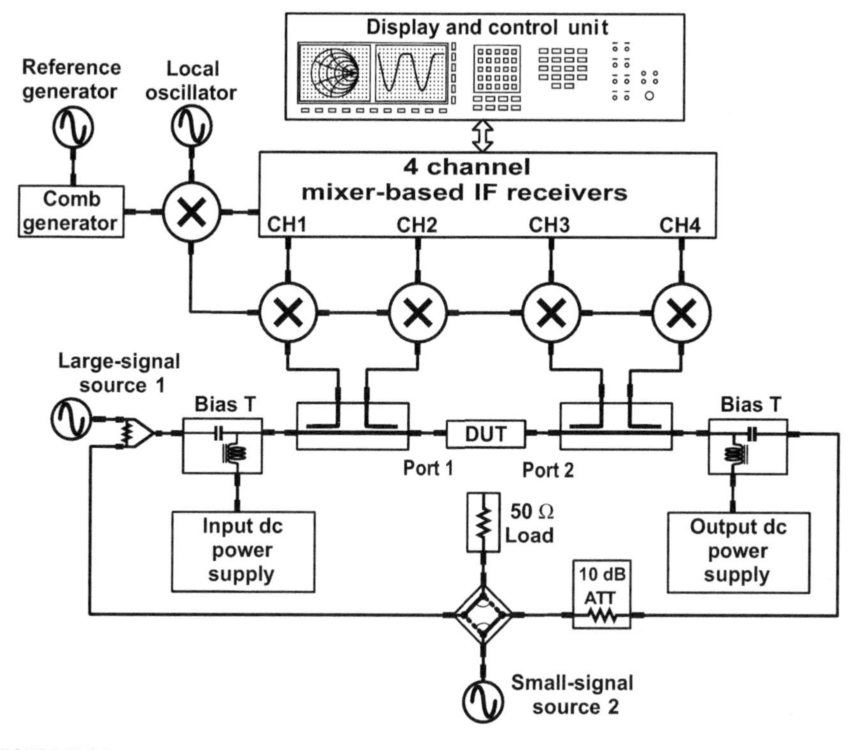

FIGURE 7.21

Simplified architecture of the nonlinear-vector network analyzer.

7.3.1 Behavioral modeling embedding and de-embedding

This and the following subsections discuss the application of embedding/de-embedding techniques to behavioral modeling. Contrary to the physical or the equivalent circuit modeling scenario in which the circuit is usually divided into a set of separate subcircuit shells that can be embedded or de-embedded, behavioral models are not conceived for such a partitioning, making these techniques much less natural and thus unusual. However, there are still some behavioral model examples that can benefit from these embedding/de-embedding procedures.

As was mentioned in Section 7.2, physical models rely on the knowledge of the internal constitution of the device to be modeled (i.e., its elementary components and their associated relationships). These components can be grouped in several subcircuits, usually as nested shells, according to their function or intended operation. This is the case, for example, of the typical equivalent circuit modeling of microwave-active devices whose elements are commonly grouped into an intrinsic subcircuit (responsible for the characteristic current–voltage behavior of the transistor) enclosed in an extrinsic subcircuit that stands for the electrical effects introduced by the parasitic semiconductor bulk resistances and package reactive elements.

Extraction of intrinsic elements' values cannot be performed looking only into the device's accessible terminals; instead, measurements at the intrinsic terminals are required. However, because this is physically impossible, what is done in practice is to measure the device's I–V characteristics at the extrinsic device terminals and then to estimate the desired measurements at the intrinsic terminals by de-embedding the effects of the (supposedly a priori known) extrinsic subcircuit.

Contrary to physical models, behavioral models are black-box representations intended to describe and predict the overall observable behavior of the device, regardless of its internal constitution. That is, a behavioral model constitutes a mapping of the system's excitation onto its response, treating the system as a whole entity whose input and output port signals are the only accessible information. Obviously, this reasoning inevitably leads to the conclusion that behavioral modeling and embedding/de-embedding can be understood as two orthogonal concepts. Hence, following this logic, we could be led to the conclusion that embedding/de-embedding makes no sense in the behavioral modeling context, which is the reason why these two topics are seldom seen together.

However, there are several cases where de-embedding can provide a significant benefit to the extraction of behavioral models, if one has more information about the internal constitution of the system and whenever it is possible to divide the behavioral model of the global system into several behavioral models of the corresponding subsystems. To understand this, let us consider an example: the extraction of two-box models.

The first two-box model to be analyzed is the one that is known to be composed of a static nonlinearity followed by a linear filter, as shown in Figure 7.22.

Supposing that the nonlinearity can be approximated by a polynomial, $\mathcal{P}[x(n)]$, and the linear filter, $\mathcal{L}[z(n)]$, by a finite-impulse response discrete convolution characteristic, we get:

$$y(n) = \mathcal{L}[z(n)] = \sum_{m=0}^{M} l(m)z(n-m) \tag{7.47}$$

and

$$z(n) = \mathcal{P}[x(n)] = \sum_{p=0}^{P} f_p[x(n)]^p \tag{7.48}$$

which leads to a simple one-dimensional Volterra series behavioral model of the form

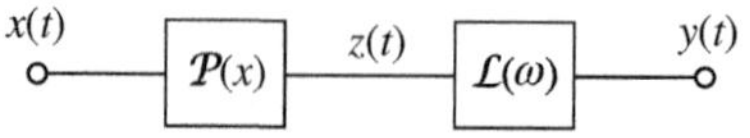

x(t) P(x) z(t) L(ω) y(t)

FIGURE 7.22

The nonlinearity-filter Hammerstein two-box behavioral model cascade.

$$y(n) = \sum_{p=0}^{P} \sum_{m=0}^{M} h_p(m, \ldots, m) x(n-m) \ldots x(n-m) = \sum_{p=0}^{P} \sum_{m=0}^{M} l(m) f_p [x(n-m)]^p \tag{7.49}$$

This expression shows that both the filter impulse response weights and the polynomial coefficients can be directly extracted using a linear least-squares (LS) procedure, except for an immaterial gain that can be overcome, such as by making $l(0) = 1$.

Now, consider what happens if we swap the filter and the static nonlinearity, creating what is known as the Wiener two-box model (Figure 7.23).

Now,

$$y(n) = \mathcal{P}[z(n)] = \sum_{p=0}^{P} f_p [z(n)]^p \tag{7.50}$$

and

$$z(n) = \mathcal{L}[z(n)] = \sum_{m=0}^{M} l(m) x(n-m) \tag{7.51}$$

which leads to a multidimensional Volterra series behavioral model of the form

$$\begin{aligned} y(n) &= \sum_{p=0}^{P} \sum_{m_1=0}^{M} \cdots \sum_{m_p=0}^{M} h_p(m_1, \ldots, m_p) x(n-m_1) \ldots x(n-m_p) \\ &= \sum_{p=0}^{P} \sum_{m_1=0}^{M} \cdots \sum_{m_p=0}^{M} f_p l(m_1) \ldots l(m_p) x(n-m_1) \ldots x(n-m_p) \end{aligned} \tag{7.52}$$

which no longer allows an extraction of the nonlinearity and filter parameters with a simple linear LS extraction. Additionally, this also means that one cannot evaluate the model storing only the filter and polynomial nonlinearity parameters, but rather all the $h_p(m_1, \ldots, m_p)$ coefficients.

In this regard, it is interesting to note that although both the Hammerstein and the Wiener cascades are completely identified with only $P + (M+1)$ parameters, the Hammerstein allows a direct linear extraction of these $P + (M+1)$ parameters, whereas the Wiener structure leads to an LS extraction of an immensely larger number of coefficients on the order of $(M+1)^P$. If they are assumed to be independent during the extraction (but in fact they are not, because they are set as combinations of the parameters of the same polynomial and linear filter), this leads to the

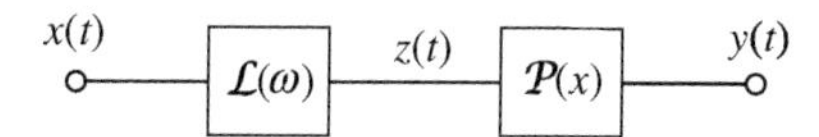

FIGURE 7.23

The filter-nonlinearity Wiener two-box behavioral model cascade.

extraction of a full Volterra series. Therefore, the knowledge, and corresponding de-embedding, of the Hammerstein linear filter would not provide any major benefit because, after knowing the filter's $M+1$ parameters, we would then still have to determine the remaining P parameters of the polynomial nonlinearity.

However, the situation is completely different for the Wiener cascade, as the knowledge of the filter's $M+1$ coefficients would not lead to a remainder of a number of parameters on the order of $(M+1)^P-(M+1)$, but to a remainder of only P parameters. In fact, if we were told the coefficients of the filter impulse response, we could immediately compute $z(n)$ from $x(n)$; from $z(n)$ and $y(n)$, we could directly determine the P polynomial nonlinearity parameters through a simple LS extraction. Therefore, the filter de-embedding of the Wiener cascade can, indeed, offer a major benefit to behavioral model formulation and extraction.

7.3.2 Considerations on model inversion

As mentioned in the previous subsection, de-embedding of behavioral models occurs whenever the DUT is composed of a set of subcircuit shells, usually in cascade, and the behavioral models of some of these subcircuits are known, whereas those of the remaining ones (usually, the inner ones) are to be determined. A typical case is that of a three-box model (three submodels in cascade, as shown in Figure 7.24) where the input and output models are, by some process, known and the one in the middle is unknown. As will be shown in the next subsection, intrinsic device modeling is one example that fits this description when one considers the input and output impact of the package extrinsic elements (electromagnetic simulations or package measurements could provide models for these) and the unknown behavior of the intrinsic device.

In this case, and assuming that only the input and output signals, $x(n)$ and $y(n)$, are accessible, it is required to estimate the signals at the terminals of the subsystem $H_2[\cdot]$, $u(n)$ and $z(n)$, for determining a behavioral model for it. Knowing the input subsystem model, it is trivial to estimate $u(n)$ because it is required only to evaluate such model excited by $x(n)$. The same is not true for the output subsystem model, $H_3[\cdot]$. Given the model from $z(n)$ to $y(n)$, it is now necessary to estimate its input $z(n)$ from the output $y(n)$, and this requires the determination of the model's inverse.

The inverse of a model is the model that generates the input signal from the measured output signal. If the model and its inverse are cascaded, then the resulting system is a unit gain static system. Depending on the complexity of the model, the calculation of its inverse can be more or less laborious, but, usually, it is not simple. Let us illustrate the model inversion process by analyzing a set of particular models.

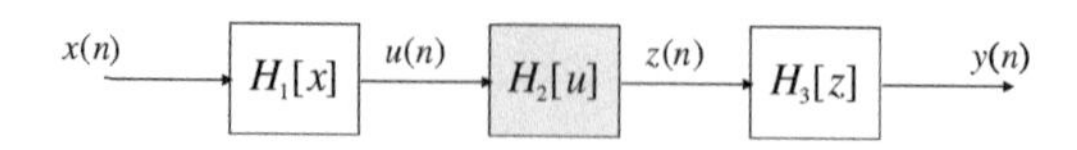

FIGURE 7.24

The three-box model.

7.3.2.1 Inversion of static nonlinear models

The inversion of a static nonlinear function is very simple, as long as the function is injective. If the forward function is implemented by an LUT, then its inversion consists of simply exchanging the variables of the LUT. If it is implemented by a static polynomial, then the inverse can also be approximated by another polynomial. Even though, in this case, it can be possible to analytically derive the inverse of the forward polynomial, generally the exact inverse polynomial is one of infinite order. Notice, for instance, that the inverse of the function $y(n) = z^2(n)$, for positive values of $z(n)$, is given by $z(n) = \sqrt{y(n)}$. As is known, the Taylor expansion of the square-root function is an infinite order polynomial. To obviate the difficulty in determining the exact analytical polynomial inverse, two approaches are usually carried out: (1) a truncation to the inverse polynomial order is selected and its coefficients are extracted from some $y(n) \rightarrow z(n)$ data, if available; (2) a truncation to the inverse polynomial order is chosen, say, to the pth order, and its coefficients are analytically determined from those of the forward polynomial so that the cascade of the inverse and the forward polynomials generates a new polynomial whose coefficients are null up to the pth order (except the first-order coefficient, which becomes equal to 1). The latter approach is the particularization (to the static case) of the denominated pth-order inverse model [37], applicable to general Volterra series, which we will revisit in Section 7.3.2.3. To exemplify the concept of the pth-order inverse, let us apply it to the static polynomial case here considered. Let the forward polynomial model be defined as

$$y(n) = \sum_{k=1}^{K} a_k z^k(n) \tag{7.53}$$

and let the inverse model be also a polynomial, limited to the pth order, as

$$\hat{z}(n) = \sum_{q=1}^{p} b_q y^q(n) \tag{7.54}$$

where $\hat{z}(n)$ is the estimate of $z(n)$. The cascade of the forward and inverse models is a polynomial with the form of

$$\hat{z}(n) = \sum_{q=1}^{p} b_q \left(\sum_{k=1}^{K} a_k z^k(n) \right)^q = \sum_{r=1}^{pK} c_r z^r(n) \tag{7.55}$$

The unknown b_q parameters are then determined to satisfy the following p equations:

$$\begin{cases} c_1 = 1 \\ c_r = 0, \quad r = 2, \ldots, p \end{cases} \tag{7.56}$$

This means that the pth-order inverse model is equal to $z(n)$, except for the remaining terms that are produced by products of orders higher than p.

The analysis presented in Eqns (7.55) and (7.56) considered the case where the inverse model is cascaded after the forward model; in this case, the inverse model is called the postinverse model. A similar conclusion can be obtained by considering the inverse model cascaded before the forward model—the preinverse model.

7.3.2.2 Inversion of linear dynamic models

Now let us concentrate on the inversion of a dynamical system, starting with the simpler case of one that is linear. If the forward model is expressed as an FIR filter, then it is known that the exact inverse model will also be a linear filter but with an infinite impulse response (IIR) form. This can be easily illustrated by observing the example of the inverse of

$$y(n) = h_0 z(n) + h_1 z(n-1) \tag{7.57}$$

which is

$$\begin{aligned} z(n) &= \frac{1}{h_0}y(n) - \frac{h_1}{h_0}z(n-1) = \frac{1}{h_0}y(n) - \frac{h_1}{h_0}\left[\frac{1}{h_0}y(n-1) - \frac{h_1}{h_0}z(n-2)\right] \\ &= \sum_{m=0}^{\infty}(-1)^m\left(\frac{h_1}{h_0}\right)^m\frac{1}{h_0}y(n-m) \end{aligned} \tag{7.58}$$

In most practical situations, it is assumed that the approximation of also considering the inverse model as one of fading memory—neglecting the past behavior beyond a certain predefined memory depth. In the case of Eqn (7.58), this is reasonable for $|h_1|<|h_0|$.

If the forward model is an IIR filter in the form of

$$y(n) = h_{D,0}z(n) + \sum_{r=1}^{R} h_{R,r}y(n-r) \tag{7.59}$$

where h_D and h_R stand for the coefficients of the direct and recursive terms, then the analytical inverse model of such IIR is an FIR filter given by

$$z(n) = -\frac{1}{h_{D,0}}\sum_{r=0}^{R} h_{R,r}y(n-r) \tag{7.60}$$

with $h_{R,0} = -1$.

Finally, for the general case where the forward model is the summation of both the convolution of the input samples and the convolution of the past output samples, as in Eqn (7.61)—a formulation usually known as autoregressive moving average (ARMA)—the inverse model will keep the configuration of an ARMA, as shown in Eqn (7.62).

$$y(n) = \sum_{m=0}^{M} h_{D,m}z(n-m) + \sum_{r=1}^{R} h_{R,r}y(n-r) \tag{7.61}$$

$$z(n) = -\frac{1}{h_{D,0}} \sum_{m=1}^{M} h_{D,m} z(n-m) - \frac{1}{h_{D,0}} \sum_{r=0}^{R} h_{R,r} y(n-r) \tag{7.62}$$

7.3.2.3 Inversion of nonlinear dynamic models

The analytical inversion of nonlinear dynamic models is, in general, much more complicated than the cases presented above. However, there are particular topologies whose inverse can be easily determined, as is the case of the Wiener and Hammerstein two-box models. For instance, the preinverse of a Wiener model is a Hammerstein whose linear filter is the inverse of the filter of the Wiener model, and whose nonlinear static function is the inverse of that of the Wiener.

For the case of the polynomial filter (Volterra series) general model formulation, Schetzen [37] has deduced the strategy to determine the kernels of the polynomial filter, which when placed in cascade (either after or before) with the forward Volterra series model, leads to null kernels of the cascade up to a specified order p, except for the first-order kernel (which is unitary). This is the pth-order (pre- or post-) inverse of the forward Volterra series model. As mentioned above, the nonlinear dynamic behavior of the forward model for polynomial orders higher than p are not accounted for in this calculation of the inverse model. Thorough details on the pth-order inverse of a Volterra series model can be found in Refs [1],[37].

7.3.3 Examples of behavioral modeling embedding and de-embedding

In this section, several examples of embedding/de-embedding applied to behavioral modeling are provided. Making use of the linear and nonlinear model inversion tools derived in the previous section, we will show how behavioral model blocks can be embedded in, or de-embedded from, the overall behavioral model that includes them.

For the sake of simplicity, and thus clarity of the presentation, we will divide the given examples into two groups. The first group deals with SISO or transfer models. These models are characterized by a direct input−output relationship, $y(n) = \mathcal{H}[x(n)]$, which is nonlinear and dynamic in general and can be associated with feed-forward, cascade, and feedback connections. Because the SISO models assume that each block does not impact the operation of the others (assuming that the external environment, even if it is determined by the other blocks, is the same as the one in which their $\mathcal{H}[x(n)]$ was extracted), these models are not used in circuit-level analysis, but only systems-level analysis.

The second group of examples deals with two-input/two-output (2I2O) models. They constitute a particular case of the more general MIMO models and are characterized by a more involved relationship, in which each of the outputs is given as a nonlinear dynamic function of both of the inputs: $y_1(n) = \mathcal{H}_1[x_1(n), x_2(n)]$ and $y_2(n) = \mathcal{H}_2[x_1(n), x_2(n)]$. When their inputs and outputs are instantiated as voltages, currents, or even incident and reflected/transmitted waves, they allow the representation of important notions of circuit analysis, such as input or output impedance or termination, being thus capable of accommodating the interaction between connected blocks. Hence, they are often used as two-port electrical network descriptions that can be arranged in series, parallel, and cascade connections.

7.3.3.1 Embedding/de-embedding SISO systems for system-level analysis

As mentioned above, SISO systems can be connected in either feed-forward, cascade, or feedback arrangements. Each of them requires a particular embedding/de-embedding process.

Starting with the feed-forward connection shown in Figure 7.25, composed of two blocks, A and B, respectively represented by $y^{A}(n) = \mathcal{H}^{A}[x^{A}(n)]$ and $y^{B}(n) = \mathcal{H}^{B}[x^{B}(n)]$, we have for the complete model, T,

$$y^{T}(n) = y^{A}(n) + y^{B}(n) \quad \Leftrightarrow \quad \mathcal{H}^{T}[x(n)] = \mathcal{H}^{A}[x(n)] + \mathcal{H}^{B}[x(n)] \tag{7.63}$$

Therefore, the de-embedding of, say, block B, from the global model T, can be readily made by subtraction, which exposes block A:

$$y^{A}(n) = y^{T}(n) - y^{B}(n) \quad \Leftrightarrow \quad \mathcal{H}^{A}[x(n)] = \mathcal{H}^{T}[x(n)] - \mathcal{H}^{B}[x(n)] \tag{7.64}$$

For example, if all models of A, B, and T are Volterra series (or their particular instances of linear FIR filters or polynomial nonlinearities), the frequency-domain nonlinear transfer function and the time-domain nonlinear impulse-response function de-embedding can be computed as follows:

$$H_p^{A}(\omega_1, \ldots, \omega_p) = H_p^{T}(\omega_1, \ldots, \omega_p) - H_p^{B}(\omega_1, \ldots, \omega_p) \tag{7.65}$$

or

$$h_p^{A}(m_1, \ldots, m_p) = h_p^{T}(m_1, \ldots, m_p) - h_p^{B}(m_1, \ldots, m_p) \tag{7.66}$$

Two particularly useful examples of this form of de-embedding are found in electrical circuits whenever one wants to de-embed the effects of a series-connected or parallel-connected element. In the series connection case, Eqns (7.63)–(7.66) can be directly applied if the elements are treated as current-controlled components:

$$y^{E}(n) = \mathcal{H}^{E}[x^{E}(n)] \quad \Leftrightarrow \quad v^{E}(n) = Z^{E}[i^{E}(n)] \tag{7.67}$$

where E stands for any of the A, B, or T blocks. In the parallel connection, the elements must be treated as voltage-controlled components:

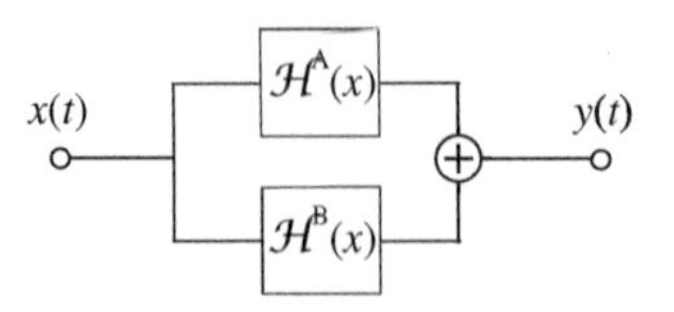

FIGURE 7.25

Feed-forward connection of two SISO models.

$$y^{E}(n) = \mathcal{H}^{E}\left[x^{E}(n)\right] \quad \Leftrightarrow \quad i^{E}(n) = Y^{E}\left[v^{E}(n)\right] \tag{7.68}$$

Now, let us turn our attention to the cascade connection shown in Figure 7.26. In this case,

$$y^{T}(n) = \mathcal{H}^{B}\left\{\mathcal{H}^{A}[x(n)]\right\} \tag{7.69}$$

and the de-embedding of block B should now be performed by

$$\mathcal{H}^{A}[x(n)] = \mathcal{H}^{-B}\left[y^{T}(n)\right] \tag{7.70}$$

where $\mathcal{H}^{-B}[.]$ represents the inverse model of $\mathcal{H}^{B}[.]$ as discussed in Section 7.3.2. This corresponds to adding the postinverse of $\mathcal{H}^{B}[.]$ at the output of the original cascade connection, because by definition of the postinverse:

$$\mathcal{H}^{A}[x(n)] = \mathcal{H}^{-B}\left[\mathcal{H}^{B}\left\{\mathcal{H}^{A}[x(n)]\right\}\right] \tag{7.71}$$

Similarly, if we wanted to de-embed block A from the cascade to expose block B, we would be led to

$$\mathcal{H}^{B}[x(n)] = \mathcal{H}^{B}\left[\mathcal{H}^{A}\left\{\mathcal{H}^{-A}[x(n)]\right\}\right] \tag{7.72}$$

which corresponds to adding the pre-inverse of block A at the input of the original cascade connection.

As a practical application, de-embedding block B from the cascade corresponds to the de-embedding of the linear filter from the Hammerstein cascade or de-embedding the static nonlinearity from the Wiener arrangement discussed in Section 7.3.1. Conversely, de-embedding port A corresponds to de-embedding the linear filter from the Wiener or the static nonlinearity from the Hammerstein cascades.

Naturally, this procedure can be extended to three-box models, the Wiener–Hammerstein filter-nonlinearity-filter or the Hammerstein–Wiener nonlinearity-filter-nonlinearity cascades, or even to more complex structures, as long as the desired de-embedding is performed from the outer to the inner blocks, successively adding the required pre- or postinverse block models.

Among these examples of block de-embedding in cascades, the nonlinear feedback system shown in Figure 7.27(b) [38] deserves special attention because it has been widely used as one of the major nonlinear distortion models of commonly used wireless power amplifiers (PAs) [39].

In Figure 7.27(b), the input and output filters, $M_i(\omega)$ and $M_o(\omega)$, represent the PA input and output matching networks from Figure 7.27(a), respectively. The static nonlinearity, $f[e(t)]$, represents the active device's actual nonlinear behavior and

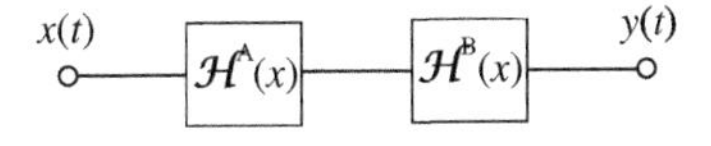

FIGURE 7.26

Cascade connection of two SISO models.

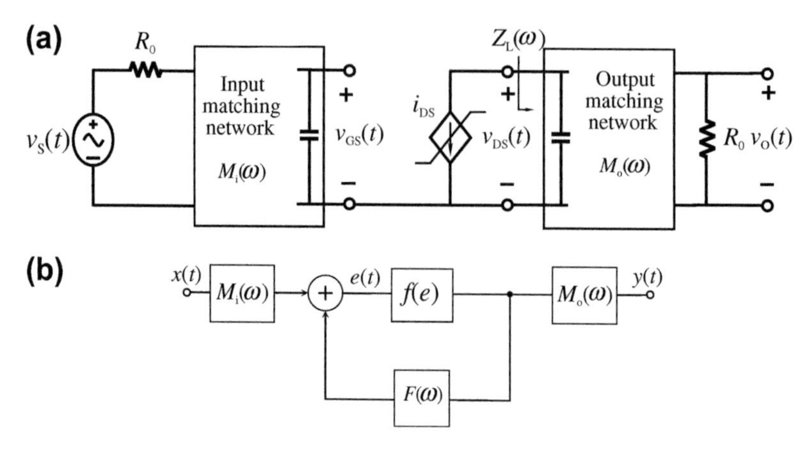

FIGURE 7.27

The wireless power amplifier nonlinear distortion model. (a) Simplified schematic diagram of the wireless amplifier circuit. (b) Equivalent system-level nonlinear dynamic feedback model.

the feedback filter, $F(\omega)$, represents the dynamic lumped effects of both the impedance seen by the output transistor terminal and the possible thermal impedance [38].

Such a model indicates that the PA dynamic effects can be understood as originating from both $M_i(\omega)$ and $M_o(\omega)$ and from the combined effects of $F(\omega)$. Because of the wide bandwidth of $M_i(\omega)$ and $M_o(\omega)$, which is necessary to accommodate the transmitted signal without significant linear distortion, it is expected that these matching networks determine memory effects of relatively short time-constants. In contrast, both the thermal impedance and the transistor bias circuitry have much smaller bandwidths and so impose memory effects of much longer time constants. So, the necessity to accurately predict the amplifiers' long-term memory effects has determined the desire to eliminate the memory contributions of the outer PA model shell of $M_i(\omega)$ and $M_o(\omega)$ and so to expose the effects of $F(\omega)$ in the inner shell.

This problem is similar to the input and output filter de-embedding just mentioned for the Wiener–Hammerstein three-box model. So, one could first identify $M_i(\omega)$ and determine, by circuit-level simulation, the transfer function between the PA voltage source to the active device's input control voltage (see Figure 7.27(a). Similarly, one could also identify $M_o(\omega)$ from the relationship between the active device's output current and the PA output voltage. Then, one could de-embed these two filters from the complete PA model adding one input linear filter of $M_i^{-1}(\omega) = 1/M_i(\omega)$ transfer function and an output filter of $M_o^{-1}(\omega) = 1/M_o(\omega)$ to the measured complete PA model.

7.3.3.2 Embedding/de-embedding of 2I2O systems for circuit-level analysis

One major difficulty faced with SISO models is that the behavior of each block within the system has to be exactly equal to that observed when the block is separately characterized. This means that the environment determined by the preceding and following blocks must equal the one determined by the source and load terminations of the characterization setup. For example, in RF and microwave circuits, where these terminations are usually normalized to 50 Ω, this means that the output port of the preceding blocks and the input port of the following blocks must be perfectly matched. Because this is a condition difficult to meet in most of the wireless circuits, these SISO system representations were substituted by 2I2O two-port network models.

In electrical circuits, 2I2O two-port network models are particular cases of MIMO n-port network models, where the inputs and outputs defined by

$$y_1(n) = \mathcal{H}_1[x_1(n), x_2(n)] \text{ and } y_2(n) = \mathcal{H}_2[x_1(n), x_2(n)] \quad (7.73)$$

are voltages and currents or incident and reflected/transmitted waves. Hence, some commonly used two-port models consist of the following:

- The admittance formulation:

$$i_1(n) = \mathcal{I}_1[v_1(n), v_2(n)] \text{ and } i_2(n) = \mathcal{I}_2[v_1(n), v_2(n)] \quad (7.74)$$

- The impedance formulation:

$$v_1(n) = \mathcal{V}_1[i_1(n), i_2(n)] \text{ and } v_2(n) = \mathcal{V}_2[i_1(n), i_2(n)] \quad (7.75)$$

- The transmission formulation:

$$v_1(n) = \mathcal{V}_1[v_2(n), -i_2(n)] \text{ and } i_1(n) = \mathcal{I}_1[v_2(n), -i_2(n)] \quad (7.76)$$

- The scattering formulation:

$$b_1(n) = \mathcal{B}_1[a_1(n), a_2(n)] \text{ and } b_2(n) = \mathcal{B}_2[a_1(n), a_2(n)] \quad (7.77)$$

- The wave-transfer formulation:

$$a_1(n) = \mathcal{A}_1[a_2(n), b_2(n)] \text{ and } b_1(n) = \mathcal{B}_1[a_2(n), b_2(n)] \quad (7.78)$$

These names were derived from the matrix formulations of their linear instances.

Because the two-port's incident and reflected waves can be given as a linear transformation from its voltages and currents, all of these formulations are equivalent and each of them can be obtained as a transformation from any of the others. Furthermore, many other representations corresponding to different selections of the excitations and responses from all the eight available variables [$v_p(n)$, $i_p(n)$, $a_p(n)$, $b_p(n)$], where $p = 1, 2$ are possible. The choice of one in detriment to the others should be made depending on the way the two-ports are connected to form the complete circuit. These linear two-port network formulations play an important role in circuit analysis and model embedding/de-embedding because they can be used to build larger models.

For example, consider the two networks, A and B, shown in Figure 7.28, which are connected in parallel to build a larger network, T.

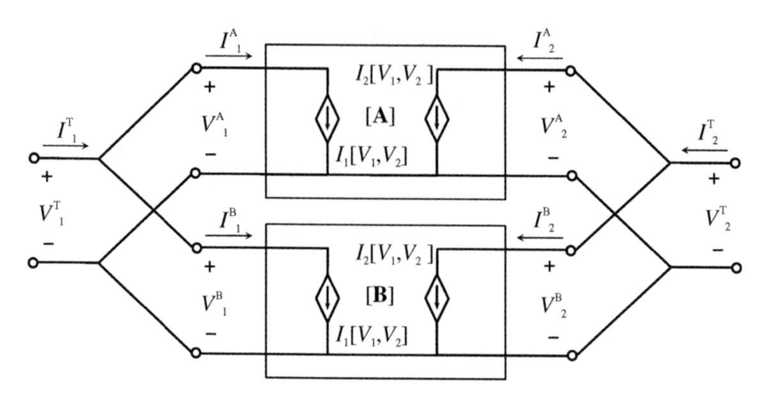

FIGURE 7.28

Two-port embedding/de-embedding through a parallel, this connection.

Because the two networks are connected in parallel, this means that

$$v_p^{\mathrm{T}}(n) = v_p^{\mathrm{A}}(n) = v_p^{\mathrm{B}}(n) \quad \text{and} \quad i_p^{\mathrm{T}}(n) = i_p^{\mathrm{A}}(n) + i_p^{\mathrm{B}}(n) \tag{7.79}$$

where p is the port number. So, selecting an admittance form allows us to write

$$\mathfrak{I}_p^{\mathrm{T}}\left[v_1^{\mathrm{T}}(n), v_2^{\mathrm{T}}(n)\right] = \mathfrak{I}_p^{\mathrm{A}}\left[v_1^{\mathrm{T}}(n), v_2^{\mathrm{T}}(n)\right] + \mathfrak{I}_p^{\mathrm{B}}\left[v_1^{\mathrm{T}}(n), v_2^{\mathrm{T}}(n)\right] \tag{7.80}$$

which shows that embedding any parallel network in, or de-embedding it from, the network T can be done by simple model operator addition or subtraction. For example, when the two-ports are linear and their currents and voltages are expressed in the frequency domain, Eqn (7.74) can be written in matrix form as

$$\begin{bmatrix} I_1(\omega) \\ I_2(\omega) \end{bmatrix} = \begin{bmatrix} Y_{11}(\omega) & Y_{12}(\omega) \\ Y_{21}(\omega) & Y_{22}(\omega) \end{bmatrix} \begin{bmatrix} V_1(\omega) \\ V_2(\omega) \end{bmatrix} \quad \text{or} \quad \mathbf{I}(\omega) = \mathbf{Y}(\omega)\mathbf{V}(\omega) \tag{7.81}$$

And Eqn (7.80) leads to:

$$\mathbf{Y}^{\mathrm{T}}(\omega) = \mathbf{Y}^{\mathrm{A}}(\omega) + \mathbf{Y}^{\mathrm{B}}(\omega) \tag{7.82}$$

meaning that $Y_{pq}^{\mathrm{T}} = Y_{pq}^{\mathrm{A}} + Y_{pq}^{\mathrm{B}}$, for any of the two port numbers p and q. Naturally, similar results can be found for the network series connection depicted in Figure 7.29, as this is an electrical dual case of the parallel connection.

Now, it is the currents through the networks A, B, and T that are equal, and it is the voltages across A and B that are added to result in the voltage across T. Therefore, for the series connection, one obtains:

$$\mathcal{V}_p^{\mathrm{T}}\left[i_1^{\mathrm{T}}(n), i_2^{\mathrm{T}}(n)\right] = \mathcal{V}_p^{\mathrm{A}}\left[i_1^{\mathrm{T}}(n), i_2^{\mathrm{T}}(n)\right] + \mathcal{V}_p^{\mathrm{B}}\left[i_1^{\mathrm{T}}(n), i_2^{\mathrm{T}}(n)\right] \tag{7.83}$$

Again, if restricted to the linear case, Eqn (7.83) corresponds to the following impedance-matrix addition:

$$\mathbf{Z}^{\mathrm{T}}(\omega) = \mathbf{Z}^{\mathrm{A}}(\omega) + \mathbf{Z}^{\mathrm{B}}(\omega) \tag{7.84}$$

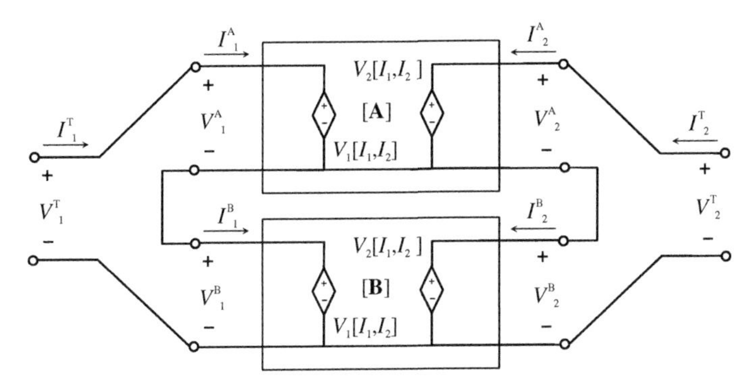

FIGURE 7.29

Two-port embedding/de-embedding through a series connection.

Expressions (7.79) and (7.80) can be understood as the relationships necessary to embed networks A or B with the other, to build network T, and so they can also be used to provide the necessary de-embedding of these blocks from the complete network T.

In the same way the admittance and impedance forms proved to be particularly useful for embedding/de-embedding two-port networks connected in parallel and series, respectively, we will now show that the transmission form can play a similar job for cascaded blocks. For that, let us consider the cascade connection shown in Figure 7.30.

When connected in cascade, the blocks' currents and voltages obey the following relationships:

$$v_1^{\mathrm{T}}(n) = v_1^{\mathrm{A}}(n), \quad i_1^{\mathrm{T}}(n) = i_1^{\mathrm{A}}(n), \quad v_2^{\mathrm{T}}(n) = v_2^{\mathrm{B}}(n), \quad i_2^{\mathrm{T}}(n) = i_2^{\mathrm{B}}(n) \quad \text{and} \quad v_1^{\mathrm{B}}(n) = v_2^{\mathrm{A}}(n), \quad i_1^{\mathrm{B}}(n) = -i_2^{\mathrm{A}}(n) \tag{7.85}$$

from which we can derive

$$v_1^{\mathrm{T}}(n) = \mathcal{V}_1^{\mathrm{A}}\left\{\mathcal{V}_1^{\mathrm{B}}\left[v_2^{\mathrm{T}}(n), -i_2^{\mathrm{T}}(n)\right], \mathcal{I}_1^{\mathrm{B}}\left[v_2^{\mathrm{T}}(n), -i_2^{\mathrm{T}}(n)\right]\right\} \tag{7.86}$$

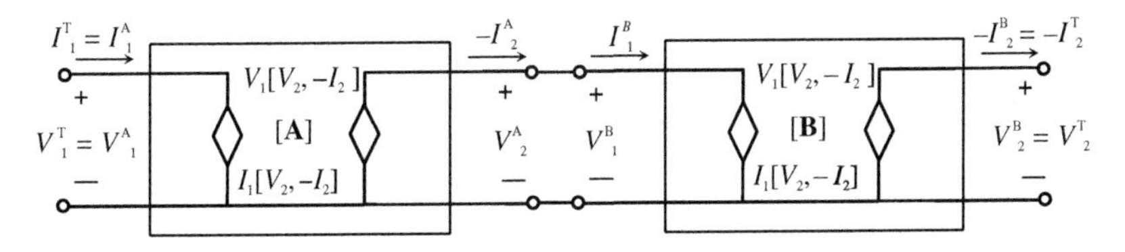

FIGURE 7.30

Two-port embedding/de-embedding through a cascade connection.

and

$$i_1^T(n) = \mathcal{I}_1^A\{\mathcal{V}_1^B[v_2^T(n), -i_2^T(n)], \mathcal{I}_1^B[v_2^T(n), -i_2^T(n)]\} \tag{7.87}$$

These two expressions show that the transmission formulation that governs the blocks' cascade is given by the composite operator rule of $\mathcal{V}_1^A\{\mathcal{V}_1^B[\cdot], \mathcal{I}_1^B[\cdot]\}$ and $\mathcal{I}_1^A\{\mathcal{V}_1^B[\cdot], \mathcal{I}_1^B[\cdot]\}$.

Similarly to what was done for the parallel and series connection, let us see what this statement implies when the blocks are linear. Under the linearity assumption and in the frequency domain, Eqn (7.76) could be written in the ABCD matrix form as follows:

$$\begin{bmatrix} V_1(\omega) \\ I_1(\omega) \end{bmatrix} = \begin{bmatrix} A(\omega) & B(\omega) \\ C(\omega) & D(\omega) \end{bmatrix} \begin{bmatrix} V_2(\omega) \\ -I_2(\omega) \end{bmatrix} \tag{7.88}$$

Equations (7.86) and (7.87) would become:

$$\begin{bmatrix} V_1(\omega) \\ I_1(\omega) \end{bmatrix}^T = \begin{bmatrix} A(\omega) & B(\omega) \\ C(\omega) & D(\omega) \end{bmatrix}^A \begin{bmatrix} A(\omega) & B(\omega) \\ C(\omega) & D(\omega) \end{bmatrix}^B \begin{bmatrix} V_2(\omega) \\ -I_2(\omega) \end{bmatrix}^T \tag{7.89}$$

which shows that the embedding of two linear two-ports in cascade corresponds to the product of their ABCD matrices. Similarly to what we had seen in the cascade of SISO systems, de-embedding one of these two-ports corresponds to the product of its pre- or postinverse ABCD matrix.

Now, let us illustrate the application of these general embedding/de-embedding tools in two of the nonlinear behavioral models addressed in the previous subsections. The first example deals with the de-embedding of the intrinsic microwave FET's I/Q model from the device package parasitics. As previously explained, the I/Q model can be used in any subcircuit whose port currents can be expressed as the summation of several current or charge components, but it fails in the presence of any series element. So, it is appropriate to model the intrinsic microwave FET active device model shown in Figure 7.31, but it cannot be used for the whole packaged device. Worse than this, it can only be applied to the intrinsic model as long as its behavior can be directly observed—something that is only possible after the required de-embedding of the package subcircuit. Therefore, we will now show how the intrinsic I/Q model can be de-embedded from the FET's extrinsic equivalent circuit model.

For analyzing the embedding correspondent to the various network shells shown in Figure 7.31(b), we will start from the intrinsic terminals of the I/Q model—G′, D′, and S′—to reach the accessible extrinsic package terminals G, D, and S. Because the first encountered network is the one identified as [R], a linear two-port connected in series with the intrinsic FET model, [I], we start from an impedance formulation of the intrinsic model frequency-domain currents and voltages of the form:

$$\mathbf{V}^I \equiv \begin{bmatrix} V_{G'} \\ V_{D'} \end{bmatrix} = \begin{bmatrix} F_g(I_{G'}, I_{D'}) \\ F_d(I_{G'}, I_{D'}) \end{bmatrix} = \mathbf{F_V}(\mathbf{I}^I) \tag{7.90}$$

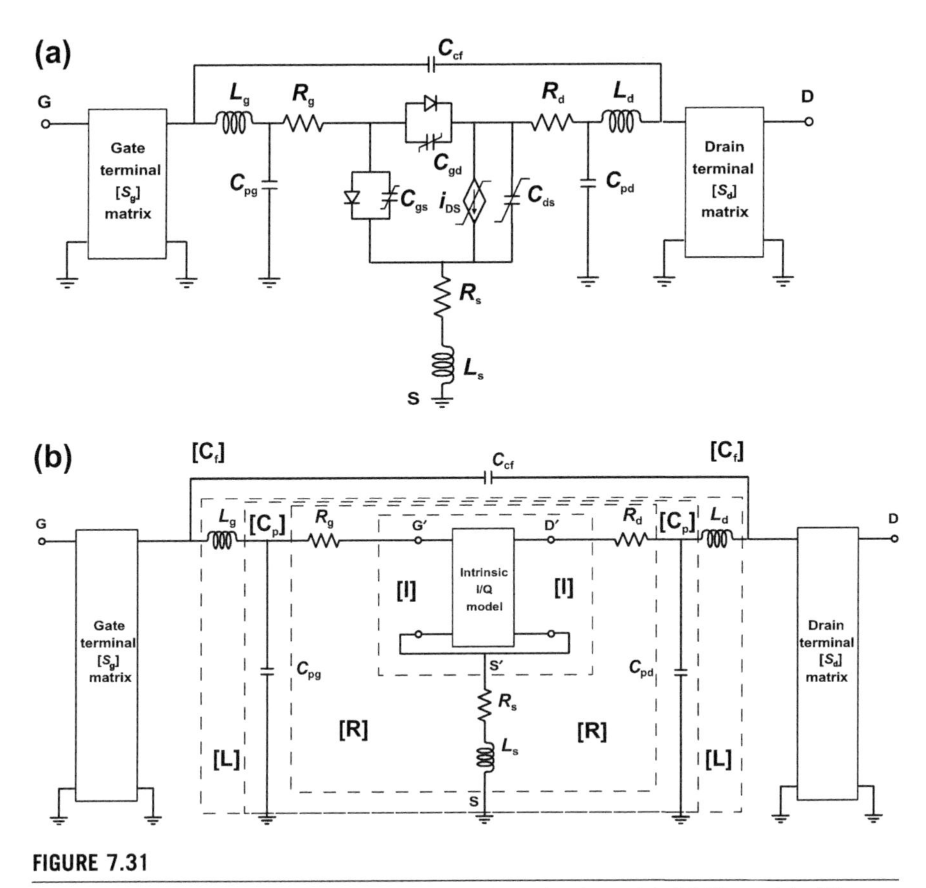

FIGURE 7.31

(a) Intrinsic and extrinsic microwave FET equivalent circuit model. (b) I/Q model of the intrinsic FET embedded in the various shells in which the extrinsic equivalent circuit was grouped.

where the explicit dependence on frequency was omitted for the sake of notation simplicity. The frequency-domain impedance matrix of [R] is such that the voltage across its outside terminals can be given by

$$\mathbf{V}^{\mathrm{T_R}} = \begin{bmatrix} R_{\mathrm{g}} + R_{\mathrm{S}} + j\omega L_{\mathrm{S}} & R_{\mathrm{S}} + j\omega L_{\mathrm{S}} \\ R_{\mathrm{S}} + j\omega L_{\mathrm{S}} & R_{\mathrm{d}} + R_{\mathrm{S}} + j\omega L_{\mathrm{S}} \end{bmatrix} \mathbf{I}^{\mathrm{R}} + \mathbf{V}^{\mathrm{I}}. \tag{7.91}$$

But, because

$$\mathbf{I}^{\mathrm{T_R}} = \mathbf{I}^{\mathrm{R}} = \mathbf{I}^{\mathrm{I}} = \begin{bmatrix} I_{\mathrm{G'}} \\ I_{\mathrm{D'}} \end{bmatrix} \tag{7.92}$$

we can write

$$\mathbf{V}^{\mathrm{T_R}} = \mathbf{Z}^{\mathrm{R}}\mathbf{I}^{\mathrm{I}} + \mathbf{V}^{\mathrm{I}} \tag{7.93}$$

The next two-port to be embedded is the one identified as $[C_p]$, connected in parallel with $[T_R]$, the series of [R] and [I], and whose admittance-matrix formulation allows the determination of the voltages and currents at its outside terminals as:

$$\mathbf{V}^{T_P} = \mathbf{V}^P = \mathbf{V}^{T_R}, \quad \text{and} \quad \mathbf{I}^{T_P} = \mathbf{Y}^P\mathbf{V}^{T_R} + \mathbf{I}^I \tag{7.94}$$

where

$$\mathbf{Y}^{T_P} = \begin{bmatrix} j\omega C_{pg} & 0 \\ 0 & j\omega C_{pd} \end{bmatrix} \tag{7.95}$$

Then, we have to embed another series network, [L], leading to:

$$\mathbf{I}^{T_L} = \mathbf{I}^L = \mathbf{I}^{T_P} \quad \text{and} \quad \mathbf{V}^{T_L} = \mathbf{Z}^L\mathbf{I}^{T_P} + \mathbf{V}^{T_P} \tag{7.96}$$

where

$$\mathbf{Z}^L = \begin{bmatrix} j\omega L_g & 0 \\ 0 & j\omega L_d \end{bmatrix} \tag{7.97}$$

The next network to be embedded is the one composed of the feedback capacitance C_{cf}, $[C_f]$, a parallel network whose admittance-matrix formulation allows one to determine the following:

$$\mathbf{V}^{T_F} = \mathbf{V}^F = \mathbf{V}^{T_L} \quad \text{and} \quad \mathbf{I}^{T_F} = \mathbf{Y}^F\mathbf{V}^{T_L} + \mathbf{I}^{T_L} = \begin{bmatrix} j\omega C_{cf} & -j\omega C_{cf} \\ -j\omega C_{cf} & j\omega C_{cf} \end{bmatrix} \mathbf{V}^{T_L} + \mathbf{I}^{T_L} \tag{7.98}$$

Up to now, the calculations corresponding to Eqns (7.93), (7.94), (7.96) and (7.98) can be summarized as follows:

$$\begin{bmatrix} \mathbf{V}^{T_F} \\ \mathbf{I}^{T_F} \end{bmatrix} = \begin{bmatrix} \mathbf{Av}_{11} & \mathbf{Z}_{12} \\ \mathbf{Y}_{21} & \mathbf{Ai}_{22} \end{bmatrix} \begin{bmatrix} \mathbf{V}^I \\ \mathbf{I}^I \end{bmatrix} \tag{7.99}$$

where

$$\mathbf{Av}_{11} = \mathbf{Z}^L\mathbf{Y}^P + \mathbf{1} \tag{7.100a}$$

$$\mathbf{Z}_{12} = \mathbf{Z}^L(\mathbf{Y}^P\mathbf{Z}^R + 1) + \mathbf{Z}^R \tag{7.100b}$$

$$\mathbf{Y}_{21} = \mathbf{Y}^F(\mathbf{Z}^L\mathbf{Y}^P + 1) + \mathbf{Y}^P \tag{7.100c}$$

$$\mathbf{Ai}_{22} = \mathbf{Y}^F[\mathbf{Z}^L(\mathbf{Y}^P\mathbf{Z}^R + 1) + \mathbf{Z}^R] + (\mathbf{Y}^P\mathbf{Z}^R + 1) \tag{7.100d}$$

and $\mathbf{1}$ stands for the 2×2 identity matrix.

At this point, we reached the distributed part of the package network, represented in Figure 7.31 by the gate and drain terminal $[S_g]$ and $[S_d]$ matrices. Because these two networks are connected in cascade with the previous ones, we first need to convert these S-parameter matrices into their corresponding ABCD matrices, say $\mathbf{T}^G$ and $\mathbf{T}^D$, to finally obtain the extrinsic FET voltages and currents:

$$\begin{bmatrix} V_G \\ I_G \end{bmatrix} = \mathbf{T}^G \begin{bmatrix} Vf_G \\ -If_G \end{bmatrix} \tag{7.101}$$

and

$$\begin{bmatrix} V_D \\ I_D \end{bmatrix} = \mathbf{T}^D \begin{bmatrix} Vf_D \\ -If_D \end{bmatrix} \tag{7.102}$$

If, on the contrary, the objective was not to start by the intrinsic voltages and currents to obtain the extrinsic ones, but rather to estimate the intrinsic voltages and currents from the measured extrinsic ones, we would have to de-embed the complete FET package. For that, the first step would be to de-embed the gate and drain distributed part by using the inverses of $\mathbf{T}^G$, $\mathbf{T}^{-G}$, and $\mathbf{T}^D$, $\mathbf{T}^{-D}$:

$$\begin{bmatrix} Vf_G \\ -If_G \end{bmatrix} = \mathbf{T}^{-G} \begin{bmatrix} V_G \\ I_G \end{bmatrix} \tag{7.103}$$

$$\begin{bmatrix} Vf_D \\ -If_D \end{bmatrix} = \mathbf{T}^{-D} \begin{bmatrix} V_D \\ I_D \end{bmatrix} \tag{7.104}$$

Then, the de-embedding of the remaining package lumped components could be directly done by inverting Eqn (7.99):

$$\begin{bmatrix} \mathbf{V}^I \\ \mathbf{I}^I \end{bmatrix} = \begin{bmatrix} \mathbf{Av}_{11} & \mathbf{Z}_{12} \\ \mathbf{Y}_{11} & \mathbf{Ai}_{22} \end{bmatrix}^{-1} \begin{bmatrix} \mathbf{V}^{T_F} \\ \mathbf{I}^{T_F} \end{bmatrix} \tag{7.105}$$

This analysis can even be extended to the de-embedding of the I/Q model current conduction functions from their associated charge functions. It is sometimes used to discriminate the dynamic effects imposed by the device's intrinsic capacitances from the quasi-static drain-source current [40]. Its basic idea is to assume that because $\mathbf{I}^I$ can be given by:

$$\mathbf{I}^I = \begin{bmatrix} I_{G'} \\ I_{D'} \end{bmatrix} = \begin{bmatrix} I_{GS}(V_{G'}, V_{D'}) + j\Omega Q_{GS} \\ I_{DS}(V_{G'}, V_{D'}) + j\Omega Q_{DS} \end{bmatrix} \tag{7.106}$$

then the $I_{DS}(V_{G'}, V_{D'})$ and also the $I_{GS}(V_{G'}, V_{D'})$ conduction currents can be obtained from the above obtained intrinsic FET currents through

$$\begin{bmatrix} I_{GS}(V_{G'}, V_{D'}) \\ I_{DS}(V_{G'}, V_{D'}) \end{bmatrix} = \begin{bmatrix} I_{G'} - j\Omega Q_{GS} \\ I_{D'} - j\Omega Q_{DS} \end{bmatrix} \tag{7.107}$$

as long as the $j\Omega Q_{GS}$ and $j\Omega Q_{DS}$ charge functions can be estimated from the previous knowledge of the FET's intrinsic nonlinear capacitances C_{gs}, C_{ds}, and C_{gd}.

The second example to illustrate the previously mentioned embedding/de-embedding procedures concerns the scaling of the X-parameter model [41,42]. The problem can be formulated as follows. Take the microwave FET model shown in Figure 7.31 and suppose that an X-parameter model was extracted for it. To reduce the size of the measurement data required to characterize an entire device family, consider that you want to measure only a unit FET cell and from it extrapolate the *X*-parameter set of the whole device family. A convenient way to do this [42] is to start by de-embedding the parasitics of the unit cell to obtain the *X*-parameter of its corresponding intrinsic unit cell, rescale the intrinsic transistor and its parasitics for any other desired cell width and number of gate fingers, and re-embed the new intrinsic device with its parasitics. This can be done using the same de-embedding procedure just discussed for the I/Q model, except that now we no longer start with measurements of voltages and currents, $[V_G V_D]^T$ and $[I_G I_D]^T$, but with incident and reflected/transmitted waves, $[A_G A_D]^T$ and $[B_G B_D]^T$ (where $\mathbf{M}^T$ represents the transpose of vector **M**). However, this difficulty can be readily overcome using the linear transformations between these two vector sets, herein rewritten in matrix form:

$$\begin{bmatrix} A \\ B \end{bmatrix} = \frac{1}{2}\begin{bmatrix} 1 & Z_0 \\ 1 & -Z_0 \end{bmatrix}\begin{bmatrix} V \\ I \end{bmatrix} \quad \text{and} \quad \begin{bmatrix} V \\ I \end{bmatrix} = 2\begin{bmatrix} 1 & Z_0 \\ 1 & -Z_0 \end{bmatrix}^{-1}\begin{bmatrix} A \\ B \end{bmatrix} = \begin{bmatrix} 1 & 1 \\ \frac{1}{Z_0} & -\frac{1}{Z_0} \end{bmatrix}\begin{bmatrix} A \\ B \end{bmatrix} \tag{7.108}$$

Moreover, Eqn (7.108) even allows us to transform directly from or to an **S**-matrix formulation to or from the **Z**- or **Y**-matrix formulations, noting that $\mathbf{V} = \mathbf{ZI}$, $\mathbf{I} = \mathbf{YV}$, and $\mathbf{B} = \mathbf{SA}$:

$$\mathbf{Y} = \frac{1}{Z_0}(\mathbf{1} + \mathbf{S})^{-1}(\mathbf{1} - \mathbf{S}) \tag{7.109}$$

$$\mathbf{Z} = Z_0(\mathbf{1} - \mathbf{S})^{-1}(\mathbf{1} + \mathbf{S}) \tag{7.110}$$

and

$$\mathbf{S} = (\mathbf{Z} + Z_0)^{-1}(\mathbf{Z} - Z_0) \tag{7.111}$$

or

$$\mathbf{S} = (Y_0 + \mathbf{Y})^{-1}(Y_0 - \mathbf{Y}) \tag{7.112}$$

References

[1] Schetzen M. The Volterra and Wiener theories of nonlinear systems (Malabar, FL, USA): Robert Krieger Publishing Company; 2006.

[2] Mathews V, Sicuranza G. Polynomial signal processing (New York, NY, USA): John Wiley & Sons, Inc; 2000.

[3] Gupta MM, Jin L, Homma N. Static and dynamic neural networks (New York, NY, USA): John Wiley & Sons, Inc; 2003.

[4] Zhang QJ, Gupta KC. Neural networks for RF and microwave design (Dedham, MA, USA): Artech House; 2000.

[5] Zhang Q-J, Gupta KC, Devabhaktuni VK. Artificial neural networks for RF and microwave design—from theory to practice. IEEE Trans Microwave Theory Tech April 2003;51(4):1339−50.
[6] Chua LO, Kang SM. Section-wise piecewise-linear functions: canonical representation, properties, and applications. Proc IEEE June 1977;65(6):915−29.
[7] Chua LO, Deng A-C. Canonical piecewise-linear modeling. IEEE Trans Circuits Syst May 1986;CAS-33(5):511−25.
[8] Kantz H, Schreiber T. Nonlinear time series analysis (Cambridge, UK): Cambridge University Press; 1997.
[9] Takens F. Detecting strange attractors in turbulence. In: Rand DA, Young L-S, editors. Dynamical systems and turbulence, lecture notes in mathematics, vol. 898. (Berlin, Germany): Springer-Verlag; 1981. pp. 366−81.
[10] Casdagli M. A dynamical systems approach to modeling input-output systems. In: Casdagli M, Eubank S, editors. Nonlinear modeling and forecasting, SFI Studies in The Sciences of Complexity, Proc, vol. XII. (Reading, MA, USA): Addison-Wesley; 1992.
[11] Schreurs D, Wood J, Tufillaro N, Usikov D, Barford L, Root DE. The construction and evaluation of behavioral models for microwave devices based on time-domain large-signal measurements. Int Electron Devices Meet IEDM'00 (San Francisco, CA, USA), December 2000:819−22.
[12] Verbeyst F, Bossche MV. VIOMAP, the *S*-parameter equivalent for weakly nonlinear RF and microwave devices. IEEE Trans Microwave Theory Tech December 1994;42(12):2531−5.
[13] Kim J, Konstantinou K. Digital predistortion of wideband signals on power amplifier model with memory. IEE Electron Lett November 2001;37(23):1417−8.
[14] Hammi O, Ghannouchi FM, Vassilakis B. A compact envelope-memory polynomial for RF transmitters modeling with application to baseband and RF-digital predistortion. IEEE Microwave Wireless Compon Lett May 2008;18(5):359−61.
[15] Morgan DR, Ma Z, Kim J, Zierdt MG, Pastalan J. A generalized memory polynomial model for digital predistortion of RF power amplifiers. IEEE Trans Signal Process October 2006;54(10):3852−60.
[16] Wang F, Zhang Q-J. Knowledge-based neural models for microwave design. IEEE Trans Microwave Theory Tech December 1997;45(12):2333−43.
[17] Cao Y, Chen X, Wang G. Dynamic behavioral modeling of nonlinear microwave devices using real-time recurrent neural network. IEEE Trans Electron Devices May 2009;56(5):1020−6.
[18] Xu J, Yagoub MCE, Ding R, Zhang QJ. Exact adjoint sensitivity analysis for neural-based microwave modeling and design. IEEE Trans Microwave Theory Tech January 2003;51(1):226−37.
[19] Roblin P, Root DE, Verspecht J, Ko Y, Teyssier JP. New trends for the nonlinear measurement and modeling of high-power RF transistors and amplifiers with memory effects. IEEE Trans Microwave Theory Tech June 2012;60(6):1964−78.
[20] W8531EP IC-CAP NeuroFET Extraction Package, Agilent Technologies Inc.
[21] Root DE, Fan S, Meyer J. Technology-independent large-signal non quasistatic FET models by direct extraction from automatically characterized device data. In: 21st Eur Microwave Conf Proc (Stüttgart, Germany), September 1991. pp. 927−32.

[22] Bosch SV, Martens L. Approximation of state functions in measurement-based transistor model. IEEE Trans Microwave Theory Tech January 1999;47(1):14–7.

[23] Koh K, Park H-M, Hong S. A spline large-signal FET model based on bias-dependent pulsed I–V measurements. IEEE Trans Microwave Theory Tech November 2002;50(11):2598–603.

[24] Wood J, Aaen PH, Bridges D, Lamey D, Guyonnet M, Chan DS, et al. A nonlinear electro-thermal scalable model for high-power RF LDMOS transistors. IEEE Trans Microwave Theory Tech February 2009;57(2):282–92.

[25] Gorissen D, Zhang L, Zhang Q-J, Dhaene T. Evolutionary neuro-space mapping technique for modeling of nonlinear microwave devices. IEEE Trans Microwave Theory Tech February 2011;59(2):213–29.

[26] Verspecht J, Root DE, Wood J, Cognata A. Broad-band, multi-harmonic frequency domain behavioral models from automated large-signal vectorial network measurements. IEEE MTT-S Int Microwave Theory Tech Symp Dig (Long Beach, CA, USA), June 2005:1975–8.

[27] Root DE, Verspecht J, Sharrit D, Wood J, Cognata A. Broad-band poly-harmonic distortion (PHD) behavioral models from fast automated simulations and large-signal vectorial network measurements. IEEE Trans Microwave Theory Tech November 2005;MTT-53(11):3656–64.

[28] Verspecht J, Root DE. Poly-harmonic distortion modeling. IEEE Microwave Mag June 2006;7(3):44–57.

[29] Application note: Solutions for characterizing and designing linear active devices. Agilent Technologies Inc.; 2009.

[30] Vendelin GD, Pavio AM, Rohde UL. Microwave circuit design using linear and nonlinear techniques. 2nd ed. (New York, NY, USA): John Wiley & Sons; 2005.

[31] Rytting DK. Network analyzer error models and calibration methods. 62nd ARFTG Conf (Boulder, CO, USA), December 2003. Short course notes.

[32] Kruppa W, Sodomsky KF. An explicit solution for the scattering parameters of a linear two-port measured with an imperfect test set. IEEE Trans Microwave Theory Tech January 1971;MTT-19(1):122–3.

[33] Engen GF, Hoer CA. Thru-reflect-line: an improved technique for calibrating the dual six-port automatic network analyzer. IEEE Trans Microwave Theory Tech December 1979;27(12):987–93.

[34] Van den Broeck T, Verspecht J. Calibrated vectorial nonlinear-network analyzers. IEEE MTT-S Int Microwave Symp Dig (San Diego, CA, USA), June 1994:1069–72.

[35] Verspecht J. Large-signal network analysis. IEEE Microwave Mag December 2005; 6(4):82–92.

[36] Blockley P, Gunyan D, Scott JB. Mixer-based, vector-corrected, vector signal/network analyzer offering 300 kHz–20 GHz bandwidth and traceable phase response. IEEE MTT-S Int Microwave Symp Dig (Long Beach, CA, USA), June 2005: 1497–500.

[37] Schetzen M. Theory of the *p*th-order inverses of nonlinear systems. IEEE Trans Circuits Syst May 1976;CAS-23(5):285–91.

[38] Pedro JC, Carvalho NB, Lavrador PM. Modeling nonlinear behavior of band-pass memoryless and dynamic systems. IEEE MTT-S Int Microwave Symp Dig (Philadelphia, PA, USA), June 2003:2133–6.

[39] Pedro JC, Maas SA. A comparative overview of microwave and wireless power-amplifier behavioral modeling approaches. IEEE Trans Microwave Theory Tech April 2005;MTT-53(4):1150–63.

[40] Vadala V, Avolio G, Raffo A, Schreurs DMM-P, Vannini G. Nonlinear embedding and de-embedding techniques for large-signal FET measurements. Microwave Optical Tech Lett December 2012;54(12):2835–8.

[41] Leckey JG. A scalable *X*-parameter model for GaAs and GaN FETs. In: 6th European Microwave Integrated Circuits Conf Proc (Manchester, UK), October 2011. pp. 13–16.

[42] Root DE, Marcu M, Horn J, Xu J, Biernacki RM, Iwamoto M. Scaling of *X*-parameters for device modeling. IEEE MTT-S Int Microwave Symp Dig (Montreal, Canada), June 2012:1–3.

CHAPTER

Electromagnetic-Analysis-Based Transistor De-embedding and Related Radio-Frequency Amplifier Design

Manuel Yarlequé[1], Dominique M.M.-P. Schreurs[2], Bart Nauwelaers[2], Davide Resca[3], Giorgio Vannini[4]

[1] *Departamento de Ingeniería Pontificia Universidad Católica del Perú, Lima, Perú,*
[2] *ESAT-TELEMIC, KU Leuven, Leuven, Belgium,*
[3] *MEC – Microwave Electronics for Communications, Bologna, Italy,*
[4] *Dipartimento di Ingegneria, Università di Ferrara, Ferrara, Italy*

8.1 Introduction

The current trend to migrate wireless applications to higher frequencies to obtain broader bandwidths and therefore higher transmission throughput has driven semiconductor companies to develop faster devices with high integration. This has come along with some complications in device characterization and modeling. Semiconductor layout dimensions are close to the wavelength of the electromagnetic (EM) wave propagating through it, and therefore the classical equivalent circuit approaches for the description of device parasitics are not representative of what is actually happening at the device and circuit level. In this sense, a relatively new way to analyze the parasitics has evolved, which is based on EM characterization of the parasitic network.

This chapter aims to describe methodologies and techniques for de-embedding device measurements from extrinsic measurements by characterizing the parasitic network surrounding the intrinsic device, through the use of a three-dimensional (3D) physical model of the network and its electromagnetic analysis. The electromagnetic assessment is carried out employing 3D EM solvers and internal ports. In the first part, four different de-embedding procedures applied to field-effect transistors (FETs) for monolithic microwave integrated circuit (MMIC) design are presented. In the second part, the de-embedding of FET devices for hybrid circuit design purposes is described.

8.2 Electromagnetic analysis of MMIC transistor layout

This section is intended as a guide to set up the EM simulation of an MMIC transistor layout, aiming at giving an accurate description of its passive parasitic

Microwave De-embedding. http://dx.doi.org/10.1016/B978-0-12-401700-9.00008-2

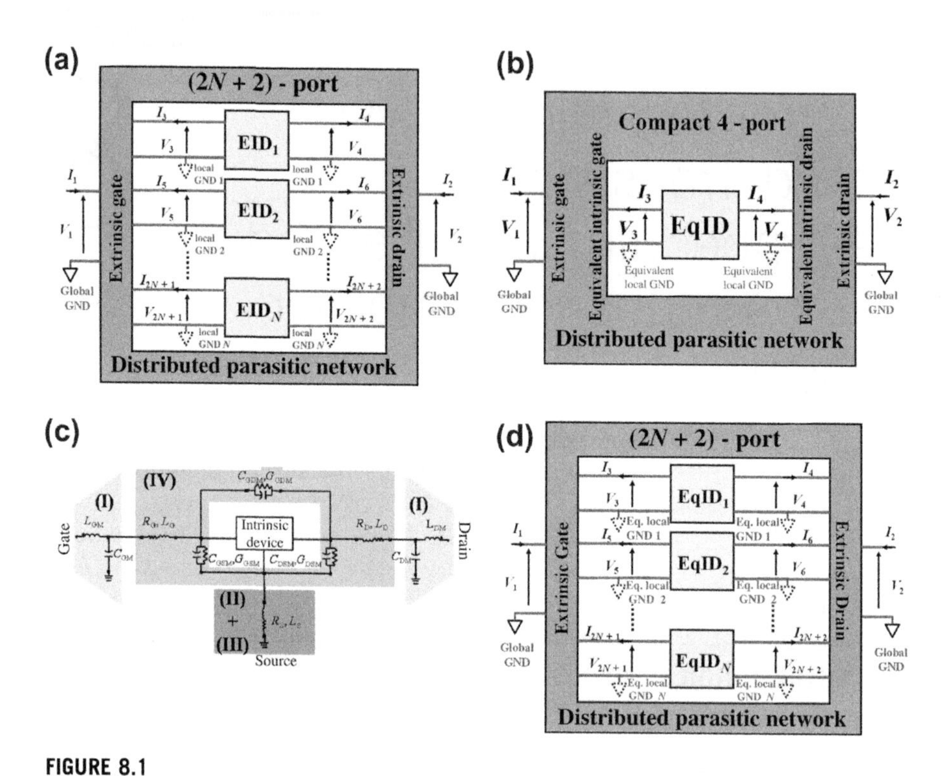

FIGURE 8.1

Different EM-based MMIC transistor models treated in this chapter. (a) Fully distributed transistor model, (b) compact distributed transistor model, (c) scalable empirical equivalent circuit model, (d) complex layout structure model. EID, elementary intrinsic device; EqID, equivalent intrinsic device.

structure. Different MMIC transistor EM-based models are shown in Figure 8.1 and will be dealt with here. In particular, Figure 8.1(a), (b) and (d) show distributed models, in which elementary intrinsic devices (EIDs) describing the active part of the transistor are suitably connected to a distributed passive network.

The EM simulation of the fully distributed model in Figure 8.1(a) [1–3], the compact distributed model in Figure 8.1(b) [4], and the complex layout structure model in Figure 8.1(d) [5] share the same EM setup. The differences between them are the different procedures used to de-embed the extrinsic measurements from the passive parasitic network to get either the EIDs or the equivalent intrinsic device (EqID). The related de-embedding procedures and examples about these different models will be described in Sections 8.3–8.5. A different topology is, instead, used for the extraction of the scalable empirical equivalent circuit model in Figure 8.1(c) [6].

Considering the structure of an MMIC transistor layout, any commercially available 3D planar EM simulator [7,8], allowing the use of internal ports, is suitable to perform the required EM simulations. In the first part of this chapter, all of the examples presented are based on the use of Sonnet EM® software [7]. Although the examples given here are all for MMIC field-effect transistors, very similar setups and techniques can be easily applied to bipolar technologies.

Before starting any simulation, the following information must be known:

- Transistor layout drawing
- Layout layers involved and their physical interconnection, according to the foundry process description
- Technological parameters
- Frequency range of the available measured data or reference device model range of validity
- Required frequency extrapolation region (if extrapolation is needed).

Fortunately, most of these data are available from the foundry manual and foundry design kit describing the MMIC process of the transistor to be modeled.

8.2.1 Dielectric stack and metal layers definition

The knowledge of the foundry process description, together with the transistor layout drawing, allows one to define the EM simulator dielectric stack of materials and the metal layers. An example is shown in Figure 8.2 with reference to a millimeter-wave low-noise gallium arsenide (GaAs) pseudomorphic high-electron-mobility transistor (pHEMT) process.

The dielectric stack comprises the GaAs substrate, the passivation oxide over the finger region, the air layer needed to define the air bridge connecting the

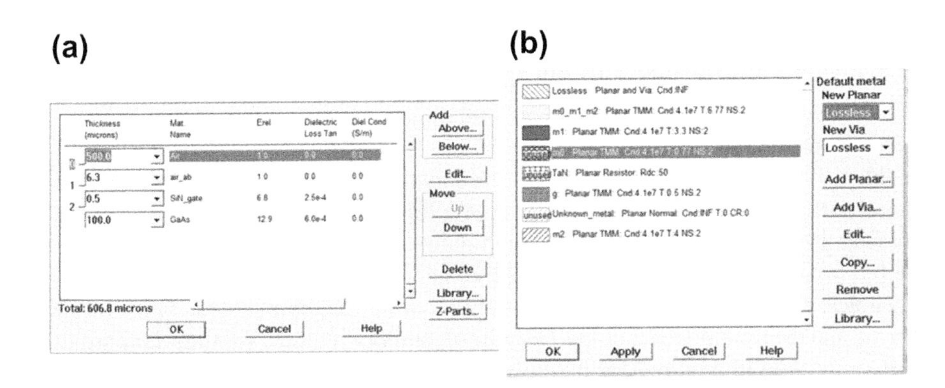

FIGURE 8.2

(a) Sonnet EM® dielectric stack and (b) metal layer definitions for a millimeter-wave low-noise GaAs pHEMT process.

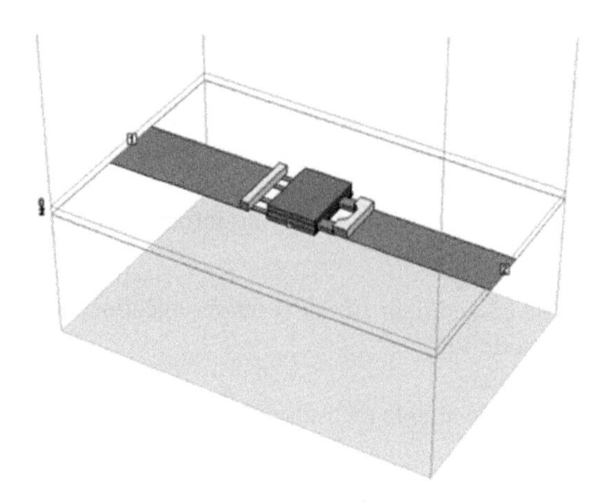

FIGURE 8.3

3D view of a four-finger low-noise GaAs pHEMT structure after layout drawing, dielectric stack definition, and metal layer mapping.

source fingers, and the air layer needed to move up the top cover of the Sonnet box, in order to reproduce a free-space condition. The metal layers are those shown in Figure 8.2(b). Because the thickness of the metals in the finger region is comparable to their distance, to simulate accurately their EM coupling, they are set as thick conductors. The foundry process description helps to set each metallization layer to the correct dielectric interface, as well as to define the via hole polygons that physically connect two metals placed at two different dielectric interfaces. After the layout drawing, dielectric definition, and metal layer mapping, the resulting 3D view of a four-finger low-noise GaAs pHEMT is shown in Figure 8.3.

8.2.2 Mesh cell dimensioning and functional partitioning

When defining the Sonnet box, the mesh cell dimension must be set to have the fairest resolution in the finger region of the device. This leads to the so-called big-small problem [9] because this minimum cell size creates a huge number of cells in the larger surrounding structures. In the case of a four-finger low-noise GaAs pHEMT (having a total gate width of 150 μm and gate length of 0.25 μm), the cell, set to be 0.5×0.125 μm, will produce a huge number of mesh cells in the source-end via holes. A functional partitioning [9] of the device to simulate separately the finger region and the via hole reduces the 10.8 GB of memory required for the whole structure to the 4.3 GB required for the finger region plus 700 MB required for one via hole (whose mesh cell size was set to be 1×1 μm). Figure 8.4 shows the finger region and via hole parts simulated.

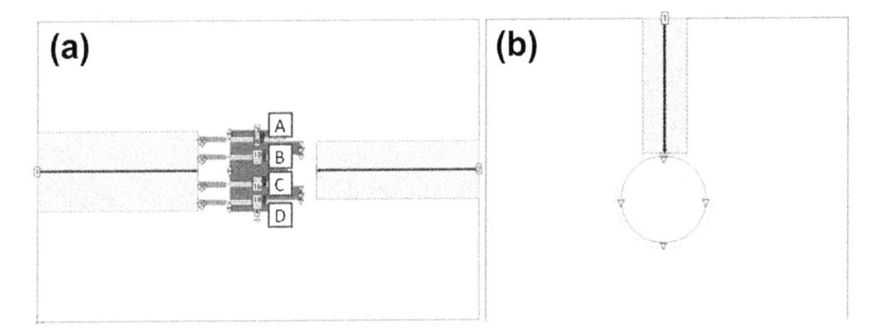

FIGURE 8.4

Functional partitioning of a 4 × 37.5 μm low-noise 0.25 μm GaAs pHEMT. EM simulator setup of the finger region (a) and of the via hole (b).

8.2.3 Port definitions

To excite the structures drawn in the Sonnet geometry editor, one has to place the EM simulator ports. Apart from the scalable empirical equivalent circuit model in Figure 8.1(c), whose setup will be addressed later, the other models will require, besides the external excitations, the definition of the attachment points for the EIDs. According to Figure 8.4(a), two Sonnet box ports are placed both at the external gate side (port 1) and at the external drain side (port 2). Because of the functional partitioning, two additional autogrounded ports are placed at the two source ends (ports 3 and 4). Then, 16 internal ports (ports 5–20) are placed in the finger region to form four different Sonnet co-calibrated port groups (the groups A, B, C, and D shown in Figure 8.4(a)); each of them is formed by four co-calibrated floating referenced ports.

To create one four-port group (e.g., group A), a tiny portion of the gate metal polygon (see Figure 8.5(a)) is cut away to create the edges where ports 5 and 7 of the group were placed. Using the "add point" tool of the Sonnet geometry editor, it is easy to define the attachment points of ports 6 and 8. By selecting ports 5–8, the co-calibrated group A is created, setting the floating ground reference, which defines the local ground in Figure 8.5(b), for each group. For the sake of the EID definition, ports 5 and 7 are an intrinsic gate attachment point (they will be connected together in the circuit simulator), port 6 is an intrinsic source attachment point, and port 8 is an intrinsic drain attachment point. It is worth noting that the intrinsic source port will be the local reference when an EID two-port representation is defined in the de-embedding procedure (local ground in Figure 8.1(a) and (b)).

In the example in Figure 8.4 and Figure 8.5, only one section of EIDs is defined in the middle of the finger region. This is exactly what is needed to extract the compact distributed model in Figure 8.1(b), whereas more sections can be defined if the fully distributed model in Figure 8.1(a) is required (when distributed, effects along the device fingers are very strong).

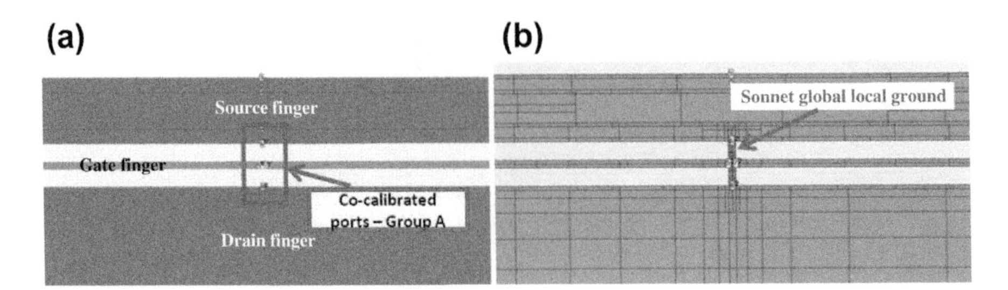

FIGURE 8.5

(a) Generation of the EID attachment points by defining sonnet co-calibrated four-port groups with floating references. (b) Sketch of the generated meshes showing the defined local grounds.

8.2.4 Setup for complex layout model

Complex layout models, such as the one in Figure 8.1(d), can be adopted when modeling relatively complex transistor structures (e.g., a cascode connection composed of the cascade of a common source [CS] and a Common Gate [CG] transistor). The EM simulator setup of the complex layout structure model is very similar to the one used for a single transistor. The only difference is that some additional functional partitioning is required. Let the cascode connection of two 4 × 37.5 μm low-noise 0.25 μm GaAs pHEMTs be the case for a complex layout structure (see Figure 8.6). There are two-finger regions of the CS and CG FETs connected in cascode fashion by a "surrounding" passive structure.

According to Figure 8.6, the model comprises a feedback line connecting the drain of the CS FET to the source of the CG FET and a tantalum nitride resistor in series to an over-via metal-insulator-metal capacitor (which provides stabilized radio-frequency [RF] shunt to ground to the CG FET). The ground connection of the CS FET source ends is obtained through the bottom plate of the over-via capacitor at the upper end and through the via hole at the bottom end. The EM simulation of the whole structure demands memory resources that might be unaffordable due to the "big-small problem." The functional partitioning helps out again: one can divide the EM simulation into smaller simulations, which are the CS and CG finger regions and the surrounding passive structure. Because in this case both CS and CG FETs have the same size, their finger region is identical and just one simulation is performed (see Figure 8.6). The related EM setup is exactly the one described in the previous sections. The surrounding structure simulation makes use of two co-calibrated port groups to create the attachment point of the CS FET (ports 1–4) and CG FET (ports 5–7) finger regions, respectively. Two additional ports (Sonnet box ports 8 and 9, not shown in Figure 8.6) provide the external gate and drain excitations to the surrounding structure simulation. It is worth noticing that the co-calibrated groups were obtained by defining two Sonnet components

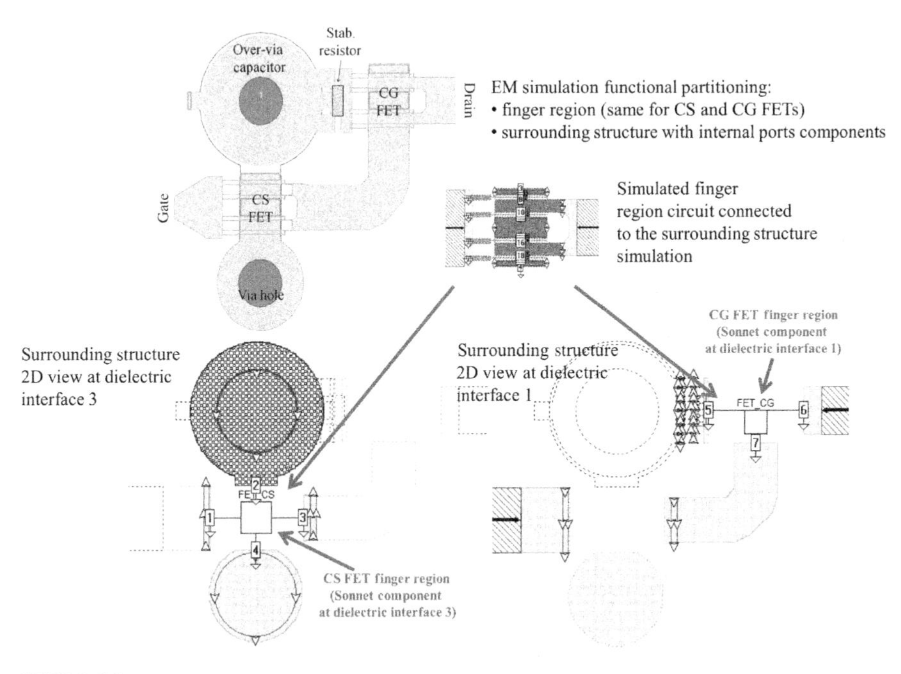

FIGURE 8.6

Complex layout model of an EM simulator setup. Case of a cascode cell made by $2 \times 4 \times 37.5\ \mu m$ low-noise 0.25 μm GaAs pHEMTs. 2D, two dimensional; CG, common gate; CS, common source; FET, field-effect transistor.

designated FET_CS and FET_CG in Figure 8.6, whose port reference is set at the Sonnet box. Due to the physical connection of the metal layers in the surrounding structure, the two components are defined in the EM setup at two different dielectric layer interfaces (interface numbers 3 and 1, respectively, in Figure 8.6).

8.2.5 Full-wave EM simulation of a transistor passive structure

The EM simulator setup suitable for the extraction of the scalable empirical equivalent circuit model in Figure 8.1(c) is simpler than the ones previously outlined. In fact, in this case the EM simulation is set to provide a two-port description of the device's passive parasitic network. The two-port data are used to extract the parasitic elements of the model according to the procedure described in Section 8.4. Figure 8.7 shows the EM simulator setup to obtain the full-wave (FW) simulation of a six-finger GaAs power pHEMT (having a total gate width of 300 μm and gate length of 0.25 μm) passive structure. The dielectric stack and metal layer definitions are very similar to those of Figure 8.2. Again, it is very useful to adopt functional partitioning of the whole structure into finger regions and via holes. There is no need for internal ports: the only ports placed are two Sonnet box excitations at the gate and

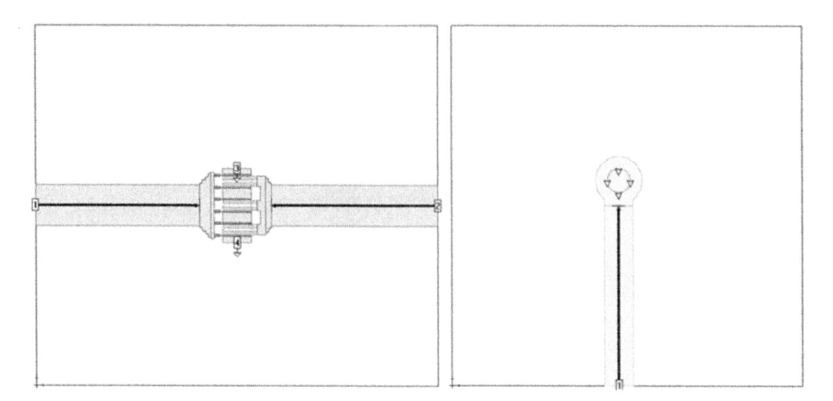

FIGURE 8.7

Functional partitioning of a 6 × 50 μm 0.25 μm GaAs power pHEMT. EM simulator setup of the finger region and the via hole.

drain sides, respectively, and two autogrounded ports to create the attachment points of the via holes.

8.2.6 Frequency range of analysis

The very last thing to do is the choice of the frequency range over which the EM analysis must be performed. Because the extraction of the models in Figure 8.1 involves the de-embedding procedures described in Sections 8.3–8.5, the EM simulation frequency range must at least match the frequency range of the measured data available (or match the reference transistor model frequency range of validity). It is also very important to add to the frequency list the direct current (DC) frequency too, especially when a nonlinear transistor model is planned to be extracted.

When the EM-based model is used for frequency extrapolation, the EM simulation must be performed at the maximum extrapolated frequency.

All of the EM simulations performed in the examples given in this section and in Sections 8.3–8.5, are obtained using the adaptive sweep feature available in Sonnet. This allows one to accurately simulate very broad frequency ranges with a small number of frequency points. (Very often, less than 10 frequency points are enough to let the adaptive sweep algorithm produce the full frequency response.)

8.2.7 EM simulator output

When the EM analysis is finished, the simulator outputs a passive scattering (**S**)-matrix. This matrix could be either a multiport matrix, as in the case of the simulations exploiting internal ports, or a two-port matrix, as in the case of the full-wave EM analysis.

It is worth noting that the data obtained from each part of a partitioned simulation have then to be properly recombined in the circuit simulation to recover the EM response of the entire structure.

A first check of the EM simulation data helps to find out if all of the previous steps were done correctly. For example, if the **S**-matrix is not passive at some frequency or some of the *S* parameters show no symmetry (when they must be symmetric because of symmetries in the simulated structure), it means that something in the setup went wrong.

Apart from those macroscopic errors, the final check of the entire EM procedure can be done only by looking at the intrinsic device frequency response, which is available only at the end of the de-embedding procedures described in Sections 8.3–8.5.

When the computational resources available allow simulating the whole structure, the functional partitioning should always be verified, because the partition procedure intrinsically neglects couplings that might be important (see for example [5]).

8.3 Transistor modeling based on a distributed parasitic network description

8.3.1 Distributed modeling approaches

For many years, research effort has been dedicated to the development of "global" design procedures [10,11] for monolithic micro- and millimeter-wave integrated circuits. Indeed, circuit design could be more effectively carried out through design approaches where, beside the values of passive (lumped or distributed) components, active device geometric parameters (e.g., gate width and number of fingers) are also used as design variables. Moreover, higher integration levels and consequently cost reduction are possible if complex coupling and distributed effects can be properly accounted for and managed in the circuit design phase. Suitable scalable FET models that accurately predict device response as a function of geometry are key elements in this context.

Often, III–V foundries provide scalable models based on equivalent circuits whose parameters scale with gate width and finger number. The scaling approaches adopted range from trivial linear rules to completely empirical expressions (usually polynomial expressions fitted on the measured electrical responses of a quite large number of different device structures).

High operating frequencies make the task of identifying a scalable model more complex. Lumped equivalent circuit descriptions of intrinsically distributed phenomena [12] (e.g., propagation along the metallization, terminal coupling) show lower accuracy at millimeter-waves and more attention must be paid to parasitic modeling and identification.

Distributed models have long been proposed [13,14] as a possible alternative to conventional equivalent circuits. They usually cascade a number of elementary cells (describing the transistor active area), which are fed by lumped passive networks that should model electromagnetic propagation and coupling phenomena related to the passive structure. These models, however, have been usually identified on the basis of experiments carried out on a relatively large number of different device structures and by using only limited information arising from simple electromagnetic simulations or analytical evaluation of the propagation and coupling phenomena.

The consistent solution of the coupled electromagnetic and electron transport problems could be dealt with also through physics-based numerical simulations [15]. Unfortunately, despite the progress in numerical algorithms and the availability of powerful workstations, these models are not sufficiently mature yet, and their application to practical problems such as device scaling and circuit simulation, among others, may be difficult [16], taking also into account their computational cost.

The progress in electromagnetic numerical analysis has led to powerful and efficient electromagnetic simulation tools (e.g., *EM*, Momentum) which can be effectively exploited in the identification of the transistor parasitic network. On this basis, quite accurate, robust, and computationally efficient distributed transistor models have been developed.

These approaches can be divided in two different classes: models in which a lumped or simplified distributed network is identified on the basis of electromagnetic simulation of transistor structures, and models in which the distributed parasitic network is characterized through a multiport matrix description directly obtained from electromagnetic simulations of the device layout. Among the latter approaches, the one proposed in [1,2] and successfully applied in [3,4] for the distributed modeling of microwave and millimeter-wave FETs will be dealt with in detail in the following section.

8.3.2 Model definition and identification

The distributed modeling approach [2] was originally proposed and validated for linear operation of transistors, although it can be (in theory) straightforwardly applied also to the more general case of nonlinear regime. Such an issue will be probed further after presenting the modeling approach.

The model [2] can be classified as a circuit-design-oriented model—that is, a model that can be effectively used for integrated circuit (e.g., MMIC) design. The basic idea is that of describing the transistor in terms of a distributed extrinsic passive network feeding a cascade of identical elementary active devices. More precisely, the active FET area is partitioned into a suitable number of "internal elementary devices" (active slices) fed by a "passive distributed network", as shown in Figure 8.8. In Figure 8.8, for simplicity, a two-finger device composed of $2N$ interconnected "active slices" has been considered (clearly, the number of fingers can be extended without limitations).

Model identification begins by characterizing the device passive network in terms of scattering parameters. To this end, an accurate electromagnetic simulation

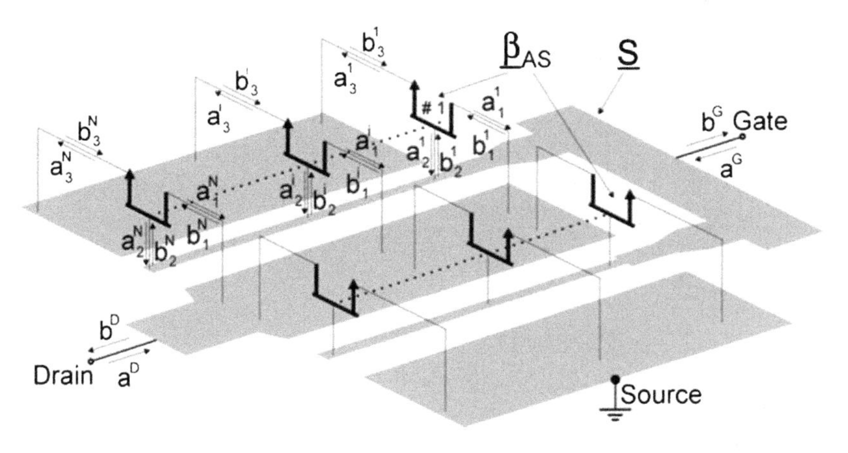

FIGURE 8.8

Structure of the distributed model.

of the device layout must be carried out. In such a way, the actual device geometry, material geometry, and characteristics, as well as losses in the dielectrics and metallizations, are accounted for any given device structure by means of a multiport scattering-matrix distributed description $\underline{\mathbf{S}}$. In this context, it is important to observe that, in the modeling of electrically distributed systems using a circuit representation, one of the main issues is the assignment of the system reference potential. In a spatially distributed structure, when the separation between two adjacent nodes is a significant fraction of the wavelength, it is not possible to define a unique common reference point. In that sense, port voltages of any elementary device are referred to as "local potential," as will be described in detail in Section 8.3.3.

The identification of the scattering matrices $\underline{\boldsymbol{\beta}}_{AS}$ associated with the active slices can be successively carried out by means of a de-embedding procedure starting from the knowledge of measured S parameters of the transistor. More precisely, the following homogeneous system of equations can be written:

$$\left[\begin{bmatrix} \underline{\mathbf{n}}_{1,1} & \underline{\mathbf{n}}_{1,2} & \vdots & \underline{\mathbf{n}}_{1,N} \\ \underline{\mathbf{n}}_{2,1} & \underline{\mathbf{n}}_{2,2} & \vdots & \underline{\mathbf{n}}_{2,N} \\ \cdots & \cdots & \ddots & \cdots \\ \underline{\mathbf{n}}_{N,1} & \underline{\mathbf{n}}_{N,2} & \vdots & \underline{\mathbf{n}}_{N,N} \end{bmatrix} - \begin{bmatrix} \underline{\boldsymbol{\beta}}_{AS} & \underline{0} & \cdots & \underline{0} \\ \underline{0} & \underline{\boldsymbol{\beta}}_{AS} & \cdots & \underline{0} \\ \vdots & \vdots & \ddots & \vdots \\ \underline{0} & \underline{0} & \cdots & \underline{\boldsymbol{\beta}}_{AS} \end{bmatrix}\right] \cdot \begin{bmatrix} \mathbf{b}^1 \\ \mathbf{b}^2 \\ \vdots \\ \mathbf{b}^{N-1} \\ \mathbf{b}^N \end{bmatrix} = \underline{0} \quad (8.1)$$

where N is the number of active slices included within each "half device." We can use this approach because a symmetric structure is very common in high-frequency devices.

In Eqn (8.1) the generic element having position p,q within the matrices $\underline{\mathbf{n}}_{h,k}$ is defined as:

$$\left(\underline{\mathbf{n}}_{h,k}\right)_{p,q} = \left(\left[\underline{\boldsymbol{\sigma}}\cdot\begin{bmatrix}\left[\underline{\boldsymbol{\alpha}}-\underline{\boldsymbol{\delta}}\right]^{-1}\cdot\underline{\boldsymbol{\gamma}}\\ \underline{\mathbf{I}}\end{bmatrix}\right]^{-1}\right)_{3h-3+p,\ 3k-3+q} \qquad \begin{aligned} h,k &= 1,\ldots,N \\ p,q &= 1,2,3 \end{aligned} \tag{8.2}$$

with:

$$\begin{aligned} \underline{\boldsymbol{\sigma}}_{i=1,\ldots,3N;\ j=1,..,2+3N} &= \underline{\tilde{\mathbf{S}}}_{i=3,\ldots,2+3N;\ j=1,..,2+3N} \\ \underline{\boldsymbol{\delta}}_{i=1,2;\ j=1,2} &= \underline{\tilde{\mathbf{S}}}_{i=1,2;\ j=1,2} \\ \underline{\boldsymbol{\gamma}}_{i=1,2;\ j=1,..,3N} &= \underline{\tilde{\mathbf{S}}}_{i=1,2;\ j=3,..,2+3N} \end{aligned}$$

$\underline{\boldsymbol{\alpha}}$ is the measured 2×2 scattering matrix of the electron device, $\underline{\mathbf{I}}$ is a $3N \times 3N$ identity matrix, and $\underline{\tilde{\mathbf{S}}}$ is the scattering matrix obtained from $\underline{\mathbf{S}}$, once the rows and columns that are linearly dependent due to the symmetry of the device structure have been eliminated.

The system of equations in Eqn (8.1) must be solved for $\underline{\boldsymbol{\beta}}_{AS}$. In the simplest case of $N = 1$ (which as we will see in the following is also the most interesting one), by performing matrix arrangements, the solution of the system leads to:

$$\underline{\boldsymbol{\beta}}_{AS} = \left[\underline{\boldsymbol{\sigma}}\cdot\begin{bmatrix}\left[\underline{\boldsymbol{\alpha}}-\underline{\boldsymbol{\delta}}\right]^{-1}\cdot\underline{\boldsymbol{\gamma}}\\ \underline{\mathbf{I}}\end{bmatrix}\right]^{-1} \tag{8.3}$$

For a generic value of N, the problem can be solved by considering that the existence of the solution of the homogenous system of equations Eqn (8.1) requires a null determinant of the coefficient matrix. This leads to an eigenvalue problem [3] that can be solved for $\underline{\boldsymbol{\beta}}_{AS}$ by means of numerical routines. A slightly different formulation, which does not exploit transistor symmetry, was used in [17], where the solution is found by solving four smaller eigenvalue problems. In the case of $N = 2$, 3, and 4, matrix arrangements of Eqn (8.1) lead to a nonlinear system of equations [3], which can be solved by means of conventional numerical routines.

It must be pointed out that, in any case, model identification (i.e., the computation of $\underline{\boldsymbol{\beta}}_{AS}$) does not require either optimization or parameter fitting. In other words, the identification procedure, although rather involved from a mathematical point of view, is straightforward and requires only conventional scattering parameters device measurements and electromagnetic simulations of the device layout. These can be carried out by exploiting commercially available, powerful EM simulation tools [7,8].

Clearly, some fundamental approximations are introduced in the formulation and identification of the above empirical model. The first and most important one is that the characterization of electromagnetic field distribution in the passive structure is not substantially modified by the current transport along the channel. A more rigorous approach should consider the use of coupled EM and carrier transport simulators. A second approximation consists of having the active part of the electron device "lumped" into a limited, finite number of active slices instead of considering a purely distributed "intrinsic" transistor. On these bases, a conventional, commercially available electromagnetic simulator for model identification can be straightforwardly adopted.

The validity of this empirical approach (and, indirectly, of the assumptions adopted) is confirmed, as will be shown in the following, by the good capabilities of this modeling technique in predicting how transistor electrical characteristics scale with gate width and number of fingers or, more generally, with device geometry variations.

8.3.3 Model scalability and frequency extrapolation capabilities

While the scaling of parasitic elements is not so straightforward, and, depending on the level of approximation adopted in parasitic description, can involve complex considerations, it is usually well accepted and confirmed by experimental evidence that the intrinsic region of an active device can be practically scaled in a proportional way. If, for the sake of convenience, we consider an admittance matrix description of the active slice, we can immediately write:

$$\underline{\mathbf{Y}}_{AS}(\omega, W_{AS}) = \underline{\hat{\mathbf{Y}}}(\omega)\cdot W_{AS} \tag{8.4}$$

where W_{AS} is the width of the active slice and $\underline{\hat{\mathbf{Y}}}(\omega)$ is the admittance matrix associated with the unitary gate width. Note that the admittance matrix $\underline{\mathbf{Y}}_{AS}$ and the corresponding scattering $\underline{\boldsymbol{\beta}}_{AS}$ matrix associated with the active slice are related by well-known transformation formulae.

Although very simple, Eqn (8.4) does not agree perfectly with experiment evidence (in a more pronounced way at high frequencies but quite evidently even at DC or low frequencies) and a better approximation with an additional degree of freedom can be obtained by adopting the following scaling rule [2]:

$$\underline{\mathbf{Y}}_{AS}(\omega, W_{AS}) = \underline{\hat{\mathbf{Y}}}(\omega)\cdot W_{AS} + \underline{\mathbf{C}}(\omega) \tag{8.5}$$

where $\underline{\mathbf{C}}(\omega)$ is a constant (width-independent) term. The use of Eqn (8.5) instead of Eqn (8.4) enables one to account for possibly "nonhomogeneous" transistor structures along the finger width (e.g., the beginning and the end of the gate finger may have some geometric differences with respect to the intermediate part of the structure) and in general for "border-like" effects. Also, a different thermal regime of the active slices might be responsible in large-periphery transistors for deviations from Eqn (8.4).

The adoption of Eqn (8.5) instead of Eqn (8.4) does not make model identification more complex. Measurements on just two transistors having different gate widths are sufficient to obtain the value of the term $\underline{\mathbf{C}}(\omega)$ and $\underline{\hat{\mathbf{Y}}}(\omega)$. It must be observed that a more sound approach can be easily obtained by identifying the parameters of the scaling rule Eqn (8.5) on the basis of measurements carried out on a larger number of transistors and applying a linear regression algorithm.

Once the matrices $\hat{\underline{\mathbf{Y}}}(\omega)$ and $\underline{\mathbf{C}}(\omega)$ have been identified, the linear rule Eqn (8.5) can be used to obtain the admittance matrix associated with active slices having an arbitrary width W_{AS}.

Accurate predictions have been obtained using the described approach and by considering only a single active slice per finger; the validity of such a choice, which substantially reduces the computational effort and simplifies the model identification procedure, has been experimentally verified in [3] by repeating the model identification procedure previously described as a function of the number of active slices per finger and comparing the simulation results with *S*-parameter measurements up to 110 GHz. The experiment, carried out by considering 0.2 μm GaAs pHEMT devices having different finger widths (namely 2 × 15 μm, 2 × 30 μm, 2 × 60 μm, and 2 × 120 μm) and a number of active slices ranging from one to four per finger, showed that the obtained $\hat{\underline{\mathbf{Y}}}(\omega)$ (the admittance matrix associated with the unitary gate width) is independent on the number of active slices adopted in model identification. It can be concluded that using a single active slice per gate finger represents a reasonable choice (giving also a good tradeoff between accuracy and complexity) for the modeling of FET devices up to millimeter-wave frequencies. Also, practical considerations of electron device design justify this conclusion. In fact, in a transistor properly designed for its operating frequency range, the propagation effects along the structure width, although important at high frequencies, cannot be so relevant as to impair the efficient exploitation of part of the active area.

As far as model scalability is concerned, the knowledge of the admittance matrix description $\underline{\mathbf{Y}}_{AS}(\omega, W_{AS})$ of the active slices, in conjunction with the electromagnetic simulation of the device layout, enables the prediction of the linear response of any given transistor structure with an arbitrary number of fingers and gate width. To this end, the distributed model can easily be implemented in any computer-aided design (CAD) environment through the simple connection of matrix-described multiport elements. Thus, device model scaling does not require an extensive experimental characterization of a large number of transistor test structures and associated identification procedures to fit empirical scaling rules. On the contrary, only a limited number of transistor structures must be experimentally characterized through *S*-parameter measurements to identify the description of the active slice. Successively, the computation of the response of a transistor having an arbitrary structure requires only the electromagnetic simulation of its layout. In other words, with respect to other empirical approaches based on equivalent circuits, the effort of expensive and wide experimental characterizations is replaced by electromagnetic simulations.

Moreover, additional advantages are associated with the described approach. For instance, such a distributed model is suitable for robust frequency extrapolation [3]. In fact, the internal active slice admittance is characterized by "regular" frequency behavior. This is consistent with the physical evidence of "short memory" conditions, which usually hold for the "intrinsic part" of microwave and millimeter-wave devices [18–20]. On the contrary, extrinsic parasitic effects cause important resonances in electron devices observed in the *Y*-domain. This can be clearly seen

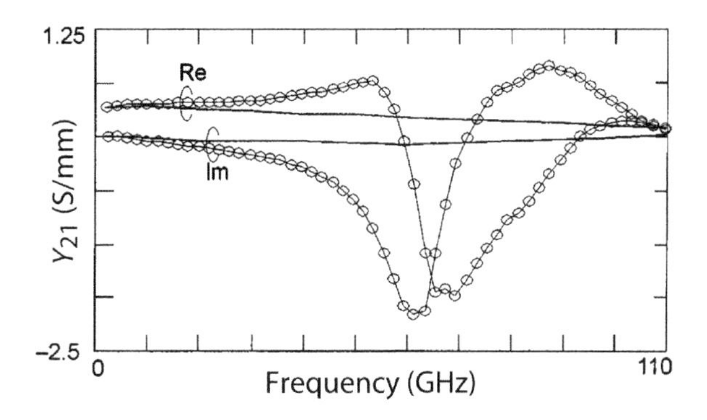

FIGURE 8.9

Measured admittance Y_{21} per unit of gate width (symbols) for a 2 × 30 μm GaAs pHEMT biased at $I_{DS} = I_{dss}$, $V_{DS} = 3$ V. Continuous lines represent the corresponding extracted admittance parameter $Y_{AS,21}$ of the internal elementary devices.

in Figure 8.9, which shows the measured admittance parameter Y_{21} (per unit of gate width) of a 2 × 30 μm GaAs pHEMT and the corresponding extracted admittance parameter $Y_{AS,21}$ of the internal elementary device obtained considering a single active slice per finger. Similar behavior is found for the other admittance matrix elements and independently of the bias conditions.

Clearly, although the direct frequency extrapolation of "extrinsic" measured device admittance parameters would be highly inaccurate, the almost linear behavior of the active slice admittance is suitable for reliable frequency extrapolation. To this end, low-order polynomial expressions can be fitted, in the frequency range used in device characterization, to the internal admittance coefficients [3]. These polynomials easily provide the frequency-extrapolated behavior of the internal elementary devices, which, in conjunction with the distributed description of the extrinsic parasitic network, are used to predict the small-signal device behavior at frequencies higher than those used in the identification measurements. To this aim, electromagnetic simulations must be obviously performed up to the highest frequency of interest. A great advantage of this procedure is that frequency limitations of instrumentation (i.e., vector network analyzer [VNA]) are overcome because the active slice identification can be carried out in a relatively "low" frequency range. Only the identification of the device passive structures requires information at higher frequencies, but such identification is carried out through only electromagnetic simulations.

8.3.4 Experimental and simulation results

The validity of the scalable, distributed model previously described was verified in [3] up to millimeter-wave frequencies. In particular, 0.2 μm GaAs pHEMTs having a different number of gate fingers and gate widths were considered. Figure 8.10 shows,

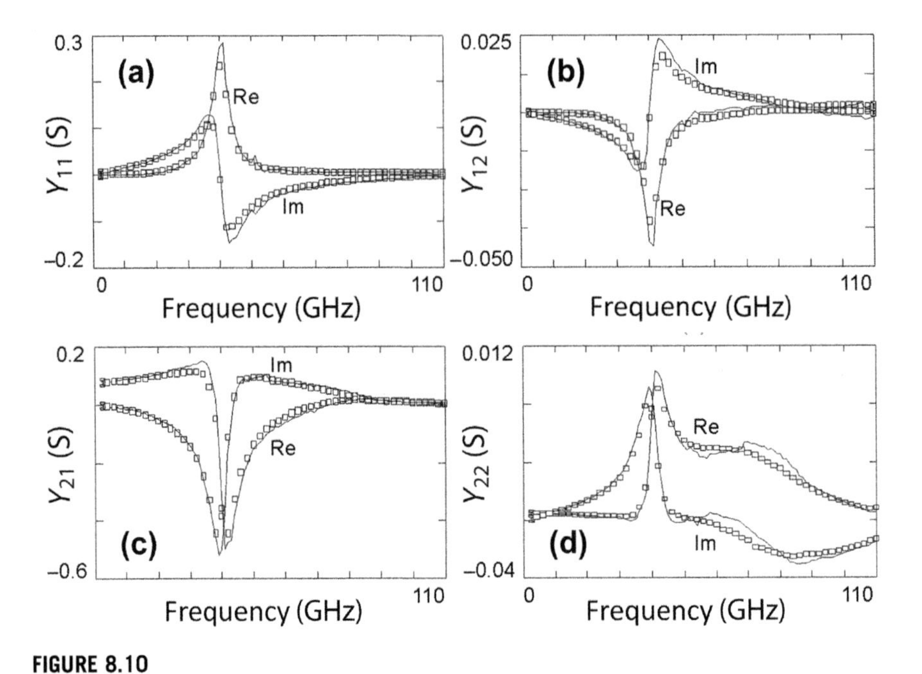

FIGURE 8.10

Measured Y parameters (symbols) and Y parameters predicted through scaling (lines) of a 6×30 μm GaAs pHEMT biased at $I_D = I_{dss}$, $V_{DS} = 3$ V: (a) Y_{11}, (b) Y_{12} (c) Y_{21}, and (d) Y_{22}.

as an example, the good capability of the model in predicting small-signal Y-parameter measurements up to 110 GHz for a 6×30 μm transistor (a structure different from the ones used in model identification).

Also, the frequency extrapolation properties of the model were verified. In particular, Y parameters obtained from scattering measurements performed on 0.2 μm GaAs pHEMTs up to 110 GHz, using an HP8510XF network analyzer, were compared with model predictions. More precisely, the admittance parameters of the active slice of the model were identified on the basis of measurements carried out only up to 50 GHz and then frequency extrapolated through a simple linear regression up to 110 GHz. Successively, these extrapolated Y parameters were embedded in the passive network of the device characterized up to 110 GHz using electromagnetic simulation. Figure 8.11 shows the comparison between measurements and predictions using frequency extrapolation for a 4×30 μm transistor. Almost pinched-off operating conditions ($V_{DS} = 3$ V and $V_{GS} = -1$ V) were considered in the experiment shown here to put in evidence that the model is capable of predicting resonant effects totally occurring in the extrapolated frequency range. Clearly, this frequency extrapolation capability can only rely on the "physical" correctness in combining the accurate electromagnetic simulation results with the almost physically consistent linear approximation/extrapolation of the internal elementary device models.

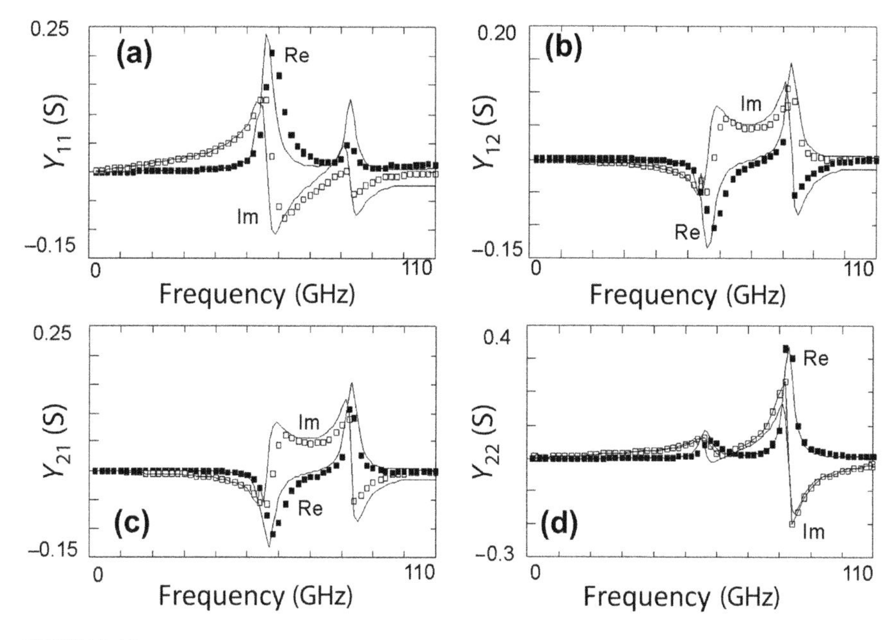

FIGURE 8.11

Measured admittance parameters (symbols) and admittance parameters predicted through frequency extrapolation above 50 GHz (lines) for a 4 × 30 μm GaAs pHEMT at $V_{GS} = -1$ V, $V_{DS} = 3$ V; (a) Y_{11}, (b) Y_{12}, (c) Y_{21}, and (d) Y_{22}.

Some authors have applied techniques similar to the one proposed in Refs. [2,3] for the distributed modeling of heterojunction bipolar transistors (HBTs) [21,22]. The approach also provides good predictive capabilities for this kind of transistor. For example, the adopted transistor partitioning and predicted scattering parameters up to 50 GHz through scaling based on electromagnetic modeling are shown in Figure 8.12 for a 6 × 3 × 40 μm HBT.

8.3.5 Complexity issues and nonlinear modeling

In the previous section, a scalable linear distributed transistor model was addressed. Such an approach could be directly extended to the nonlinear case. To this end, different techniques [23] for the identification of a nonlinear model on the basis of current–voltage (I/V) characteristics and bias-dependent alternating current small-signal parameters, could be applied to "substitute" the linear description of the active slices with a nonlinear one. However, a relevant problem arising in the context of nonlinear distributed transistor modeling is related to computational efficiency. In fact, nonlinear circuit analysis (based either on iterative harmonic-balance or time domain algorithms) requires repeated evaluations of the device response.

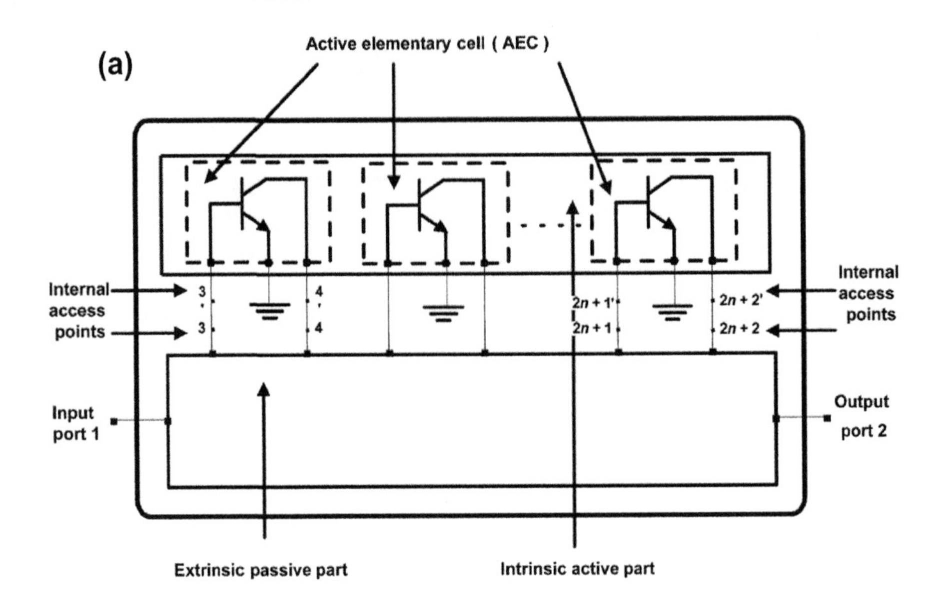

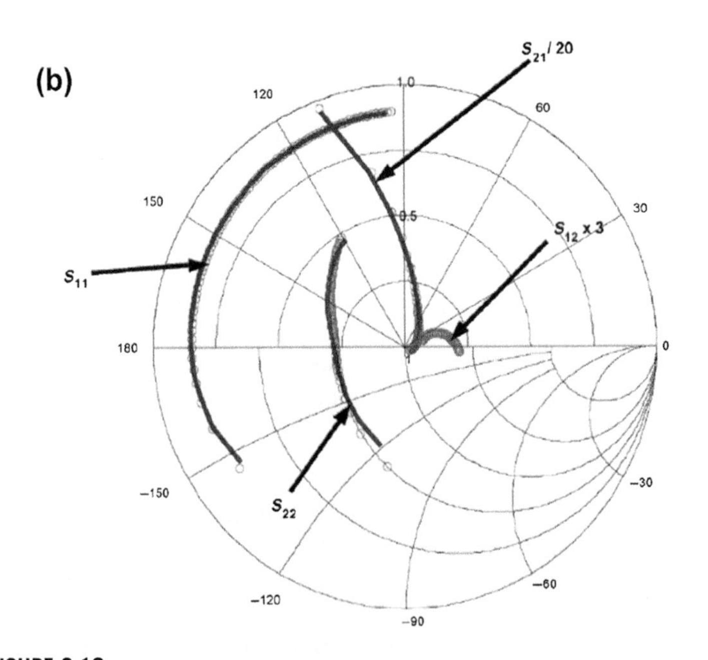

FIGURE 8.12

(a) Partitioning of the HBT structure. (b) Measured (circles) and predicted (solid line) S parameters of a $6 \times 3 \times 40\,\mu\text{m}$ HBT.

Thus, a distributed model based on a large number of nonlinear active slices would require a huge amount of computer processing time and memory occupation (in fact, computing effort grows more than linearly with the number of nonlinear nodes). This consideration suggests that some kind of simplification with respect to a "fully distributed" approach must be introduced if computational efficiency must be guaranteed.

To cope with this problem, an original approach was proposed in Ref. [4] to define a nonlinear distributed model based on an "equivalent" two-port intrinsic nonlinear block connected to a linear distributed four-port passive parasitic network. The starting point is, as in the previously described approach, the partitioning of the transistor active area in a given number N of active slices or EIDs. These ones are interconnected through a passive distributed $2N+2$-port network (see Figure 8.13).

Accurate electromagnetic simulations can be carried out to characterize the $2N+2$-port network in terms of its admittance matrix $\mathbf{Y}_{\mathrm{EM}}$ accounting for the parasitic effects due to the gate and drain accesses to the active area, along the device fingers and possible transverse couplings between fingers. The transistor structure shown in Figure 8.13 is the same adopted in Ref. [1–3] to deal with linear distributed modeling. In Ref. [4], with the aim of developing a distributed nonlinear model,

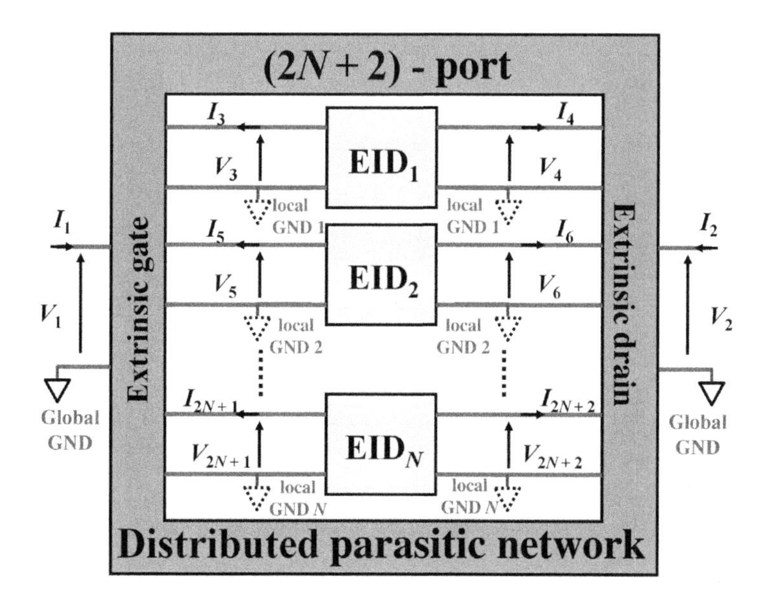

FIGURE 8.13

Distributed parasitic network (gray pattern) directly obtained from the EM simulation of the device passive structure (described by the $\mathbf{Y}_{\mathrm{EM}}$ admittance matrix). Voltage and current phasors at the external gate and drain terminals and at the gate and drain elementary intrinsic device (EID) terminals are also shown.

the "equivalent compact" structure in Figure 8.14 was introduced, where a single EqID is considered jointly with a corresponding, suitable definition of a compact distributed parasitic four-port network, described by the admittance matrix $\mathbf{Y_C}\,[4 \times 4]$.

A quite simple procedure can be adopted to make the structure in Figure 8.14 equivalent to the one in Figure 8.13. In fact, under the simplifying hypothesis that every EID is equal to each other and is fed by identical excitations, the following conditions can be imposed:

$$\begin{aligned} \mathrm{V}_1 &\doteq V_1 \\ \mathrm{V}_2 &\doteq V_2 \\ \mathrm{V}_3 &\doteq V_3 = V_5 = \ldots = V_{2N+1} \\ \mathrm{V}_4 &\doteq V_4 = V_6 = \ldots = V_{2N+2} \end{aligned} \tag{8.6}$$

and

$$\begin{aligned} \mathrm{I}_1 &\doteq I_1 \\ \mathrm{I}_2 &\doteq I_2 \\ \mathrm{I}_3/N &\doteq I_3 = I_5 = \ldots = I_{2N+1} \\ \mathrm{I}_4/N &\doteq I_4 = I_6 = \ldots = I_{2N+2} \end{aligned} \tag{8.7}$$

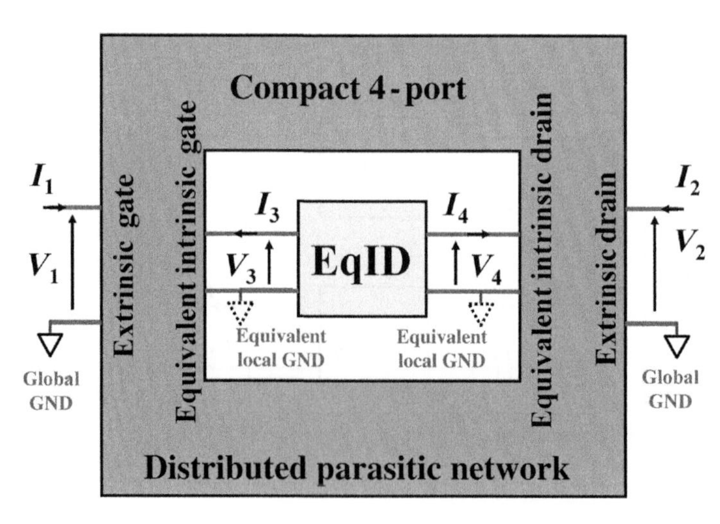

FIGURE 8.14

Electron device model composed by the single equivalent intrinsic device (EqID) and the compact four-port distributed parasitic network identified directly from the electromagnetic simulation through Eqn (8.8).

where V_1, V_2, I_1, I_2 and V_j, V_{j+1}, I_j, I_{j+1} ($j = 3, 5, \ldots, 2N+1$) are the phasors of extrinsic gate-source, drain-source, and EID voltages and currents in Figure 8.6, whereas V_i, I_i ($i = 1, \ldots, 4$) are the phasors of voltages and currents at the ports of the yet-unknown compact parasitic network in Figure 8.14.

On the basis of Eqns (8.6) and (8.7), the admittance matrix $\mathbf{Y}_\mathrm{C}$ of the compact distributed parasitic network can be easily evaluated:

$$\mathbf{Y}_\mathrm{C} = \begin{pmatrix} y_{11} & y_{12} & \sum_{j=2}^{N+1} y_{1,2j-1} & \sum_{j=2}^{N+1} y_{1,2j} \\ y_{21} & y_{22} & \sum_{j=2}^{N+1} y_{2,2j-1} & \sum_{j=2}^{N+1} y_{2,2j} \\ \sum_{i=2}^{N+1} y_{2i-1,1} & \sum_{i=2}^{N+1} y_{2i-1,2} & \sum_{i=2}^{N+1}\sum_{j=2}^{N+1} y_{2i-1,2j-1} & \sum_{i=2}^{N+1}\sum_{j=2}^{N+1} y_{2i-1,2j} \\ \sum_{i=2}^{N+1} y_{2i,1} & \sum_{i=2}^{N+1} y_{2i,2} & \sum_{i=2}^{N+1}\sum_{j=2}^{N+1} y_{2i,2j-1} & \sum_{i=2}^{N+1}\sum_{j=2}^{N+1} y_{2i,2j} \end{pmatrix} \tag{8.8}$$

where y_{ij} ($i, j = 1, \ldots, 2N+2$) are the elements of the $\mathbf{Y}_\mathrm{EM}$ matrix.

Multifrequency closed-form de-embedding of the parasitic network described by Eqn (8.8) from small-signal device measurements directly leads to the multibias, multifrequency small-signal model of the EqID.

The procedure described previously is simple and elegant but relies on some hypotheses that must be probed further. In particular, the hypothesis stating that every EID is fed by identical excitations could be questionable because, especially at high frequencies, important attenuations or delays could be present in the excitations of different EIDs. In order to verify if the approximations involved by this hypothesis is acceptable, in Ref. [4] a different approach was adopted for model identification. In particular, the admittance matrix $\mathbf{Y}_\mathrm{C}$ of the compact distributed parasitic network and the admittance matrix $\mathbf{Y}_\mathrm{EqID}$ of the EqID were identified by applying a quite complex algorithm where a "least-square congruence" is imposed between the response of the EqID in Figure 8.14 and the response of every EID in Figure 8.13. The adopted algorithm requires multibias measurements to have a well-conditioned problem and does not impose equal excitations for the EIDs. The model obtained with this latter approach, even for relatively large transistor structures (e.g., $10 \times 48\ \mu\mathrm{m}$ and $12 \times 75\ \mu\mathrm{m}$ transistors) and frequencies up to 65 GHz, was practically identical to the one obtained with the simpler approach corresponding to Eqn (8.8). This result can be justified by considering that if an electron device is accurately designed for its operating frequency range, the propagation effects along the structure width—in particular, signal attenuation—should not be too relevant; otherwise, there will be some parts of the active area not efficiently exploited. Thus, the approximation of EIDs practically fed by the same excitations does not prove so limiting.

Once the parasitic description $\mathbf{Y}_\mathrm{C}$ and the frequency- and bias-dependent small-signal description of the EqID have been obtained, any possible approach to

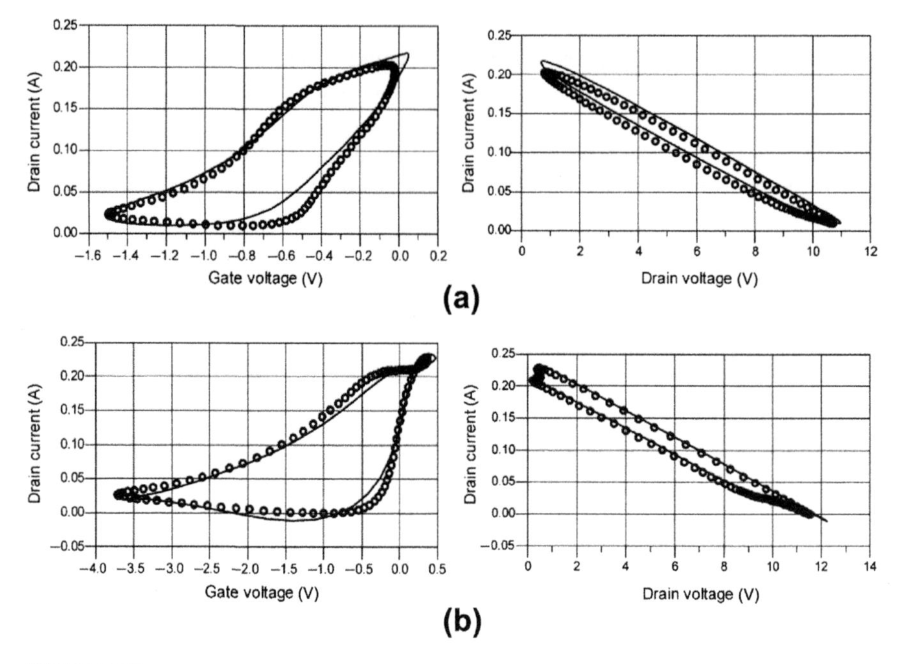

FIGURE 8.15

The dynamic trans- and output characteristics of a 12 × 75 μm pHEMT at 5 GHz (50 Ω source and load terminations). Comparison between scaled model predictions (lines) and measurements (circles) for the device biased at (a) $V_{DS} = -0.6$ V, $V_{DS} = 6$ V, and (b) $V_{GS} = -1.1$ V, $V_{DS} = 6$ V.

transistor nonlinear modeling could be applied. In particular, in [4] the nonlinear discrete convolution model [19] was adopted jointly with a suitable I/V model to account for dispersive effects [24]. As an example of the predictive capabilities of this nonlinear distributed model, Figure 8.15 shows dynamic trans- and output characteristics under class-A and class-B bias for a 12 × 75 μm GaAs pHEMT working at 5 GHz fundamental frequency.

8.4 Full-wave EM analysis for transistor equivalent circuit parasitic element extraction

As previously discussed, optimum device periphery selection is crucial in MMIC design. Having available a scalable model, the designer is able to evaluate suitably defined figures of merit and design criteria such as those defined in [25,26]. In addition, scalable models also allow one to save time during the empirical characterization of foundry processes because they require fewer measurements and model extractions to be carried out.

When microwave and millimeter-wave applications are involved, the overall quality of a scalable model is strictly dependent on the rules applied to the scaling of its extrinsic parasitic network, which takes into account the parasitic phenomena occurring in the physical interconnections of the intrinsic device to the external terminals.

Because simple linear scaling rules [12] and [27,28] applied to conventional lumped parasitic networks are not very accurate at high frequencies, complex, technology-dependent scaling rules [29–32] or quasi-distributed models [3] and [33–41] have been recently proposed. In different approaches (e.g., [3] and [41]), commercial EM simulators have been also exploited for the analysis of extrinsic parasitic phenomena and the definition of distributed models, where the entire active area is divided into a suitable number of elementary "active slices." Such approaches have proven very suitable for scaling, providing extremely accurate electrical response predictions, but they suffer from relatively poor numerical efficiency in harmonic-balance analyses due to the proliferation in the number of internal nonlinear ports (as explained in Section 8.3.4).

In this section, an almost conventional lumped parasitic network is adopted where the parasitic equivalent circuit parameters (ECPs) are identified on the basis of a full-wave EM simulation of the device layout [6]. Suitable scaling rules are provided, leading to a critical improvement in prediction accuracy of scaled devices. This is proved by the experimental validation of the scalable model applied to a family of 0.25 μm GaAs power pHEMTs for millimeter-wave applications.

8.4.1 Model definition and identification

Having to model the extrinsic parasitic phenomena associated with a typical microwave FET layout through a scalable lumped component network, the device layout in Figure 8.16 is separated into four different regions. This allows one to maintain the link between the device layout drawing and the device size, making it possible to define very accurate and simple linear scaling rules.

Region (I) accounts for the parasitic effects due to the gate and drain manifolds, which feed the signals to each finger. Region (II) accounts for the parasitic effects due to via holes, which provide source connections to ground. Region (III) accounts for the air bridge, which provides the ground connection to the source fingers through the via hole pads. Region (IV) accounts for the parasitic effects due to the interactions between the finger metallizations (i.e., ohmic losses along the gate and drain fingers and capacitive-like couplings both between the gate and drain fingers themselves and between the gate/drain fingers and the air bridge).

This somehow intuitive separation leads to the lumped extrinsic parasitic network shown in Figure 8.17.

The four elements L_{GM}, C_{GM}, L_{DM}, and C_{DM} (where M stands here for *manifold*) model the parasitic effects of region (I). The elements R_S and L_S model the effects of via holes and air bridge in regions (II) and (III). The parasitic effects of the two via holes of region (II) are accounted for separately by means of a standing alone EM

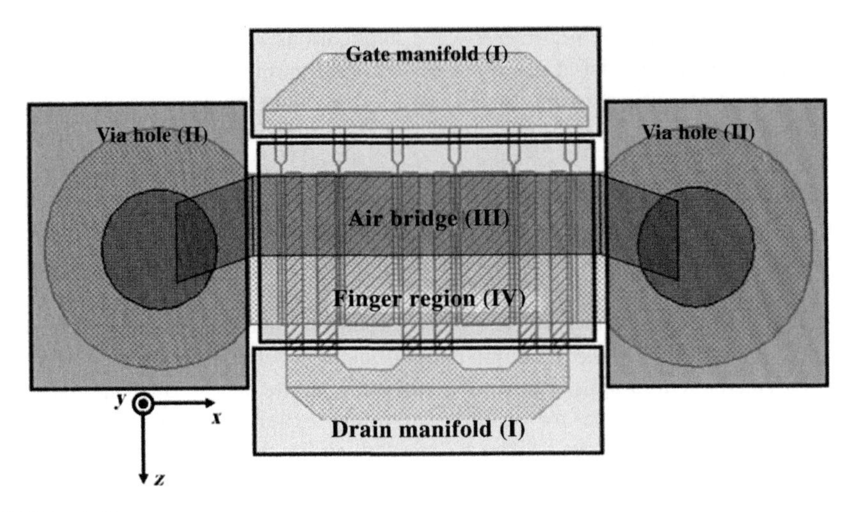

FIGURE 8.16

Device parasitic effects subdivision through device layout considerations. Gate and drain manifolds (I), source via holes + air bridge (II + III), and finger region (IV).

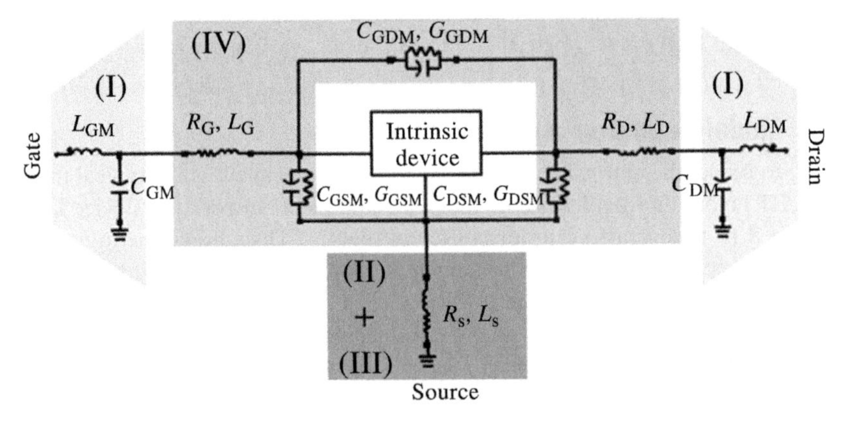

FIGURE 8.17

Lumped component network describing the device extrinsic parasitic phenomena. Schematic regions (I), (II), (III), and (IV) are strictly related to the corresponding layout regions shown in Figure 8.16.

simulation of the via structure: this choice is twice useful because it leads to the functional partitioning of the full-wave EM simulation discussed in Section 8.2.2, allowing the via holes to be removed when floating source electrodes are needed in the specific application. The elements R_G, L_G, R_D, and L_D describe the series parasitic effects due to the multi-gate/drain-finger region (IV), whereas the elements

C_{GSM}, G_{GSM}, C_{GDM}, G_{GDM}, C_{DSM}, and G_{DSM} (where M stands here for *metallization*) account for the capacitive-like coupling and dielectric losses both between the metallization structures of the gate and drain fingers and between the gate/drain fingers and the air bridge.

According to the axes defined in Figure 8.16, the following geometry dependent scaling factors [42] are defined:

$$\begin{aligned} SF_Z &= \frac{W_g^{sc}}{W_g^{ref}} \\ SF_X &= \frac{N_g^{sc}}{N_g^{ref}}, \end{aligned} \tag{8.9}$$

where W_g^{sc}, W_g^{ref}, N_g^{sc}, N_g^{ref} are the gate width and the number of gate fingers of the scaled and reference device, respectively. The subscript Z represents the direction along the gate/drain fingers (z-axis), whereas X denotes the direction along the channel (x-axis), respectively. Thus, the factors SF_Z and SF_X will be applied whenever a scaling of the gate width and/or a different number of paralleled fingers are involved.

Considering the parasitic phenomena associated to region (IV), simple scaling rules are adopted for the lumped components of the corresponding schematic region:

$$(R_i, L_i)^{sc} = (R_i, L_i)^{ref} \cdot \frac{SF_Z}{SF_X}, \quad i = G, D \tag{8.10}$$

$$(C_i, G_i)^{sc} = (C_i, G_i)^{ref} \cdot SF_Z \cdot SF_X \cdot \quad i = GSM, GDM, DSM \tag{8.11}$$

Considering region (I), the gate and drain manifolds widening (or shrinkage) effects in the presence of a varying number N of fingers have been investigated through simple EM simulations of the different gate and drain manifold layouts drawn by modifying the reference device layout on the basis of the minimum distance along the x-axis between two successive gate fingers. The study is carried out for a number of fingers ranging from 4 to 12, which corresponds to the set of electron devices made available in the foundry process considered. A simple LC network, as shown in the gate and drain regions (I) in Figure 8.17, perfectly fits the obtained EM description up to 100 GHz [6] for both the gate and drain manifolds. The extracted values of the LC model parameters concerning the gate and drain manifolds are shown in Table 8.1. The values of L_{GM} and L_{DM} are nearly constant versus the number of fingers, whereas the values of C_{GM} and C_{DM} increase almost linearly with N, according to the scaling factor SF_X. This is also clearly seen in Figure 8.18, where the extracted inductances and capacitances are plotted versus the number of fingers.

These results justify the following scaling rules for the parasitic lumped components in the schematic region (I):

$$\begin{aligned} (L_{GM}, L_{DM})^{sc} &= (L_{GM}, L_{DM})^{ref} \\ (C_{GM}, C_{DM})^{sc} &= (C_{GM}, C_{DM})^{ref} \cdot SF_X. \end{aligned} \tag{8.12}$$

Table 8.1 LC model of the gate and drain manifolds

	4 Fingers	6 Fingers	8 Fingers	10 Fingers	12 Fingers
C_{GM} [fF]	7.5	10.6	14.4	19.81	25
L_{GM} [pH]	12.5	11.3	10.7	10.7	10.3
C_{DM} [fF]	12.5	14	20.3	25.5	34.4
L_{DM} [pH]	12	10.6	10.2	12.2	11.3

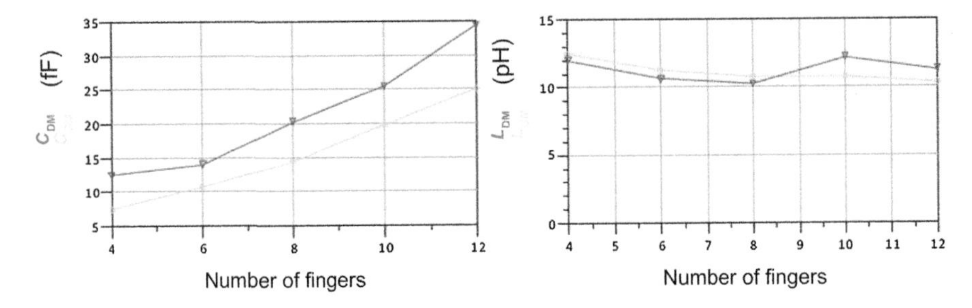

FIGURE 8.18

Variation of the L and C parameters in the gate and drain manifold models with the number of gate fingers: L_{GM} (circles), L_{DM} (triangles), $C_{GM}SF_X$ (circles), $C_{DM}SF_X$ (triangles). SF_X is evaluated by considering six gate fingers in the reference device.

A thorough analysis of regions (II) and (III), taking into account the presence of the air bridge, as shown in Figure 8.19, suggests that R_S and L_S should not scale with the device periphery. Because the internal source fingers are connected to the ground terminal through the air bridge that is running along the x-axis, negligible parasitic effects actually take place in the Z-direction (the source finger opposite ends almost correspond to open impedance terminations).

The series parasitic phenomena associated with the air bridge along the x-axis can be considered almost negligible with respect to those introduced by the via holes, as confirmed by a dedicated EM analysis of a stand-alone air bridge.

Thus, since the via holes (actually two vias connected in parallel) are the same for all the device peripheries, the R_S and L_S elements of Figure 8.17 may be thought of as not requiring any scaling at all:

$$(R_S, L_S)^{sc} = (R_S, L_S)^{ref}. \tag{8.13}$$

Particular care must be paid when the source structure is either different than that in Figure 8.19 or the air bridge has significant parasitic effects, because the source parasitic elements scaling rule, Eqn (8.13), might change.

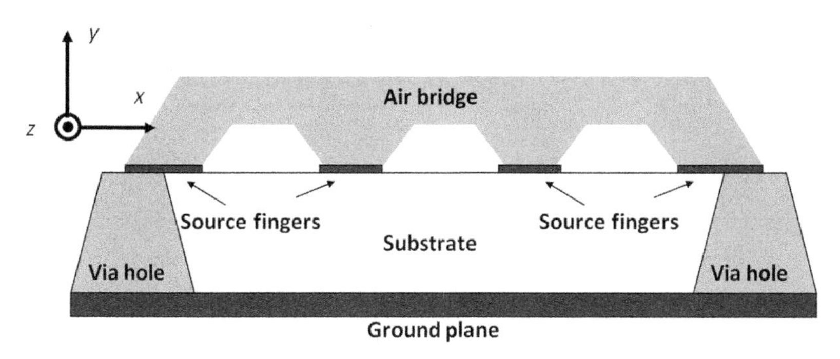

FIGURE 8.19

Cross-section of the air bridge providing ground connection to the internal source fingers (six gate-finger device). The entire extrinsic structure associated with the source is composed by the via holes in the layout region (II) connected to the air bridge representing the layout region (III) (see Figure 8.16).

The identification of the extrinsic lumped components involves just a few full-wave EM simulations of a single reference device structure. Measurements under off-state conditions (or forward-gate, cold-FET) and characterization of multiple different-in-size devices are not required for the identification of the extrinsic elements using this approach. The full-wave EM analyses are set accordingly to the description of Section 8.2.5. A $6 \times 50\ \mu m$ pHEMT ($L = 0.25\ \mu m$) belonging to the power millimeter-wave GaAs process mentioned previously was selected as the reference device.

According to the detailed description found in Ref. [6], the model identification procedure consists of three different steps, which are here briefly recalled. The first step consists of the EM simulations that are the full-wave EM analysis of the four-port device extrinsic structure (without via holes), the full-wave EM simulation of the conic via hole, and the full-wave EM simulations of the gate and drain manifolds (the functional partitioning is described in Section 8.2.5).

In the second step, the simulated data are used to extract the extrinsic equivalent circuit parameters. First, R_S and L_S in regions (II) and (III) are extracted by fitting the via hole simulation while L_{GM}, C_{GM}, L_{DM}, and C_{DM} in region (I) are extracted by fitting the gate and drain manifold simulations (these are separate simulations: see Ref. [6]).

Then, the distributed two-port description associated with the entire extrinsic layout is used (by means of a standard optimization routine) for the identification of the remaining extrinsic circuit parameters: R_i, L_i (i = G, D) and C_i, G_i (i = GSM, GDM, DSM).

The extraction procedure of the extrinsic elements involves a very well-conditioned optimization problem in 10 unknowns. Fast convergence to the same solution is observed by considering very different initial guesses of the parameter values.

Table 8.2 Extracted extrinsic parasitic elements

C_{GSM}	C_{GDM}	C_{DSM}	G_{GSM}	G_{GDM}	G_{DSM}
27.1 fF	25.2 fF	32.2 fF	3.0e-8 S	2.5e-8 S	2.8e-9 S
L_G	L_D	L_S	R_G	R_D	R_S
14.3 pH	13.0 pH	10.63 pH	0.603 Ω	0.085 Ω	0.056 Ω

At the end of the procedure, the measured behavior of the reference device is de-embedded from the extracted extrinsic equivalent circuit. The obtained description is eventually used for the identification of the preferred intrinsic device model. It is worth noting that just a single set of *S*-parameter measurements at the operating bias on a single device periphery is needed for the identification of this linear scalable model.

The 6 × 50 μm pHEMT reference device full-wave EM simulations are performed in the frequency range of 0–100 GHz. The extracted values of the gate and drain manifolds parasitic elements have already been reported in Table 8.1 for the six-finger reference device. The extracted values of the remaining parasitic elements are listed in Table 8.2.

8.4.2 Model scalability and frequency extrapolation capabilities

In order to verify the scaling rules, Eqns (8.10)–(8.13), full-wave EM analyses of a 10 × 50 μm and a 12 × 75 μm pHEMT of the same process as the reference device were performed. The extracted extrinsic equivalent circuit model predictions are compared to the simulated data in Figure 8.20. The results highlight that the scaled parasitic network is capable of predicting the simulated behavior of the scaled device, not only up to the maximum optimization frequency of 100 GHz but even further, putting in evidence the excellent frequency extrapolation capabilities of the extracted parasitic network (not only for the reference device but even for the scaled devices).

8.4.3 Experimental and simulation results

Experimental validation of the proposed modeling approach has been carried out for a family of 0.25 μm GaAs pHEMTs. As already said, the chosen reference device is a 6 × 50 μm pHEMT.

S-parameter measurements in the frequency range of 5–65 GHz have been performed by biasing the device in class-A operation at $V_{GS} = -0.5$ V, $V_{DS} = 5$ V ($I_{DS} = 30$ mA).

Those measurements are de-embedded from the parasitic elements of Table 8.1 and Table 8.2 in order to obtain the intrinsic measured admittance parameters. These data can be used for the identification of any intrinsic linear (or nonlinear) device model. Aiming at investigating just the scalability of the proposed extrinsic parasitic

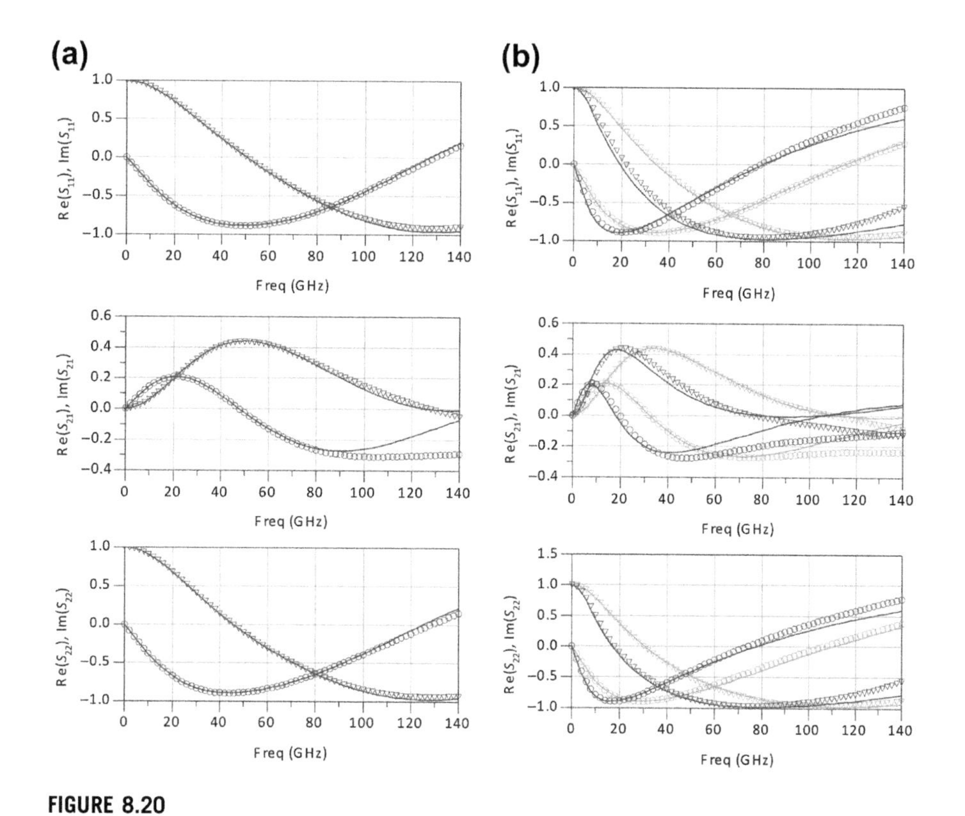

FIGURE 8.20

(a) Parasitic network equivalent circuit model fitting (lines) to the simulated data (symbols) of the 6 × 50 μm reference phemt. Real parts (triangles) and imaginary parts (circles) of the *S* parameters were in the frequency range of 0.1–140 GHz. (b) Scaled parasitic network equivalent circuit model fitting (lines) to the simulated data (symbols) of the scaled 10 × 50 μm and 12 × 75 μm pHEMTs. Real parts (triangles) and imaginary parts (circles) of the *S* parameters were in the frequency range of 0.1–140 GHz.

modeling approach, a linear table-based description of the intrinsic device, in terms of admittance matrix, is used hereinafter. Thus, the intrinsic device behavior scales following a simple linear rule (i.e., $Y_{\text{int}}^{\text{sc}}(j\omega) = Y_{\text{int}}^{\text{sc}}(j\omega)\ \text{SF}_Z\ \text{SF}_X$).

The complete linear scalable model has been experimentally validated in terms of prediction capabilities of both *S* parameters and global design parameters in [6] [42], such as the stability factor μ [43], the maximum available gain (MAG), or the maximum stable gain (MSG), whenever the MAG is not defined [27]. This represents a rather heavy experimental validation because it also takes into account the way the discrepancies in the prediction of each *S* parameter combine within such important global figures of merit. Four devices have been considered, having remarkable differences in number of gate fingers and gate width, namely the

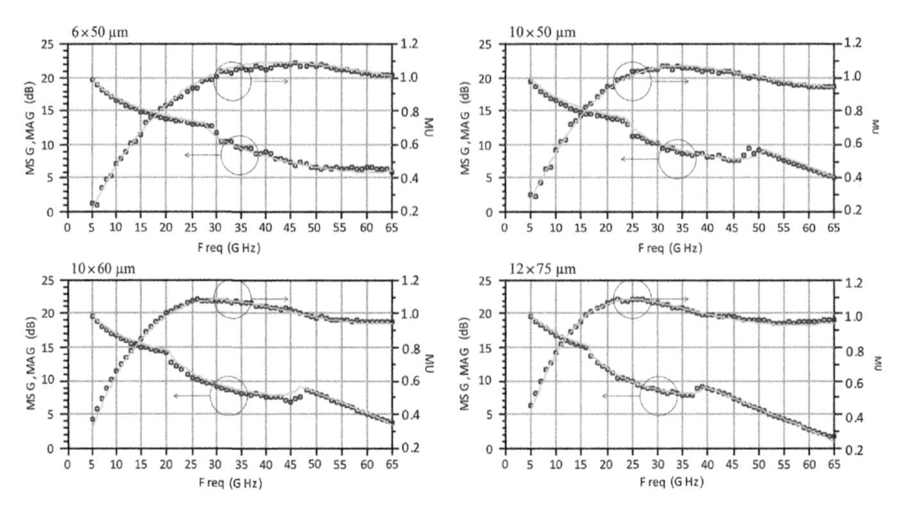

FIGURE 8.21

MAG/MSG and the stability factor μ versus frequency of 6 × 50 μm, 10 × 50 μm, 10 × 60 μm, and 12 × 75 μm pHEMTs. Measurements (circles) versus predictions obtained through the empirical equivalent circuit scalable model. The devices are biased in typical class-a operation. Analogous results are obtained in different bias conditions.

6 × 50 μm pHEMT reference device plus a 10 × 50 μm, a 10 × 60 μm, and a 12 × 75 μm pHEMT.

Some of those experimental results are plotted in Figure 8.21, which shows the sound model prediction capability of scaled devices. For the complete experimental validation, see Ref. [6].

8.5 Examples of application to MMIC design

The modeling techniques described in the previous sections are illustrated here for the final application, namely MMIC designs.

8.5.1 A more flexible use of transistor layout

In Section 8.3 and Section 8.4, it was shown how accurate electromagnetic simulations of transistor layout can be effectively exploited to identify robust and flexible scalable transistor models up to millimeter-wave frequencies. The modeling approaches that have been described make number of fingers and finger width available as design variables. This undoubtedly is a mandatory starting point for a more flexible use of transistor layout in circuit design.

In particular, the approach proposed in Ref. [6] has the important advantage of not requiring an electromagnetic simulation of each transistor structure for obtaining the corresponding model. Moreover, gate width and number of gate fingers can be directly, continuously, and conveniently swept in circuit design and/or optimization. This is not always the case with foundry models for which it is usually possible to continuously scale the finger width, whereas separate models are often required for transistor structures having different finger numbers. As an example, Figure 8.22 shows how source and load reflection coefficients, which provide simultaneous input/output conjugate match, can be easily computed versus the number of gate fingers and finger width by means of the model [6].

On the contrary, the approaches adopted in Refs. [1−44] require the electromagnetic simulation of each transistor structure to identify the corresponding model (however, also in this case, a limited number of scattering parameter measurements is needed). Although such a recurring use of electromagnetic simulations can be considered time and resource consuming, it is well assessed that their adoption significantly improves MMIC design accuracy while simultaneously reducing development and, possibly, manufacturing costs. Indeed, the design of low-cost, high-performance monolithic microwave and millimeter-wave integrated circuits requires strong and important coupling phenomena, usually existing in complex and dense structures, to be taken into account. As a matter of fact, basic approaches based

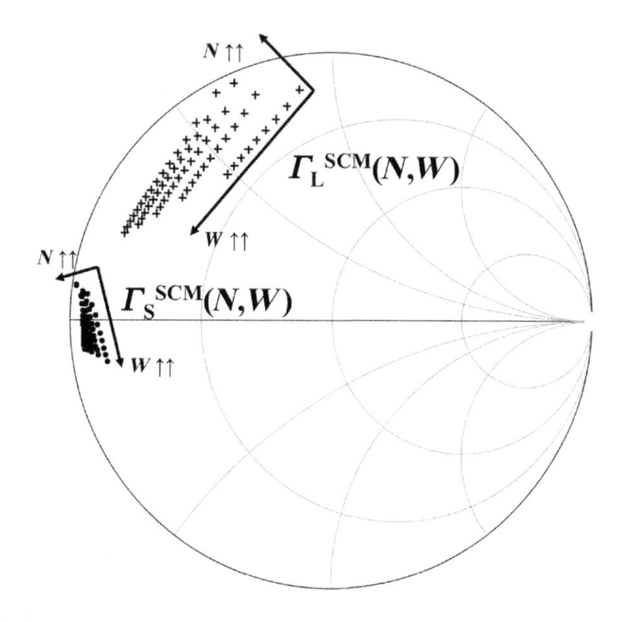

FIGURE 8.22

Load and source reflection coefficients (Γ_L and Γ_S) providing input/output simultaneous conjugate match versus number of gate fingers and finger widths (N and W) at 30 GHz. N varies from 4 to 12 numbers of fingers (even numbers), whereas W varies from 30 μm to 100 μm (Step 5 μm). The arrows show how the reflection coefficients vary when the independent variables increase.

on conventional circuit simulations by means of component libraries supplied by the foundry allow limited accuracy in performance prediction.

MMIC design would definitely benefit from "global" design procedures where the effects of the passive parts as well as the interactions with active devices are consistently taken into account. This goal can be obtained by using accurate electromagnetic simulations of the complete passive parts of the circuit, including the FET metallizations, and a proper characterization of the intrinsic part of the active device. The electromagnetic simulations are based on the effective layout and substrate parameters, while the intrinsic part of the transistors is identified by means of the modeling approaches presented in Refs [1—44]. In the next section, some MMIC design examples will be described with the aim of highlighting how circuit designers can take advantage of a "global" design approach.

8.5.2 MMIC design based on distributed transistor models

The approaches previously described have been adopted to accurately predict the response of transistors having a given number of fingers and finger widths, as well as to successfully design microwave and millimeter-wave monolithic integrated circuits [5],[45]. Moreover, modeling techniques like the ones proposed in Refs [1—44] can be also effectively exploited to design high-density integrated circuits, which may consist of unconventional configurations where active and passive devices are closely spaced in order to get compact size and consequent reduction of production costs. For instance, a "global" design approach based on the work in Refs [1—3] allows the performance prediction of MMIC structures where interconnecting lines between transistors are removed by realizing with a single electrode the ohmic electrodes of adjacent FETs (i.e., ohmic electrode-sharing technology [OEST]) [46]. This enables a drastic reduction of the chip size as well as a decrease of the transmission losses to be obtained. Clearly, such circuits cannot be simulated by simply using conventional CAD tools and standard libraries.

In Ref. [47], the design of the 14 GHz amplifier shown in Figure 8.23 has been carried out as an example to demonstrate the application of the approach [2] in designing OEST structures.

A conventional design approach would require, for the circuit in Figure 8.23, a layout implementation based on four different FETs interconnected through transmission lines. As an alternative, the amplifier can be designed by adopting an "unconventional" OEST-based layout, where all the interconnecting lines between cascaded FETs are removed because the ohmic electrodes of two adjacent FETs with the same electric potential are realized with a single electrode. The typical OEST implementation of such a topology is shown in Figure 8.24, where the labels have the same meaning as in Figure 8.23.

A limited number of transistor structures have been used to identify the admittance matrix associated with the unitary gate width of the active devices by using the technique proposed in Ref. [3]. Successively, circuit performance has been optimized by using the admittance matrix description of the active part and accurate

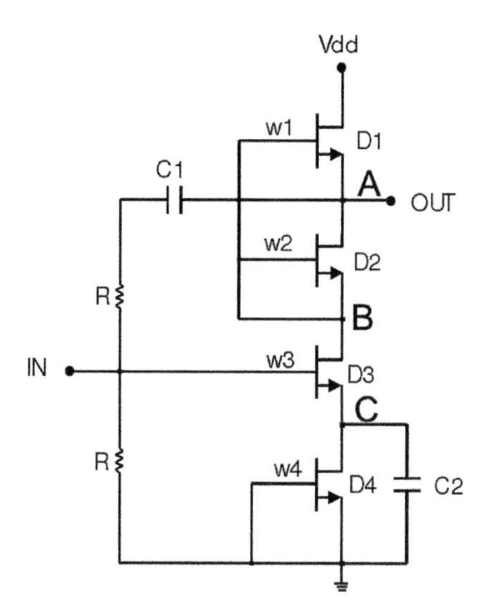

FIGURE 8.23

Schematic of a 14 GHz GaAs FET amplifier designed by using the OMMIC ED02AH pHEMT process and the OEST approach.

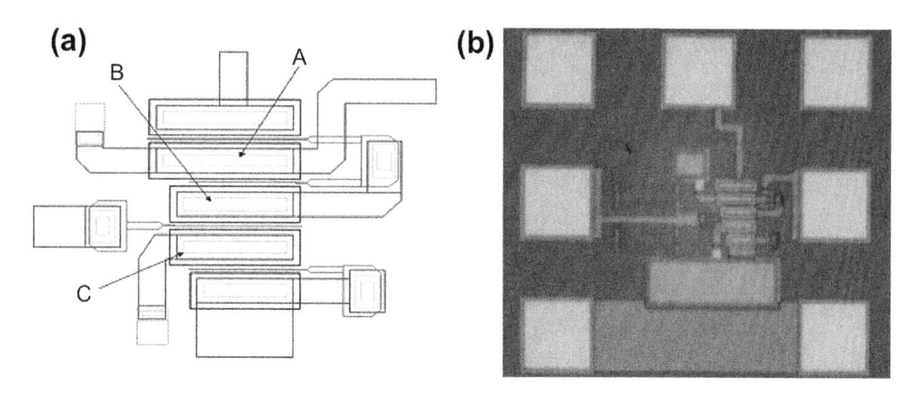

FIGURE 8.24

(a) Layout and (b) chip photograph of a 14 GHz GaAs FET amplifier designed by using the OMMIC ED02AH GaAs 0.2 μm pHEMT process and the OEST approach.

electromagnetic simulations of the layout, thus taking into account the complex electromagnetic effects and interactions between active devices due to such a compact and unconventional implementation. The manufactured MMIC is shown in Figure 8.24(b). The chip area, excluding contacting ground-signal-ground test pads, is 0.12×0.10 mm^2. The conventional layout of the same amplifier, designed by using four "separated" active devices, would occupy an integrated area of $0.3 \times 0.2\ mm^2$. Good agreement was obtained between measurements and performance predicted through simulations, as can be seen in Figure 8.25.

The same approach can be applied to other kinds of circuits. For instance, in Figure 8.26, the circuit schematic of a single-pole double-throw (SPDT) switch is depicted. The OEST layout implementation and the manufactured MMIC are shown in Figure 8.27(a) and (b), respectively.

Another interesting example of application to MMIC design can be found in Ref. [5], where the distributed EM-based empirical modeling technique proposed in Ref. [4] and recalled in Section 8.3.4 has been applied to build an accurate model of a cascode FET configuration, namely cascaded CS (common source) and CG (common gate) transistors.

Design kits provided by the foundries often do not include either a cascode FET model or a CG model. Moreover, cascode and CG transistor samples are not usually readily available for characterization and identification of a suitable model [48]. As an alternative, the cascode model could be obtained by interconnecting a CS FET model and a CG model obtained from the latter by applying simple admittance matrix transformations or "rotation" once the source electrode has been floated. However, this procedure is usually scarcely accurate.

To overcome the above problems, the cascode FET compact model shown in Figure 8.28(b) was adopted in Ref. [5], in which two equivalent intrinsic devices, one for the CS stage and the other for the CG stage ($EqID^{CS}$, $EqID^{CG}$), are connected

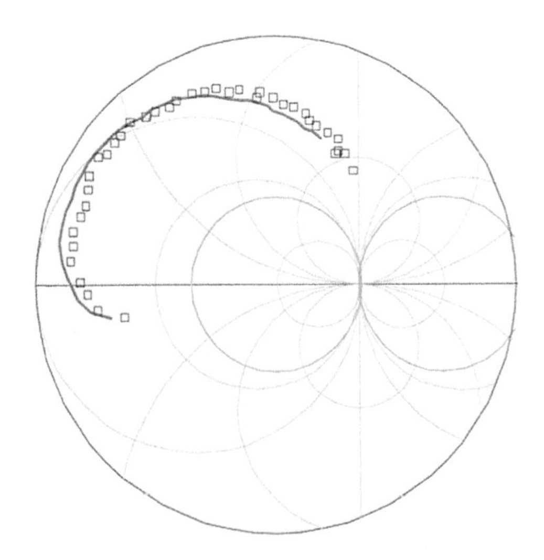

FIGURE 8.25

Comparison of S_{21} (1–20 GHz) measurements (squares) and simulations for the amplifier in Figure 8.24(b).

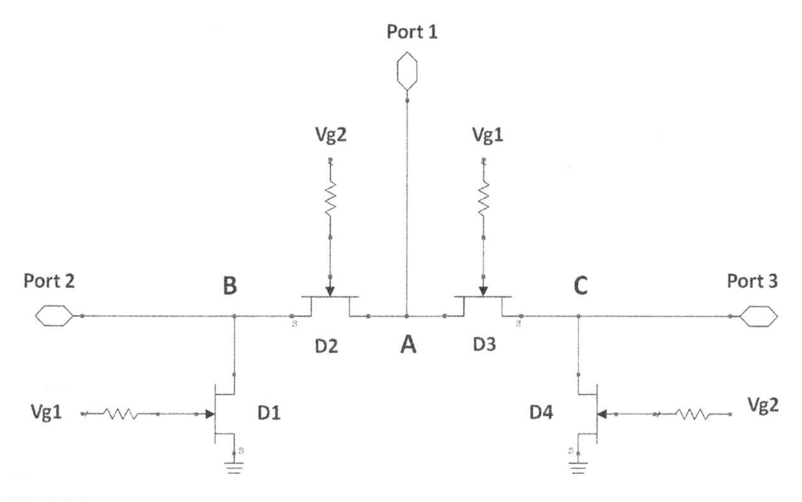

FIGURE 8.26

Schematic of a SPDT switch designed by using the OEST approach.

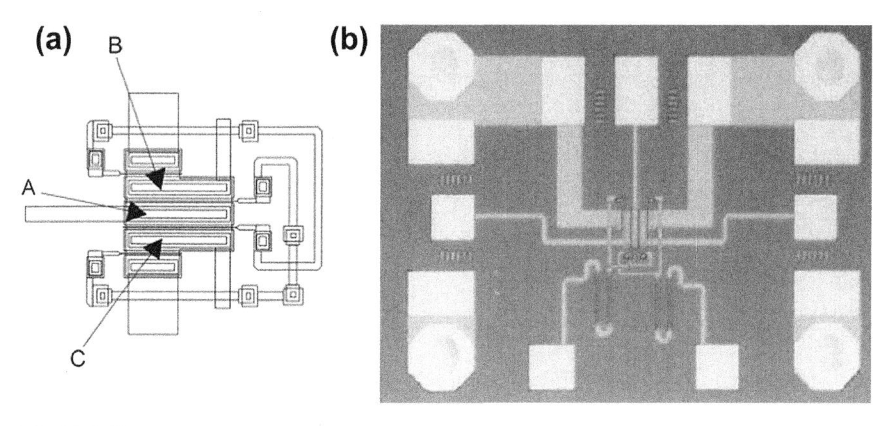

FIGURE 8.27

(a) Layout and (b) chip photograph of a SPDT switch designed by using the OEST approach.

to a linear distributed six-port passive network characterized by the admittance matrix $\mathbf{Y}_{6P}$ and describing the extrinsic parasitics related to the metallization layout of the cascade overall cell.

The identification of the compact model begins by partitioning the CS and CG device geometries into a suitable number (N^{CS} and N^{CG}) of EIDs [4], placed along the layout fingers (see Figure 8.28(a)). The EIDs are interconnected by a linear passive distributed structure, which is described in terms of an admittance matrix $\mathbf{Y}_{EM}$

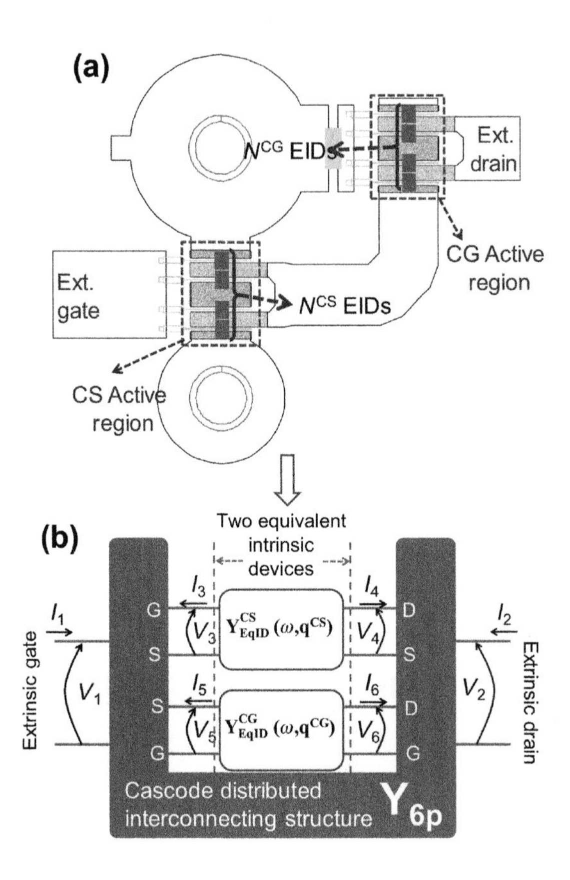

FIGURE 8.28

Cascode transistor EM-based empirical model. (a) Device layout EM simulation setup where ($N^{CS} + N^{CG}$) EIDs are defined. (b) Empirical model composed of a compact six-port distributed parasitic network connected to two eqids—one for the CS FET and the other for the CG FET.

$[2(N^{CS} + N^{CG}) + 2 \times 2(N^{CS} + N^{CG}) + 2]$, which can be computed on the basis of accurate electromagnetic simulations of the transistor layout.

The compacting procedure of the linear distributed multiport network described by $\mathbf{Y}_{EM}$ into the minimal six-port network in Figure 8.28(b) described by $\mathbf{Y}_{6P}$ has been carried out by applying the same assumptions made for the single transistor [4]—that is, by considering the EIDs inside the CS or CG stage to be equal and equally excited. On this basis, the $\mathbf{Y}_{6P}$ matrix can be identified by means of closed-form relationships [5].

Following the procedure [4] recalled in Section 8.3.4, the $\mathbf{Y}_{EqID}^{CS}$ and $\mathbf{Y}_{EqID}^{CG}$ admittance matrixes associated to the equivalent intrinsic devices have been identified by using a reference sample (a CS FET provided by the foundry). Note that the

CG admittance matrix can be obtained from the CS one by means of well-known transformations.

It was demonstrated in Ref. [5] that the obtained cascode FET model in Figure 8.28(b) is capable of accurately predicting the performance of a 0.8–20 GHz MMIC distributed amplifier exploiting six cascode stages and, in particular, instabilities that are not detectable by using conventional equivalent circuit models.

Finally, in Ref. [5], the same approach has been used to design the symmetric-layout cascode cell shown in Figure 8.29. Excellent agreement between performance simulation and measurements has been obtained up to 45 GHz (see Figure 8.30), whereas the accuracy of a conventional model is definitely lower.

8.6 De-embedding for bare-die transistor

The previous sections described the modeling and de-embedding of on-wafer devices with MMIC design as the final application. The remainder of the chapter focuses on modeling and de-embedding of bare-die devices, employed in hybrid circuit design.

Newly processed transistors can be measured and characterized directly, using high-frequency probes, as they are embedded in the fabricated wafer. This is the easiest way to identify all the extrinsic elements that surround the intrinsic transistor

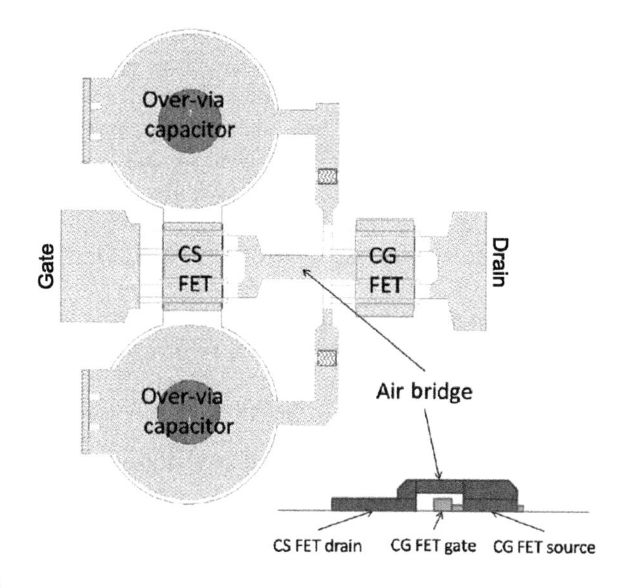

FIGURE 8.29

Layout of a symmetric cascode cell. CG, common gate; CS, common source; FET, field-effect transistor.

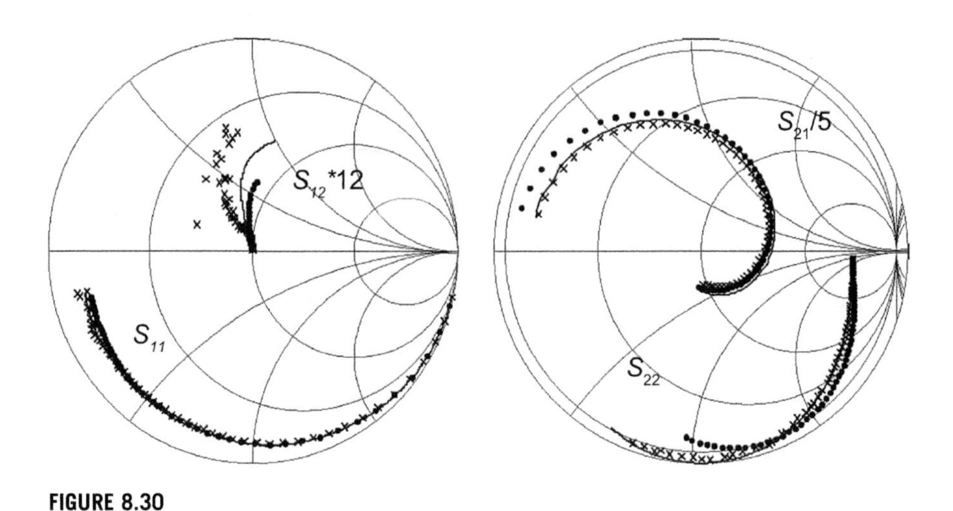

FIGURE 8.30

The 1.5–45 GHz *S*-parameter predictions compared to device measurements for a designed symmetric cascode cell. Crosses: Device measurements. Lines: EM-based model. Dots: conventional model.

and subsequently to de-embed them. Another type of characterization is when the device is diced out of the wafer and is available as a single die. In this case, a test fixture, on which the device is wire bonded, is necessary for its measurement and characterization. A third case is when the transistor bare-die is packaged. In this case, a test fixture with transmission lines for connecting the package leads is required for characterization. The latter turns out to be the most difficult characterization case because of the high number of extrinsic parasitics and the difficulty in obtaining their values. On the other hand, it is known that high-frequency devices aimed for power applications are difficult to be characterized by on-wafer measurements due to power dissipation constraints; therefore, characterization using test fixtures is necessary.

A bare-die transistor mounted in a test fixture will be analyzed in this section following the classical equivalent circuit modeling approach. Next, a method based on the four-port model concept, which employs EM analysis, will be explained.

8.6.1 Case study: bare-die transistor in test fixture

The device to be characterized and de-embedded is a Filtronic bare-die GaAs HEMT. This device is wire bonded to the test fixture as shown in Figure 8.31. For this fixture, an aluminum oxide (Al_2O_3) ceramic substrate is used. It exhibits a dielectric constant, ε_r, of 9.8 and loss tangent of 0.0001. The thickness of the substrate is 600 μm. Coplanar waveguide (CPW) transmission lines are used at the input and output ports. The grounds of the input and output CPW transmission lines are

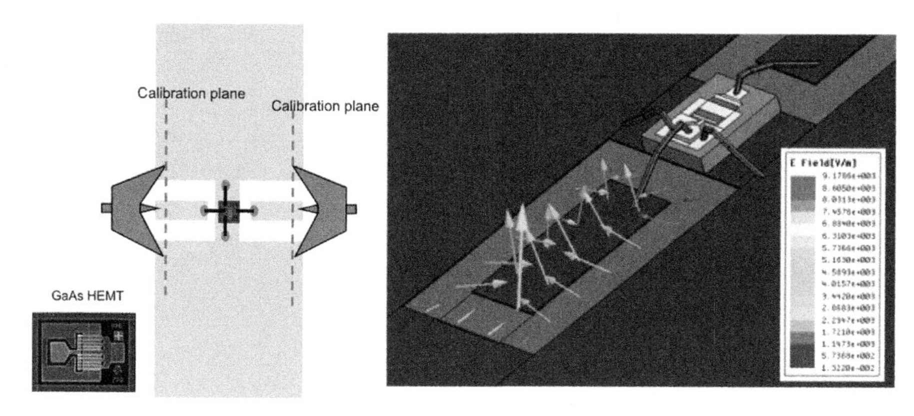

FIGURE 8.31

GaAs HEMT bare-die and test fixture.

joined together beneath the device. This ground also serves to conduct the device heat to the wide fin areas (CPW grounds). The CPW transmission lines are tailored for 50 Ω and their lengths are set to 2.5° at 1 GHz. For testing, microwave probes of 800 μm pitch are used.

Once the test fixture has been verified to work properly (i.e., the transistor did not oscillate when biased), the measurement, de-embedding, and characterization of the GaAs transistor can begin. The extrinsic network that includes the effect of the CPW transmission lines, bonding wires, and parasitic capacitances has to be determined. These elements should be subtracted from the measurements. Conventional procedures such as short-open-line-thru, thru-reflect-line (TRL), and multi-TRL are not possible in this case because they require de-embedding standards (open, short, lines) processed in a GaAs material with similar characteristics as the device under test, which is very cumbersome. Therefore, two de-embedding procedures based on equivalent circuit modeling and on four-port EM analysis, respectively, will be presented.

8.6.2 Equivalent circuit model for extrinsic networks

An identification of the parasitic elements is necessary to establish a topology for the extrinsic network. The topology defined for the test fixture under consideration is reported in Figure 8.32. The following elements can be identified:

- CPW_g and CPW_d correspond to the CPW transmission lines at ports 1 and 2, respectively.
- L_{gp}, L_{dp}, and L_{sp} correspond to the parasitic inductances of the bonding wires, including the transistor metallization.
- C_{pg} and C_{pd} correspond to the parasitic capacitances between the transistor contact pad (gate and drain) and the metallization underneath the bare die.

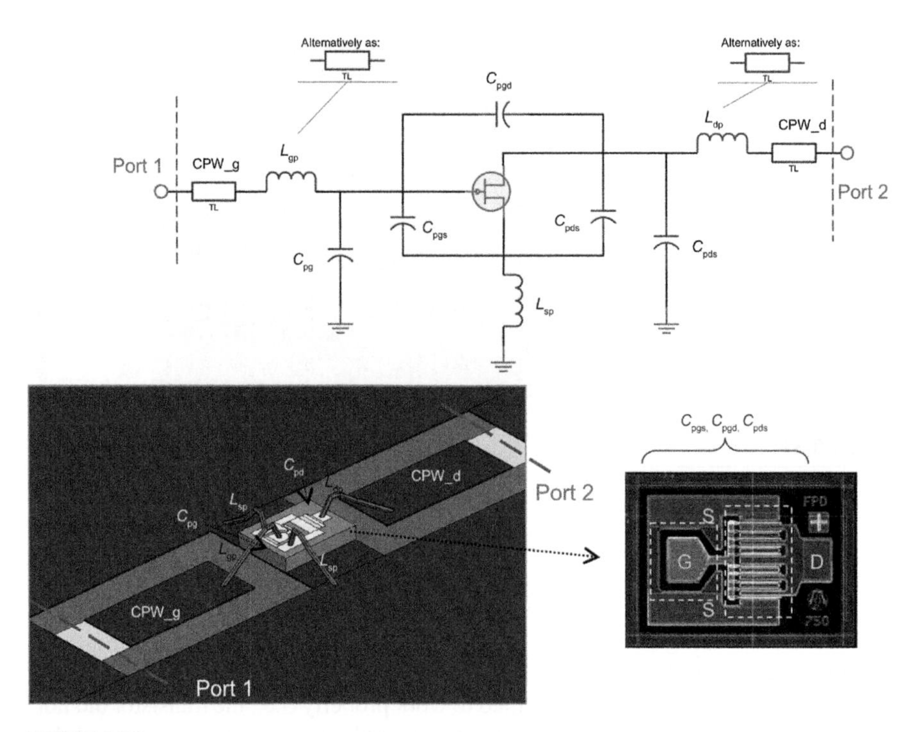

FIGURE 8.32

Equivalent circuit of the FPD750 test fixture.

- C_{pgs}, C_{pgd}, and C_{pds} correspond to the parasitic intrafinger capacitance as well as the capacitance between pads (gate, source and drain).

For completeness, the short CPW lines and bonding wires also have associated parasitic resistances. Nevertheless, these are negligible in comparison with the access or contact resistances. These resistances are not indicated in the test fixture because they are physically inside the bare-die transistor. They are dependent on the technology and processing. For clarification, they are shown in Figure 8.33 together with the schematic of the final extrinsic network.

Different topologies with larger numbers of elements could be used as well. Although this would lead to higher precision, the determination of the element values would be more cumbersome. On the other hand, the reported topology was shown to be sufficient for the intended frequency range (up to 10 GHz) of the model.

Bonding wires can be modeled as pure inductances or transmission lines. For a wire inductance, the rule of thumb of 1 nH/mm can be used to determine its value, once the physical length is estimated. The average length of the bonding wires in this case study is 500 μm; therefore, the estimated inductance values would be 0.5 nH for the drain and gate bonding wires and 0.25 nH for the source bonding wire. However,

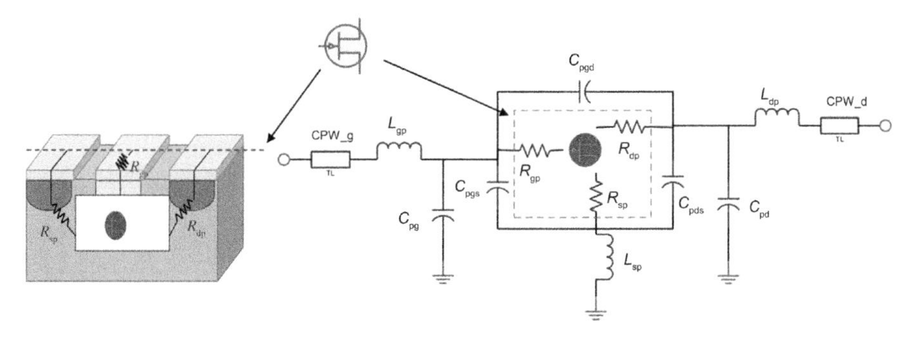

FIGURE 8.33

Complete equivalent circuit including contact resistances for the FPD750 mounted on the test fixture from Figure 8.32.

these values are not accurate enough for the de-embedding process. A better approximation can be obtained if they are modeled as transmission lines. They can be considered as lines suspended in air and in parallel to a ground plane. The distributed capacitance and inductance can be calculated, followed by the characteristic impedance and electrical length. Values of 145 Ω and 0.77° are calculated for the bonding wires at the gate and drain, respectively. Because the bonding wire at the source terminal is much shorter, it can be kept as a pure inductance. Moreover, this simplifies the de-embedding calculation. Its value can be calculated based on the distributed inductance of the transmission line model instead of the rule of thumb. Given a calculated distributed inductance of 0.46 nH/mm, then its value would be about 0.115 nH.

The parasitic resistances can be determined based on the classical procedure found in Ref. [49], which is widely known as the cold-FET method, in which the device is driven in forward mode ($V_{DS} = 0$ V and $V_{GS} = 0.8$ V). The intrinsic model to be considered under this condition is reported in Figure 8.34. Unlike what is established in Ref. [49], the gate capacitance C_{gate} is not neglected and is included in the model, as suggested in Refs [50,51], for weak forward conduction (C_{gate} plus R_{gate}). Besides the extraction of the resistance values, this cold FET condition can be used to verify and correct the values of the bonding wires modeled either as transmission lines (L_{gp}, L_{dp}) or as pure inductance (L_{sp}). The following expressions can be established in terms of Z parameters:

$$Z_{11} - Z_{12} = R_{gp} + \frac{R_{gate}}{1 + j\omega C_{gate} R_{gate}} + j\omega L_{gp} \tag{8.14}$$

$$Z_{12} = R_{ch1} + R_{sp} + j\omega L_{sp} \tag{8.15}$$

$$Z_{22} - Z_{12} = R_{ch2} + R_{dp} + j\omega L_{dp} \tag{8.16}$$

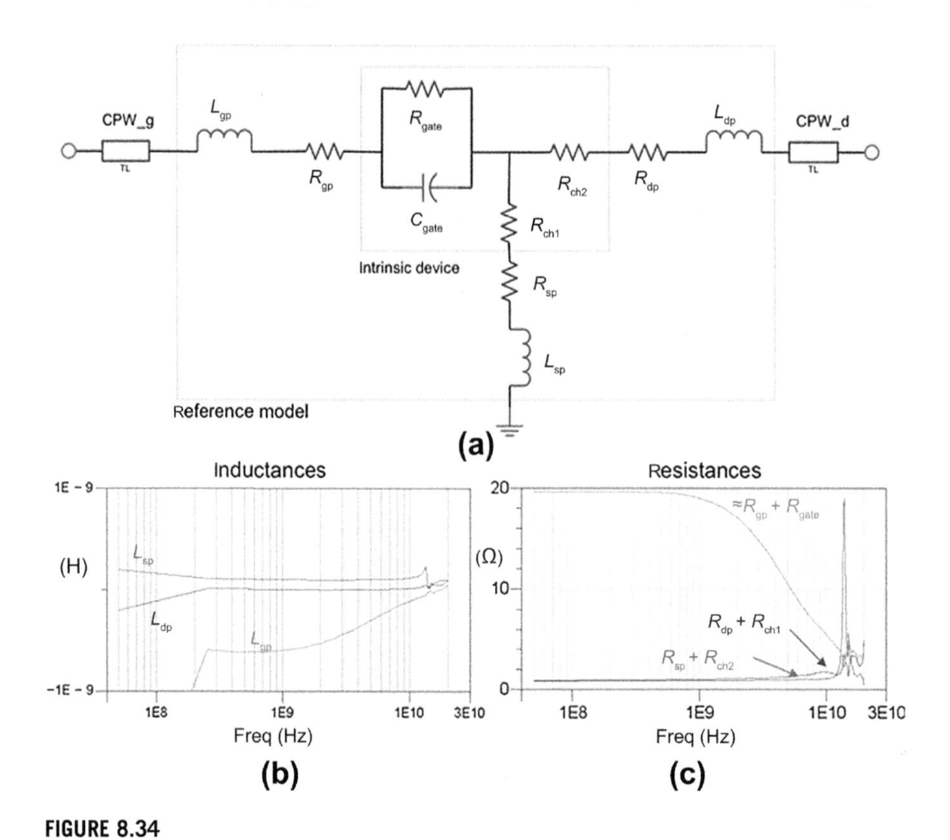

FIGURE 8.34

(a) Equivalent circuit and extracted (b) inductance and (c) resistance values of FPD750 under weak forward cold FET conditions.

As can be deduced from Eqns (8.15) and (8.16), inductances L_{sp} and L_{dp} can be solved straightforwardly. Therefore, this is a well-conditioned problem.

As mentioned, the initial characterization of these inductances has already been performed as transmission lines (L_{gp}, L_{dp}) and as lumped inductance (L_{sp}). In fact, de-embedding of the CPW transmission lines and inductances L_{gp} and L_{dp} (as transmission lines) is performed before the application of Eqns (8.14)–(8.16). As a result, if the bonding wires L_{gp} and L_{dp} are well modeled, the inductance values should be zero when these formulas are applied; otherwise, the characteristics of the transmission lines should be modified until zero values are obtained. This is possible for L_{dp}, but not for L_{gp} due to the influence of C_{gate}. However, this is not so relevant because the characteristics of the bonding wires L_{dp} and L_{gp} can be considered similar in most cases because most test fixtures are symmetric. The inductance L_{sp} can be extracted straightforwardly and its value

Table 8.3 Extrinsic network elements values extracted for the FPD750 transistor

Characteristic	Symbol	Value
CPW characteristic impedance at gate (Ω)	$Z_{o,g}$	52
CPW characteristic impedance at drain (Ω)	$Z_{o,d}$	52
CPW phase delay at gate (degrees)	E_g	2.55
CPW phase delay at drain (degrees)	E_d	2.55
Bonding wire characteristic impedance at gate (Ω)	$Z_{o,bg}$	155
Bonding wire characteristic impedance at drain (Ω)	$Z_{o,bd}$	155
Bonding wire phase delay at gate (degrees)	E_{bg}	0.9
Bonding wire phase delay at drain (degrees)	E_{bd}	0.8
Pad-ground parasitic capacitance at gate (fF)	C_{pg}	32
Pad-ground parasitic capacitance at drain (fF)	C_{pd}	32
Bonding wire inductance at source (nH)	L_{sp}	0.085
Interdigital capacitance gate-source (fF)	C_{pgs}	64
Interdigital capacitance gate-drain (fF)	C_{pgd}	44
Interdigital capacitance drain-source (fF)	C_{pds}	60
Contact resistance at gate (Ω)	R_{gp}	0.27
Contact resistance at drain (Ω)	R_{dp}	0.6
Contact resistance at source (Ω)	R_{sp}	0.6

(0.085 nH) in this case study is close to the estimated value indicated before (0.115 nH). The corrected values for the transmission lines and bonding wires are reported in Table 8.3.

Unlike the inductance case, the resistance cannot be solved in a direct way because it is an ill-conditioned problem. There are six unknowns for a set of three equations. Then, some assumptions or alternative approaches should be taken. The channel resistances corresponding to the gate-source (R_{ch1}) and gate-drain (R_{ch2}) junctions could be considered similar; therefore, both are taken as half of the total channel resistance [49,51]. A rough estimate of R_{sp}, R_{dp}, and R_{gp} can be obtained using another cold FET condition in which the effect of the channel resistance and forward conduction impedance are removed. For instance, a cold FET condition corresponding to the bias $V_{DS}=0$ V and $V_{GS}=0$ V (open-channel condition) can be used to determine R_{gp}. The corresponding equivalent circuit is reported in Figure 8.35 [50]. It can be inferred that, by taking the real part of the factor $Z_{11}-Z_{12}$ of this network, an initial value of R_{gp} can be obtained. On the other hand, R_{sp} and R_{dp} can be considered approximately equal as a first approach, given that the geometry of the physical contacts is approximately the same. This can be also verified by observing Figure 8.34. The values of $R_{sp}+R_{ch1}$ ($R_{ch}/2$) and $R_{dp}+R_{ch2}$ ($R_{ch}/2$) coincide up to about 3 GHz. Finally, to determine separate values for either R_{sp} or R_{dp}, a cold FET condition with

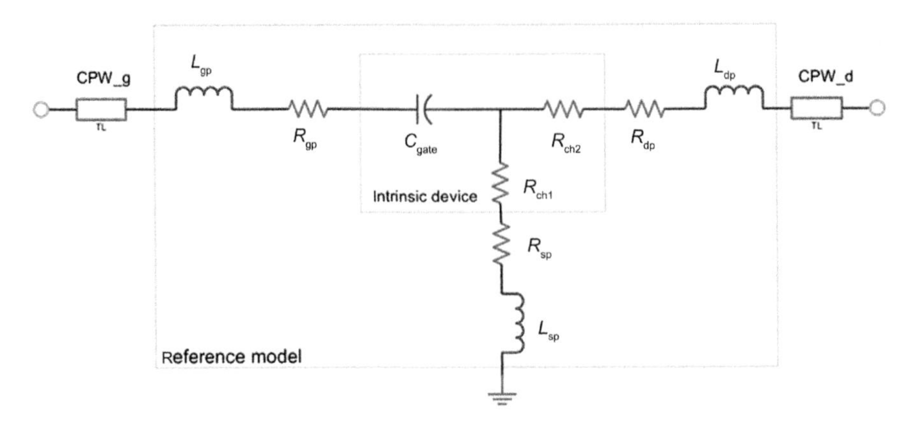

FIGURE 8.35

Cold FET condition for $V_{GS}=0$ V and $V_{DS}=0$ V.

$V_{DS}=0$ V and $V_{GS}=-1.4$ V is used. The equivalent circuit for this condition can be observed in Figure 8.36. Based on this, it can be deduced that the real part of the Z parameters could be used for extracting a first estimate of the resistances R_{sp} or R_{dp}. It is important to remark that this value is not accurate because it changes with V_{GS} lower than pinch-off. However, the chosen bias point gives a coherent value that is constant over the frequency range. Finally, the flyback method [52,53] is a more accurate method to determine the value of the resistances R_{sp} and R_{dp}. The complete set of values of these parasitic extrinsic resistors is reported in Table 8.3.

Finally, in order to determine proper values for the extrinsic capacitances, the devices have to be driven again in cold FET conditions—that is, $V_{DS}=0$ V and V_{GS} equal to or lower than the pinch-off voltage. Under this condition, the intrinsic device can be represented by its depletion capacitances C_{gs} and C_{gd} plus an additional capacitance C_{ds}. These elements form a delta network as shown in Figure 8.36. The imaginary parts of the Y parameters are used for the determination of these capacitances. At low frequencies, the following relationships can be established:

$$\mathrm{Im}(Y_{11}-Y_{12}) = \omega\left(C_{pg}+C_{pgs}+C_{gs}\right) \tag{8.17}$$

$$\mathrm{Im}(Y_{12}) = \omega\left(C_{pgd}+C_{gd}\right) \tag{8.18}$$

$$\mathrm{Im}(Y_{22}-Y_{12}) = \omega\left(C_{pd}+C_{pds}+C_{ds}\right) \tag{8.19}$$

However, these equations are insufficient in number to obtain all the capacitance values. There are eight unknowns for only three equations. Furthermore, it is not possible to separate the values of the parallel capacitances C_{pg}, C_{pgs}, and C_{gs}. Therefore, additional concepts or approaches have to be used to sort out these

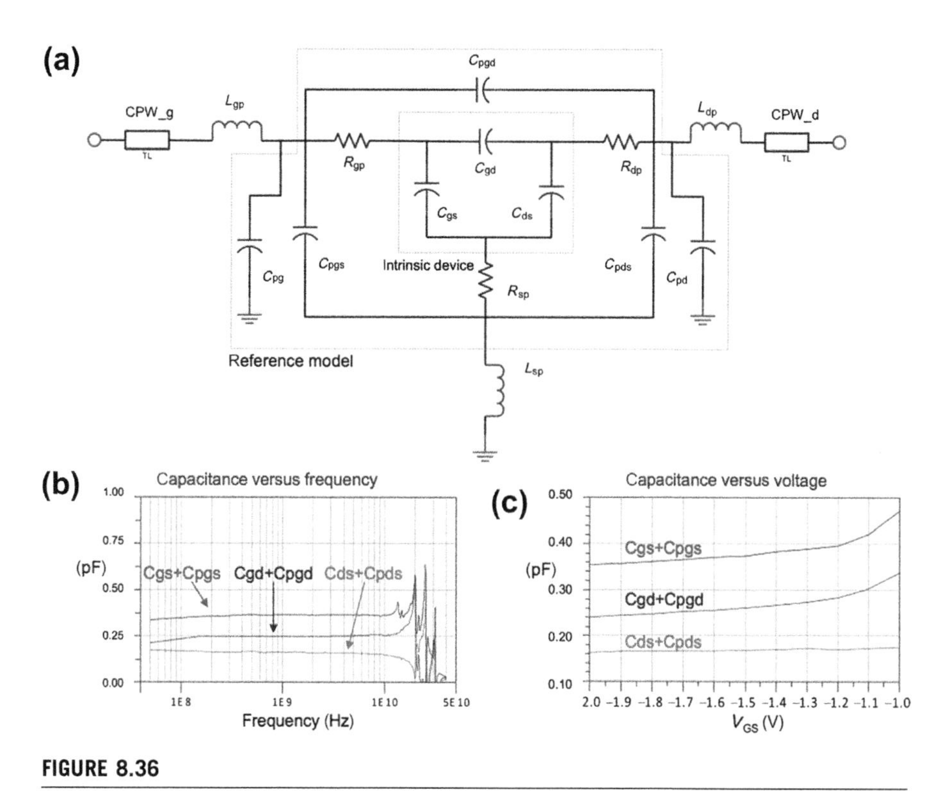

FIGURE 8.36

(a) Equivalent circuit and (b, c) extracted capacitances for the FPD750 device under cold FET conditions for voltages V_{GS} lower than pinch-off voltage.

nontrivial conditions. In particular, the capacitances C_{pg} and C_{pd} could be characterized by using the parallel plate capacitance concept and formula. However, this has to be corrected due to the presence of fringing fields, which increase the apparent width of the capacitor plates by an amount proportional to the thickness of the dielectric [54], in this case a GaAs substrate. The final values obtained for C_{pg} and C_{pd} are reported in Table 8.3. Although the determination of C_{pg} relaxes the number of unknowns in Eqn (8.17), the capacitances C_{pgs} and C_{gs} still cannot be solved separately, as well as C_{pgd} and C_{gd} in Eqn (8.19). In order to separate these variables, other tests have to be performed [50]. If the gate-source voltage is decreased, the corresponding depletion capacitance C_{gs} will decrease as well. Then in the limit when $V_{GS} = -\infty$, C_{gs} becomes equal to zero. As a result, the only remaining capacitance will be C_{pgs}. In practice, a plot of the extracted capacitance $C_{pgs}-C_{gs}$ versus V_{GS} is extrapolated for $V_{GS} = -\infty$. In fact, curve-fitting procedures are performed in order to find this extrapolated value. Another important factor to take into account when this fitting is performed is that C_{gs} is equal to C_{gd} for these cold FET conditions. The graphs reported in Figure 8.36 shows this characteristic for $C_{pgs}-C_{gs}$ and $C_{pgd}-C_{gd}$. On the other hand, the separation of C_{pds} and C_{ds} is more difficult because this does not follow such characteristics, but it remains approximately constant. Then this value can be assumed to be a constant and unique parasitic value. The values for all these extrinsic capacitances are reported in Table 8.3.

8.6.3 Four-port representation of extrinsic network

Another methodology to perform the characterization of the test fixture is by using a four-port network definition [55,56]. In this case, most of the extrinsic parasitics, studied in the previous section, are embedded in a "black-box" four-port network or matrix. The positions of the ports in the test fixture are indicated in Figure 8.37. Ports 1 and 2 correspond to the measurement planes, whereas ports 3 and 4 correspond to the intrinsic device plane. In this case, the intrinsic device plane cannot go deeper and surpass the contact resistances, and therefore these resistances have to be de-embedded later. In previous research [55,56] applying to on-wafer devices, the four-port network matrix was successfully determined by using a 3D EM simulator or a set of standard structures. Nevertheless, the determination of the four-port matrix based on 3D EM simulation fails when it is applied to large structures such as the present test fixture. This is mainly due to the effect of the local ground plane [57]; that is, the ground of ports 1 and 2 is different from the ground of ports 3 and 4, whereas the general definition of a four-port network considers only one common ground (see Figure 8.37(b)). The effect of the local ground can be overcome by using a five-port network characterization [58], where the local ground becomes a fifth port. However, this approach is not feasible or applicable with a 3D EM characterization.

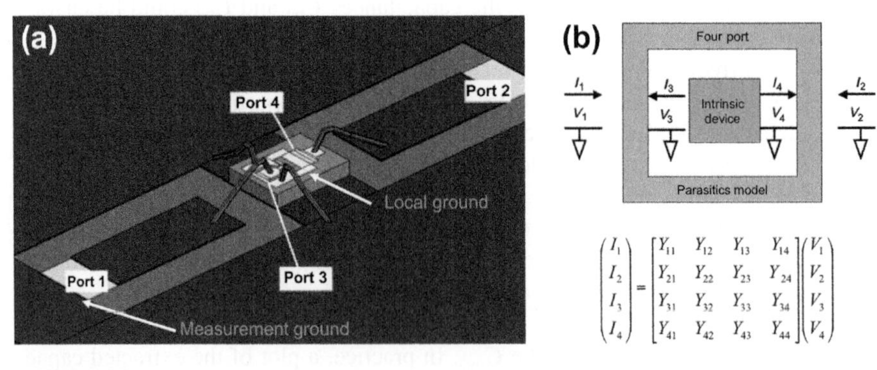

FIGURE 8.37

(a) Device and test fixtures with four-port network definition. Local ground is associated with ports 3 and 4. (b) Conventional four-port definition of an extrinsic parasitic. Common ground is used for all the ports.

Furthermore, EM calibration and de-embedding techniques such as double-delay or short-open-calibration (SOC) [58,59] may not provide good results in this case due to the complexity of the test fixture. An alternative four-port characterization of the text fixture that corrects the effect of the local ground, without the recourse to a five-port characterization, will be described next [60,61].

As was already described, the four-port network matrix can be determined by using a 3D EM simulator, such as a high-frequency structural simulator. However, the internal ports used for this 3D EM analysis have their own local ground, which is not a common ground. Then, this four-port network matrix solution has to be corrected in order to transform it into one that has a common ground.

Using the definition of variables indicated in Ref. [55], the network can be represented by its **Y**-matrix, with a common ground reference (see Figure 8.37(b)):

$$\begin{pmatrix} I_1 \\ I_2 \\ I_3 \\ I_4 \end{pmatrix} = \begin{bmatrix} Y_{11} & Y_{12} & Y_{13} & Y_{14} \\ Y_{21} & Y_{22} & Y_{23} & Y_{24} \\ Y_{31} & Y_{32} & Y_{33} & Y_{34} \\ Y_{41} & Y_{42} & Y_{43} & Y_{44} \end{bmatrix} \begin{pmatrix} V_1 \\ V_2 \\ V_3 \\ V_4 \end{pmatrix} \tag{8.20}$$

By defining vectors $\mathbf{V}_e$ and $\mathbf{I}_e$ as the extrinsic vectors and $\mathbf{V}_i$ and $\mathbf{I}_i$ as the intrinsic ones,

$$\begin{pmatrix} \mathbf{I}_e \\ \mathbf{I}_i \end{pmatrix} = \begin{pmatrix} I_1 \\ I_2 \\ - \\ I_3 \\ I_4 \end{pmatrix} \text{ and } \begin{pmatrix} \mathbf{V}_e \\ \mathbf{V}_i \end{pmatrix} = \begin{pmatrix} V_1 \\ V_2 \\ - \\ V_3 \\ V_4 \end{pmatrix} \tag{8.22}$$

Equation (8.7) can be rewritten as

$$\begin{pmatrix} \mathbf{I}_e \\ \mathbf{I}_i \end{pmatrix} = \begin{bmatrix} Y_{11} & Y_{12} & \vdots & Y_{13} & Y_{14} \\ Y_{21} & Y_{22} & \vdots & Y_{23} & Y_{24} \\ \cdots & \cdots & \vdots & \cdots & \cdots \\ Y_{31} & Y_{32} & \vdots & Y_{33} & Y_{34} \\ Y_{41} & Y_{42} & \vdots & Y_{43} & Y_{44} \end{bmatrix} \begin{pmatrix} \mathbf{V}_e \\ \mathbf{V}_i \end{pmatrix} \tag{8.21}$$

$$\begin{pmatrix} \mathbf{I}_e \\ \mathbf{I}_i \end{pmatrix} = \begin{bmatrix} \mathbf{Y}_{ee} & \mathbf{Y}_{ei} \\ \mathbf{Y}_{ie} & \mathbf{Y}_{ii} \end{bmatrix} \begin{pmatrix} \mathbf{V}_e \\ \mathbf{V}_i \end{pmatrix} \tag{8.22}$$

where $\mathbf{Y}_{ee}$, $\mathbf{Y}_{ei}$, $\mathbf{Y}_{ie}$, and $\mathbf{Y}_{ii}$ are four 2×2 matrices.

Noting that $\mathbf{I}_e = Y^{EXT} \cdot \mathbf{V}_e$ and $\mathbf{I}_i = -Y^{INT} \cdot \mathbf{V}_i$, where Y^{EXT} and Y^{INT} are the extrinsic and intrinsic device's Y parameters, respectively, the following relationships can be established from Eqn (8.22):

$$\left.\begin{aligned} Y^{EXT}\mathbf{V}_e &= \mathbf{Y}_{ee}\mathbf{V}_e + \mathbf{Y}_{ei}\mathbf{V}_i \\ -Y^{INT}\mathbf{V}_i &= \mathbf{Y}_{ie}\mathbf{V}_e + \mathbf{Y}_{ii}\mathbf{V}_i \end{aligned}\right\} \tag{8.23}$$

Solving the simple set of Eqn (8.23) for Y^{EXT} provides

$$Y^{EXT} = \mathbf{Y}_{ee} - \mathbf{Y}_{ei}\left(Y^{INT} + \mathbf{Y}_{ii}\right)^{-1}\mathbf{Y}_{ie} \tag{8.24}$$

As a matter of fact, Y^{EXT} coincides with the device measurements (including the test fixture) at the calibration plane, whereas Y^{INT} corresponds to the intrinsic device Y parameters. Then, the idea of this four-port technique [60,61] is to have input data for Y^{EXT} and Y^{INT} at different bias conditions in order to solve for $\mathbf{Y}_{ij}$ ($\mathbf{Y}_{ee}$, $\mathbf{Y}_{ei}$, $\mathbf{Y}_{ie}$, and $\mathbf{Y}_{ii}$), which are the Y parameters of the four-port parasitic model. An initial solution for the four-port network has already been obtained from a 3D EM simulator. However, this needs to be corrected in order to eliminate the local ground effect. Y^{EXT} input data are collected by doing S-parameter measurements at ports 1 and 2, which are the only physically accessible ports. Y^{INT} input data can be obtained by driving the device in bias conditions where its behavior can be easily defined and modeled, and these correspond to the cold FET bias conditions ($V_{DS} = 0$ V).

On the other hand, in order to solve for the 16 unknown variables of the four-port network $\mathbf{Y}_{ij}$ (Y_{11}, Y_{12}, Y_{13}, ... Y_{44}), 16 equations are needed, or four different conditions for Y^{EXT} and Y^{INT} have to be known. Each condition provides four equations. These four conditions can be reduced to two conditions or eight equations, taking into account the following observations:

- The four-port network corresponds to a passive network, implying that the matrix should be reciprocal (i.e., $\mathbf{Y}_{ij} = \mathbf{Y}_{ji}$).
- Because the losses are minimal, the phase of the Y parameters should be very close to 90°.
- Due to the physical layout, a symmetry approach can be considered for Y_{11} and Y_{22} ($Y_{11} = Y_{22}$), which is not true for Y_{33} and Y_{44}. The same argument holds for Y_{23} and Y_{14} ($Y_{23} = Y_{14}$).

Therefore, only two conditions are needed to solve the problem. The two conditions used in this technique are two cold FET conditions: cutoff and forward conduction condition. These two conditions are chosen because the determination of the intrinsic parameter values (Y^{INT}) and the Y^{EXT} measurements can be realized straightforwardly.

The intrinsic device models used for obtaining Y^{INT} input data at these bias conditions are indicated in Figure 8.38 (see also Figure 8.34 and Figure 8.36). These models are extracted at low frequencies where the influence of the device's inductive parasitics can be neglected.

For these two cold FET conditions, two matrix equations can be established from Eqn (8.24):

$$Y^{EXT}_{\text{pinch-off}} = \mathbf{Y}_{ee} - \mathbf{Y}_{ei}\left(Y^{INT}_{\text{pinch-off}} + \mathbf{Y}_{ii}\right)^{-1}\mathbf{Y}_{ie} \tag{8.25}$$

$$Y^{EXT}_{\text{forward}} = \mathbf{Y}_{ee} - \mathbf{Y}_{ei}\left(Y^{INT}_{\text{forward}} + \mathbf{Y}_{ii}\right)^{-1}\mathbf{Y}_{ie} \tag{8.26}$$

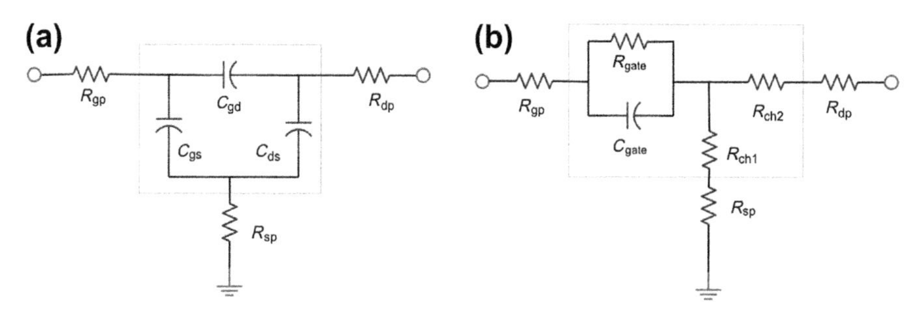

FIGURE 8.38

Intrinsic device models used to obtain Y^{INT} data at two cold FET bias conditions: (a) pinch-off bias condition and (b) forward conduction condition.

This set of equations cannot be solved straightforwardly due to the matrix inversion operation that includes one set of variables. Therefore, this has to be solved by numerical methods. Equations (8.25) and (8.26) plus the considerations indicated above are used to solve the Y parameters by using a differential evolution optimization process [62] and by taking as initial values the ones given by the EM simulator. In order to explain the process followed to solve this problem, a flow diagram of the algorithm is shown in Figure 8.39.

The algorithm starts by initializing the values of the variables $\mathbf{Y}_{\mathrm{ij}}$ $(Y_{11},Y_{12},Y_{13},\ldots Y_{44})$ with the values given by the 3D EM simulator. At the same time, the values for $Y^{\mathrm{INT}}_{\text{pinch-off}}$ and $Y^{\mathrm{INT}}_{\text{forward}}$ are obtained from the simulation performed with the model indicated in Figure 8.38. Then, all of these values are combined to calculate $Y^{\mathrm{EXT}}_{\text{pinch-off}}$ and $Y^{\mathrm{EXT}}_{\text{forward}}$ using Eqns (8.25) and (8.26). If the latter values coincide with the values obtained from measurements, then the algorithm exits. Actually, what is calculated is the difference between the theoretical values and the measured ones using the following formula:

$$\mathrm{Error} = \sum_{j=1}^{2}\sum_{i=1}^{2}\left|\frac{Y_{\mathrm{DUT,labmeas}}(i,j) - Y_{\mathrm{DUT,extrinsic}}(i,j)}{Y_{\mathrm{DUT,labmeas}}(i,j)}\right| \tag{8.27}$$

where $Y_{\mathrm{DUT,labmeas}}$ corresponds to the measured Y parameters and $Y_{\mathrm{DUT,extrinsic}}$ corresponds to the calculated Y parameters based on the four-port network. Both $Y_{\mathrm{DUT,labmeas}}$ and $Y_{\mathrm{DUT,extrinsic}}$ are set at the calibration plane.

This error is calculated for each bias condition. Then $Y_{\mathrm{DUT,extrinsic}}$ of Eqn (8.27) can be either $Y^{\mathrm{EXT}}_{\text{pinch-off}}$ or $Y^{\mathrm{EXT}}_{\text{forward}}$. The sum of the errors for both bias conditions is the total error, which should be minimized. If this total error is greater than a threshold value, then the assumed solution is not valid and a new set of values for $\mathbf{Y}_{\mathrm{ij}}$ is calculated based on the differential evolution (DE) optimization algorithm. With the latter operation, the algorithm returns to the beginning of the program, establishing in that way an iterative loop that will end when the error is lower

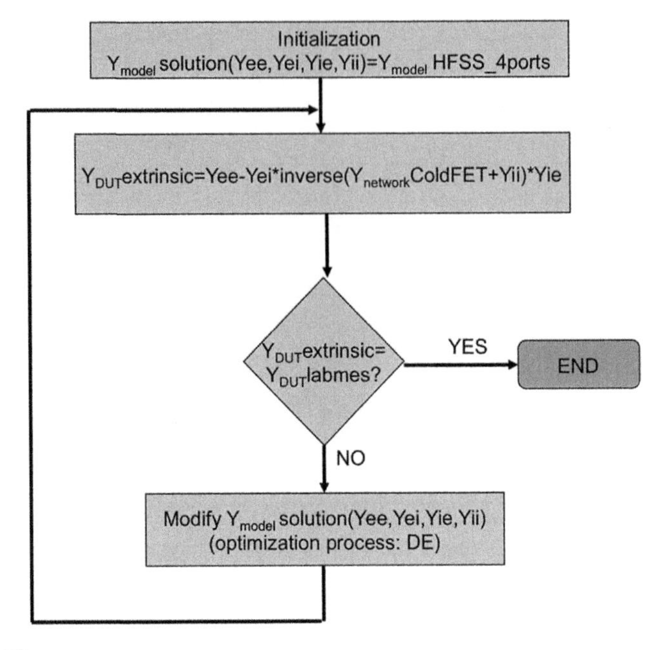

FIGURE 8.39

Flow diagram of the algorithm used to correct the *Y* parameters of the four-port de-embedding circuit (at a single frequency).

than or equal to the threshold value. This algorithm is repeated for all the analysis frequencies.

To illustrate the validity of the method, measurement data of the device mounted on the test fixture are collected at different bias conditions. Additionally, a 3D EM model of the structure is realized, according to Figure 8.37(a). Simulations performed on this 3D EM model provide a set of four-port *Y* parameters that define the extrinsic model of the test fixture [56]. Simple models of the intrinsic device are extracted at the two cold FET conditions as explained above. *Y* parameters of these models can be obtained straightforwardly; by applying Eqn (8.24), the corresponding extrinsic *Y* parameters can be obtained and compared with the measurements. This is shown in Figures 8.40(a) and 8.41(a). As can be seen, the extrinsic *S* parameters based on the 3D EM four-port extrinsic model start to diverge from measurements at high frequencies (around 6 GHz), and this difference increases with frequency. As mentioned, this is due to the local ground effect. The described technique corrects this problem. By realizing the algorithm described above, a new set of four-port *Y* parameters that models the extrinsic parasitics is obtained. The calculated extrinsic *Y* parameters, and corresponding *S* parameters, are compared with the measurements in Figures 8.40(b) and 8.41(b). It can be observed that the agreement with the measurements is remarkably better in the second case, up to 20 GHz.

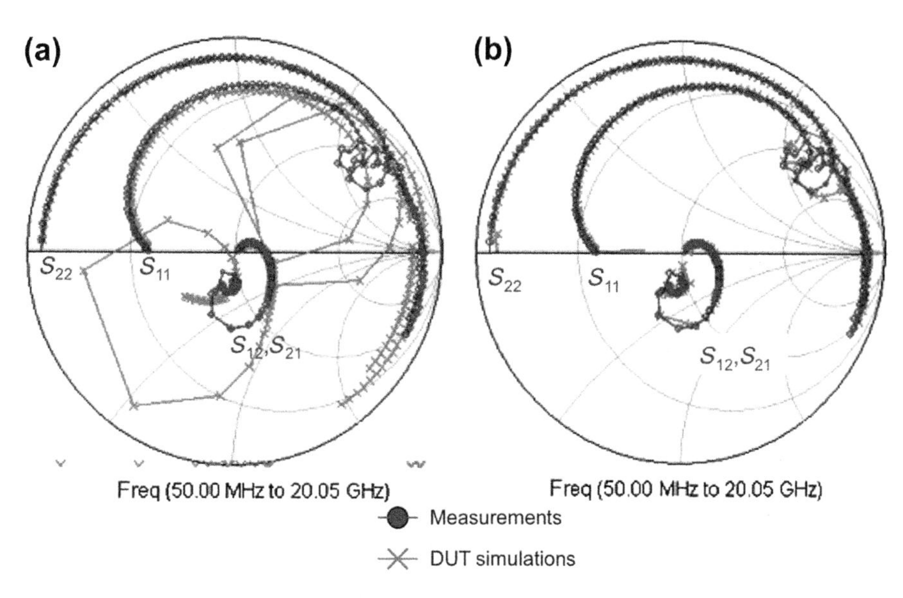

FIGURE 8.40

(a) S parameters with 3D EM four-port values and (b) with the proposed technique at cold-FET forward condition. DUT, device under test.

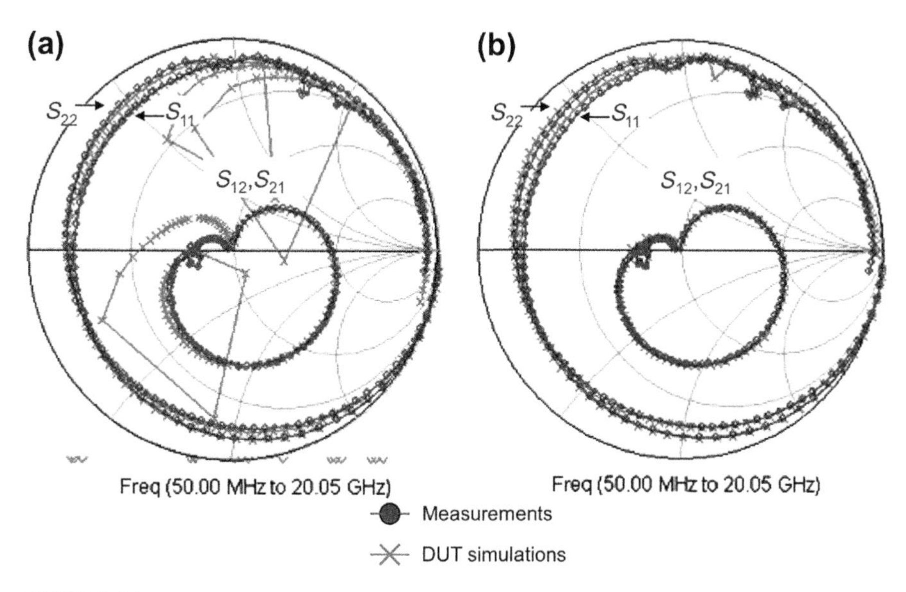

FIGURE 8.41

(a) S Parameters with 3D EM four-port values and (b) with the proposed technique at cold-FET pinch-off condition. DUT, device under test.

8.6.4 De-embedding of transistor *S* parameters

Once a model of the extrinsic network has been defined by either an equivalent circuit or four-port model, the intrinsic *S* parameters of the device can be obtained through de-embedding. To this purpose, *S*-parameter measurements are performed on the test fixture under different bias points. These are chosen in order to cover the saturation, cutoff, and linear regions. If the model is intended for power amplifiers, *S*-parameter measurements under forward conduction bias have to be collected as well. The final set of bias points for the considered device is the following:

- V_{GS} from -1.4 to 0.8 V with 0.1 V steps and V_{DS} from 0 to 2 V with 0.2 V steps.
- V_{GS} from-1.4 to 0 V with 0.1 V steps and V_{DS} from 2 to 6 V with 0.5 V steps.

As mentioned, the *S* parameters collected at these bias points have to be de-embedded in order to translate them to the intrinsic device plane. This de-embedding process can be realized in two ways, depending on how the extrinsic network (test fixture plus additional parasitics) is modeled. If the extrinsic network is modeled as an equivalent circuit such as the one already described in Section 8.6.2 and reported in Figure 8.42 for further explanation, then the sequence of de-embedding is as follows:

1. For moving from calibration plane 1 to plane 2, first a transformation of measured *S* parameters into ABCD parameters (ABCD1) is performed. Then the following operation to de-embed the CPW transmission lines is done:

$$\mathrm{ABCD}_2 = \mathrm{ABCD}_{\mathrm{g}}^{-1} * \mathrm{ABCD}_1 * \mathrm{ABCD}_{\mathrm{d}}^{-1} \tag{8.28}$$

where

$$\mathrm{ABCD}_{\mathrm{g}} = \begin{bmatrix} \cos(E_{\mathrm{g}}) & jZ_{\mathrm{o,g}}\sin(E_{\mathrm{g}}) \\ \dfrac{j}{Z_{\mathrm{o,g}}}\sin(E_{\mathrm{g}}) & \cos(E_{\mathrm{g}}) \end{bmatrix} \tag{8.29}$$

$$\mathrm{ABCD}_{\mathrm{d}} = \begin{bmatrix} \cos(E_{\mathrm{d}}) & jZ_{\mathrm{o,g}}\sin(E_{\mathrm{d}}) \\ \dfrac{j}{Z_{\mathrm{o,g}}}\sin(E_{\mathrm{d}}) & \cos(E_{\mathrm{d}}) \end{bmatrix} \tag{8.30}$$

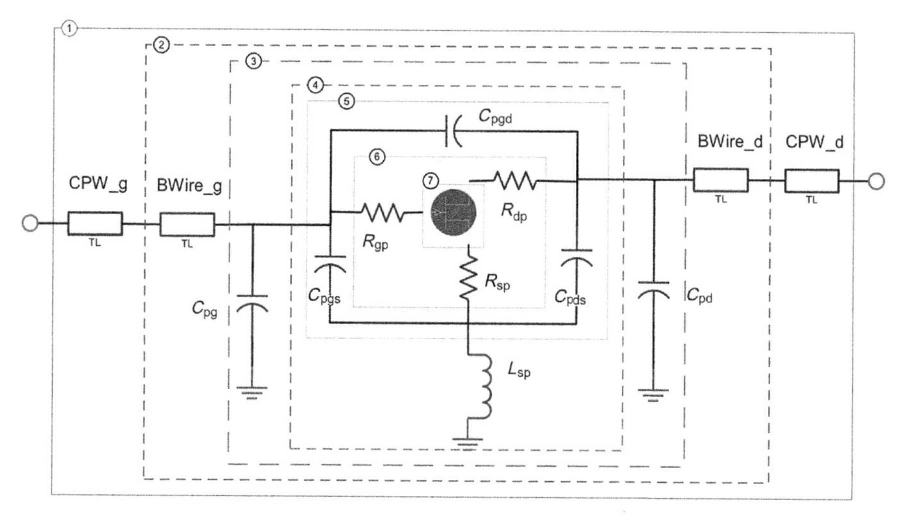

FIGURE 8.42

Test fixture equivalent circuit and de-embedding planes.

2. From plane 2 to plane 3, the following operation to de-embed the bonding wires at gate and drain is performed:

$$\mathrm{ABCD}_3 = \mathrm{ABCD}_{\mathrm{bg}}^{-1} * \mathrm{ABCD}_2 * \mathrm{ABCD}_{\mathrm{bd}}^{-1} \tag{8.31}$$

where

$$\mathrm{ABCD}_{\mathrm{bg}} = \begin{bmatrix} \cos(E_{\mathrm{bg}}) & jZ_{\mathrm{o,g}}\sin(E_{\mathrm{bg}}) \\ \dfrac{j}{Z_{\mathrm{o,g}}}\sin(E_{\mathrm{bg}}) & \cos(E_{\mathrm{bg}}) \end{bmatrix} \tag{8.32}$$

$$\mathrm{ABCD}_{\mathrm{bd}} = \begin{bmatrix} \cos(E_{\mathrm{bd}}) & jZ_{\mathrm{o,g}}\sin(E_{\mathrm{bd}}) \\ \dfrac{j}{Z_{\mathrm{o,g}}}\sin(E_{\mathrm{bd}}) & \cos(E_{\mathrm{bd}}) \end{bmatrix} \tag{8.33}$$

3. From plane 3 to plane 4, first a transformation of ABCD parameters (ABCD_3) to Y parameters (Y_3) is done, followed by an operation to de-embed capacitances C_{pg} and C_{pd}.

$$Y_4 = Y_3 - j\omega\begin{bmatrix} C_{pg} & 0 \\ 0 & C_{pd} \end{bmatrix} \quad (8.34)$$

4. From plane 4 to plane 5, first a transformation of Y parameters (Y_4) into Z parameters (Z_4) is done; then the following operation is performed to de-embed the source inductance:

$$Z_5 = Z_4 - j\omega\begin{bmatrix} L_{sp} & L_{sp} \\ L_{sp} & L_{sp} \end{bmatrix} \quad (8.35)$$

5. From plane 5 to plane 6, first Z parameters (Z_5) are transformed to Y parameters (Y_5); then the operation to de-embed capacitances C_{pgs}, C_{pgd}, and C_{pds} is performed:

$$Y_6 = Y_5 - j\omega\begin{bmatrix} C_{pgs} + C_{pgd} & -C_{pgd} \\ -C_{pgd} & C_{pds} + C_{pgd} \end{bmatrix} \quad (8.36)$$

6. From plane 6 to plane 7, first Y parameters (Y_6) are converted to Z parameters (Z_6); then resistances R_{gp}, R_{dp} and R_{sp} are de-embedded:

$$Z_7 = Z_6 - \begin{bmatrix} R_{gp} + R_{sp} & R_{sp} \\ R_{sp} & R_{dp} + R_{sp} \end{bmatrix} \quad (8.37)$$

This last set of Z parameters (Z_7) can be transformed into S parameters (S_7) again, which would become the intrinsic device S parameters.

This de-embedding procedure can be realized using the four-port network approach as well. In this case, steps 1–5 are replaced by the following single operation (based on Eqn (8.23)):

$$Y_6 = -\mathrm{Y}_{ie}(Y_1 - \mathrm{Y}_{ee})^{-1}\mathrm{Y}_{ei} - \mathrm{Y}_{ii} \quad (8.38)$$

where Y_{ie}, Y_{ee}, Y_{ei}, and Y_{ii} are the 2×2 matrices of the four-port network, as defined above (see Eqns (8.21) and (8.22)).

Then, in order to complete the de-embedding process using the four-port network, step 6 of the equivalent circuit procedure has to be applied as well.

8.7 Bare-die transistor modeling and power amplifier design

High-frequency circuit design relies on a well-constructed device model. Up to this point, what has been obtained is the transistor S parameters after de-embedding. With these S parameters, it is possible to begin transistor modeling for small-signal regime and subsequently a large-signal model. These aspects will be covered in this section, as well as how this model is utilized to perform basic RF power amplifier designs.

8.7.1 Transistor modeling

Once the intrinsic device S parameters are obtained, the elements of a basic small-signal model, as shown in Figure 8.43(a), can be determined. For this purpose, closed-form relationships already exist to determine each of these elements based on the transistor's Y parameters (which can be straightforwardly calculated from the transistor's S parameters). These equations can be found in Ref. [63]; however, more complete relationships will be described next. The typical small-signal model reported in Figure 8.43(a) is valid only for negative V_{GS}, which is an insufficient description for a device aimed to operate under very large-signal conditions (e.g., power amplifier).

As a result, the characterization and modeling of the device for positive V_{GS} is required as well. A small-signal model covering negative and positive V_{GS} bias conditions is reported in Figure 8.43(b). The major difference with the one reported in Figure 8.43(a) is the addition of two resistances, which account for the weak/mild/strong conduction of the gate junctions (gate-source and gate-drain) when a positive V_{GS} is applied. The determination of these resistances is not straightforward and it is better to obtain them from S parameters at low frequencies (e.g., 50 MHz), instead of from DC measurements. Close relationships to determine these resistances along with the typical intrinsic elements can be found in Ref. [64].

After performing these operations, the extracted intrinsic parameters C_{gs}, C_{gd}, and g_m of the small-signal model are determined. The results are shown in Figure 8.44. These parameters are constant over frequencies up to 12 GHz. Above this frequency, board resonances take place and distort their behavior. Using

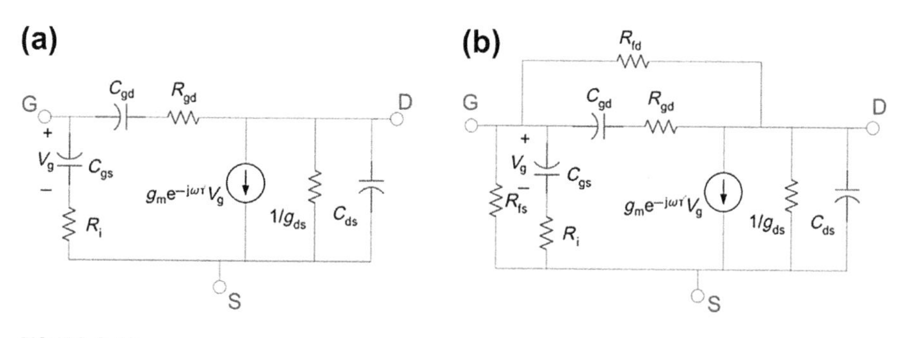

FIGURE 8.43

(a) Basic small-signal equivalent circuit and (b) extended circuit including forward conduction resistances.

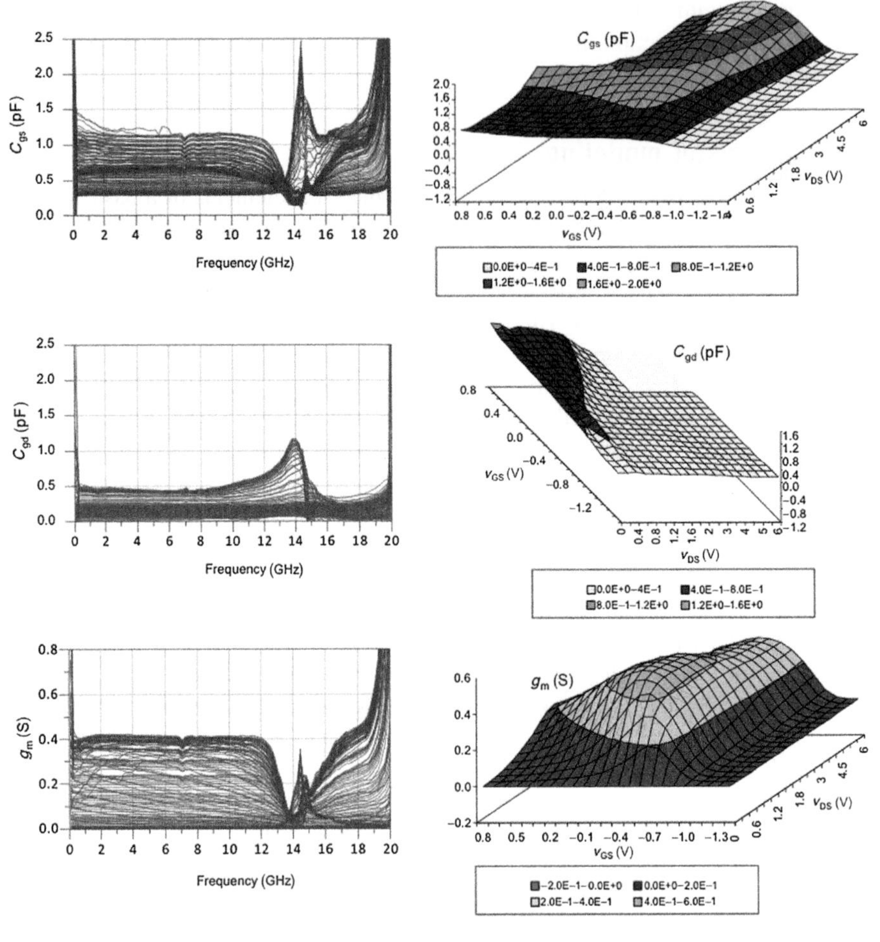

FIGURE 8.44

Extracted device intrinsic values for C_{gs}, C_{gd}, and g_m versus frequency and versus bias voltages V_{GS} and V_{DS}.

four-port techniques, the influence of these board resonances can be reduced [60] and make the extracted values constant over almost the whole frequency range (20 GHz). At the end, this is not critical because only one value, taken at one specific frequency or averaged over a frequency range, is necessary. The variation of these intrinsic values with the bias voltages V_{GS} and V_{DS} is more relevant for nonlinear modeling, and this is also reported in Figure 8.44.

The small-signal model, shown in Figure 8.43(b), can be used as a reference to build the basic configuration of the large-signal model. The first step is to identify the main elements that change their behavior with the voltages applied to the

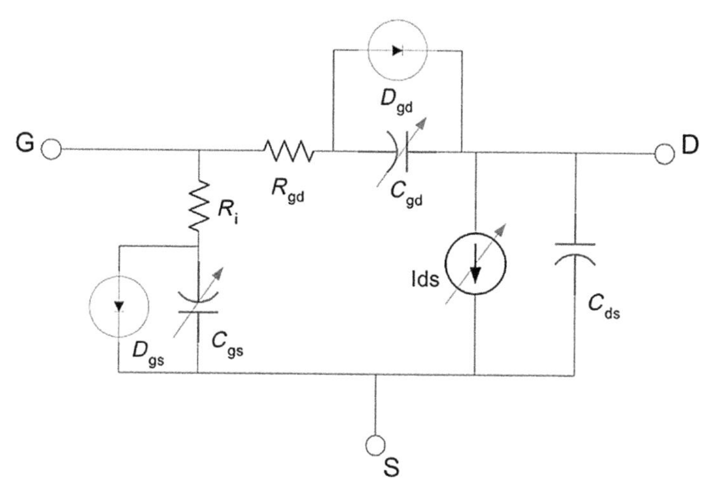

FIGURE 8.45

Basic configuration of a large-signal model.

device. The forward conduction resistances R_{fs} and R_{fd} depend on the bias voltages, and this dependency can be modeled properly by two diodes (see Figure 8.45), which corresponds to the physical characteristic of the device gate. The drain current is related to the transconductance g_m. As g_m varies with V_{GS} and V_{DS} (see Figure 8.44), the drain current has to be modeled with a non-linear dependency on V_{GS} and V_{DS} (see Figure 8.45). As can be observed from Figure 8.44 as well, capacitances C_{gs} and C_{gd} are also dependent on the voltages V_{GS} and V_{GD}, and therefore they have to be modeled as such (see Figure 8.45).

Although correct nonlinear current and capacitances modeling is fundamental for large-signal operation, this is insufficient to make the model robust under different signal excitations. In particular, one of the most difficult tasks is to make the model work at high output powers for both low and high frequencies without major convergence issues. To perform a finer tuning, thermal effects and cross-capacitances may have to be taken into account.

The mathematical equations that represent the nonlinear behavior of these elements are what make a specific model unique and different from others. For the identification of such a model, we refer the readers to Chapter 5. This chapter describes in detail the extraction of the Angelov large-signal model [65–67].

8.7.2 Class-AB and class-E power amplifier design

In this section, we illustrate that the model of the bare-die device, obtained after de-embedding of the extrinsic network, can be successfully applied to power amplifier design.

We first have to introduce the concept of equivalent capacitance before we can elaborate on the actual design examples. The reason is that simplified transistor models, such as a current source (class-AB) or switch (class-E), widely used in low-frequency design methodologies, can no longer be applied at high frequencies because the device's intrinsic nonlinear capacitances, nonlinear currents, and dispersion effects play a crucial role at high frequencies [67]. Such equivalent capacitance gathers most of the device's nonlinearities and nonidealities [68–70]. Once this value is determined, the procedure to design the amplifier is straightforward, and classic class-AB or class-E operation with their unique characteristics is guaranteed at high frequencies without recourse to optimization procedures. For this purpose, the nonlinear model, constructed earlier, and a harmonic-balance simulator are employed.

First of all, the device is biased for class-AB or class-E operation and is driven such that the maximum excursion is reached for class-AB (or cutoff and saturation region for class E) by applying a high-power sinusoidal waveform at the input (see Figure 8.46). It has to be remarked that all the extrinsic parasitics have to be disabled during this simulation; otherwise, this will influence the value of the equivalent capacitance. Then shunt loads, consisting of a resistive plus an inductive part, are presented at the intrinsic lead of the device. For class AB, the initial value of the resistive part is calculated in order to reach the maximum excursion. For class-E operation, this value is chosen such that cutoff and saturation regions are reached. Next, the inductive part values are chosen such that they eliminate the reactance of the output capacitance of the device.

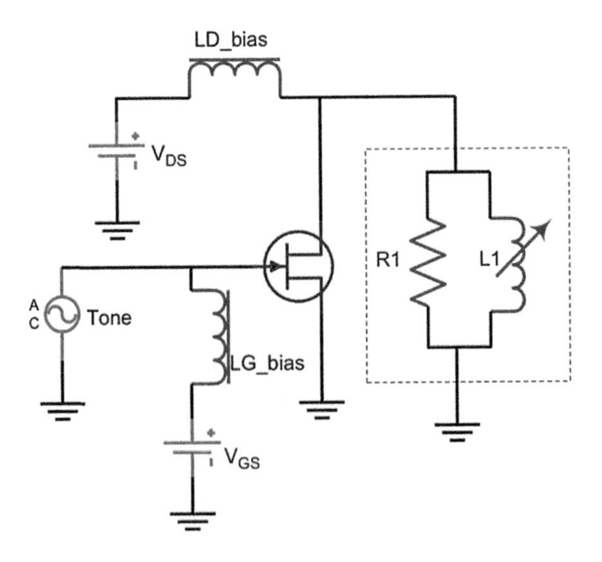

FIGURE 8.46

Test bench to determine equivalent capacitance.

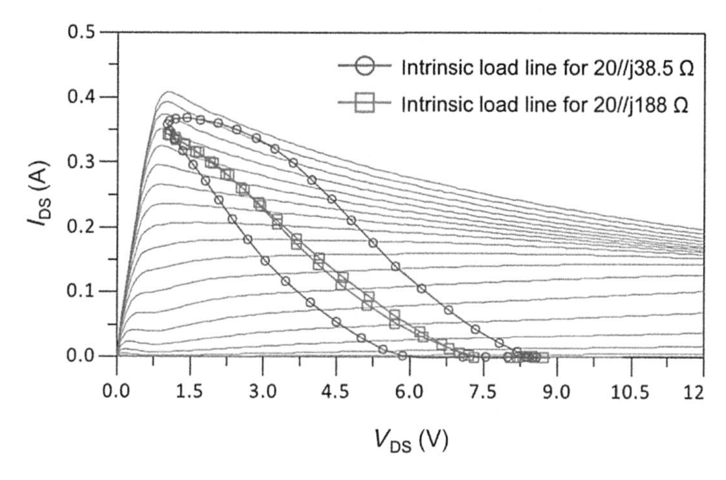

FIGURE 8.47

Load line at the intrinsic drain for class-E operation and for a starting load (circles) and the optimal load (squares).

First trial values give the load line as indicated in Figure 8.47 (line with circles), which looks like an ellipsoid. As can be observed, this value does not compensate completely the reactive part of the device. If the value of the shunt inductive part is increased, a point will be reached where the load line will no longer show hysteresis, indicating that the global nonlinear output capacitance has been compensated by a linear inductive reactance (line with squares in Figure 8.47). Notice that the looping may not be canceled completely by just adding a linear inductance. Then, the following relationship can be established:

$$X_{C_{eq}} = X_L \Rightarrow \frac{1}{\omega_o C_{eq}} = \omega_o L \tag{8.42}$$

In this way, the equivalent capacitance can be determined and used later for designing a class-AB or class-E power amplifier.

The described method is applied to the FPD750 GaAs HEMT to design a class-AB and class-E power amplifier at 3.5 GHz. The Angelov nonlinear model developed for this device will be used here.

For class-AB operation, the real part of the optimal load is calculated as [71]:

$$R_{opt-AB} = \frac{\Delta v}{\Delta i} = \frac{V_{breakdown} - V_{knee}}{I_{max}} = \frac{15 - 1\ \text{V}}{0.36\ \text{A}} = 38.8\ \Omega \tag{8.43}$$

As indicated in [71], this value has to be corrected by a factor of 0.95, giving the value of 37 Ω. Then, in order to calculate the equivalent capacitance, different inductive loads are presented until the reactive part is canceled. This process can be observed for two different loads in Figure 8.48(a), by which the device has been

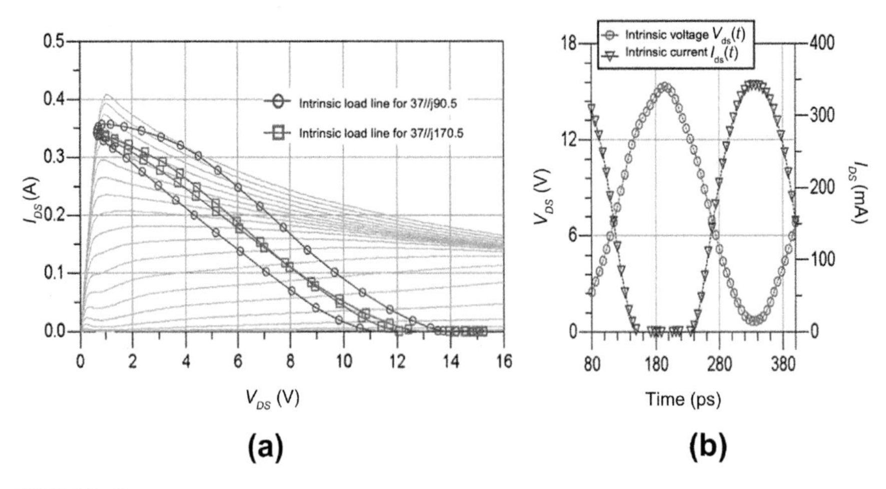

FIGURE 8.48

(a) Dynamic load line and (b) time domain voltage and current waveforms at the intrinsic drain for the class-AB amplifier.

biased for class-AB operation. The equivalent reactance turned out to be $-170\text{j}\,\Omega$. As a result, the optimal load can be defined as $37\,\Omega/170\text{j}\,\Omega$, which converted in series impedance, becomes $36+8\text{j}\,\Omega$. Providing this load at the fundamental frequency plus short circuits at the harmonics, the typical class-AB time domain waveform (intrinsic sinusoidal drain voltage and truncated sinusoidal drain-source current) can be observed as shown in Figure 8.48(b).

For class-E amplifier design, a similar test is performed as described above with the device biased for class-E operation, obtaining $X_{C_{\text{eq}}} = 188\,\Omega$ at 3.5 GHz, and therefore $C_{\text{eq}} = 0.24$ pF. A graph showing this procedure is given in Figure 8.47. Based on this value, the optimal load for class-E operation can be determined using the following [72,73]:

$$Z_{\text{opt}} = \frac{0.28}{\omega_0 C_{\text{out}}} e^{\text{j}49} = 0.28\left(X_{C_{\text{eq}}} @ \omega_0\right) e^{\text{j}49} = 34.7 + \text{j}40\,\Omega \tag{8.44}$$

Providing this optimal load at the intrinsic drain (i.e., without extrinsic parasitics), the time domain voltage and current waveforms shown in Figure 8.49(a) are obtained. Unlike in class-AB operation, in class E, the harmonics are kept as open circuits as stated in the original work [72,73].

As we can see, the ideal class-E waveforms are not completely realized. This is due to the fact that the exciting source generates a sinusoidal signal, whose rise time is not short enough to provide the right excitation for a class-E waveform. This can be improved by increasing the power, which therefore shortens the rise time, but at the expense of degrading the power-added efficiency (PAE). If the excitation signal is changed from sinusoidal to a trapezoidal input signal, better class-E waveforms

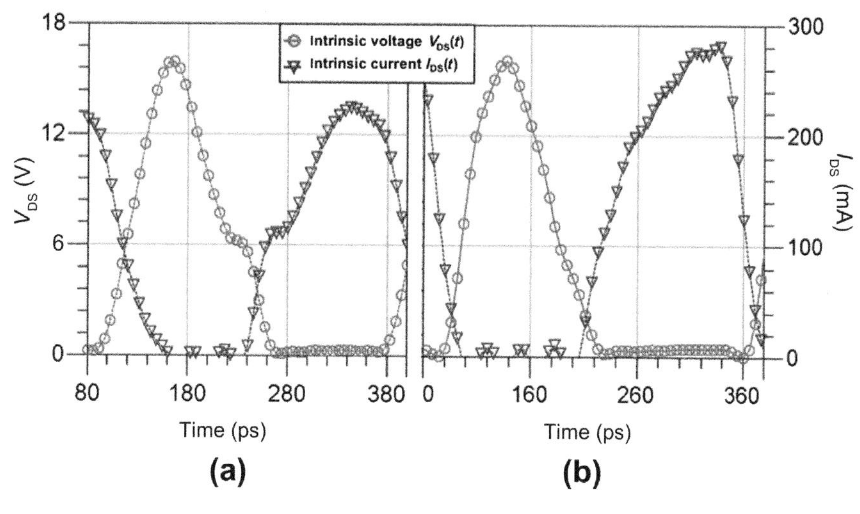

FIGURE 8.49

Time-domain voltage and current waveforms at the intrinsic drain for the class-E amplifier using (a) a sinusoidal excitation signal and (b) a trapezoidal excitation.

can be obtained. A trapezoidal excitation signal is used with rise and fall time of 14 ps (theoretical class E assumes transition times equal to zero). As it can be observed in Figure 8.49(b), this resembles more the ideal class-E waveforms. The major difference is due to the fall-time effect in the current waveform and the on-resistance in the drain voltage.

The following step is to implement this load with distributed elements. To realize this, matching networks providing the loads for class-AB and class-E operation, indicated in Table 8.4, are designed.

Two harmonics are sufficient to guarantee class-AB and class-E waveforms. Higher harmonics are inherently short-circuited due to the intrinsic/extrinsic capacitances. It has to be indicated that in the process of designing the matching networks, the effect of all the extrinsic parasitics have to be absorbed in the matching network. For these particular designs, an additional stabilization network was defined in order to make the amplifiers unconditionally stable over a broad frequency range.

Table 8.4 Class-AB and class-E Design load values

Operation Mode	Load at f_0	Load at $2f_0$	Load at $3f_0$
Class-AB	$36+j8\,\Omega$	Zero (short-circuit)	Zero (short-circuit)
Class-E	$34.7+j40\,\Omega$	Infinite (open-circuit)	Infinite (open-circuit)

The stabilizations of the device and amplifier are realized using series and shunt gate resistors. Unfortunately, these networks introduce losses at the input that degrade the PAE. The input matching is then determined, which is basically the complex conjugate of the load seen at the gate lead. It is very important also to include in this calculation the effect of the stabilization network.

The complete circuit, including all the parasitics and matching/biasing/stabilizing networks, is initially simulated using the Agilent design system. The wave-

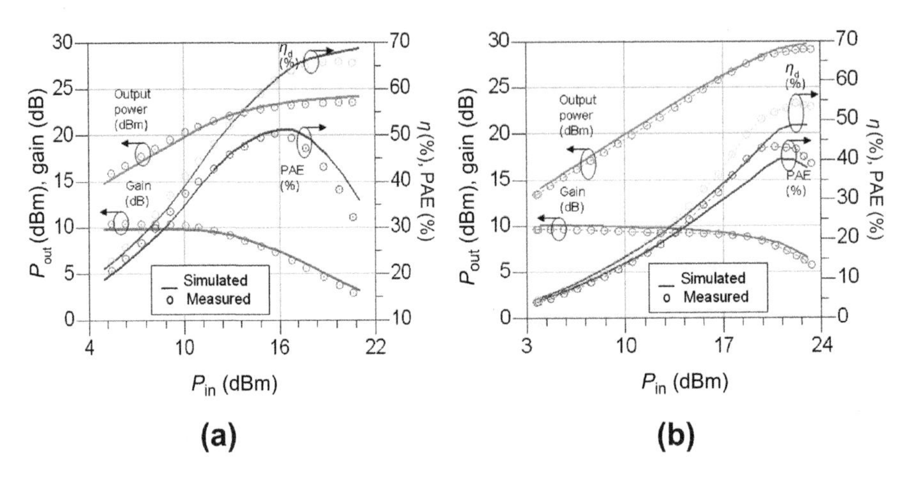

FIGURE 8.50

Comparison of measured and simulated (a) class-E ($V_{DS} = 5.15$ V, $V_{GS} = -0.8$ V) and (b) class-AB ($V_{DS} = 8.3$ V, $V_{GS} = -0.8$ V) amplifier performance figures.

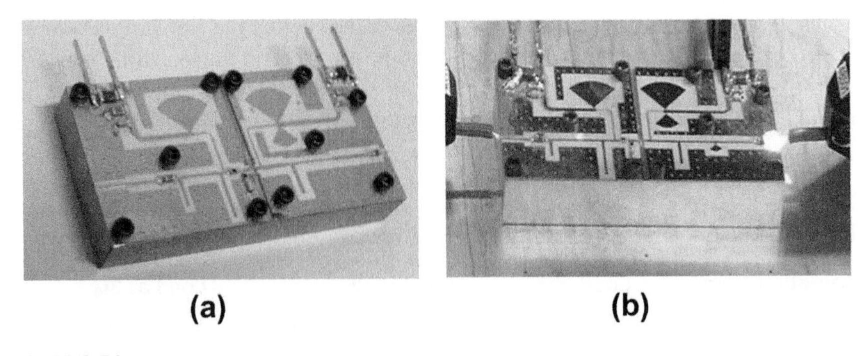

FIGURE 8.51

(a) 3.5 GHz class-AB power amplifier and (b) class-E power amplifier during LSNA testing.

forms are verified and are very much like the ones calculated at the intrinsic drain, which are given in Figure 8.48(b) and Figure 8.49. The performance figures, based on this simulation, are presented in Figure 8.50(a) and (b), for class E and class AB, respectively. On the other hand, the actual implementation of the complete amplifier including biasing networks and input/output matching can be observed in Figure 8.51(a) for class AB and in Figure 8.51(b) for class E during large signal network analyzer (LSNA) measurements. The measurement results, which can also be observed in Figure 8.50(a) and (b), present good agreement in output power, gain, drain efficiency, and PAE with simulation results for class E. As expected from theory, the drain efficiency of the class-E amplifier (65.2%) outperforms the drain efficiency of the class-AB amplifier (42%) at the same input power (17.5 dBm). These results act as a demonstrator for the large-signal model developed and thus also for the accuracy of the de-embedding step.

References

[1] Cidronali A, Collodi G, Vannini G, Santarelli A. Small-signal distributed FET model consistent with device scaling. Electron Lett March 1999;35(5):371–2.

[2] Cidronali A, Collodi G, Vannini G, Santarelli A, Manes G. A new approach to FET model scaling and MMIC design based on electromagnetic analysis. IEEE Trans Microwave Theory Tech June 1999;47(6):900–7.

[3] Cidronali A, Collodi G, Santarelli A, Vannini G, Manes G. Millimeter-wave FET modeling using on-wafer measurements and EM simulation. IEEE Trans Microwave Theory Tech February 2002;50(2):425–32.

[4] Resca D, Santarelli A, Raffo A, Cignani R, Vannini G, Filicori F, et al. Scalable nonlinear FET model based on a distributed parasitic network description. IEEE Trans Microwave Theory Tech April 2008;56(4):755–66.

[5] Resca D, Lonac JA, Cignani R, Raffo A, Santarelli A, Vannini G, et al. Accurate EM-based modeling of cascode FETs. IEEE Trans Microwave Theory Tech April 2010;58(4):719–29.

[6] Resca D, Raffo A, Santarelli A, Vannini G, Filicori F. Scalable equivalent circuit FET model for MMIC design identified through FW-EM-analyses. IEEE Trans Microwave Theory Tech February 2009;57(2):245–53.

[7] Sonnet em user's manual. (North Syracuse, NY, USA): Sonnet software.

[8] Momentum (Santa Clara, CA, USA): Agilent EEsof EDA.

[9] Rautio JC. Shortening the design cycle. IEEE Microwave Mag December 2008;9(6):86–96.

[10] Imtiaz SMS, El-Ghazaly SM. Global modeling of millimeter-wave circuits: electromagnetic simulation of amplifiers. IEEE Trans Microwave Theory Tech December 1997;45(12):2208–16.

[11] Cetiner BA, Coccioli R, Housmand B, De Flaviis F, Itoh T. Global modeling approach for prematched multifinger FET. Microwave Opt Technol Lett February 2002;32(3):174–8.

[12] Heinrich W. On the limits of FET modeling by lumped elements. Electron Lett June 1986;22(12):630–2.

[13] Ghione G, Naldi CU. Modelling and simulation of wave propagation effects in MESFET devices based on physical models. In: Solid State Device Res Conf (ESS-DERC) (Bologna, Italy), September 1987. pp. 317–20.
[14] Xiong A, Charbonniaud C, Gatard E, Dellier S. A scalable and distributed electro-thermal model of AlGaN/GaN HEMT dedicated to multi-fingers transistors. In: IEEE Comp Semiconductor Integ Circuit Symp (CSICS) (Costa Mesa, CA, USA), October 2010. pp. 1–4.
[15] Grondin RO, El-Ghazaly S, Goodnick S. A review of global modeling of charge transport in semiconductors and full-wave electromagnetic. IEEE Trans Microwave Theory Tech June 1999;47(6):817–29.
[16] Denis D, Snowden CM, Hunter IC. Coupled electrothermal, electromagnetic, and physical modeling of microwave power FETs. IEEE Trans Microwave Theory Tech June 2006;54(6):2465–70.
[17] Laloue A, David JB, Quere R, Mallet-Guy B, Laporte E, Villemazet JF, et al. Extrapolation of a measurement-based millimeter-wave nonlinear model of pHEMT to arbitrary-shaped transistors through electromagnetic simulations. IEEE Trans Microwave Theory Tech June 1999;47(6):908–14.
[18] Filicori F, Vannini G, Monaco VA. A nonlinear integral model of electron devices for HB circuit analysis. IEEE Trans Microwave Theory Tech July 1992;40(7): 1456–65.
[19] Filicori F, Santarelli A, Traverso PA, Raffo A, Vannini G, Pagani M. Nonlinear RF device modelling in the presence of low-frequency dispersive phenomena. Int J RF Microwave Comput-Aided Eng January 2006;16(1):81–94.
[20] Mirri D, Iuculano G, Filicori F, Pasini G, Vannini G, Pellegrini G. A modified Volterra series approach for nonlinear dynamic system modeling. IEEE Trans Circuits Syst I Fundam Theory Appl August 2002;49(8):1118–28.
[21] Ooi BL, Zhong Z, Wang Y, Shan XC, Lu A. A distributed millimeter-wave small-signal HBT model based on electromagnetic simulation. IEEE Trans Veh Technol September 2008;57(5):2667–74.
[22] Ooi BL, Zhou T, Kooi PS, Lin FJ, Hu SC. A distributed small-signal HBT model for millimeter-wave applications. In: 3rd Int Conf Digital Object Identifier (ICMMT) (Beijing, China), August 2002. pp. 318–21.
[23] Rudolph M, Fager C, Root DE, editors. Nonlinear transistor model parameter extraction techniques (Cambridge, UK): Cambridge University Press; 2011.
[24] Santarelli A, Filicori F, Vannini G, Rinaldi P. Backgating' model including self-heating for low-frequency dispersive effects in III-V FETs. Electron Lett October 1998;34(20):1974–6.
[25] Ladbrooke PH, Turner J. MMIC design GaAs FETs and HEMTs (Boston, MA, USA): Artech House; 1989.
[26] Gonzalez G. Microwave transistor amplifiers: analysis and design. 2nd ed. (Upper Saddle River, NJ, USA): Prentice Hall; 1996.
[27] Golio JM. Microwave MESFETs and HEMTs (Boston, MA, USA): Artech House; 1991.
[28] Resca D, Santarelli A, Raffo A, Cignani R, Vannini G, Filicori F. Scalable equivalent circuit PHEMT modelling using an EM-based parasitic network description. In: Eur Microwave Integr Circuits Conf (Munich, Germany), October 2007. pp. 60–3.

[29] Chen SW, Aina O, Li W, Phelps L, Lee T. An accurately scaled small-signal model for interdigitated power P-HEMT up to 50 GHz. IEEE Trans Microwave Theory Tech May 1997;45(5):700–3.

[30] Cojocaru VI, Brazil TJ. A scalable general-purpose model for microwave FETs including DC/AC dispersion effects. IEEE Trans Microwave Theory Tech December 1997;45(12):2248–55.

[31] Wood J, Root DE. Bias-dependent linear scalable millimeter-wave FET model. IEEE Trans Microwave Theory Tech December 2000;48(12):1381–4.

[32] Jarndal A, Kompa G. A new small signal model parameter extraction method applied to GaN devices. In: IEEE MTT-s Int Microwave Symp (Long Beach, CA, USA), June 2005. pp. 1423–6.

[33] Kuvas RL. Equivalent circuit model of FET including distributed gate effects. IEEE Trans Electron Devices June 1980;27(6):1193–5.

[34] LaRue R, Yuen C, Zdasiuk G. Distributed GaAs FET circuit model for broadband and millimeter wave applications. In: IEEE MTT-s Int Microwave Symp (San Francisco, CA, USA), May 1984. pp. 164–6.

[35] Heinrich W. Distributed equivalent-circuit model for travelling-wave FET design. IEEE Trans Microwave Theory Tech May 1987;35(5):487–91.

[36] Martin-Guerrero TM, Camacho-Penalosa C. Nonlinearities in a MESFET distributed model. Int J Microwave Millimeter-Wave Comput Aided Eng July 1996;6(4):243–8.

[37] Martin-Guerrero TM, Camacho-Peñalosa C. Simulation of the small-signal performance of a HEMT using a distributed model. In: IEEE Mediterranean Electrotech Conf (MELECON) (Bari, Italy), May 1996. pp. 567–70.

[38] Abdipour A, Pacaud A. Complete sliced model of microwave FET's and comparison with lumped model and experimental results. IEEE Trans Microwave Theory Tech January 1996;44(1):4–9.

[39] Nash SJ, Platzker A, Struble W. Distributed small signal model for multi-fingered GaAs PHEMT/MESFET devices. In: IEEE MTT-s Int Microwave Symp (San Francisco, CA, USA), June 1996. pp. 1075–8.

[40] Masuda S, Hirose T, Watanabe Y. An accurate distributed small signal FET model for millimeter-wave applications. In: IEEE MTT-s Int Microwave Symp (Anaheim, CA, USA), June 1999. pp. 157–60.

[41] Cetiner B, Coccioli R, Housmand B, Itoh T. Combination of circuit and full wave analysis for pre-matched multifinger FET. In: 30th Eur Microwave Conf (Paris, France), October 2000. pp. 1–4.

[42] Resca D, Raffo A, Santarelli A, Vannini G, Filicori F. Extraction of an extrinsic parasitic network for accurate mm-wave FET scalable modeling on the basis of full-wave EM simulation. In: IEEE MTT-s Int Microwave Symp (Atlanta, GA, USA), June 2008. pp. 1405–8.

[43] Edwards ML, Sinsky JH. A new criterion for linear 2-port stability using geometrically derived parameters. IEEE Trans Microwave Theory Tech December 1992;40(12):2303–11.

[44] McMacken J, Nedeljkovic S, Gering J. HBT modeling for cellular handset applications. In: Wireless and Microwave Tech Conf (WAMICON) (Clearwater, FL, USA), September 2009. pp. 1–5.

[45] Choi W, Jung G, Kim J, Kwon Y. Scalable small-signal modeling of RF CMOS FET based on 3-D EM-based extraction of parasitic effects and its application to

millimeter-wave amplifier design. IEEE Trans Microwave Theory Tech December 2009;57(12):3345–53.

[46] Mizutani H, Funabashi M, Kuzuhara M, Takayama Y. Compact DC-60GHz HJFET MMIC switches using ohmic electrode-sharing technology. IEEE Trans Microwave Theory Tech November 1998;46(11):1597–603.

[47] Collodi G, Cidronali A, Toccafondi C, Santarelli A, Vannini G. Global modeling approach to the design of an MMIC amplifier using ohmic electrode-sharing technology. In: Gallium Arsenide Appl Symp (GAAS) (Munich, Germany), October 2003. pp. 529–32.

[48] Martin A, Reveyrand T, Campovecchio M, Aubry R, Piotrowicz S, Floriot D, et al. Design method of balanced AlGaN/GaN HEMT cascode cells for wideband distributed power amplifiers. Proc Eur Microwave Assoc December 2008;4(12):261–7.

[49] Dambrine G, Cappy A, Heliodore F, Playez E. A new method for determining the FET small-signal equivalent circuit. IEEE Trans Microwave Theory Tech July 1998;36(7):1151–9.

[50] Giannini F, Leuzzi G. Nonlinear microwave circuit design (Chichester, UK): Wiley; June 2004.

[51] Wood J, Root DE. Bias-dependent linear scalable millimeter-wave FET model. IEEE Trans Microwave Theory Tech December 2000;48(12):2352–60.

[52] Lee K, Shur M, Lee K, Vu T, Roberts P, Helix M. A new interpretation of "End" resistance measurements. IEEE Electron Device Lett January 1984;5(1):5–7.

[53] Debie P, Martens L. Fast and accurate extraction of parasitic resistances for nonlinear GaAs MESFET device models. IEEE Trans Electron Devices December 1995; 42(12):2239–42.

[54] Hastings A. The art of analog layout (Upper Saddle River, NJ, USA): Prentice Hall; 2001.

[55] Liang Q, Cressler J, Niu G, Lu Y, Freeman G, Ahlgren D, et al. A simple four-port parasitic deembedding methodology for high-frequency scattering parameter and noise characterization of SiGe HBTs. IEEE Trans Microwave Theory Tech November 2003;51(11):2165–74.

[56] Bousnina S, Falt C, Mandeville P, Kouki A, Ghannouchi F. An accurate on-wafer deembedding technique with application to HBT devices characterization. IEEE Trans Microwave Theory Tech February 2002;50(2):420–4.

[57] Mahmoudi R, Tauritz J. A five-port deembedding method for floating two-port networks. IEEE Trans Instrum Meas April 1998;47(2):482–8.

[58] Rautio J. Deembedding the effect of a local ground plane in electromagnetic analysis. IEEE Trans Microwave Theory Tech February 2005;53(2):770–6.

[59] Rautio J, Okhmatovski V. Unification of double-delay and SOC electro-magnetic deembedding. IEEE Trans Microwave Theory Tech September 2005;53(9):2892–8.

[60] Yarleque Medina M, Schreurs D, Nauwelaers B. Medium-power RF FET intrinsic parameter extraction based on four-port extrinsic model. In: URSI Benelux Meeting (Eindhoven, The Netherlands), May 2006.

[61] Yarleque Medina M, Schreurs D, Nauwelaers B. Four-port deembedding technique for FET devices mounted in hybrid test fixture. In: Eur Microwave Integr Circuits Conf (Manchester, UK), September 2006. pp. 464–7.

[62] Price KV, Storn RM, Lampinen JA. Differential evolution: a practical approach to global optimization. 1st ed. Natural Computing Series: Springer; December 2005.

[63] Berroth M, Bosch R. Broad-band determination of the FET small-signal equivalent circuit. IEEE Trans Microwave Theory Tech July 1990;38(7):891–5.

[64] Berroth M, Bosch R. High-frequency equivalent circuit of GaAs FETs for large-signal applications. IEEE Trans Microwave Theory Tech February 1991;39(2):224–9.

[65] Angelov I, Zirath H, Rorsman N. A new empirical nonlinear model for HEMT-devices. IEEE MTT-s Int Microwave Symp (Albuquerque, NM, USA), June 1992;3:1583–6.

[66] Angelov I, Bengtsson L, Garcia M. Extensions of the Chalmers nonlinear HEMT and MESFET model. IEEE Trans Microwave Theory Tech October 1996;44(10):1664–74.

[67] Grabinski W, Nauwelaers B, Schreurs D, editors. Transistor level modeling for analog/RF IC design. 1st ed. (Heidelberg, Berlin, Germany): Springer; May 2006.

[68] Yarlequé Medina M. RF power amplifiers for wireless communications [PhD thesis]. Supervised by Nauwelaers B, Schreurs D; 24 June, 2008.

[69] Yarlequé Medina M, Schreurs D, Nauwelaers B. WiMAX class AB and class E power amplifier design using equivalent capacitance concept. Int J RF Microwave Comput-Aided Eng November 2008;18(6):543–51.

[70] Yarlequé Medina M, Schreurs D, Nauwelaers B. RF class-E power amplifier design based on a load line-equivalent capacitance method. IEEE Microwave Wireless Compon Lett March 2008;18(3):206–8.

[71] Cripps SC. RF power amplifiers for wireless communications (Norwood, MA, USA): Artech House; 2006.

[72] Mader T, Popovic Z. The transmission-line high-efficiency class-E amplifier. IEEE Microwave Guided Wave Lett September 1995;5(9):290–2.

[73] Mader T, Bryerton E, Markovic M, Forman M, Popovic Z. Switched-mode high-efficiency microwave power amplifiers in a free-space power-combiner array. IEEE Trans Microwave Theory Tech October 1998;46(10):1391–8.

CHAPTER

9

Nonlinear Embedding and De-embedding: Theory and Applications

Antonio Raffo, Valeria Vadalà, Giorgio Vannini

Dipartimento di Ingegneria, University of Ferrara, Ferrara, Italy

9.1 Introduction

This chapter introduces two new design techniques for power amplifiers (PAs), which are based on both measurement and modeling concepts. The idea is very intuitive: to measure what is simple to measure and to model what is simple to model.

From a measurement point of view, the meaning of the word "simple" is twofold: it can refer to the measurement setup or to the uncertainty evaluation. High-frequency and high-power operation requires complex setups and, as a natural consequence, complex procedures to evaluate the accuracy of the measurement. However, time-domain nonlinear measurement setups are limited to 67 GHz [1−4]; thus, when the spectral components of the electrical quantities contain harmonic contributions at higher frequencies, these measurement systems are not able to provide a complete characterization, despite their high costs. Moreover, although the calibration procedure of these setups is traceable, a verification device does not exist for quantifying the quality of the calibration [5,6]. Measurement uncertainty evaluation under nonlinear dynamic operation is a new topic, in which the microwave community is in the earliest stages [7−9].

On the other hand, from a modeling point of view, "simple" indicates a model formulation that can be identified and validated, possibly under actual device operation, with reasonable confidence, by exploiting standard and reliable procedures. It is clear that the two facets are inherently related: measurement uncertainty induces model parameter uncertainty in the identification phase and questionable model validation. The study of how measurement uncertainty influences model parameters represents another minefield for microwave device modelers [10].

Starting from these assumptions, at microwave and millimeter-wave frequencies, a simple, well-known, and commonly adopted characterization technique is able to assess device performance to the upper limit of the millimeter band: *S* parameter measurements [11−14]. The design procedures described in this chapter (at least in their original formulations) use *S* parameters to accurately assess a device's high-frequency behavior. In fact, bias- and frequency-dependent *S* parameter

Microwave De-embedding. http://dx.doi.org/10.1016/B978-0-12-401700-9.00009-4

measurements are able to capture the strictly nonlinear dynamic behavior of the transistor that is, the nonlinear dynamic contribution related to intrinsic reactive elements.

In addition, the methodologies described in this chapter may be extended to the design of mixers and oscillators, thanks to the generality of their theoretical formulation.

9.2 Waveform engineering at the current-generator plane

The concept of waveform engineering was introduced by Tasker [15–17] as "the ability to modify in a quantified manner the time-varying voltage and current present at the terminals of the device under test (DUT)." The definition is obviously correct; nevertheless, it contains some nontrivial facets to be analyzed.

At microwave frequencies, trying to find the optimum working condition by observing the waveforms of electrical quantities at DUT terminals is like searching for a small object in a large, dark room. The set of operating conditions in which the optimum has to be found is ideally infinite, and no information is given on how the waveforms have to be modified in order to reach the optimum operation. A theory that explains how to modify waveforms at the device terminals to guarantee the DUT's optimum performance does not exist, and it probably never will, no matter what the selected class of operation is. Different design methodologies have been proposed in recent decades to define the waveform shapes that guarantee optimum device performance: from class A (e.g., Cripp's load line theory) to switching classes (e.g., Refs [18–21]). Nevertheless, all of these theoretical analyses do not refer to the device terminals; they define the waveform shapes at the transistor current-generator plane (CGP), which are inaccessible due to the presence of linear and nonlinear reactive parasitic elements, especially at microwave frequencies. As the frequency increases, the problem becomes more and more challenging. In this chapter, we try to shed some light on microwave frequencies.

9.2.1 Modeling hints

PA design techniques define the optimum device operation at the transistor's current-generator plane. To clarify this fundamental aspect, Figure 9.1 shows the generic nonlinear equivalent circuit representation of a microwave transistor. In this chapter, we focus on field-effect transistor (FET) electron devices (EDs) because they are most commonly exploited in microwave and millimeter-wave PA design; nevertheless, the following considerations are also valid for bipolar transistors.

The linear extrinsic parasitic network in Figure 9.1 describes the passive structure in the device's active area, accounting for metallization and dielectric losses as well as associated inductive and capacitive effects. From a modeling perspective, parasitic elements can be characterized by conventional lumped descriptions [23–25], which can be identified by using small-signal measurements, or, alternatively, by adopting electromagnetic simulations of the device layout [26–28].

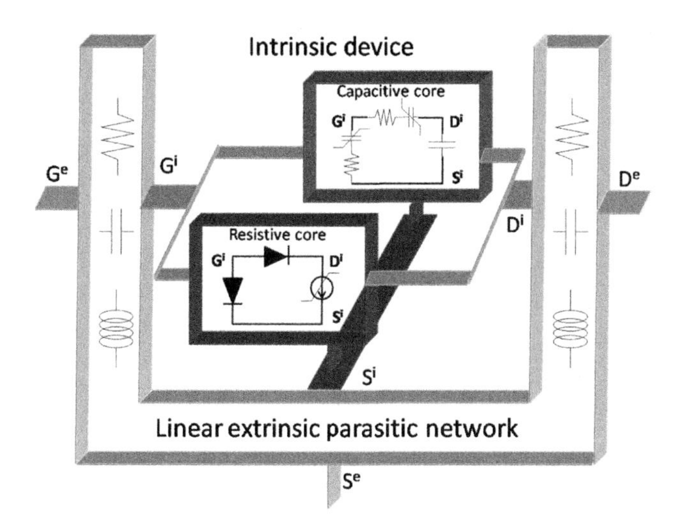

FIGURE 9.1

Nonlinear equivalent circuit of a FET. G, gate; D, drain; S, source; e, extrinsic; and i, intrinsic.

From [22]

As clearly shown in Figure 9.1, the intrinsic device is divided into two parts, which can be considered in parallel: a capacitive core, which describes the nonlinear dynamic phenomena, and a resistive core, which accounts for the direct current (DC) and low-frequency (LF) current–voltage (I/V) device characteristics. The latter differs from the DC response because of surface state densities, deep-level traps, and thermal phenomena [22,29–33]. A dynamic term should also be considered for the LF device behavior, when harmonic components at very low frequency (i.e., spectral components below the megahertz range) are present. Nevertheless, design techniques commonly used at microwave frequencies do not take into account such dynamics, assuming that all the spectral components (apart from the DC component) lie above the cutoff frequency of dispersive phenomena (i.e., hundreds of kilohertz). For this reason, the LF dynamic term will be ignored in our discussion.

The modeling of an EDs resistive core is extremely complex. A number of important nonlinear phenomena must be considered (e.g., Schottky junction forward conduction and breakdown, knee of I/V characteristics). In addition, the presence of dispersive effects [29–33] requires nonlinear measurements in the model identification phase to guarantee accurate predictions.

Figure 9.2 schematically shows the deviations existing between static and dynamic device output characteristics, measured at operating frequencies where the nonlinear dynamic effects associated with charge storage variations can be totally neglected. In this figure, the knee walkout (KW) of the dynamic characteristics, the negative slope (NS) of the DC characteristics in the saturation region, and

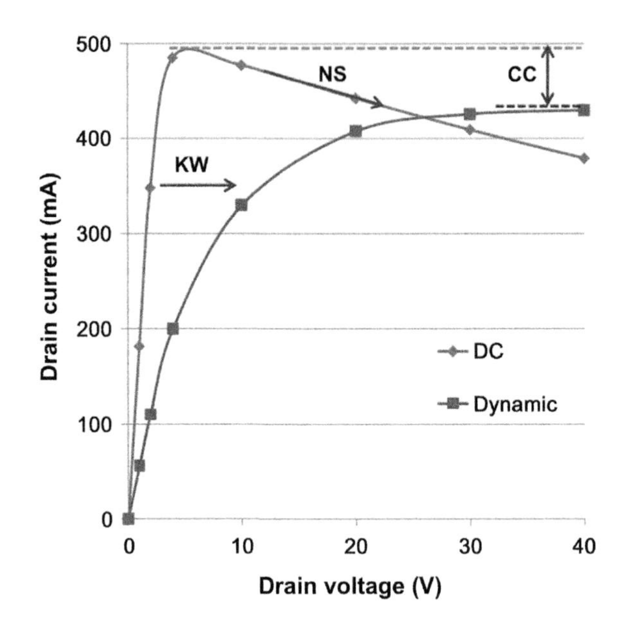

FIGURE 9.2

Schematic representation of the differences between DC and dynamic device characteristics.

From [32]

the saturation-current collapse (CC) are shown (the current collapse also can be defined with respect to a pulsed characteristic measured by exploiting a bias condition, which guarantees both no power dissipation and channel formation (e.g., $V_{g0} = 0$ V and $V_{d0} = 0$ V)). All of these effects globally contribute to decrease device performance under dynamic operation and, as a consequence, have to be correctly accounted for.

The KW can be accurately predicted by modeling the different slopes shown by the dynamic characteristics in the linear region. It has been largely demonstrated (e.g., [34–36]) that the KW monotonically increases by increasing the drain bias voltage V_{d0}. In Figure 9.3, different measurement sets, carried out on a $0.7 \times 800\ \mu m^2$ gallium nitride (GaN) high-electron-mobility transistor (HEMT) at 2 MHz, demonstrate this phenomenon. In particular, measurements were performed with different drain voltage quiescent conditions by setting the amplitude of the input incident signals in order to dynamically reach the value $v_g = 0$ V. It is evident how the knee region is subject to a shift as the average value of the drain voltage increases.

As is well known, when dynamic operation is considered, the dynamic I/V characteristics do not show any NS in the saturation region. Nevertheless, from a modeling perspective, this is not the major problem. Indeed, the complexity derives from the fact that dynamic characteristics measured under the same thermal state

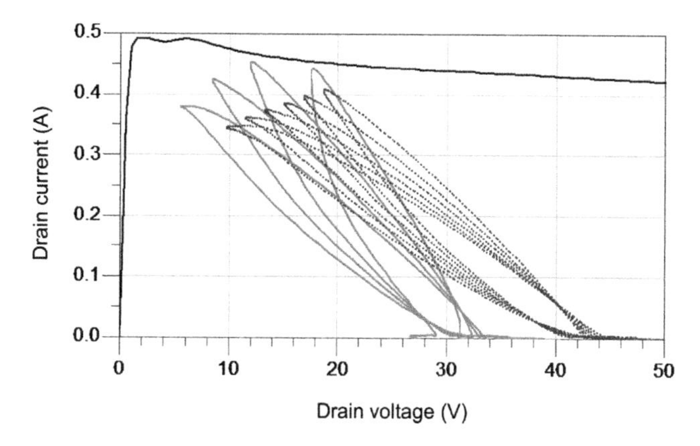

FIGURE 9.3

Measurements performed on a 0.7 × 800-μm² GaN HEMT at 2 MHz by exploiting the large-signal measurement system [36], bias condition $V_{g0} = -3$ V and $V_{d0} = 25$ V (continuous lines) or $V_{d0} = 35$ V (dotted lines). Measurements are superimposed on the DC characteristic at $V_{g0} = 0$ V. Different measurement sets are obtained by varying the amplitude of the output incident signal phasors.

From [36]

(e.g., under no power dissipation), but starting from different bias conditions, show different slopes in the saturation region [34,35]. In other words, dynamic characteristics show a positive slope in the saturation region, but the slope depends on the bias and thermal device state. The CC essentially arises from the presence of surface trap states [34,35]. Such a phenomenon is strongly dependent on both the bias condition and the device's thermal state.

The LF measurements shown in Figure 9.4, carried out on a 0.7 × 800 μm² GaN HEMT at 2 MHz, highlight the CC dependence on the average value of the gate voltage V_{g0}. The crosses in the enlarged view within the inset correspond to the same values of the instantaneous voltages ($v_g = 0$ V, $v_d = 5.4$ V); if LF dispersion were not present, the same instantaneous current value (corresponding to the DC one) should be measured. Instead, it is well evident that, as the quiescent gate voltage V_{g0} moves into the off-state region, the instantaneous drain current goes down. Finally, it is worth noting that traps and thermal effects cannot be separately dealt with because the device thermal state influences the trapping state [37–39] and time constants can be of the same order of magnitude.

Following the above considerations, the gate and drain currents of the resistive core can be expressed as follows:

$$\begin{aligned} i_g(t) &= h\left(\underline{\mathbf{v}}(t), P_0, \theta_{\text{case}}\right) \\ i_d(t) &= f\left(\underline{\mathbf{v}}(t), \underline{\mathbf{V}}_0, P_0, \theta_{\text{case}}\right) \end{aligned} \tag{9.1}$$

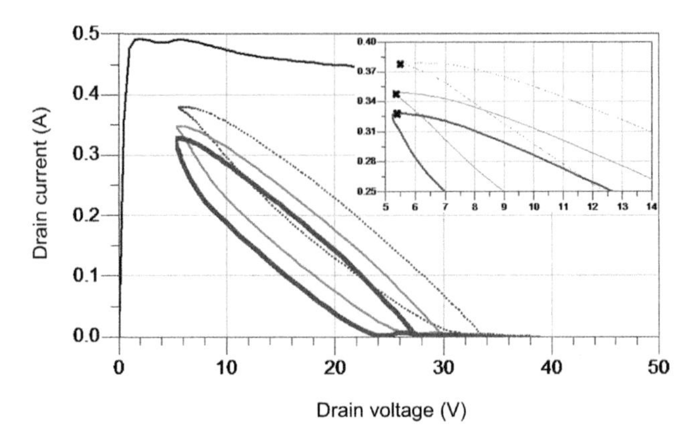

FIGURE 9.4

Measurements performed on a 0.7 × 800-μm^2 GaN HEMT at 2 MHz by exploiting the large-signal measurement system [36], bias condition $V_{d0} = 25$ V and $V_{g0} = -3$ V (dotted line), $V_{g0} = -4$ V (continuous thin line), or $V_{g0} = -5$ V (continuous thick line). The measured load lines are superimposed on the DC characteristics at $V_{g0} = 0$ V. The crosses in the enlarged view within the inset correspond to the same instantaneous voltage pair ($v_g = 0$ V, $v_d = 5.4$ V).

From [36]

In Eqn (9.1), h and f are two algebraic (i.e., memoryless) functions, $\underline{\mathbf{v}}$ is the vector of the intrinsic voltages, $\underline{\mathbf{V}}_0$ is its average value, P_0 is the average dissipated power, and θ_{case} is the device's case temperature. The dependence on $\underline{\mathbf{V}}_0$ accounts for the influence of traps and surface state densities [29–33], whereas P_0 and θ_{case} determine, through the thermal resistance, the device's I/V characteristic dependence on the junction temperature.

Identification of Eqn (9.1), particularly the drain current equation, is quite a prohibitive task. It requires the introduction of suitable approximations to make the problem affordable. A number of LF modeling approaches have been proposed in the literature [29–33], based on lookup tables or analytical expressions. Some of them introduce assumptions that enable the models to be identified on the basis of bias-dependent DC and alternating current (AC) small-signal differential measurements carried out above the cutoff of LF dispersion. However, in practice, model accuracy is commonly improved by exploiting in the identification phase, such as AC and DC measurements and large-signal dynamic measurements, including pulsed I/V characteristics [40–42]. Despite the use of quite expensive, special-purpose pulsed I/V setups, the identification of an accurate, *global* model for the resistive core still remains a very complex and hard task. This explains why foundry models often properly work in a limited number of given quiescent bias conditions; typically, a limited set of pulsed I/V measurements is fitted. Such an approach inevitably leads to *local* models, which cannot provide accurate information outside the range of the few quiescent bias conditions considered.

As far as the capacitive core is concerned, dispersive phenomena due to traps and thermal behavior represent second-order effects (whose evidence has rarely been dealt with in the literature [43,44]), which are regularly neglected in ED models oriented to PA design [45–54]. A π model of capacitors (usually assuming C_{gs} and C_{gd} to be nonlinearly dependent on the intrinsic device voltages and C_{ds} constant) is often adopted. Better prediction capabilities can be obtained at higher frequencies by introducing gate-source and gate-drain resistor–capacitor (RC) series, as shown in Figure 9.1, to describe non-quasi-static effects, which accounts for a finite device memory time [44–47]. The capacitive core can be equivalently and correctly described by adopting nonlinear charge sources (e.g., [48]) instead of nonlinear capacitors. Nevertheless, the procedures described in the present chapter can be applied for any approach adopted for the capacitive-core description.

Provided a careful de-embedding of the parasitic network is carried out, small-signal, bias/frequency-dependent S parameter measurements are usually sufficient to accurately determine the voltage-dependent capacitive-core parameters. These parameters can be fitted by means of suitable analytical expressions (e.g., [47]) or directly stored into lookup tables [48–51] to build a nonlinear dynamic model. Even problems related to charge conservation, although dealt with in the literature, do not seem to represent a major problem once suitable expedients are adopted [55–58].

To summarize, mostly due to LF dispersion phenomena and important nonlinear effects, the accurate modeling of the resistive core is the most complex issue in nonlinear ED modeling; the linear parasitic elements and the capacitive core can be more easily identified. This assumption is the basis of the design approaches described in the following sections.

9.2.2 Measurement hints

In Chapter 6, the complexity and the limitations (in terms of power and bandwidth capability) related to nonlinear measurement setups operating at microwave frequencies have been extensively addressed. Nevertheless, as explained in the previous paragraphs, PA design techniques require the characterization of the resistive core behavior, which is, by definition, frequency independent. So, if this assumption is true, a natural question comes to mind: is it really necessary to characterize the resistive core at microwave frequencies? The answer is *no*. In fact, the resistive core characterization can be conveniently carried out under LF operation. In particular, the frequency of operation has to be sufficiently low for neglecting the (linear and nonlinear) dynamic contributions, but sufficiently high for operating above the cutoff of LF dispersion (i.e., some hundreds of kilohertz). Therefore, a time-domain nonlinear measurement setup operating at low frequency represents the optimum solution for the current generator behavior characterization.

A possible architecture for such a measurement system is shown in Figure 9.5. In particular, the function generator has two 50-Ω channels that independently provide arbitrary waveforms in the frequency range of 1 mHz–120 MHz. To overcome

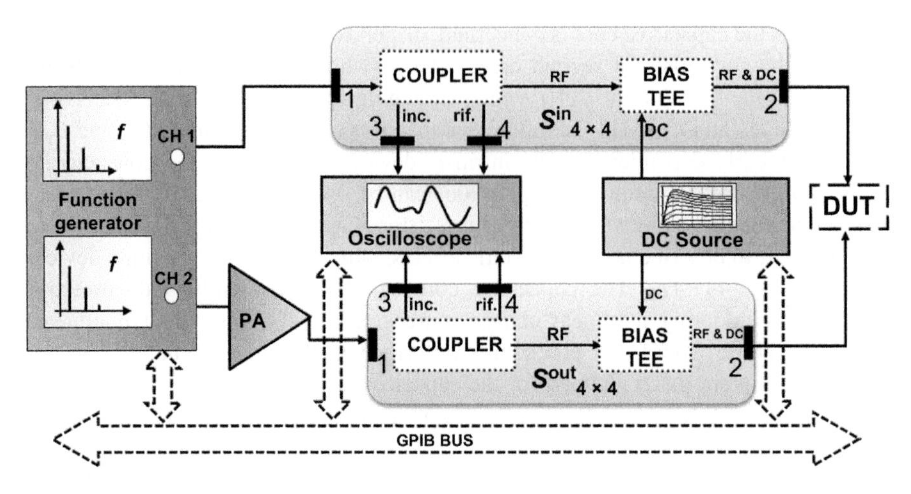

FIGURE 9.5

Block diagram of a low-frequency nonlinear measurement system.

From [36]

power limitations, a 30-W PA is cascaded to the channel devoted to the device's output port excitation. Two wideband (10 kHz–400 MHz) dual-directional couplers monitor the DUT incident and reflected waves, which are acquired by means of a four-channel digital oscilloscope (4 GSa/s). A high-resolution (4 μV, 20 fA) and accurate (*V*: 0.05%, *I*: 0.2%) DC source provides the bias for the DUT. To ensure DC and radio frequency (RF) path isolation as well as DUT stability, two wideband (200 kHz–12 GHz) bias tees are used.

In order to characterize only the device resistive core, it is convenient to operate at frequencies where the dynamic effects associated with charge storage variations and/or finite transit times can be neglected. Moreover, it is of primary interest to characterize device response above the cutoff of LF dispersion (this is why pulsed setups exploit very short pulse durations). To this end, a 2-MHz fundamental frequency is usually adequate for III–V EDs. The validity of such a choice can be verified on the basis of *S* parameter measurements carried out, in the frequency range, of 300 kHz–98 MHz, by exploiting a LF vector network analyzer (VNA; HP4195A).

Figure 9.6 shows the measured output conductance, for a 0.25-μm GaN HEMT device having a periphery of 400 μm, under two different bias conditions corresponding respectively to class-A and -AB operation. In the same figure, the output conductance DC values are reported to highlight that frequency variations in the range of 2–98 MHz, although present, are completely negligible. This confirms that a 2-MHz frequency is sufficient to operate above the cutoff of LF dispersion, which is in line with the large number of papers devoted to microwave device characterization in which pulsed measurements with pulse width of 500 ns or longer were adopted. In recent years, different papers have addressed the presence of

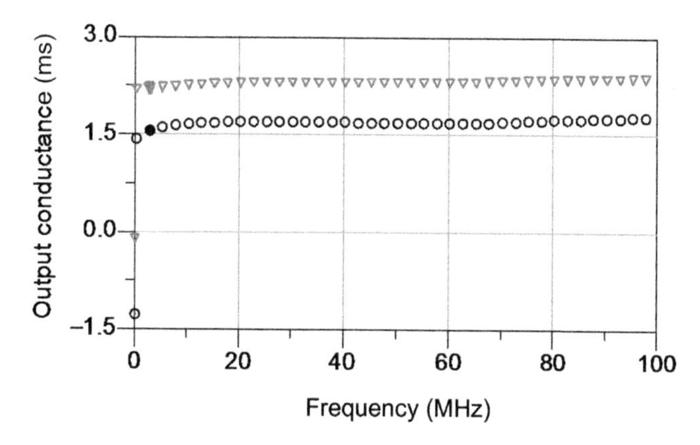

FIGURE 9.6

The 0.25 × 400-μm² GaN HEMT output conductance versus frequency for two different bias conditions: $V_{gO} = -3$ V, $V_{dO} = 20$ V (triangles), and $V_{gO} = -2$ V, $V_{dO} = 25$ V (circles). The filled symbols represent the output conductance value at 2 MHz.

From [36]

very short time constants associated to trapping effects in GaN HEMTs, essentially related to the electron capture rate (e.g., Chapter 7 in Ref. [59]). Nevertheless, despite a growing interest in these additional dispersive effects, their influence on the device performance under nonlinear operation has not yet been clearly demonstrated. As a matter of fact, model formulations exploited by the most important foundries (e.g., UMS, Triquint) do not account for fast trapping effects.

In the selected frequency range, all the measurement setup components satisfy linear nondistortion conditions. This greatly simplifies the setup calibration procedure, which practically consists of the experimental characterization of the two four-port networks ($S^{in}{}_{4\times4}$ and $S^{out}{}_{4\times4}$), shown in Figure 9.5. The characterization can be carried out by adopting the same procedure that has been fully detailed in [60] or, alternatively, by simply exploiting a LF VNA (HP4195A).

By observing the block diagram shown in Figure 9.5, it is evident that the proposed setup looks similar to a large-signal network analyzer (LSNA) [1–4]. Nevertheless, the setup here described is specifically oriented to the characterization of the resistive core. In this context, the LF operation makes the required instrumentation inexpensive and avoids complex calibration procedures, which are essentially related to microwave operation as reported in Chapter 6.

The measurement setup has been automated via an IEEE488 standard interface by means of commercial instrument automation software. The graphical user interface (see Figure 9.7) of the control software enables measurements to be carried out in an automated manner: the user can define the DC parameters, in terms of a bias grid (voltage or current) and related compliances, in accordance with the device's safe operating area. The two channels of the signal generator, which set the incident

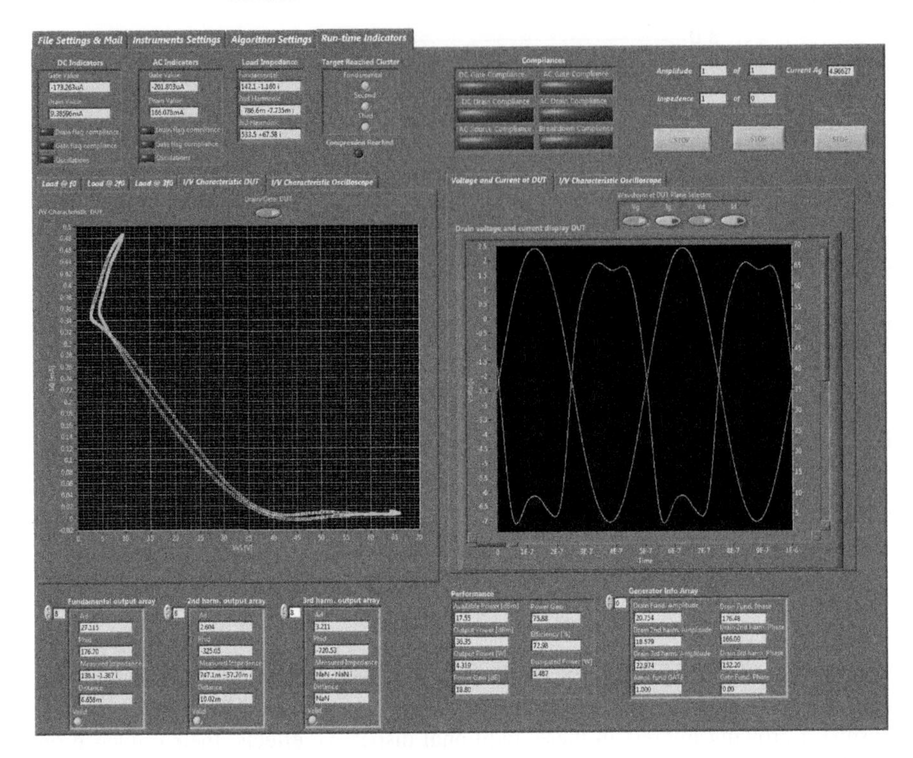

FIGURE 9.7

Graphical user interface of the control software.

signals applied to the device ports, are controlled in an independent way: for each one, the user can define the number of harmonics to be controlled. In particular, for each spectral component, on the input channel the user defines amplitude and phase of the incident signal, whereas on the output channel the impedance values to be synthesized are indicated.

Computer-aided design (CAD)-based amplifier design techniques exploit the accurate knowledge of the device's intrinsic resistive core. A clear example is provided by Cripps' load-line theory, which identifies the optimum device operation by analyzing the device's DC characteristics and defining the optimum loading condition as the resistance that maximizes voltage and current excursions. As a matter of fact, all amplifier design techniques, from class-A to high efficiency (e.g., class-F), are de facto based on the definition of the optimum waveforms of the electrical quantities at the device's intrinsic current source (i.e., the device's resistive core). However, due to traps and thermal effects [29–33], under dynamic operation the transistor resistive core shows a behavior that is very different than under static operation. The proposed setup has the unique capability of directly and exhaustively characterizing such a behavior.

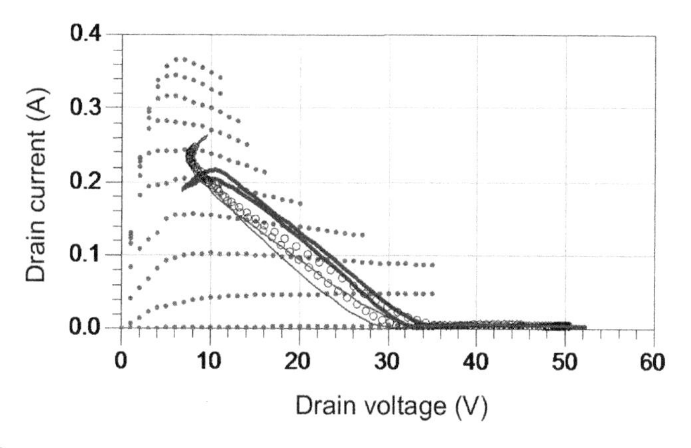

FIGURE 9.8

Different load lines synthesized by exploiting the large-signal measurement system [36] on a $0.25 \times 400\text{-}\mu m^2$ GaN HEMT device at 2 MHz, bias condition ($V_{g0} = -4$ V, $V_{d0} = 30$ V). Amplitude of the gate incident signal fundamental phasor is $A_g = 2.5$ V. The measured load lines are superimposed on DC characteristics ($-6\ V \leq V_{g0} \leq 1$ V, step 0.5 V).

From [36]

As a case study, we investigate the behavior of a $0.25 \times 400\ \mu m^2$ GaN HEMT under high-efficiency operation. In particular, Figure 9.8 shows three different load lines providing the different performance levels reported in Table 9.1. Also in this case, the impact of LF dispersion on the device's performance is well evident. The device is not able to dynamically reach the DC characteristic at $V_{g0} = 1$ V, and the I/V knee under dynamic operation is very far from the DC one. By limiting the drain current and voltage excursions, both of these phenomena reduce the deliverable output power compared to what is predicted for DC characteristics.

By observing the load lines in Figure 9.8, at first glance it is evident which is the best. Efficiency and power are worse for the thick load line because it has the highest dissipation path and minimizes the drain current excursion. Similar considerations indicate that the thin load line is the best one. Such a simple statement of the best operating condition has to be regarded as peculiar to the described characterization technique, because similar information cannot be drawn by observing extrinsic load lines carried out at microwave frequencies.

Table 9.1 Performance related to the different load lines shown in Figure 9.8

Load Line	Output Power (W)	Efficiency (%)
Thick	1.25	54
Circles	1.64	60
Thin	1.72	61

Theoretical considerations can be also carried out by observing the intrinsic drain voltage waveforms corresponding to the considered load lines; they are shown in Figure 9.9 with their harmonic components. Figure 9.9(a) refers to the thick load line; in this case, only the impedance at the fundamental frequency has been controlled. This is evident by looking at the second and third harmonics, which assume very low amplitudes. Figure 9.9(b) refers to the load line with circles; in this case, the impedance at the third harmonic also has been manipulated. It is evident that the contribution of the third harmonic, being out-of-phase with respect to the fundamental one, raises the amplifier performance according to class-F operation [18,19]. Nevertheless, class-F operation requires also a short loading condition at the second harmonic; this condition can be simply obtained by means of the considered setup. The result is shown in Figure 9.9(c), where the amplitude of the second harmonic has been halved.

In Table 9.2, the load impedances synthesized for the different load lines are reported. The impedances at the second and third harmonics differ from 50 Ω, although, for the thick load line, only the load impedance at the fundamental frequency has been controlled. This can be explained by considering that the values in Table 9.2 refer to the DUT planes and account for the nonidealities of the measurement setup (e.g., 30-W PA output impedance, attenuation and delay due to the signal paths). By looking at the impedance values in Table 9.2, it is evident how the impedances synthesized for the thin load-line are the nearest ones to the ideal class-F amplifier behavior (i.e., a short circuit at the second harmonic and an open circuit at the third harmonic).

9.3 Nonlinear embedding design technique

Nonlinear embedding is an original approach that is oriented to the design of PAs. It overcomes the major problems mentioned in the previous paragraph. In particular, nonlinear embedding is mainly based on LF nonlinear experimental ED characterization and enables the same level of accuracy provided by load-pull-based design. Moreover, the proposed technique overcomes power and frequency limitations related to nonlinear microwave measurement techniques. Finally, device currents and voltages compatible with reliability requirements can be directly monitored.

9.3.1 Theoretical formulation

To explain how the LF load line measured at the extrinsic ED ports can be used for PA design, it is convenient to express intrinsic and extrinsic voltages and currents in terms of their practically finite number M of spectral components:

$$x(t) = \sum_{k=-M}^{M} X(k\omega)e^{jk\omega t} \tag{9.2}$$

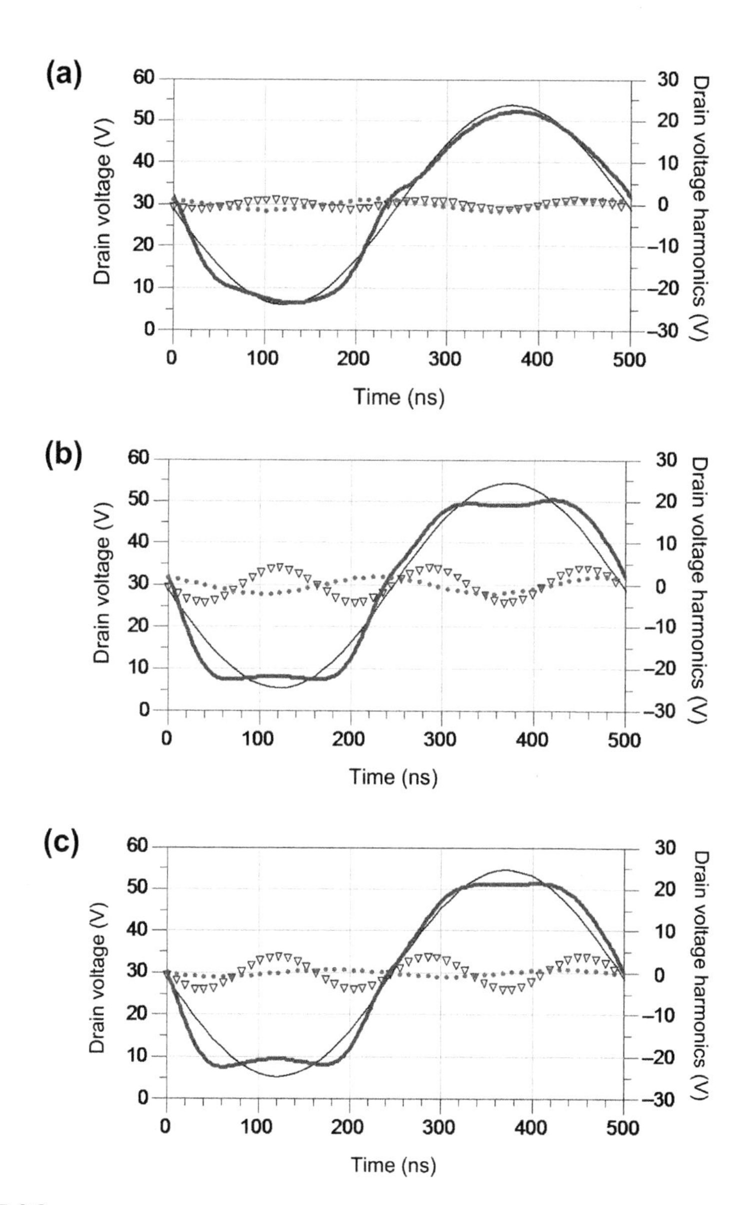

FIGURE 9.9

Measurements performed on a $0.25 \times 400\text{-}\mu\text{m}^2$ GaN HEMT at 2 MHz by exploiting the large-signal measurement system, bias condition ($V_{g0} = -4$ V, $V_{d0} = 30$ V). Time-domain voltage waveform (bold line) and its harmonic components (fundamental: fine line, second harmonic: dots, third harmonic: triangles) corresponding to the three load lines reported in Figure 9.8. (a) Thick line, (b) circles, (c) thin line.

From [36]

Table 9.2 Synthesized load terminations for the different load lines shown in Figure 9.8

Load Line	Fundamental (Ω)	Second Harmonic (Ω)	Third Harmonic (Ω)
Thick	219 – i*19	36 + i*13	47 + i*20
Circles	182 – i*11	37 + i*13	349 – i*47
Thin	177 – i*10	1 + i*15	481 – i*94

By considering any possible topology (based on lumped or distributed elements) for the parasitic network in Figure 9.1, intrinsic and extrinsic electrical variables are conveniently related by the following equations in the frequency domain:

$$\begin{bmatrix} V_{gs}^{i}(k\omega) \\ V_{ds}^{i}(k\omega) \\ I_{g}^{i}(k\omega) \\ I_{d}^{i}(k\omega) \end{bmatrix} = \underline{\mathbf{H}}(k\omega) \begin{bmatrix} V_{gs}^{e}(k\omega) \\ V_{ds}^{e}(k\omega) \\ I_{g}^{e}(k\omega) \\ I_{d}^{e}(k\omega) \end{bmatrix}, k = -M, \ldots, M \tag{9.3}$$

where $\underline{\mathbf{H}}(\omega)$ is a suitable hybrid-matrix description of the linear extrinsic parasitic network. At microwave frequencies, the harmonic components of the global intrinsic currents are composed of the sum of the conduction and displacement currents denoted with the superscripts R and C, respectively:

$$\begin{bmatrix} I_{g}^{i}(k\omega_{RF}) \\ I_{d}^{i}(k\omega_{RF}) \end{bmatrix} = \begin{bmatrix} I_{g}^{i,R}(k\omega_{RF}) + I_{g}^{i,C}(k\omega_{RF}) \\ I_{d}^{i,R}(k\omega_{RF}) + I_{d}^{i,C}(k\omega_{RF}) \end{bmatrix}, k = -M, \ldots, M \tag{9.4}$$

Due to the frequency independence of the conduction current, the phasors of its harmonic RF components coincide with the LF ones—that is $I_{x}^{i,R}(k\omega_{RF}) = I_{x}^{i,R}(k\omega_{LF})$.

When considering the LF load-line characterization carried out at the fundamental frequency $\omega = \omega_{LF}$ through the setup in Figure 9.5, the displacement current can be totally neglected. Moreover, $\underline{\mathbf{H}}(\omega)$ practically becomes a real- and frequency-independent matrix (which, when considering the most common case of lumped parasitic description, reduces to the series parasitic resistors). In such a case, Eqn (9.3) enables one to directly compute, starting from the knowledge of the LF harmonic components of the extrinsic voltages and currents, the intrinsic electrical variables (and, as consequence, the intrinsic load line) at the ED resistive core: $V_{gs}^{i}(k\omega_{LF}), V_{ds}^{i}(k\omega_{LF}), I_{g}^{i,R}(k\omega_{LF}), I_{d}^{i,R}(k\omega_{LF})$.

To exploit the LF load-line characterization for microwave PA design, the extrinsic device load and source conditions must be computed at the design frequency, which enables the electrical regime corresponding to the chosen intrinsic load line to be imposed. To this end, the displacement currents related to the electron device capacitive core must be evaluated according to the following explicit equations:

$$\begin{bmatrix} i_g^{i,C}(t) \\ i_d^{i,C}(t) \end{bmatrix} = \begin{bmatrix} \sum_{k=-M}^{M} I_g^{i,C}(k\omega_{RF})e^{jk\omega_{RF}t} \\ \sum_{k=-M}^{M} I_d^{i,C}(k\omega_{RF})e^{jk\omega_{RF}t} \end{bmatrix}$$

$$= \sum_{k=-M}^{M} jk\omega_{RF}\underline{\mathbf{C}}\left(v_{gs}^{i}(t), v_{ds}^{i}(t)\right) \begin{bmatrix} V_{gs}^{i}(k\omega_{LF})e^{jk\omega_{RF}t} \\ V_{ds}^{i}(k\omega_{LF})e^{jk\omega_{RF}t} \end{bmatrix} \tag{9.5}$$

ω_{RF} is the fundamental operating frequency of the PA with:

$$\begin{aligned} v_{gs}^{i}(t) &= \sum_{k=-M}^{M} V_{gs}^{i}(k\omega_{LF})e^{jk\omega_{RF}t} \\ v_{ds}^{i}(t) &= \sum_{k=-M}^{M} V_{ds}^{i}(k\omega_{LF})e^{jk\omega_{RF}t} \end{aligned} \tag{9.6}$$

The capacitance matrix $\underline{\mathbf{C}}$ in Eqn (9.5) can be identified on the basis of frequency- and bias-dependent S parameter measurements. Alternatively, the capacitive core of a suitable, already available, nonlinear model can be used. The difference existing, given a bias condition, among the imaginary parts of the admittance parameters Y_{12} and Y_{21} (typically modeled by a transcapacitance) can be accounted for by Eqn (9.5) without any approximation.

When high-frequency non-quasi-static effects are not negligible, the displacement currents cannot be explicitly evaluated in terms of a capacitance matrix only, as in Eqn (9.5), but nonlinear circuit analysis is required. To this end, any frequency- or time-domain available CAD tool for nonlinear circuit analysis can be easily adopted.

Once the harmonic components of the global intrinsic currents have been evaluated by use of Eqn (9.4), the extrinsic electrical variables that define the load and source extrinsic regime can be computed by:

$$\begin{bmatrix} V_{gs}^{e}(k\omega_{RF}) \\ V_{ds}^{e}(k\omega_{RF}) \\ I_g^{e}(k\omega_{RF}) \\ I_d^{e}(k\omega_{RF}) \end{bmatrix} = \underline{\mathbf{H}}^{-1}(k\omega_{RF}) \begin{bmatrix} V_{gs}^{i}(k\omega_{RF}) \\ V_{ds}^{i}(k\omega_{RF}) \\ I_g^{i}(k\omega_{RF}) \\ I_d^{i}(k\omega_{RF}) \end{bmatrix}, k = -M, \ldots, M \tag{9.7}$$

and finally, the load impedance at the fundamental and harmonic frequencies can be obtained:

$$Z_L(k\omega_{RF}) = -\frac{V_{ds}^{e}(k\omega_{RF})}{I_d^{e}(k\omega_{RF})} \tag{9.8}$$

In addition, the input device's large-signal impedance can be easily computed:

$$Z_{IN}(k\omega_{RF}) = \frac{V_{gs}^{e}(k\omega_{RF})}{I_g^{e}(k\omega_{RF})} \tag{9.9}$$

This can be used to synthesize the optimum source impedance (i.e., $Z_s = \text{conj}(Z_{IN})$), which provides a matching condition under large-signal operation. This information is not obtainable through scalar load-pull systems, but only by adopting a time-domain load-pull measurement setup [1–4].

To summarize, the proposed approach, which is based on LF, large-signal measurements and bias/frequency-dependent small-signal measurements performed in the frequency range of interest for the considered design, is fully able to provide the same kind of information obtainable by means of expensive nonlinear measurement setups operating at microwave frequencies. The only assumption is that a negligible uncertainty can be achieved in the description of the intrinsic ED capacitive core and parasitic elements, whose accuracy ultimately defines the frequency limitations. As will be demonstrated in the following section, such a hypothesis is more than reasonable from a practical point of view.

The flowchart in Figure 9.10 summarizes the fundamental steps of the described design techniques.

9.3.2 Design examples

As a first example, the design and realization of a hybrid L-band high-power amplifier (HPA) exploiting a 0.7-μm GaN HEMT process is considered. In particular, a discrete GaN HEMT power bar composed of six 2-mm cells for a total gate periphery of 12 mm was used.

The described design procedure was applied to the 2-mm elementary cell in order to define the optimum load impedance. The bias condition was chosen for class-AB operation ($V_{g0} = -3$ V, $V_{d0} = 35$ V, $I_{d0} = 140$ mA), and the load line shown in Figure 9.11 was selected, which provides a single-cell output power of 38.8 dBm (3.8 W/mm) and a 67% drain efficiency.

The capacitive core of the foundry models, accounting also for non-quasi-static phenomena, was exploited in order to compute the displacement currents. The load impedance was then evaluated according to the described procedure, while the source impedance was chosen as equal to the conjugate of the device input large-signal impedance (see Table 9.3).

In Figure 9.12, the total device gate current and its resistive component are shown. It is well evident that the latter is negligible with respect to the capacitive

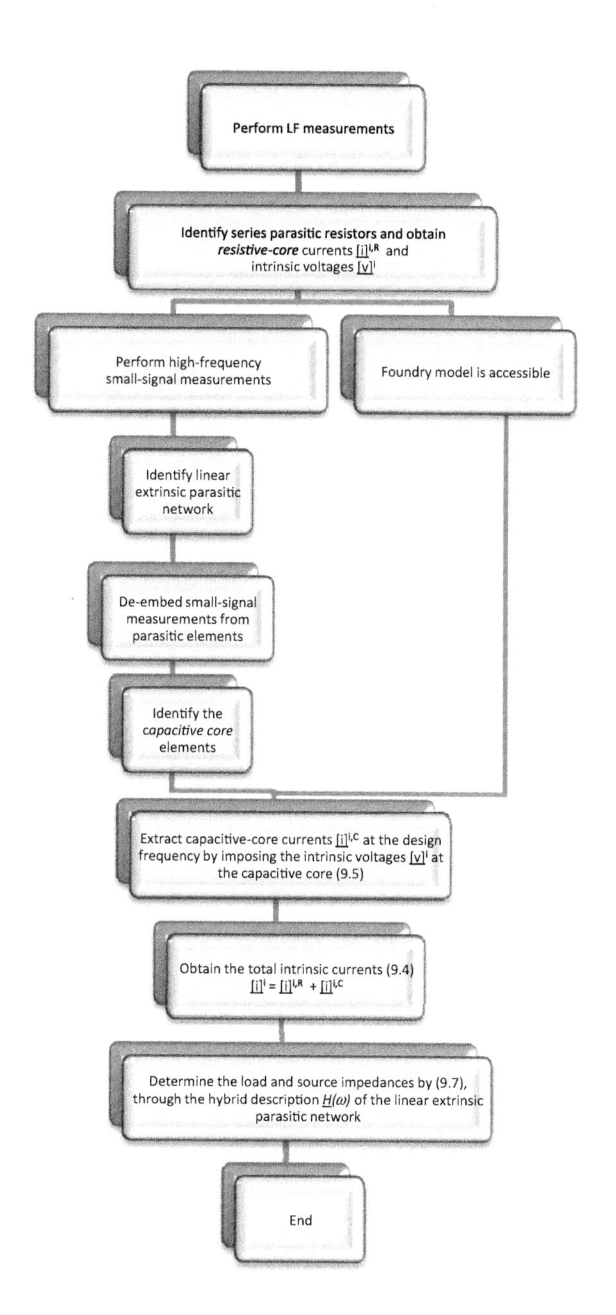

FIGURE 9.10

Flowchart describing the described design technique.

From [22]

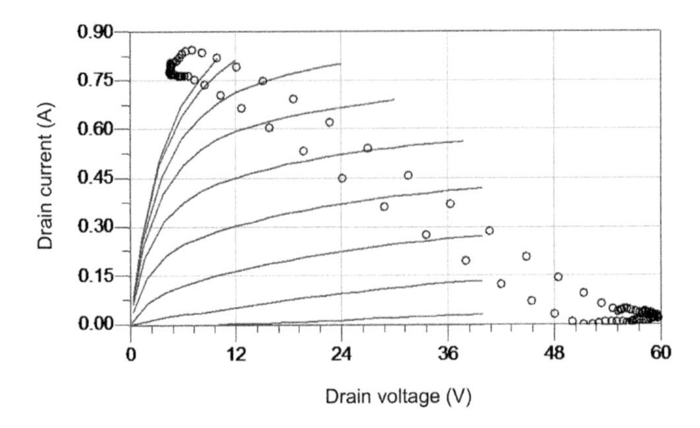

FIGURE 9.11

Load line of a 0.7 × 2000-μm^2 GaN HEMT device biased at $V_{g0} = -3$ V and $V_{d0} = 35$ V, chosen by exploiting the LF measurement system shown in Figure 9.5 at 2 MHz. The load line is superimposed to pulsed characteristics (-5.5 V $\leq V_g \leq 1.5$ V, Step 0.5 V) carried out from the considered bias condition.

From [22]

Table 9.3 Synthesized source and load impedances for the 2-mm GaN HEMT

Source Impedance (Ω)	Frequency (GHz)	Load Impedance (Ω)
9.31 + i*23.58	1.275	58.32 + i*23.06
	2.550	31.05 + i*25.80
	3.825	26.10 + i*22.52

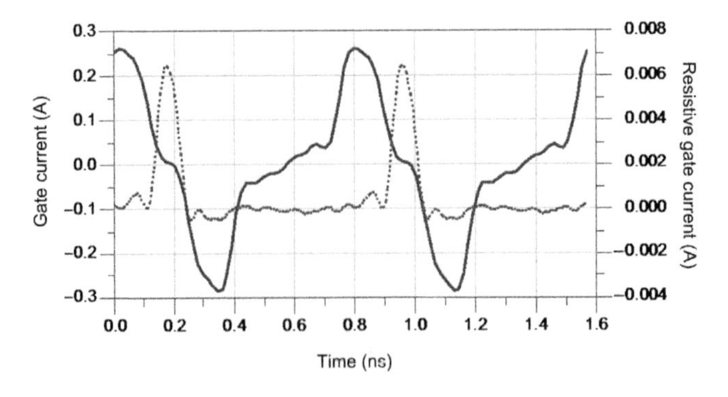

FIGURE 9.12

Gate current of a 0.7 × 2000-μm^2 GaN HEMT device biased at $V_{g0} = -3$ V and $V_{d0} = 35$ V, at the intrinsic device I_g^i (solid line) and its resistive portion (dots). $f_{RF} = 1.275$ GHz.

From [22]

current, also under high-compressed amplifier operation. Nevertheless, the average value of the gate current, which is dependent only on the resistive current, is a widely adopted marker for device reliability issues (i.e., gate-source diode conduction and gate-drain diode breakdown). Average gate current values are easily used for this purpose because they are easily put in a relationship with reliability characterization based on static device measurements (e.g., [61−63]).

The proposed approach enables the resistive gate current to be directly controlled when choosing the device operating conditions. Moreover, the time-dependent waveform of the gate current is also directly measured, which can be useful for a more in-depth analysis of reliability issues [64].

The L-band PA has been designed by synthesizing, on a high-frequency laminate, an output network that provides the chosen load impedance (Table 9.3) to each 2-mm cell of the power bar. A photograph of the realized amplifier is shown in Figure 9.13.

The high load impedance required by the GaN devices allows for impedance transformation through a simple microstrip step-impedance network connected to the six elementary cells by means of six bonding wires. An input network composed of a radial stub and a step-impedance solution optimizes the transducer gain. Furthermore, a combined series-shunt RC stabilizing network, implemented by means of surface-mount device (SMD) resistors and ceramic capacitors, makes the device unconditionally stable from DC to the cutoff frequency. Gate and drain bias networks employ high impedance $\lambda/4$ microstrip lines with shunted SMD chip capacitors.

Figure 9.14 shows the main performance of the amplifier: a saturated output power of approximately 46.75 dBm and 63% efficiency fully comply with the expected performance of the technology.

In Table 9.4, the L-band PA measured data are compared with the predictions obtained through the described technique for the same level of output power (46.64 dBm). Considering the unavoidable dispersion in the realization phase and that only the load impedance at the fundamental frequency has been accurately

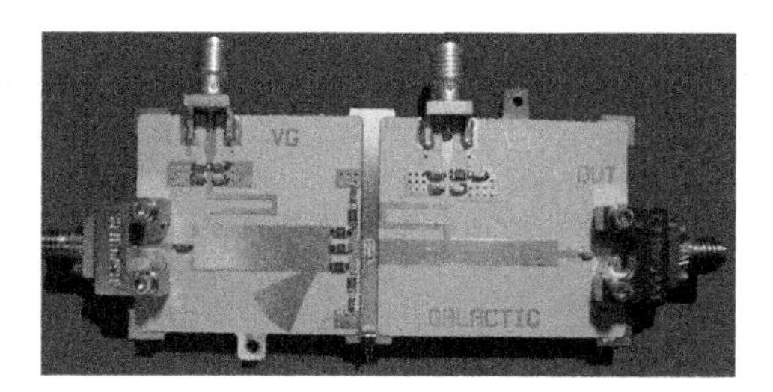

FIGURE 9.13

GaN amplifier for L-band application exploiting a planar technology integrated circuit. The active device is a 12-mm GaN power bar combining six 2-mm elementary cells.

From [22]

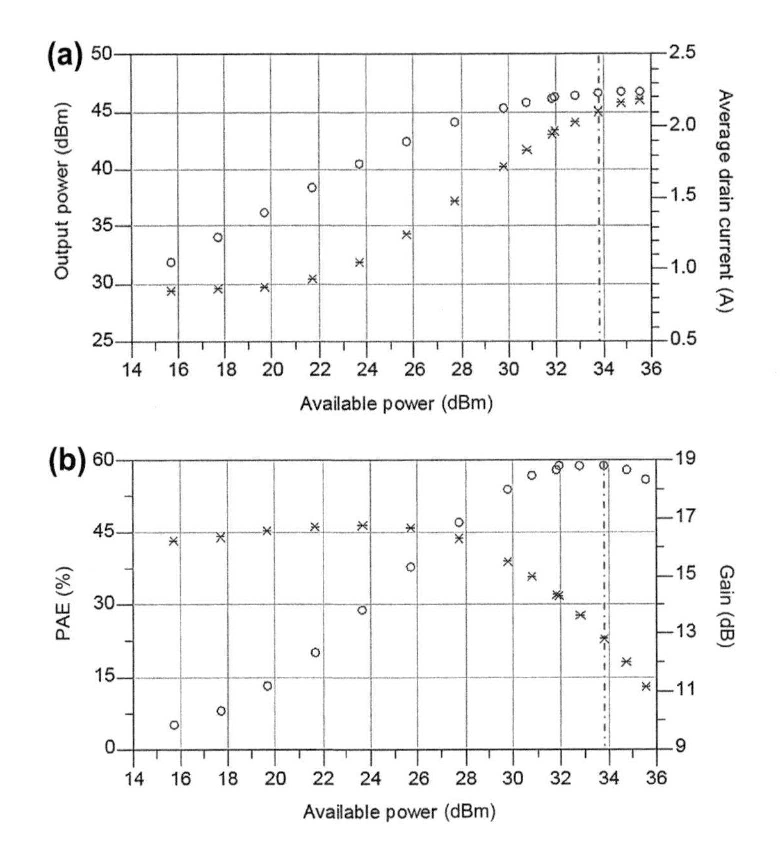

FIGURE 9.14

Measured performance of the hybrid I-band high-power amplifier of Figure 9.13. The dotted vertical lines identify the values corresponding to the output power level of interest. (a) Output power (circles) and average drain current (stars). (b) Efficiency (circles) and transducer power gain (stars).

From [22]

Table 9.4 Comparison between device performance predicted by the described technique and measurement data (P_{out} = 46.64 dBm, design frequency = 1.275 GHz)

Device Performance	Predicted	Actual
Average drain current (A)	1.95	2.1
Drain efficiency (%)	67	63
Gain (dB)	13.4	12.8
Power added efficiency (%)	64	59

Table 9.5 Obtained low-frequency drain impedances

Frequency (MHz)	Load Impedance (Ω)
2	104.1 − j*92.0
4	1.2 + j*103.2
6	299.8 + j*181.3

synthesized, the results confirm the effectiveness of the proposed approach for the design of PAs.

Also, high-efficiency amplifiers can be simply designed in accordance with the embedding methodology. As an example, in Ref. [65], a discrete C-band $0.25 \times 1250\ \mu m^2$ GaN on SiC HEMT was adopted to design a class-E amplifier. The device was initially biased under a class-B condition ($V_{g0} = -4$ V, $V_{d0} = 21$ V, $I_{d0} = 0$ A). More precisely, the drain bias voltage was chosen in order to not exceed the breakdown limitation of the selected technology under dynamic operation. Successively, appropriate LF drain and gate incident signals, at fundamental and harmonic frequencies, were imposed on the ED in order to synthesize a class-E-shaped load line. Table 9.5 summarizes the obtained LF drain impedances, whereas Figure 9.15 reports the trajectory of the synthesized load line. The dynamic drain voltage was kept lower than 60 V, well below the device breakdown limit (i.e., 70 V). An output power of 33.7 dBm and a drain efficiency of 81.1% were determined.

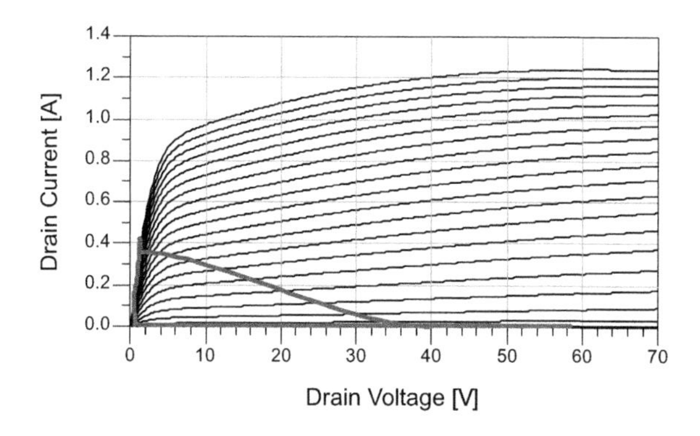

FIGURE 9.15

Measurements performed on a $0.25 \times 1250\text{-}\mu m^2$ GaN HEMT biased at $V_{d0} = 21$ V and $I_{d0} = 0$ A by exploiting the LF load-pull setup at 2 MHz. The measured load-line is superimposed on simulated DC characteristics ($-4\ V \leq V_{gs} \leq 0$ V, step 0.2 V).

From [65]

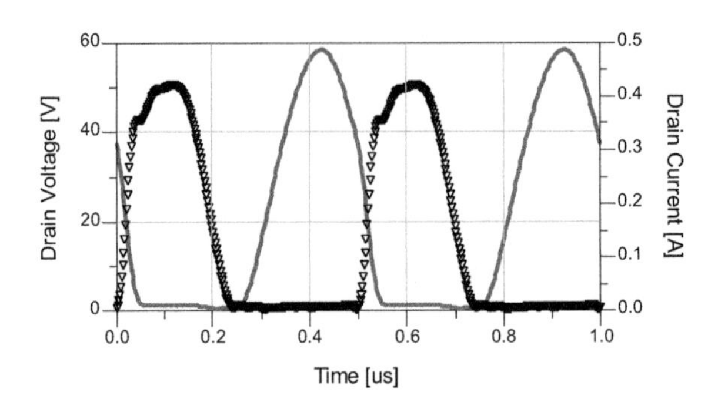

FIGURE 9.16

Low-frequency time-domain voltage (continuous line) and current (triangles) waveforms at the device current-generator plane, corresponding to the synthesized class-E operation mode shown in Figure 9.15.

From [65]

Figure 9.16 shows the intrinsic time-domain voltage and current waveforms. It is evident that the selected operating mode is typical of a class-E PA and satisfies the minimal-overlapping condition imposed by switching mode PAs.

To find out load and input impedances at the design frequency of 1.2 GHz starting from the LF data, in accordance with the previously described procedure, a foundry-model-based description for both the ED capacitances and parasitic elements was adopted. Table 9.6 summarizes the computed source and load terminations at the fundamental frequency of 1.2 GHz and related harmonics. In particular, the source termination was chosen equal to the conjugate of the ED input large signal impedance.

The L-band class-E GaN PA was designed by synthesizing, on a high-frequency laminate, suitable input and output networks that provided the chosen source and load impedances (Table 9.6) to the discrete 1.25-mm ED. A picture of the realized amplifier is shown in Figure 9.17.

Impedance transformations related to both input and output matching networks (OMNs) have been implemented by means of radial stubs in conjunction with step-impedance structures, which guarantee the predicted terminations at the ED plane. In particular, the input matching network (IMN) was designed to synthesize the

Table 9.6 Computed source and load impedances

Frequency (GHz)	Source Impedance (Ω)	Load Impedance (Ω)
1.2	16.7 + j*72.9	175.5 – j*31.2
2.4	15.6 – j*2.3	1.0 + j*54.5
3.6	1.8 – j*5.1	11.5 + j*55.3

FIGURE 9.17

Realized GaN class-E hybrid power amplifier for I-band application.

From [65]

predicted impedance only at the fundamental frequency. For the OMN, two radial stubs with an electrical length of $\lambda/4$ at the second and third harmonic, respectively, were added close to the drain of the ED (see Figure 9.17) to match the harmonic impedances.

Decoupling ceramic capacitors were chosen for both the matching networks to separate RF and DC signals. Moreover, gate and drain bias networks were realized by employing high impedance $\lambda/4$ microstrip lines with shunted SMD chip capacitors. Finally, the ED was connected to the input and OMNs by means of gold bonding wires.

Figure 9.18(a) shows the main performance of the realized PA: an output power of 32.7 dBm with a drain efficiency of 76% were measured. Moreover, Figure 9.18(b) shows the performance of the amplifier referred to the device plane. As confirmed by Table 9.7, output power and drain efficiency at this plane are in excellent agreement with the performance predicted by the LF characterization.

The OMN of a PA is generally designed to reach optimal performance under large-signal operations. When dealing with predriver or driver amplification stages, IMN design is often based on small-signal measured or simulated data to optimize both small-signal gain and input return loss [18,19]. Indeed, input power levels of the first stages in the amplification lineup are definitely lower than the output power levels, thus justifying the described design paradigm. On the contrary, in the final amplification stage, the input power levels become significantly high. As a consequence, IMN design based on small-signal operation could be inappropriate because the small-signal optimal impedance is not necessarily related to the optimal impedance under large-signal operation. In any case, the large-signal matching condition does not coincide with the small-signal one.

In the design example [66], a hybrid HPA based on GaN technology synthesizes broadband terminations for both the IMN and OMN optimized for large-signal

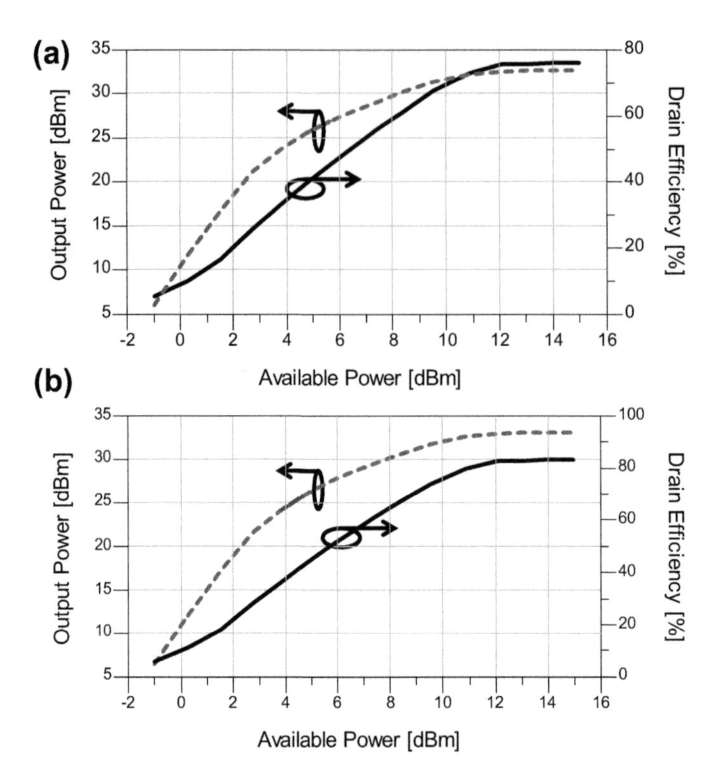

FIGURE 9.18

Measured performance of the hybrid L-band class-E PA from Figure 9.17(a): output power (dashed line) and drain efficiency (solid line) of the PA; (b) output power (dashed line) and drain efficiency (solid line) referred to the ED plane.

From [65]

Table 9.7 Comparison between device performance predicted by the proposed technique and measurement data

Performance Measurement	Predicted	Actual
Output power	33.7 dBm	33.4 dBm
Drain efficiency	81.1%	83%

operation. In particular, as far as the IMN is concerned, a conjugate matching condition was reached under large-signal operation over the whole design bandwidth. By exploiting the described LF time-domain load-pull system (Figure 9.5), a preliminary large-signal characterization was carried out on a $0.5 \times 2000\text{-}\mu m^2$ ED, representative of the 20-mm power bar composed of 10 elementary cells electrically in

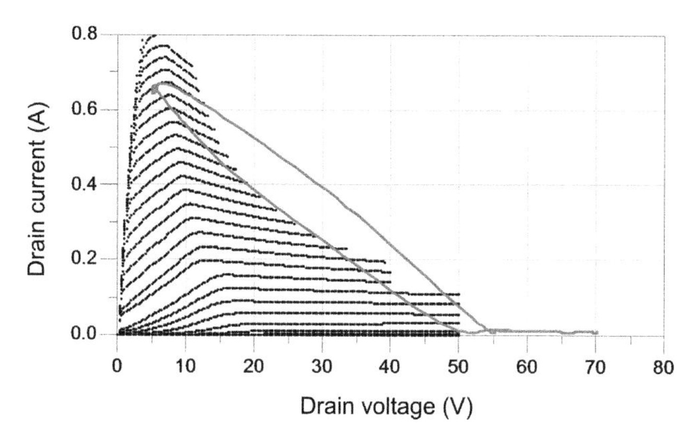

FIGURE 9.19

Measurements performed on a $0.5 \times 2000\text{-}\mu m^2$ GaN HEMT device biased at $V_{d0} = 40$ V and $I_{d0} = 57$ mA at 2 MHz. The load line is superimposed on the measured DC characteristics ($-2\ V \leq V_{gs} \leq 1$ V, step 0.1 V).

From [66]

parallel. Such experimental characterization was performed to find input and output broadband terminations that ensured adequate PA performance in the design band of 400 MHz centered at 1.2 GHz. In particular, the measured broadband load impedances correspond to the dynamic load line referred to the CGP shown in Figure 9.19.

The current-generator load line was embedded by considering nonlinear dynamic phenomena and parasitic effects to obtain the optimum impedances at the design frequencies. As a matter of fact, such a load line provides an output power of 38.1 dBm with a drain efficiency of 63.8%. The elementary cell is biased under deep class-AB conditions ($V_{d0} = 40$ V, $I_{d0} = 57$ mA). On the other hand, source termination was chosen in order to guarantee in the whole design bandwidth, a conjugate matching with respect to the large-signal input ED impedance; such a condition allows the maximum power transfer under large-signal operations.

Figure 9.20 shows the trajectories of the predicted source and load impedances in the frequency range of 1–1.4 GHz, whereas Table 9.8 reports the synthesized source and load terminations at the center frequency (i.e., 1.2 GHz) and at the extremes of the selected band related to the 2-mm elementary cell.

A picture of the realized amplifier is shown in Figure 9.21. The high load impedance required by the GaN devices allows for impedance transformation through a simple microstrip step-impedance structure together with open- and short-circuit stubs. IMN transformation was obtained by means of a step impedance transmission line in conjunction with two radial open-circuit stubs. Furthermore, a combined series-shunt RC stabilizing network, implemented by means of SMD resistors and ceramic capacitors, makes the device unconditionally stable from DC to the cutoff frequency. The power bar was connected to both IMN and OMN boards by means of gold bonding wires.

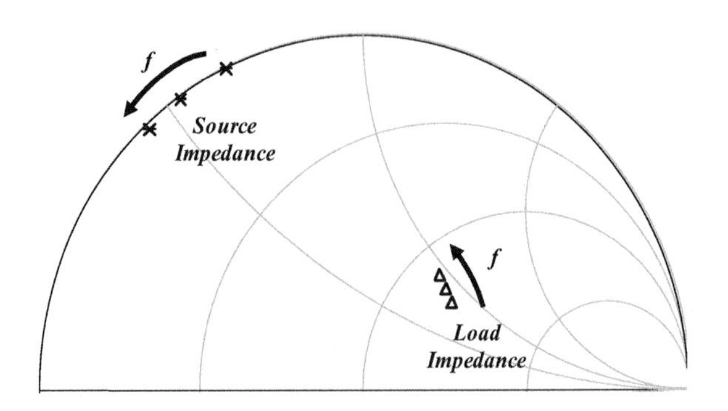

FIGURE 9.20

Broadband source (stars) and load (triangles) impedance trajectories evaluated in the frequency range of 1–1.4 GHz.

From [66]

Table 9.8 Synthesized LF load terminations in the considered frequency range

Frequency (GHz)	Source Impedance (Ω)	Load Impedance (Ω)
1.0	0.1 + j*32.0	73.0 + j*41.2
1.2	0.3 + j*26.4	66.7 + j*44.5
1.4	0.5 + j*22.4	60.6 + j*46.7

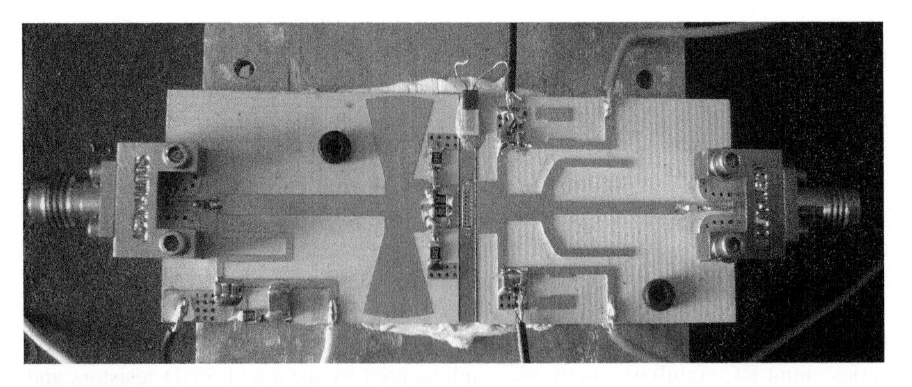

FIGURE 9.21

Realized GaN hybrid HPA.

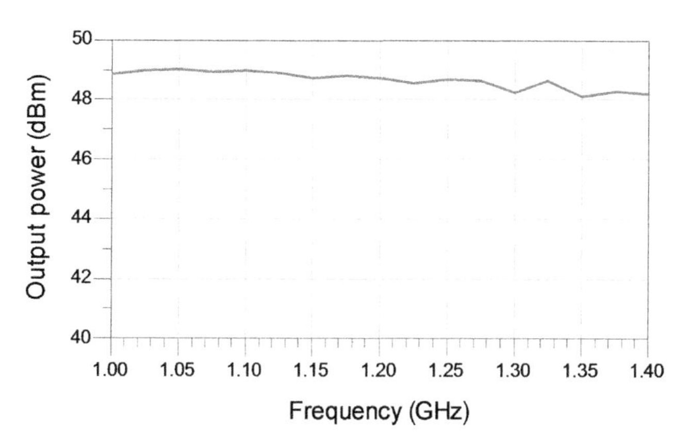

FIGURE 9.22

Saturated output power of the realized HPA of Figure 9.21 across the bandwidth of 1–1.4 GHz.

From [66]

As shown in Figure 9.22, the PA, driven with a constant available power of 38 dBm over the whole bandwidth, delivers an average saturated output power of 48.5 dBm over the frequency range of 1–1.4 GHz, with a power variation lower than 1 dB. This measurement condition corresponds to 3.5 dB of compression with respect to the measured linear gain of 14 dB.

Moreover, the measured power added efficiency (PAE) shown in Figure 9.23 is greater than 55% in the whole design bandwidth, with a peak value of 65.2% measured at 1.05 GHz. The measured performance of the designed PA is perfectly in agreement with the one predicted by the LF time-domain load-pull characterization, performed on the 2-mm elementary cell. To guarantee the same load line at the CGP over the entire bandwidth, the three load impedances in Table 9.8 were equally weighted in the IMN and OMN optimization. As a consequence, it is not surprising to have slightly better behavior at the lower bound of the bandwidth (Figure 9.23).

The last design example concerns an X-band GaN monolithic microwave integrated circuit (MMIC) HPA suitable for future-generation Synthetic-Aperture Radar (SAR) systems [67]. The technology selected for the design is the UMS 0.25-μm gate length HEMT AlGaN–GaN on an SiC process, featuring low-loss microstrip lines, monolithic microwave integrated capacitors, spiral inductors, TaN resistors, air bridges, and via holes.

LF load-pull measurements were performed by means of the LF measurement setup previously described. The intrinsic load line shown in Figure 9.24 is found to be the most suitable in terms of output power and drain efficiency.

Following the described design technique, the measured LF waveforms were embedded with the nonlinear dynamic effects occurring at high frequency and

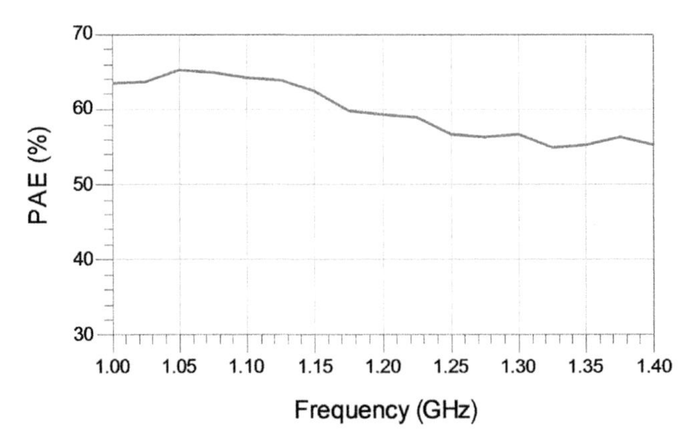

FIGURE 9.23

PAE of the realized HPA across the bandwidth 1–1.4 GHz.

From [66]

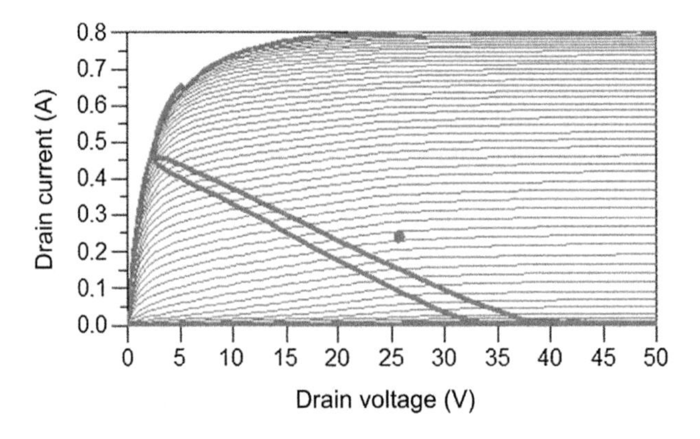

FIGURE 9.24

Selected measured intrinsic load line of the $0.25 \times 1000\ \mu m^2$ elementary power cell, with $V_{d0} = 26$ V and $I_{d0} = 80$ mA/mm. The load line is superimposed on the simulated DC characteristics.

From [67]

scaled to the frequency band of interest. The calculated performance values of the elementary power cell are listed in Table 9.9.

Two stages of amplification are required to obtain a high linear gain of 25 dB at X-band with the selected technology. According to Table 9.9, an output stage of $4 \times 8 \times 125\ \mu m$ active periphery is required to get 14 W of output power, whereas a first stage of $2 \times 8 \times 125\ \mu m$ of active periphery gives a good driving margin. The load and source impedances of the elementary cell listed in Table 9.9 are

Table 9.9 8 × 125 μm Device impedances and performance in the frequency band of interest

	Frequency		
	8.8 GHz	**9.6 GHz**	**10.4 GHz**
Z_L^{fund} (Ω)	18.4 + j*25.8	16.0 + j*23.8	14.0 + j*21.8
Z_S^{fund} (Ω)	1.9 − j*12.7	1.9 − j*10.3	1.9 − j*8.2
P_{out} (W)	3.9		
η_D (%)	62		
I_{D0} (mA)	242		

then scaled, accordingly to the chosen total active periphery of each stage, to synthesize the output, interstage, and IMNs.

Careful electromagnetic simulations were performed to recover the asymmetries within the matching networks. In addition, the interstage and the IMNs include resistive elements that guarantee the overall stability of the amplifier and prevent odd mode oscillations. A photograph of the manufactured 4.5 × 4 mm^2 MMIC HPA is shown in Figure 9.25.

The manufactured PA was characterized on a test fixture to include the external bonding wire connections. The quiescent bias point corresponds to $V_{d0} = 26$ V and

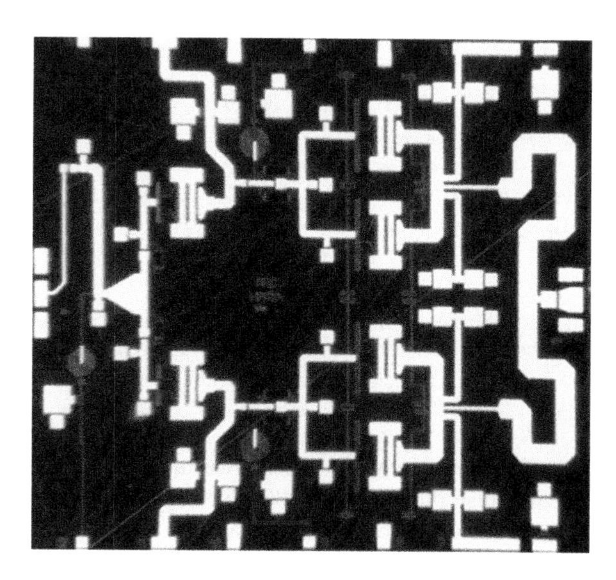

FIGURE 9.25

Photograph of the X-band GaN MMIC PA.

From [67]

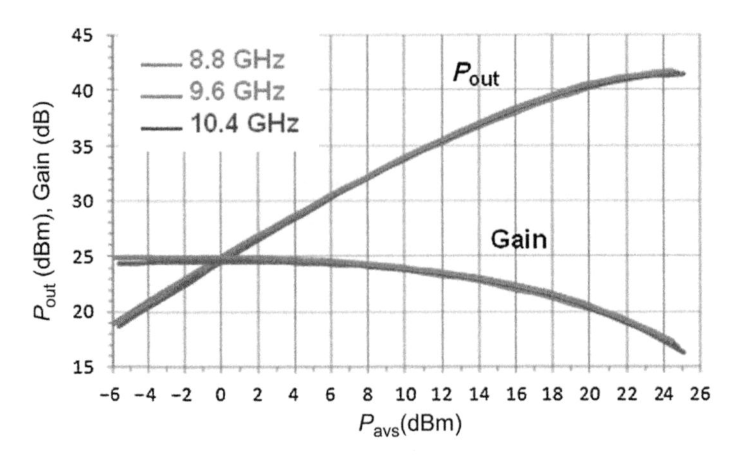

FIGURE 9.26

HPA measured output power and gain as a function of the power available from the source for the three different frequencies of 8.8, 9.6, and 10.4 GHz.

From [67]

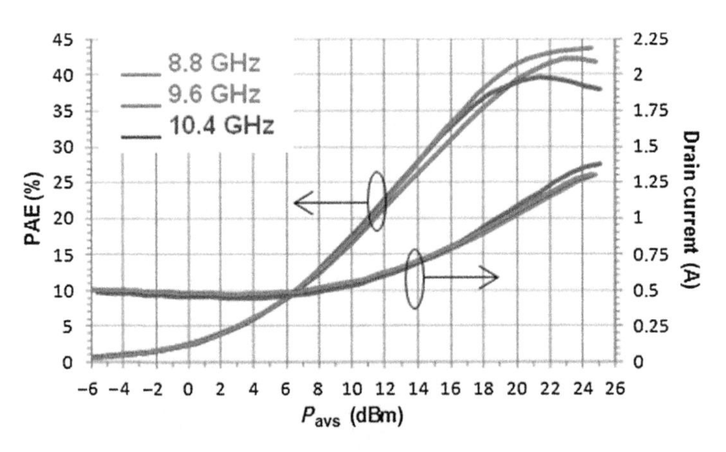

FIGURE 9.27

HPA measured PAE and DC drain current as a function of the power available from the source for the three different frequencies of 8.8, 9.6, and 10.4 GHz.

From [67]

$I_{d0} = 480$ mA/mm. The measurements were performed under pulsed quiescent conditions with 33 μs pulse duration and 10% duty cycle. Figures 9.26 and 9.27 show the measured PA output power, gain, PAE, and DC drain current as a function of the power available from the source and for three different frequencies: 8.8, 9.6, and 10.4 GHz. Considering that measurements refer to the entire two-stage amplifier, also in this case the agreement with respect to the performance predicted by the proposed design technique is excellent.

9.4 Nonlinear de-embedding design technique

Microwave nonlinear measurement systems [1–4] enable optimum loading conditions at the extrinsic ports of the ED to be found. This information is essential for the synthesis of IMN and OMN of the PA. Nevertheless, from such a setup, no information about the electrical variables at the CGP is immediately available. As a consequence, de-embedding techniques are required for retrieving this information. In fact, as previously discussed, without the knowledge of the load line at the CGP, it is not possible to state in which class of operation the device is actually working.

Starting from high-frequency nonlinear measurements and exploiting standard approaches for the parasitic network and capacitive-core models (see Figure 9.1), the nonlinear de-embedding procedure [68–70] is able to give all the information in terms of time-domain waveforms, load impedance, output power, etc., at the CGP. The only assumption is that the models of the parasitic network and the capacitive core guarantee the required level of accuracy.

Figure 9.28 shows a simplified flowchart describing the steps of the nonlinear de-embedding design technique. First of all, the extrinsic parasitic network and the capacitive-core models have to be identified. The parasitic network model allows shifting of the electrical variables from the extrinsic plane (EP) of the ED (i.e., the measurement plane) to the intrinsic plane (IP), whereas the capacitive-core model allows one to calculate the capacitive current contribution that is essential to correctly move from the IP to the CGP.

Despite that the problem of retrieving electrical variables at the current-generator plane is clearly nonlinear, linear de-embedding procedures have been typically adopted in literature. Actually, in manuscripts devoted to PA design based on microwave nonlinear measurements, often a "gray zone" exists, and it is not clear at which plane electrical waveforms are referred to. Clearly, it is not possible to draw any theoretical considerations on the basis of waveforms at the extrinsic planes.

Linear de-embedding procedures consider the capacitive core as being composed of linear components; they differ from each other because of the approximations made in the model exploited for the parasitic network description [71]. With the aim of clarifying this aspect, a complete nonlinear model topology is shown in Figure 9.29(a). The linear de-embedding procedures introduce two additional strong approximations with respect to that model: the gate-drain capacitance C_{gd} and the

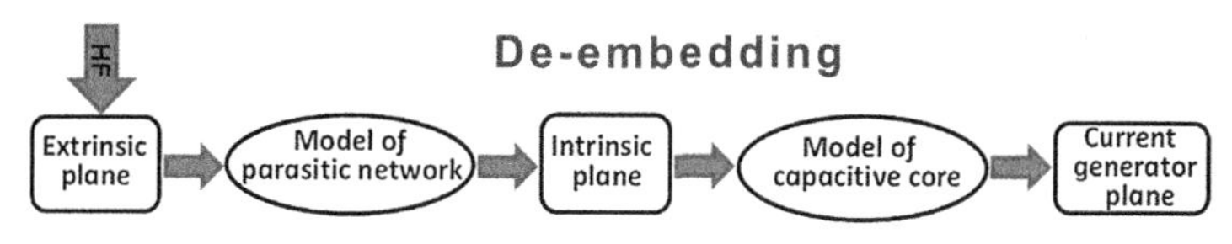

FIGURE 9.28

Flowchart describing the nonlinear de-embedding procedure.

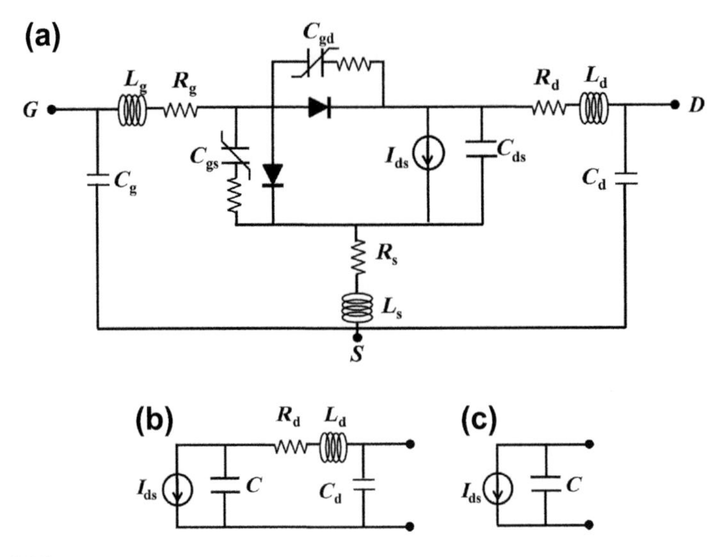

FIGURE 9.29

Nonlinear equivalent circuit of an FET electron device (a), simplified model (drain side) considering only drain parasitic network and a single capacitance (b), and a further simplified description (drain side) considering only a capacitance (c).

source parasitic effects are completely neglected. In this way, gate and drain ports can be treated separately, and as a consequence the drain can be modeled as shown in Figure 9.29(b) or, with a piratical coup, as reported in Figure 9.29(c). The question is: how many designers would choose to deactivate the gate-drain capacitor and the source parasitic elements in their nonlinear model? Or, looking at the other side of the microwave coin, why do modelers love to complicate their work by including these elements in their model formulations? To clarify this point, three de-embedding procedures are presented, showing the differences in their waveform prediction at the current-generator plane. A GaN HEMT device ($0.7 \times 800\ \mu m^2$) operating under class-F operation is chosen and its model (see Figure 9.29(a)) is exploited to draw some considerations in the general case where harmonic terminations also have to be identified. The electrical variables corresponding to this condition are considered to be the starting point to apply the three different de-embedding procedures.

In the first example, referred to as procedure A, a complete model (model A), such as the one in Figure 9.29(a), is considered, where both parasitic extrinsic network and capacitive-core models are extracted through commonly used procedures based on multi-bias S parameters. The second procedure, referred to as procedure B, considers the same extrinsic parasitic network for the drain port but the model of the capacitive core is limited to a linear drain-source capacitance (designated model B; see Figure 9.29(b)). The value of this capacitance can be obtained

Table 9.10 Load impedance synthesized at fundamental frequency and harmonics

Frequency (GHz)	EP Terminations (Ω)	CGP Terminations (Ω)
4	**32.9 + j*44.2**	**100**
8	3.4 – j*2.5	10
12	1.2 + j*19.4	600

by the small-signal model extracted for the selected bias condition (as in the following example) or, alternatively, read from the foundry manual.

The third de-embedding procedure, referred to as procedure C, exploits just a single capacitance that includes all the dynamic effects of the ED at the drain port (model C; see Figure 9.29(c)). The value of this capacitance can be obtained from the current and voltage at the EP by forcing a purely resistive load at the fundamental frequency at the CGP.

By exploiting a full model of the selected GaN ED, the extrinsic loading condition corresponding to class-F operation at the current-generator plane is found at 4 GHz; this condition is considered as the reference for the successive comparisons.

The impedance values synthesized at the EP and at the CGP at the fundamental frequency of 4 GHz and its harmonics are reported in Table 9.10. The terminations at the EP and at the CGP, obtained through the full model, are also shown in the Smith chart in Figure 9.30.

In agreement with class-F theory, a low resistance value (ideally a short circuit) is synthesized at the second harmonic frequency and a high resistance value (ideally an open circuit) is synthesized at the third harmonic at the CGP.

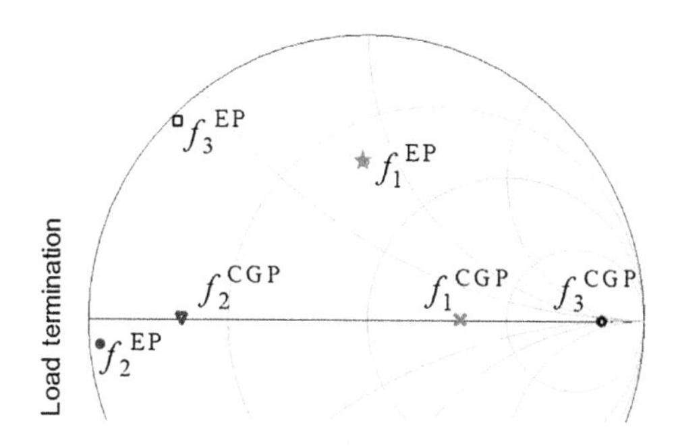

FIGURE 9.30

Terminations synthesized at 4 GHz at the EP at the fundamental frequency (star), the second harmonic (dot), and the third harmonic (square) and at CGP at fundamental (cross), second harmonic (triangle), and third harmonic (circle).

Figure 9.31 shows the load line synthesized at the extrinsic transistor plane and the associated current and voltage waveforms corresponding to the loading condition shown in Table 9.10.

Figure 9.32 shows the load line synthesized at the transistor CGP and the associated current and voltage waveforms corresponding to the loading condition in Table 9.10. By comparing Figure 9.31(a) and Figure 9.32(a), it is evident that the EP load line is very different from the CGP load line. As a consequence, by looking only at the EP, it is not possible to immediately state class-F operation. The capacitive core and parasitic network are responsible for these differences because their contributions hide the actual voltage and current shapes at the current generator.

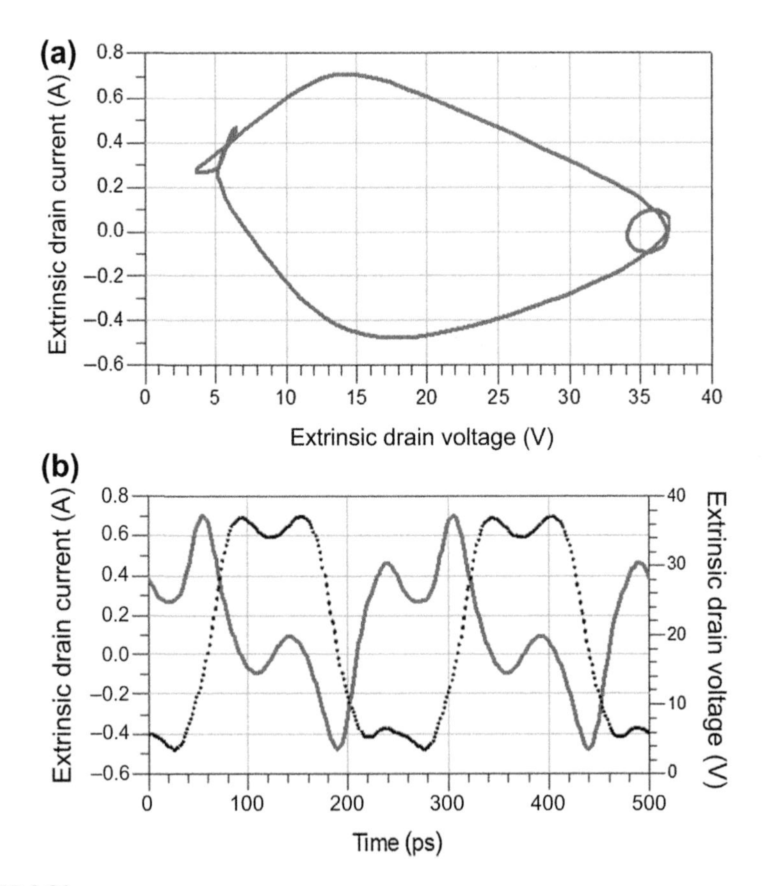

FIGURE 9.31

Load line (a) and time-domain voltage (dotted line) and current (solid line) waveforms (b) simulated at 4 GHz at the extrinsic plane for the load impedance defined in Table 9.10. Bias condition $V_{g0} = -3$ V, $V_{d0} = 20$ V, $I_{d0} = 60$ mA.

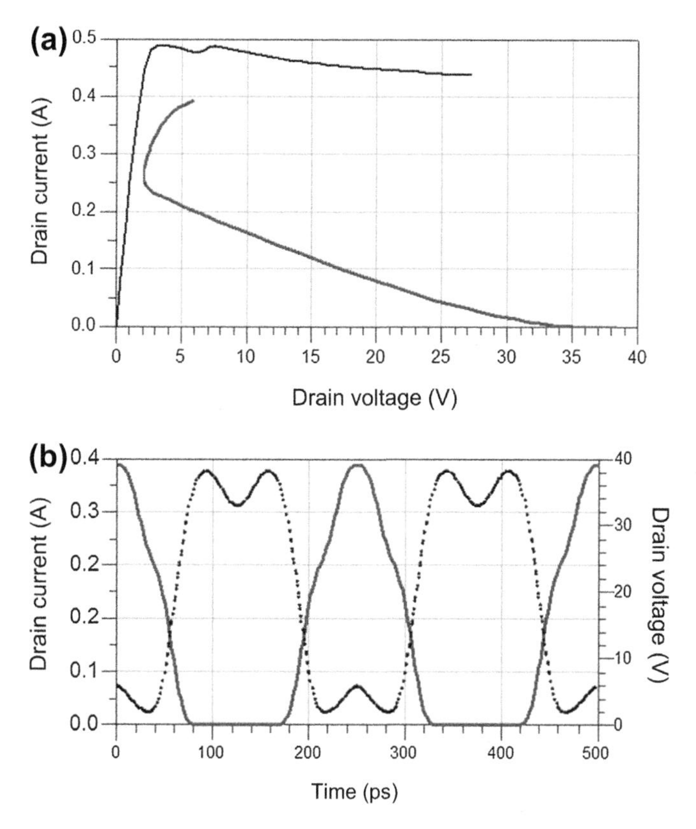

FIGURE 9.32

Load line (a) and time-domain voltage (dots) and current (solid line) waveforms (b) at the current-generator plane exploiting full model for the load impedance defined in Table 9.10. Bias condition $V_{g0} = -3$ V, $V_{d0} = 20$ V, $I_{d0} = 60$ mA; frequency 4 GHz. In (a), the load line is superimposed on DC characteristic at $V_{gs} = 0$ V.

The extrinsic electrical variables obtained by means of the full nonlinear model are now exploited to apply the three different de-embedding procedures.

Figure 9.33 shows the load line at the transistor CGP and the associated current and voltage obtained after the de-embedding based on procedure A (i.e., nonlinear de-embedding). As expected, because the nonlinear de-embedding procedure is exact by definition in the simulation environment, the de-embedded load line and current and voltage waveforms perfectly match the reference ones.

The next step consists of de-embedding the waveforms of Figure 9.31 by exploiting procedure B (see Figure 9.29(b)). The resulting waveforms at the CGP are shown in Figure 9.34, where the waveforms obtained by exploiting the full nonlinear model are also reported to make the comparison easier. As can be seen, the load line

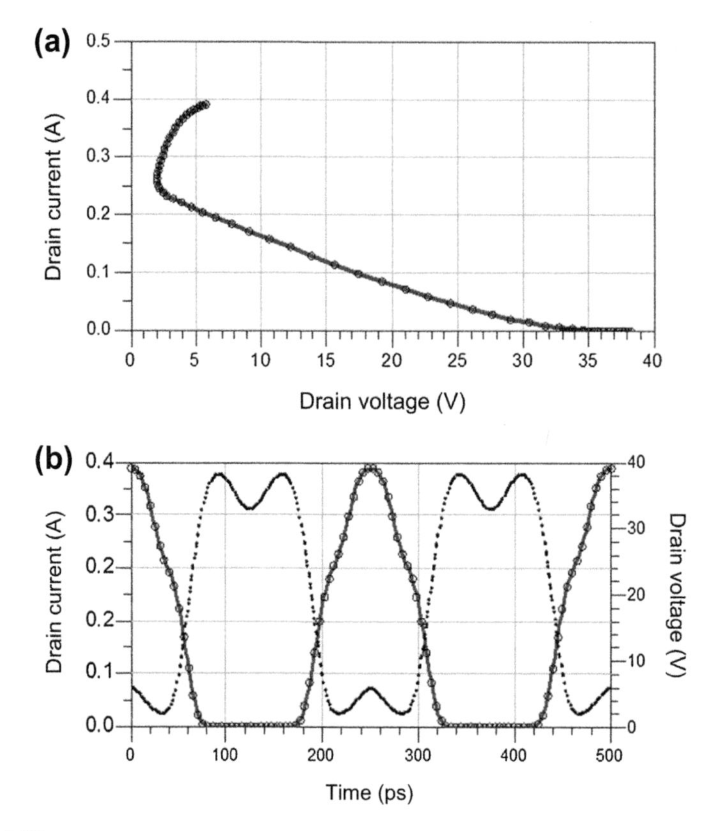

FIGURE 9.33

Load line (a) and time-domain voltage (solid line) and current (symbols) waveforms (b) de-embedded at the current-generator plane exploiting procedure A for the load impedance defined in Table 9.10. Bias condition $V_{g0} = -3$ V, $V_{d0} = 20$ V, $I_{d0} = 60$ mA. In (a), the load-line is superimposed on DC characteristic at $V_{gs} = 0$ V.

obtained in this way is significantly different from the reference one. The impedance values corresponding to this load line are reported in Table 9.11. Note that the first harmonic moves away from the reference value, whereas the second harmonic components are close to the reference value. Finally, the third harmonic is far from the open condition defined at the CGP for class-F operation.

Finally, the same extrinsic waveforms in Figure 9.31 are de-embedded to obtain CGP waveforms using procedure C (see Figure 9.29(c)), which uses a single capacitance to model both the parasitic network and capacitive core contributions.

The de-embedding results are shown in Figure 9.35, where the waveforms obtained by exploiting the full model are reported to make the comparison easier. As can be seen, the load line obtained in this way is again different from the

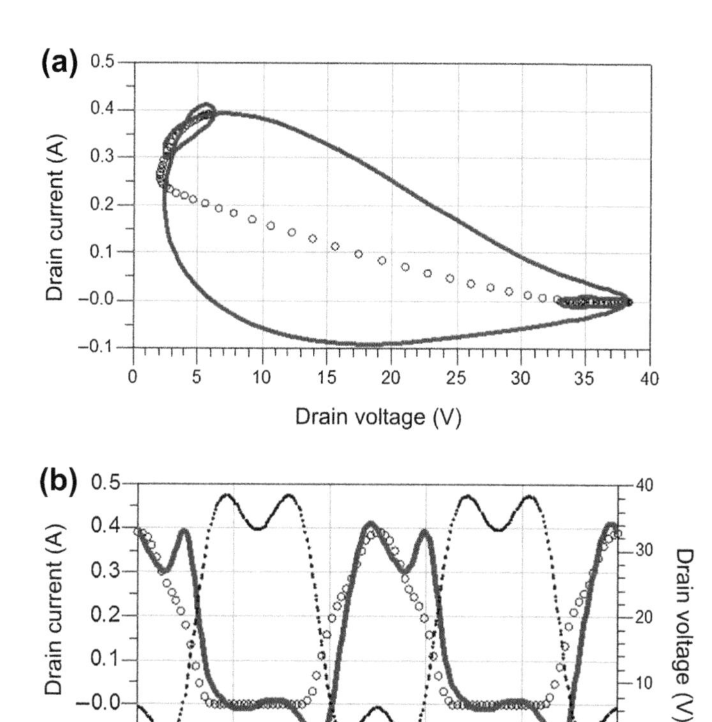

FIGURE 9.34

(a) Load line obtained at the current-generator plane for procedure B (solid line) and the procedure from Figure 9.32 (symbols). (b) Time-domain voltage (dotted line) and current (thick solid line) obtained from procedure B and time-domain current of Figure 9.32 (circles).

Table 9.11 Load impedance retrieved at the current generator plane

Frequency (GHz)	Full Model	Procedure A (Ω)	Procedure B (Ω)	Procedure C (Ω)
4	100	100	80 + j*39	92.2
8	10	10	6.9 + j*3.4	3 − j*2.6
12	600	600	12.2 + j*102.1	−47.8 + j*111.2

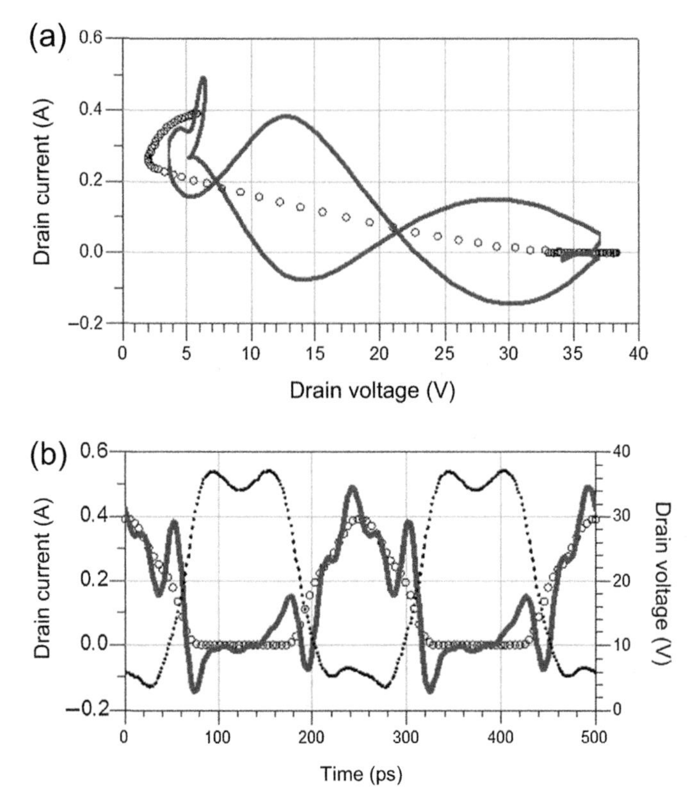

FIGURE 9.35

(a) Load line obtained at the current-generator plane exploiting procedure C (solid line) and the procedure from Figure 9.32 (symbols). (b) Time-domain voltage (dotted line) and current (thick solid line) obtained exploiting procedure C and time-domain current from Figure 9.32 (circles).

reference one. The impedance values corresponding to this load line are reported in Table 9.11. Note that the second harmonic component is close to the reference value, whereas the first harmonic moves away from the reference value and the third harmonic assumes a negative value.

Obviously, the differences between the three de-embedding techniques explode when the operating frequency increases, as is evident by looking at the second and third harmonic values in the previous examples. On the other hand, when the operating frequency decreases, the differences are less evident. As an example, the extrinsic waveforms corresponding to the same current generator termination defined in Table 9.10 are found by using the full nonlinear model for the operating frequency of 1 GHz, as reported in Table 9.12.

Table 9.12 Load impedance synthesized at fundamental frequency and harmonics

Frequency (GHz)	EP Terminations (Ω)	CGP Terminations (Ω)
1	86.1 + j*29	100
2	5.9 − j*0.2	10
3	17.9 + j*106.3	600

The EP load line synthesized at 1 GHz and the current and voltage waveforms corresponding to the loading condition of Table 9.12 are shown in Figure 9.36.

Figure 9.37 shows the load line synthesized at the transistor CGP and the associated current and voltage waveforms corresponding to the loading condition in Table 9.12. Also in this case, the EP load line is different from the CGP one because

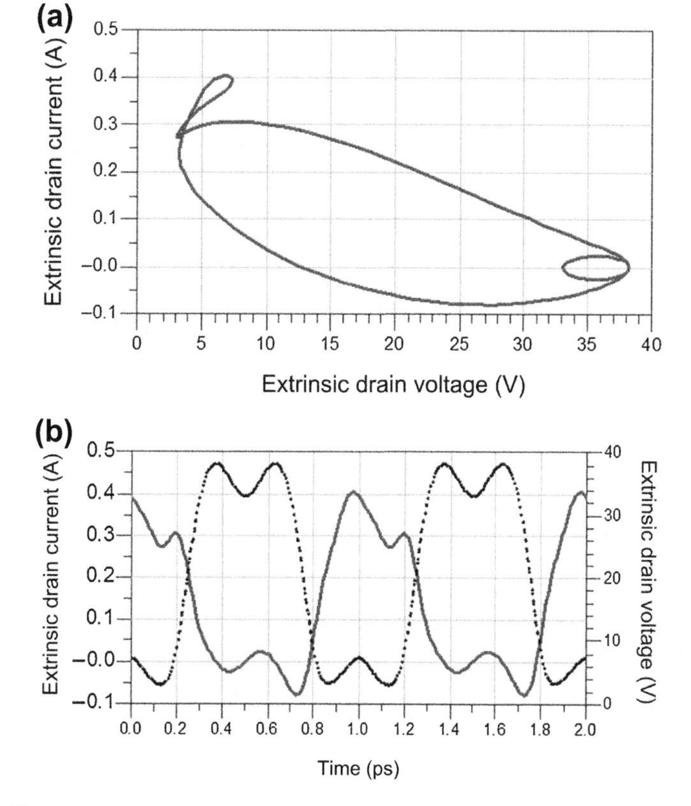

FIGURE 9.36

Load line (a) and time-domain voltage (dotted line) and current (solid line) waveforms (b) simulated at 1 GHz at the extrinsic plane for the load impedance defined in Table 9.12. Bias condition $V_{g0} = -3$ V, $V_{d0} = 20$ V, $I_{d0} = 60$ mA.

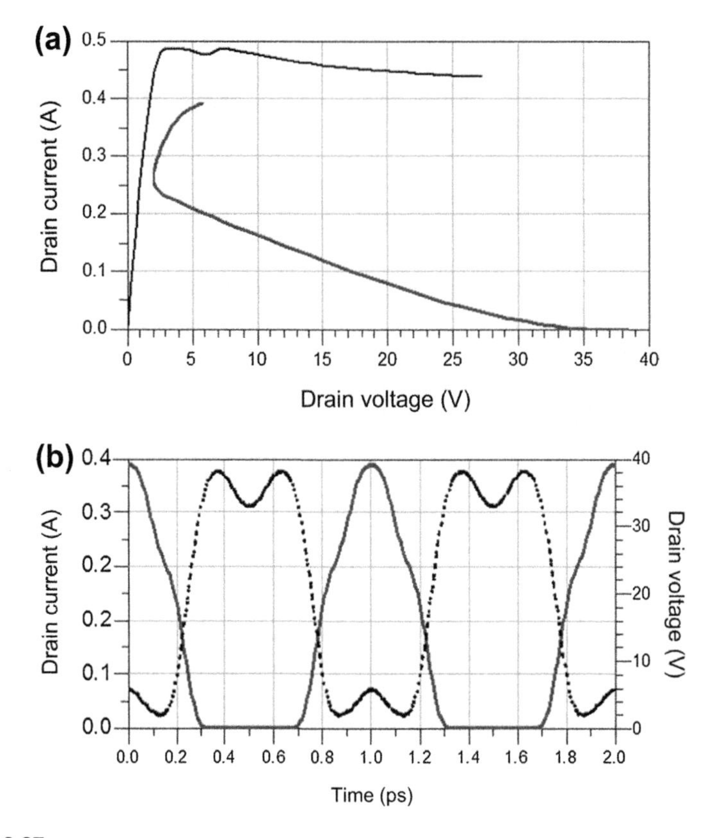

FIGURE 9.37

Load line (a) and time-domain voltage (symbols) and current (solid line) waveforms (b) at the current-generator plane exploiting full model for the load impedance defined in Table 9.12. Bias condition $V_{g0} = -3$ V, $V_{d0} = 20$ V, $I_{d0} = 60$ mA; frequency 1 GHz. In (a), the load line is superimposed on DC characteristic at $V_{gs} = 0$ V.

the capacitive core and parasitic network contributions hide the actual voltage and current shapes at the current generator.

The extrinsic waveforms at 1 GHz are de-embedded by using procedures A, B, and C in order to obtain the waveforms at the current-generator plane. The results are shown in Figures 9.38–9.40, respectively. The impedance values corresponding to these load lines are reported in Table 9.13. The proposed results confirm the lower discrepancies introduced at lower frequencies.

Another important aspect to be considered is the limited number of harmonics that can be gathered by microwave time-domain measurement setups due to their bandwidth limitation. The cutoff of harmonic components adversely affects the

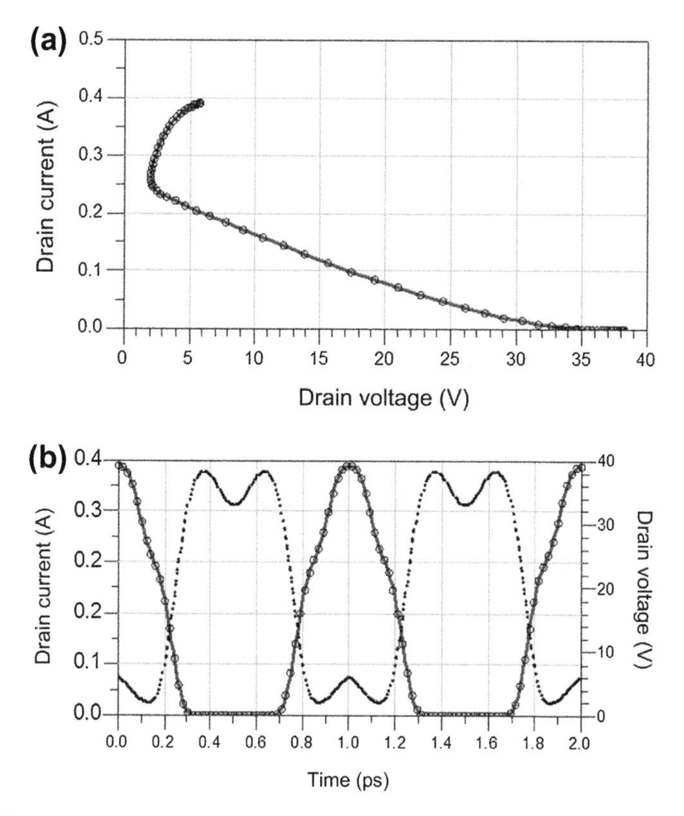

FIGURE 9.38

(a) Load line obtained at the current-generator plane for procedure A (solid line) and the load line from Figure 9.37 (symbols). (b) Time-domain voltage (dotted line) and current (thick solid line) obtained for procedure A and the time-domain current of Figure 9.37 (circles).

de-embedded waveforms at the current-generator plane. For instance, also in this case, a negative current is computed at the current-generator plane when harmonics above the third harmonic are neglected. Figure 9.41 shows an example of the effect of harmonic cutoff on the current waveforms drawn in Figure 9.33. The solid line represents the drain current waveform de-embedded by exploiting procedure A (i.e., nonlinear de-embedding). Circles and dashed lines represent the same waveforms considering three and five harmonic components, respectively. As shown, there is a negative contribution of few milliamps, equal to approximately 2.2% of the maximum drain current (i.e., 400 mA). When nonlinear de-embedding is considered, the impact of the cutoff of harmonic components is small. When simplified procedures B or C are used for the de-embedding (i.e., linear de-embedding), the problem becomes critical.

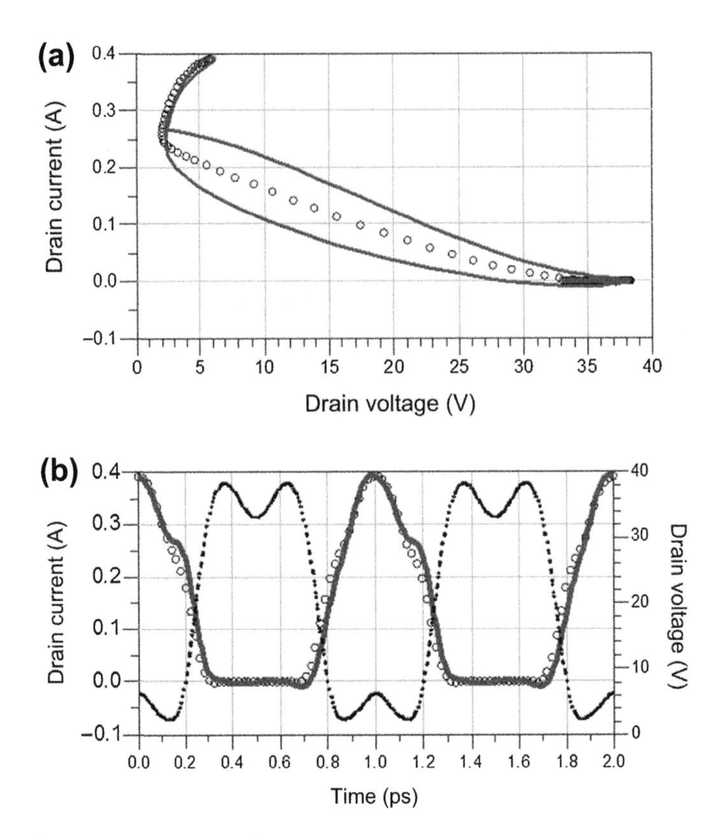

FIGURE 9.39

(a) Load line obtained at the current-generator plane for procedure B (solid line) and the load line from Figure 9.37 (symbols). (b) Time-domain voltage (dotted line) and current (thick solid line) obtained for procedure B and time-domain current of Figure 9.37 (circles).

With the aim of demonstrating this assumption, Figure 9.42 shows an example of the effect of harmonic cutoff on the load line when procedure B and C are used. The solid line represents the drain current waveform de-embedded by exploiting model A and considering 10 harmonics. Circles and dotted lines refer to the same waveforms de-embedded by exploiting procedures B and C, respectively, and considering only three harmonics. As shown for procedure B, the negative current achieves a negative peak of 95 mA (i.e., 24% of the maximum drain current); for procedure C, the negative current achieves a negative peak of 71 mA (i.e., 18% of the maximum drain current).

To summarize, harmonic truncation leads to evident nonphysical behavior when linear de-embedding is adopted whereas it affects the results of nonlinear de-embedding to a lesser extent.

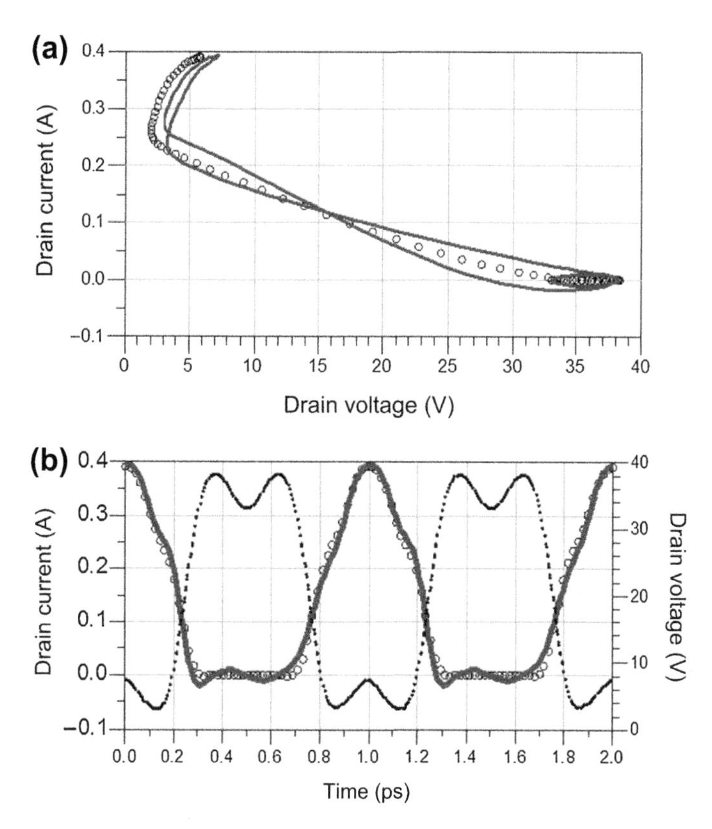

FIGURE 9.40

(a) Load line obtained at the current-generator plane for procedure C (solid line) and the load line from Figure 9.37 (symbols). (b) Time-domain voltage (dotted line) and current (thick solid line) obtained for procedure C and time-domain current from Figure 9.37 (circles).

Table 9.13 Load impedance retrieved at the current generator plane

Frequency (GHz)	Full Model (Ω)	Procedure A (Ω)	Procedure B (Ω)	Procedure C (Ω)
1	100	100	99.8 + j*3.6	95.9
2	10	10	9.9 + j*1.2	5.8 + j*0.6
3	600	600	588.4 – j*35	362 – j*322

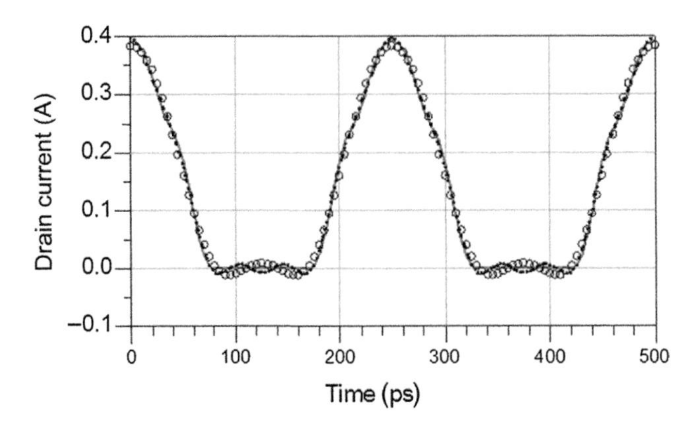

FIGURE 9.41

Drain current waveform de-embedded from the load line of Figure 9.31 using procedure A, considering 10 harmonics (solid line), five harmonics (dashed line) and three harmonics (circles).

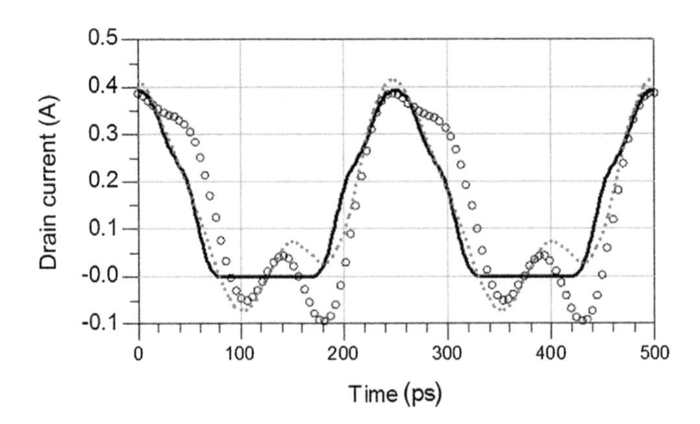

FIGURE 9.42

Drain current waveform de-embedded from the load line of Figure 9.31 using procedure A (solid line), procedure B (circles), and procedure C (dotted line).

9.4.1 Theoretical formulation

The nonlinear de-embedding technique reported in this paragraph relies on the model topology shown in Figure 9.1. To explain how the current and voltage waveforms at the extrinsic ED ports can be used to obtain the waveforms at the current-generator plane, it is convenient to express intrinsic and extrinsic voltages and currents in terms of their practically finite number M of spectral components as in Eqn (9.2).

After carrying out high-frequency time-domain measurements, the first step of the nonlinear de-embedding procedure consists of shifting the measured spectra up to the IPs. This step requires an accurate description of the parasitic networks (including the device package), which can be obtained through *S* parameter measurements by widely known techniques based on lumped elements approximation (e.g., [23–25]) or, alternatively, by adopting electromagnetic simulations of the device layout (e.g., [26–28]). This step can be accomplished straightforwardly by applying a linear transformation in frequency domain, as expressed by Eqn (9.3). The currents obtained after transformation in Eqn (9.3) can be simply expressed as the vector sum of the contributions coming from the resistive and the capacitive cores, which are assumed to be strictly in parallel, as expressed by Eqn (9.4).

The second step of the procedure consists of separating the reactive component from the resistive one in Eqn (9.4). This step is fundamental to obtaining the device load line at the resistive core plane (i.e., the CGP). Due to the presence of the capacitive core, the knowledge of the electrical quantities at the IP does not allow one to correctly evaluate the current, and consequently the load line, synthesized at the CGP.

The de-embedding of the parasitic elements can give just an idea of the resistive core load line, which becomes adequately accurate when the capacitive core contribution can be neglected; as previously discussed, this is the case in a sufficiently LF operation. It is necessary to identify the displacement current contribution in Eqn (9.4). To this aim, a model-based description of the nonlinear capacitances has to be exploited, which, under the hypothesis of negligible dispersion of the capacitive core, can be simply identified by adopting standard multi-bias *S* parameter measurements or, alternatively, by exploiting large-signal time-domain measurements. When non-quasi-static effects can be neglected, the intrinsic capacitive currents can be expressed as in Eqn (9.5).

Once this operation has been carried out, the resistive core electrical variables are available. In particular, the resistive currents can be obtained by simply subtracting the capacitive currents from the total intrinsic currents by exploiting Eqn (9.4), whereas the voltages coincide with the intrinsic ones because the capacitive and resistive cores are in parallel.

These operations allow one to retrieve the load line and the synthesized impedances at the current generator at the fundamental and harmonic frequencies. The described design technique can be successfully used to perform waveform engineering in the measurement phase by monitoring the dynamic load line at the resistive core, while optimum source and loading conditions are achieved at the accessible EPs [72]. The flowchart in Figure 9.43 summarizes the fundamental steps of the proposed nonlinear de-embedding technique.

9.4.2 Experimental examples

As a first example, the whole de-embedding procedure is applied on a GaN HEMT ($0.7 \times 800\ \mu m^2$). For this device, the model of the parasitic network was extracted

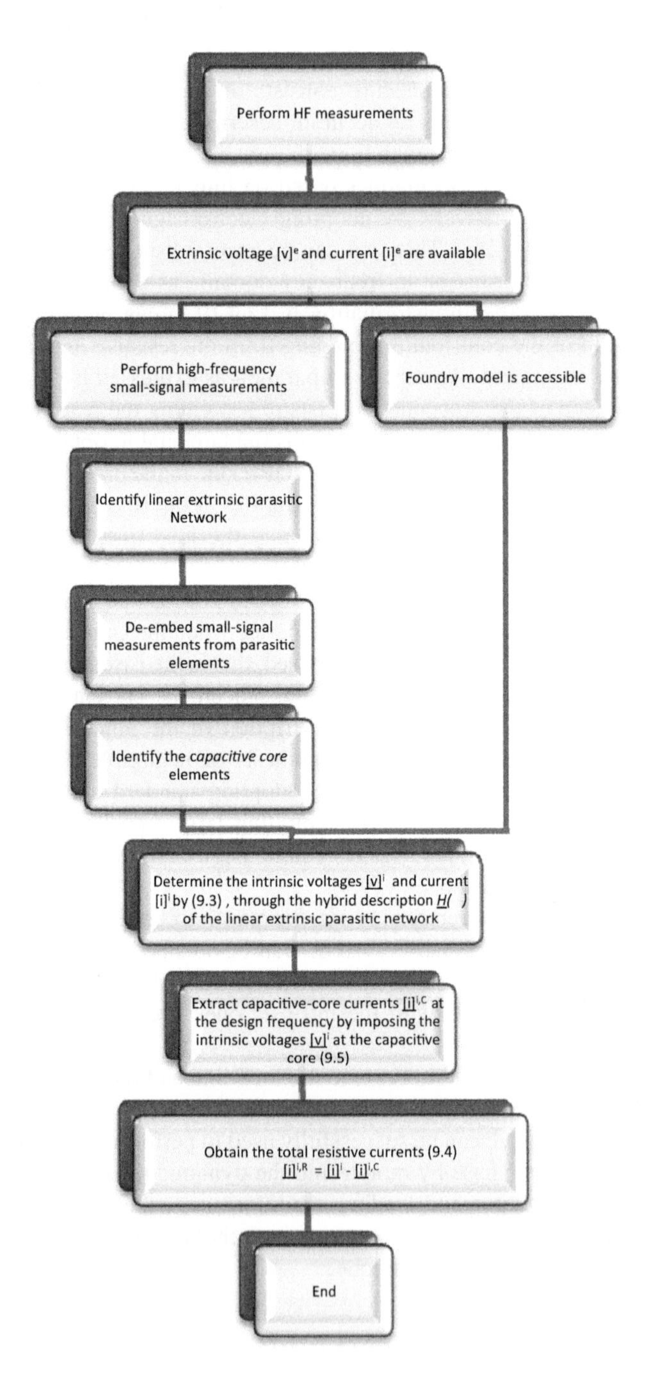

FIGURE 9.43

Flowchart describing the nonlinear de-embedding technique.

by exploiting a conventional technique based on cold scattering parameter measurements, while the capacitive-core model was extracted from multi-bias S parameter measurements and then stored into a lookup table. In particular, a quasi-static description was found to be adequate for the capacitive core.

With the aim of applying a nonlinear de-embedding procedure, active load-pull time-domain measurements were carried out by exploiting a LSNA, which provides vector calibrated scattered waves in the 600 MHz−50 GHz frequency bandwidth. High-frequency sources combined with PAs are used to reproduce realistic trajectories and to actively control the termination at the fundamental frequency (f_0). The latter is set equal to 4 GHz and the device is biased at $V_{g0} = -2$ V and $V_{d0} = 25$ V, which corresponds to class-A operation. Different loading trajectories are obtained by varying the phase of the signal injected at the output port, thus changing the phase relationship between input and output signals.

The nonlinear de-embedding procedure was applied to the measured extrinsic load line to obtain the waveforms at the intrinsic and current-generator planes. The results of de-embedding are reported in Figure 9.44 for three different values of the output power. As shown in the figures, the extrinsic load lines are affected by a strong reactive component, whose major contribution is due to the nonlinear capacitive core. In fact, for the selected measurement frequency, the linear parasitic network slightly rotates and shrinks the measured load line, as confirmed by comparing the extrinsic load line and the transformed one after applying Eqn (9.3). Finally, in Figure 9.44, the load lines at the CGP, which actually determine the active device performance and represent the theoretical basis of PA design techniques, are shown. It is clear that the combined effect of the parasitic network in conjunction with the capacitive nonlinearities impacts significantly the shape of the load line, masking the information relative to the solely resistive core.

To demonstrate the impact of the capacitive core on the impedances synthesized at the CGP when microwave operation is involved, a comparison between the synthesized impedances at the three different considered planes (i.e., extrinsic, intrinsic, and CGPs) is reported in Table 9.14. The numbers in the table further highlight the importance of shifting the measurements at the current-generator plane, where the synthesized impedances strongly differ from the one obtained at the measurement plane. It must be underlined that the impedances at the current-generator plane determine the voltage and current waveforms, which can be compared with the theoretical waveforms in order to correctly perform waveform engineering [15−17].

In Figure 9.45, the three load lines at the CGP are drawn together with the static characteristic measured at $V_{GS0} = 0$ V. The marked point on the load lines corresponds to an instantaneous value of the intrinsic gate voltage of 0 V; if LF dispersion was not present, these points should belong to the depicted static characteristic. However, as clearly observable, the dynamic current significantly deviates from the static characteristics, which is an indication of the influence of thermal effects and traps [29−36].

Another strategy to identify the capacitive-core model, which can be used in rigorous nonlinear de-embedding, is to carry out LSNA vector large-signal

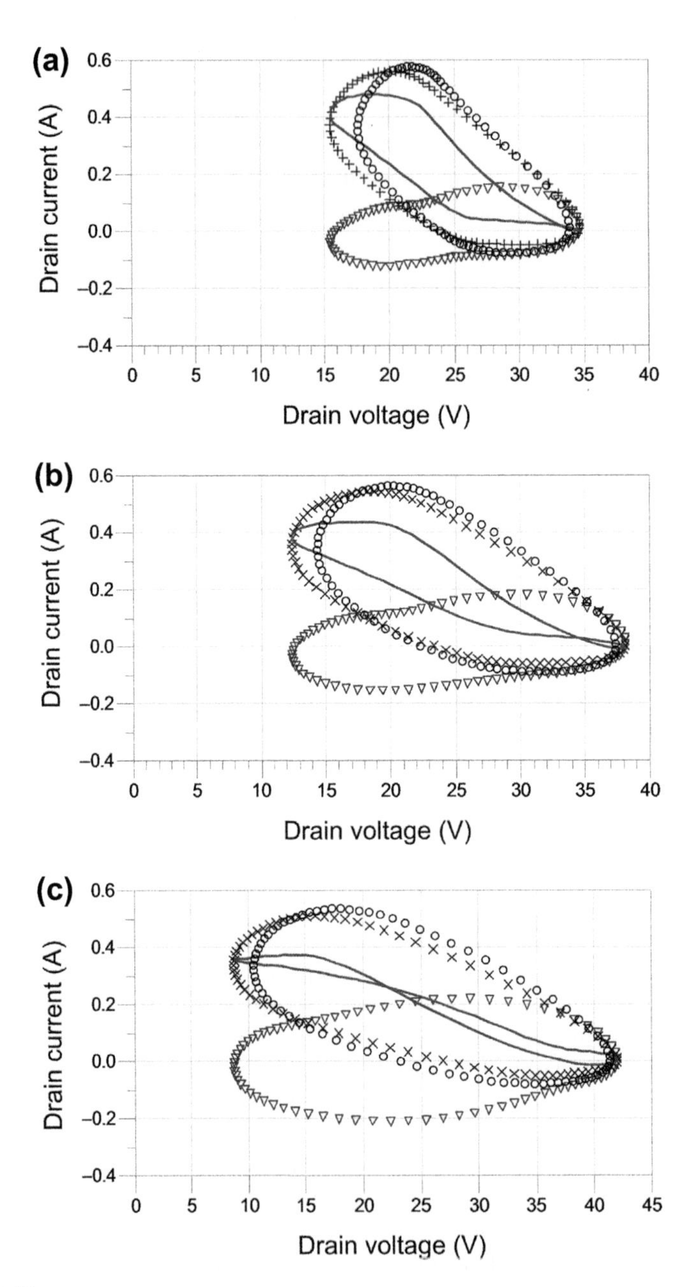

FIGURE 9.44

Measurements performed on GaN HEMT 0.7 × 800 μm^2 biased at $V_{g0} = -2$ V and $V_{d0} = 25$ V at $f_0 = 4$ GHz. Measured extrinsic load line (circles), load line after parasitic network de-embedding (crosses), displacement current (triangles), and load line at the intrinsic resistive core (continuous line) for $P_{out} = 0.86$ W (a); $P_{out} = 1.26$ W (b); and $P_{out} = 1.7$ W (c).

From [68]

Table 9.14 Load impedances corresponding to the synthesized load lines for three power levels and up to the third harmonic

	Frequency (GHz)	Extrinsic Plane (Ω)	Intrinsic Plane (Ω)	Current-Generator Plane (Ω)
$P_{out} = 0.86$ W	4	12.0 + j*16.9	16.3 + j*20.9	29.1 + j*16.8
	8	28.3 − j*16.5	28.9 − j*10.9	20.5 − j*25.1
	12	58.0 + j*18.5	70.4 + j*12.4	18.5 − j*27.8
$P_{out} = 1.26$ W	4	18.4 + j*25.2	23.5 + j*29.5	47.4 + j*20.3
	8	28.1 − j*16.8	28.8 − j*11.5	25.0 − j*23.7
	12	54.6 + j*18.2	67.0 + j*14.0	17.4 − j*26.4
$P_{out} = 1.7$ W	4	31.4 + j*37.7	38.8 + j*42.2	82.0 + j*5.6
	8	327.8 − j*17.2	28.7 − j*12.3	29.1 − j*16.9
	12	53.9 + j*23.3	69.4 + j*19.5	20.4 − j*26.6

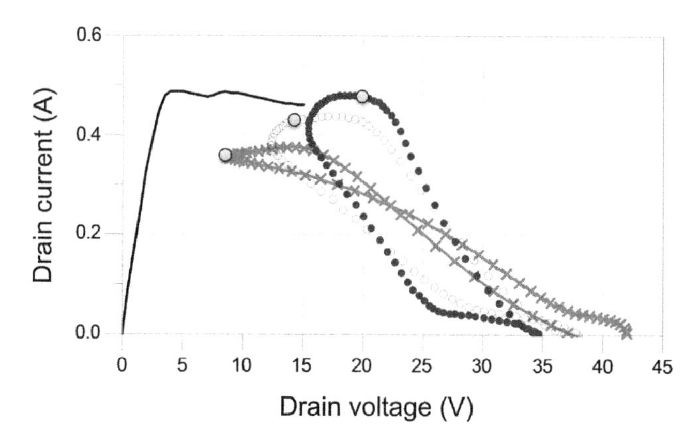

FIGURE 9.45

Load lines (symbols) at the intrinsic resistive core for the output power values reported in Figure 9.44: $P_{out} = 0.86$ W (crosses); $P_{out} = 1.26$ W (circles); and $P_{out} = 1.7$ W (dots). Load lines are superimposed on the static characteristic at $V_{GS0} = 0$ V (continuous line). The marked point on each load line corresponds to the instantaneous gate voltage value equal to 0 V.

From [68]

measurements in the low- (<30 MHz) and high-frequency range. Different from multi-bias S parameter based identification, the use of large-signal excitations allows one to extract a model under realistic conditions and with a reduced number of experiments, covering a wide region of operation instantaneously, without incurring degradation of the device. The parasitic network is here extracted by exploiting a conventional technique based on cold scattering parameter measurements; then, the obtained values are further optimized by exploiting the LSNA measurements. To better understand the capacitive-core model extraction procedure, some details are given in the following.

First of all, LF measurements are carried out with the aim of extracting a local model for the resistive core and to optimize the values of the resistive elements of the parasitic network. The fundamental frequency of the excitations in this phase is set around 2 MHz, which allows simplification of the circuit in Figure 9.29(a) to only the resistive parasitic elements and the I_{ds} generator, totally neglecting the contribution of the capacitive core and the reactive elements of the parasitic network. Next, large-signal high-frequency measurements are carried out with the aim of extracting the model of the capacitive core and optimizing the values of the reactive parasitic network elements. At such an operating frequency, the complete circuit in Figure 9.29(a) is considered wherein R_s, R_d, R_g, and I_{ds} are known from the first optimization step based on the LF measurements, whereas the capacitive-core model has to be extracted. Any analytical or empirical model formulation could be used in this phase for the capacitive core. After the identification steps, all the information is available in order to perform a full nonlinear de-embedding of any high-frequency load line measured in different bias conditions from the one exploited in the optimization process. This procedure was applied to an AlGaN/GaN HEMT device ($0.7 \times 800\ \mu m^2$) on an SiC substrate. The capacitive core and parasitic network model extractions followed the two-step procedure just explained. First, a set of LF large-signal measurements, shown in Figure 9.46(a), is performed at 2 MHz under a fixed bias condition $V_{g0} = -2$ V and $V_{d0} = 20$ V. These measurements were used for the extraction of the nonlinear Angelov's model and the parameters were obtained for the selected bias condition, through least-squares based numerical optimization combined with a harmonic balance solver. Only the resistive parasitic elements R_S and R_D were considered in this step. R_g is neglected in this phase because there is no conduction at the gate side for the selected operating conditions. In the second step, a set of high-frequency load lines measured at 4 GHz, shown in Figure 9.46(b), in the same bias condition as before are employed to determine the model of the capacitive core and to optimize the values of the remaining parasitic elements (i.e., R_g, L_g, L_s, L_d, C_g, C_d). For the capacitive core, the charge equations of Angelov's model were exploited [47]. Different from the I_{ds} source, the capacitive core is assumed to be nondispersive; as such, its instantaneous response depends only on the instantaneous values of the intrinsic voltages, $v_{gs}(t)$ and $v_{ds}(t)$. As a consequence, the descriptions obtained for the charge sources continue to be valid regardless of the selected bias point and the device thermal state [32]. As before, both the parameters of the capacitive-core model and the value of the parasitic are obtained by numerical optimization.

After the identification steps, active load-pull measurements were performed at $f_0 = 4$ GHz, $V_{g0} = -3$ V, and $V_{d0} = 25$ V to perform a full nonlinear de-embedding. The de-embedded transcharacteristic and load line are shown in Figure 9.47. The fact that the de-embedded transcharacteristic shows clipping to zero ampere when

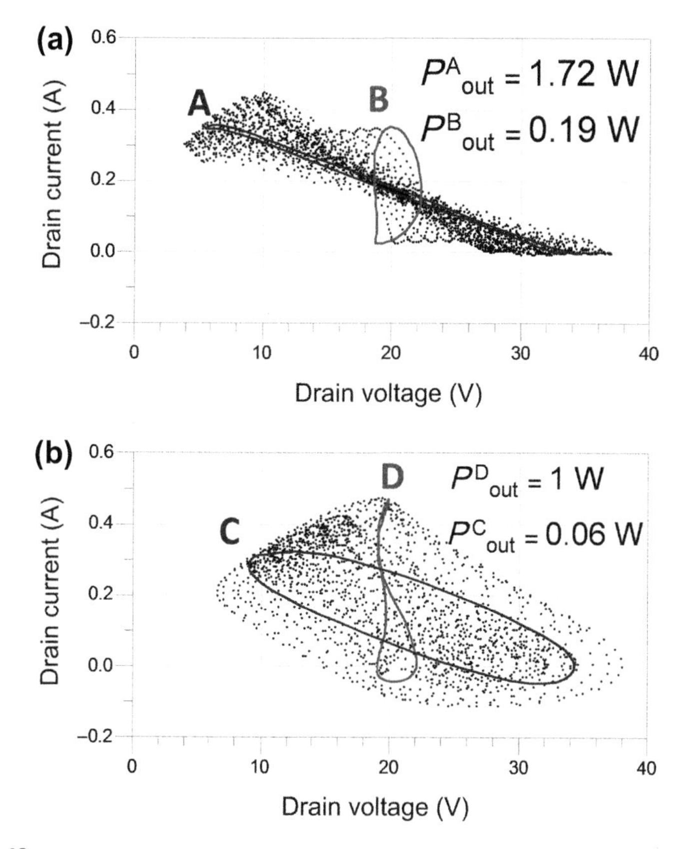

FIGURE 9.46

Measurements performed on a GaN HEMT 0.7 × 800 μm^2 biased at $V_{g0} = -2$ V and $V_{d0} = 20$ V. Measured load lines were at the extrinsic plane at $f_0 = 2$ MHz (a) and $f_0 = 4$ GHz (b). These data are used during the numerical optimization. The continuous line marks load lines corresponding to different power levels.

From [69]

the gate voltage stays below the threshold voltage constitutes a qualitative indication of the validity of the applied procedure.

9.5 Nonlinear embedding versus de-embedding: a comparative analysis

In this section, a comparative study is presented for the two nonlinear procedures described in this chapter: embedding and de-embedding. The flowchart in Figure 9.48 summarizes and links the fundamental steps of the described

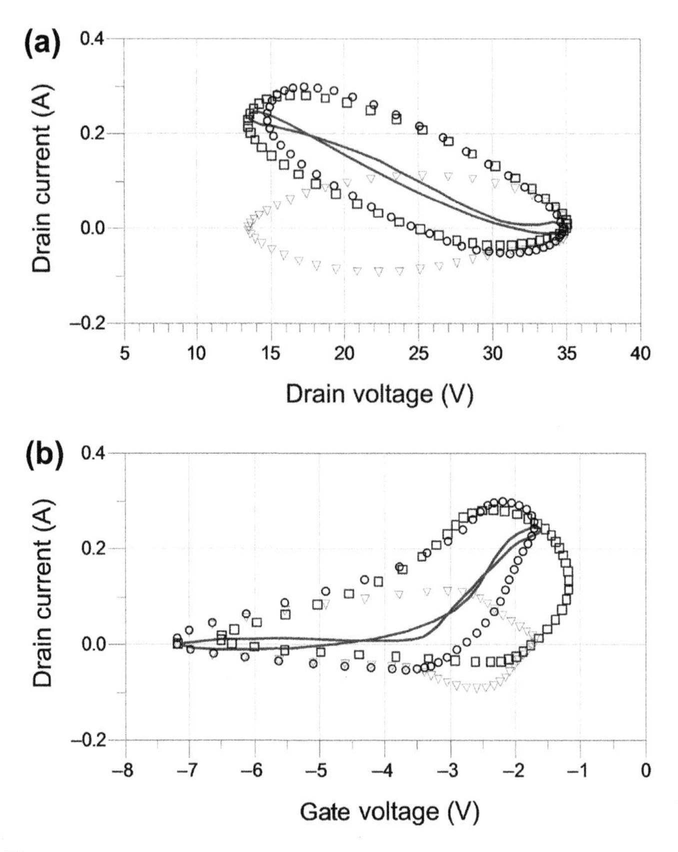

FIGURE 9.47

Measurements performed on GaN HEMT 0.7 × 800 μm² biased at $V_{g0} = -3$ V and $V_{d0} = 25$ V at $f_0 = 4$ GHz. Load line (a) and transcharacteristic (b) are shown. Measurements were at the extrinsic plane (circles); after parasitic de-embedding (squares); at the drain-source CGP (continuous line); and at the displacement current (triangles). $P_{out} = 0.701$ W.

From [69]

procedures. The de-embedding procedure exploits high-frequency nonlinear measurements performed by means of an LSNA setup. These measurements fully characterize the device behavior at the EP. As shown by following the solid arrows in Figure 9.48, to move from the EP to the IP, Eqn (9.3) has to be exploited for obtaining voltage and current phasors de-embedded from the parasitic network effects. Next, electrical variables at the CGP can be computed by simply subtracting the capacitive currents, obtained from the model by means of Eqn (9.5), from the intrinsic currents in Eqn (9.4). In this way, the load line and thus the synthesized impedance at the current-generator plane at the fundamental and harmonic frequencies are derived.

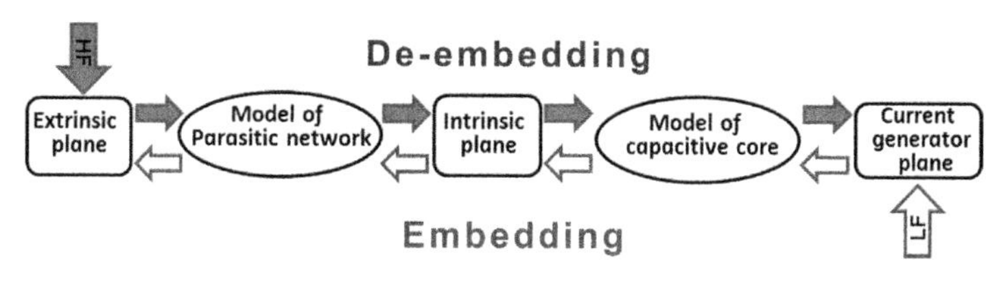

FIGURE 9.48

Flowchart describing the nonlinear de-embedding and embedding procedures.

From [70]

The embedding procedure uses LF nonlinear measurements, performed by means of the measurement setup described in Ref. [36], to directly characterize the ED resistive core. Thus, these measurements are defined at the CGP (see Figure 9.10). As shown by following the outlined arrows in Figure 9.48, to move from the CGP to the IP, the capacitive current contribution, obtained by Eqn (9.5), has to be summed to the measured resistive contribution by exploiting Eqn (9.4). In this way, the intrinsic current is derived. Then, the extrinsic quantities are available by simply applying Eqn (9.3), which allows one to embed the effects of the parasitic network. At this point, the load line and the corresponding impedances to be synthesized at the EP at the fundamental and harmonic frequencies are available.

The two techniques were compared on a GaN HEMT ($0.7 \times 800\ \mu m^2$) biased under class-A operation at the fundamental frequency of 4 GHz. High-linearity operation was chosen in order to comply with the power capability of the adopted LSNA. As described, the two procedures require a description of the parasitic network and capacitive core. The parasitic elements were extracted by exploiting a conventional technique based on cold scattering parameter measurements. Regarding the capacitive core, at the operating frequency of 4 GHz, a quasi-static description exploiting multi-bias S parameters was found adequate to extract the bias-dependent capacitances.

Once the capacitive core and parasitic network models were identified, measurements were carried out in order to compare the two approaches. Nonlinear measurements at 4 GHz were carried out by exploiting an LSNA, thus obtaining waveforms at the EP. Then, by applying the de-embedding procedure, the load line at the CGP was derived.

To compare the two procedures, nonlinear measurements at 2 MHz were carried out using the LF measurement setup proposed in Ref. [36] and the embedding procedure was applied. In particular, with the aim of setting the same operating condition for the two procedures, the intrinsic voltages found by de-embedding were applied at the ED resistive core by means of the LF measurement setup, obtaining waveforms at the CGP. Successively, by applying the embedding procedure, the waveforms at the EP were derived.

Figure 9.49(a) shows the comparison between the load lines measured at low frequency and the ones obtained at the current-generator plane through the nonlinear

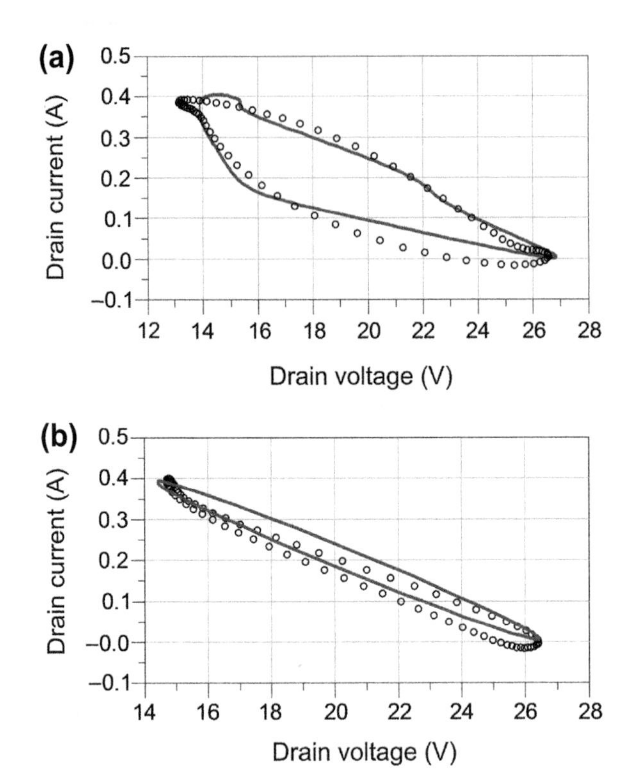

FIGURE 9.49

Measurements performed on a GaN HEMT 0.7 × 800 μm² biased at $V_{g0} = -2$ V and $V_{d0} = 20$ V. Load lines are at the current-generator (a) and extrinsic (b) planes. In (a), symbols indicate the load line de-embedded from high frequency. The continuous line shows the measurement at 2 MHz. In (b), symbols represent the measurement at 4 GHz. The continuous line is the load line embedded from low frequency.

From [70]

de-embedding. Figure 9.49(b) shows the comparison between the load lines measured at high frequency and the one obtained at the EP through the embedding procedure. The agreement between the two approaches is good both at the extrinsic and current-generator planes.

As shown in Table 9.15, the output impedances at the fundamental frequency obtained through the two approaches are in excellent agreement. For both procedures, it is important to consider the uncertainty related to the capacitive-core model; in the de-embedding procedure, it affects the variables at the CGP, whereas in the embedding procedure, it affects the variables at the EP. In PA design, it is clearly preferable to be more accurate in the impedance to be synthesized at the EP, although it affects the cost of the measurement system. As previously mentioned,

Table 9.15 Load impedances at extrinsic and current generator planes

	Current-Generator Plane (Ω)	Extrinsic Plane (Ω)
Embedding	29.8 – i*11.2	30.6 + i*4.4
De-embedding	30.3 – i*10.7	28.9 + i*2.8

high-frequency measurement systems are more expensive and have frequency and power limitations that can be simply overcome by using the embedding procedure. In fact, the power limit at low frequency is not a critical issue. High power levels can be easily managed without expensive instrumentation. In addition, increasing the operating frequency leads necessarily to increased measurement uncertainty.

The two procedures are not equivalent in terms of uncertainty associated with the electrical quantities of interest. As an example, the resistive gate current (i.e., gate-source diode conduction and gate-drain diode breakdown) is of great interest for reliability issues. In the embedding procedure, the current $I_g^{i,R}$, which is in the micro-ampere order of magnitude, is directly measured; in the de-embedding procedure, it is obtained by subtracting the capacitive current to the total intrinsic one by applying Eqn (9.4). Because these two components (i.e., $I_g^{i,C}$ and I_g^i) are of the same order of magnitude, the uncertainty on the estimated $I_G^{i,R}$ is clearly unacceptable.

Acknowledgments

The authors wish to thank Prof. Dominique Schreurs, Dr Gustavo Avolio, and Dr Giovanni Crupi for fruitful discussions on measurement and modeling techniques.

References

[1] Schreurs D, Van der Zanden K, Verspecht J, De Raedt W, Nauwelaers B. Real-time measurement of InP HEMTs during large-signal RF overdrive stress. In: Proc Eur Gallium Arsenide Related III–V Comp Applic Symp (Amsterdam, The Netherlands), October 1998. pp. 545–50.

[2] Horn JM, Verspecht J, Gunyan D, Betts L, Root DE, Eriksson J. X-parameter measurement and simulation of a GSM handset amplifier. In: Proc Microwave Integ Circuit Conf (Amsterdam, The Netherlands), October 2008. pp. 135–8.

[3] Verspecht J, Gunyan D, Horn J, Xu J, Cognata A, Root DE. Multi-tone, multi-port, and dynamic memory enhancements to PHD nonlinear behavioral models from large-signal measurements and simulations. In: Proc. IEEE MTT-S Int. Microwave Symp (Honolulu, HI, USA), June 2007. pp. 969–72.

[4] Van Moer W, Rolain Y. A large-signal network analyzer: Why is it needed? IEEE Microwave Mag December 2006;7(6):46–62.

[5] Lott U. Measurement of magnitude and phase of harmonics generated in nonlinear microwave two-ports. IEEE Trans Microw Theory Tech October 1989;37(10):1506–11.
[6] Adhikari S, Ghiotto A, Wang K, Wu K. Development of a large-signal network-analyzer round-robin artifact. IEEE Microwave Mag January 2013;14(1):140–5.
[7] Blockley PS, Scott JB, Gunyan D, Parker AE. The random component of mixer-based nonlinear vector network analyzer measurement uncertainty. IEEE Trans Microwave Theory Tech October 2007;55(10):2231–9.
[8] Lin M, Zhang Y. Covariance-matrix-based uncertainty analysis for NVNA measurements. IEEE Trans Instrum Meas January 2012;61(1):93–102.
[9] Ferrero A, Teppati V, Carullo A. Accuracy evaluation of on-wafer load-pull measurements. IEEE Trans Microwave Theory Tech January 2001;49(1):39–43.
[10] Fager C, Linner LJP, Pedro JC. Optimal parameter extraction and uncertainty estimation in intrinsic FET small-signal models. IEEE Trans Microwave Theory Tech December 2002;50(12):2797–803.
[11] Lewandowski A, Williams DF, Hale PD, Wang JCM, Dienstfrey A. Covariance-based vector-network-analyzer uncertainty analysis for time- and frequency-domain measurements. IEEE Trans Microwave Theory Tech July 2010;58(7):1877–86.
[12] Fung A, Samoska L, Pukala D, Dawson D, Kangaslahti P, Varonen M, et al. On-wafer *S*-parameter measurements in the 325–508 GHz band. IEEE Trans Terahertz Sci Tech March 2012;2(2):186–92.
[13] Agilent Technologies. Available: http://www.home.agilent.com/.
[14] Anritsu. Available: http://www.anritsu.com/.
[15] Tasker PJ. Practical waveform engineering. IEEE Microwave Mag December 2009;10(7):65–76.
[16] Benedikt J, Gaddi R, Tasker P, Goss M. High-power time-domain measurement system with active harmonic load-pull for high-efficiency base-station amplifier design. IEEE Trans Microwave Theory Tech December 2000;48(12):2617–24.
[17] Roff C, Benedikt J, Tasker PJ, Wallis DJ, Hilton KP, Maclean JO, et al. Analysis of DC–RF dispersion in AlGaN/GaN HFETs using RF waveform engineering. IEEE Trans Electron Dev January 2009;56(1):13–19.
[18] Cripps SC. RF power amplifiers for wireless communication (Norwood, MA, USA): Artech House; March 1999.
[19] Colantonio P, Giannini F, Limiti E. High efficiency RF and microwave solid state power amplifiers (New York, NY, USA): J. Wiley & Sons; July 2009.
[20] Colantonio P, Giannini F, Leuzzi G, Limiti E. On the class-F power amplifiers. Int J RF Microwave Comput Aided Eng February 1999;9(2):129–49.
[21] Xu H, Gao S, Heikman S, Long SI, Mishra UK, York RA. A high-efficiency class-E GaN HEMT power amplifier at 1.9 GHz. IEEE Microwave Wireless Compon Lett January 2006;16(1):22–4.
[22] Raffo A, Scappaviva F, Vannini G. A new approach to microwave power amplifier design based on the experimental characterization of the intrinsic electron-device load line. IEEE Trans Microwave Theory Tech July 2009;57(7):1743–52.
[23] Dambrine G, Cappy A, Heliodore F, Playez E. A new method for determining the FET small-signal equivalent circuit. IEEE Trans Microwave Theory Tech March 1988;36(7):1151–9.

[24] Rorsman N, Garcia M, Karlsson C, Zirath H. Accurate small signal modeling of HFET's for millimeter-wave applications. IEEE Trans Microwave Theory Tech October 1996;44(3):432–7.

[25] Crupi G, Xiao D, Schreurs DMM-P, Limiti E, Caddemi A, De Raedt W, et al. Accurate multibias equivalent circuit extraction for GaN HEMTs. IEEE Trans Microwave Theory Tech October 2006;54(10):3616–22.

[26] Cetiner B, Coccioli R, Housmand B, Itoh T. Combination of circuit and full wave analysis for pre-matched multifinger FET. In: Proc 30th Eur Microwave Conf (Paris, France), October 2000. pp. 1–4.

[27] Resca D, Santarelli A, Raffo A, Cignani R, Vannini G, Filicori F, et al. Scalable nonlinear FET model based on a distributed parasitic network description. IEEE Trans Microwave Theory Tech February 2008;56(4):755–66.

[28] Resca D, Raffo A, Santarelli A, Vannini G, Filicori F. Scalable equivalent circuit FET model for MMIC design identified through FW–EM analyses. IEEE Trans Microwave Theory Tech July 2009;57(2):245–53.

[29] Santarelli A, Vannini G, Filicori F, Rinaldi P. Backgating model including self-heating for low-frequency dispersive effects in III–V FETs. Electron Lett October 1998; 34(20):1974–6.

[30] Chaibi M, Fernandez T, Rodriguez-Tellez J, Cano JL, Aghoutane M. Accurate large-signal single current source thermal model for GaAs MESFET/HEMT. Electron Lett July 2007;43(14):775–7.

[31] Raffo A, Santarelli A, Traverso PA, Pagani M, Palomba F, Scappaviva F, et al. Accurate PHEMT nonlinear modeling in the presence of low-frequency dispersive effects. IEEE Trans Microwave Theory Tech November 2005;53(11):3449–59.

[32] Raffo A, Vadalà V, Schreurs DMM-P, Crupi G, Avolio G, Caddemi A, et al. Nonlinear dispersive modeling of electron devices oriented to GaN power amplifier design. IEEE Trans Microwave Theory Tech April 2010;58(4):710–18.

[33] Jardel O, De Groote F, Reveyrand T, Jacquet JC, Charbonniaud C, Teyssier JP, et al. An electrothermal model for AlGaN/GaN power HEMTs including trapping effects to improve large-signal simulation results on high VSWR. IEEE Trans Microwave Theory Tech December 2007;55(12):2660–9.

[34] McGovern P, Benedikt J, Tasker PJ, Powell J, Hilton KP, Glasper JL, et al. Analysis of DC–RF dispersion in AlGaN/GaN HFETs using pulsed I–V and time-domain waveform measurements. In: Proc IEEE MTT-S Int Microwave Symp (Long Beach, CA, USA), June 2005. pp. 503–6.

[35] Ciccognani W, Giannini F, Limiti E, Longhi PE, Nanni MA, Serino A, et al. GaN device technology: manufacturing, characterization, modelling and verification. In: Proc IEEE 14th Conf on Microwave Tech (Prague, Czech Republic), April 2008. pp. 1–6.

[36] Raffo A, Di Falco S, Vadalà V, Vannini G. Characterization of GaN HEMT low-frequency dispersion through a multiharmonic measurement system. IEEE Trans Microwave Theory Tech September 2010;58(9):2490–6.

[37] Sze SM, Ng KK. Physics of semiconductor devices. New York: J. Wiley & Sons; 2006.

[38] Raffo A, Vadalà V, Vannini G, Santarelli A. A new empirical model for the characterization of low-frequency dispersive effects in FET electron devices accounting for thermal influence on the trapping state. In: Proc IEEE MTT-S Int Microwave Symp (Atlanta, GA, USA), June 2008. pp. 1421–4.

[39] Augaudy S, Quere R, Teyssier JP, Di Forte-Poisson MA, Cassette S, Dessertenne B, et al. Pulse characterization of trapping and thermal effects of microwave GaN power FETs. In: Proc IEEE MTT-S Int Microwave Symp (Phoenix, AZ, USA), May 2001. pp. 427–30.
[40] Scott J, Rathmell JG, Parker A, Sayed M. Pulsed device measurements and applications. IEEE Trans Microwave Theory Tech December 1996;44(12):2718–23.
[41] Rodriguez-Tellez J, Fernandez T, Mediavilla A, Tazon A. Characterization of thermal and frequency-dispersion effects in GaAs MESFET devices. IEEE Trans Microwave Theory Tech July 2001;49(7):1352–5.
[42] Santarelli A, Cignani R, Niessen D, Traverso PA, Filicori F. New pulsed measurement setup for GaN and GaAs FETs characterization. Int J Microwave Wireless Tech April 2012;4(3):387–97.
[43] Graffeuil J, Hadjoub Z, Fortea JP, Pouysegur M. Analysis of capacitance and transconductance frequency dispersion in MESFETs for surface characterization. Solid State Electron October 1986;29(10):1087–97.
[44] Golio JM. Microwave MESFETs and HEMTs. Norwood, MA: Artech House; 1991.
[45] Jarndal A, Kompa G. Large-signal model for AlGaN/GaN HEMTs accurately predicts trapping- and self-heating-induced dispersion and intermodulation distortion. IEEE Trans Electron Dev November 2007;54(11):2830–6.
[46] Crupi G, Raffo A, Schreurs D, Avolio G, Vadalà V, Di Falco S, et al. Accurate GaN HEMT nonquasi-static large-signal model including dispersive effects. Microwave Opt Technol Lett March 2011;53(3):692–7.
[47] Angelov I, Rorsman N, Stenarson J, Garcia M, Zirath H. An empirical table-based FET model. IEEE Trans Microwave Theory Tech December 1999;47(12):2350–7.
[48] Crupi G, Schreurs DMM-P, Xiao D, Caddemi A, Parvais B, Mercha A, et al. Determination and validation of new nonlinear FinFET model based on lookup tables. IEEE Microwave Wireless Compon Lett May 2007;17(5):361–3.
[49] Santarelli A, Di Giacomo V, Raffo A, Traverso PA, Vannini G, Filicori F. A nonquasi-static empirical model of electron devices. IEEE Trans Microwave Theory Tech December 2006;54(12):4021–31.
[50] Fernández-Barciela M, Tasker PJ, Campos-Roca Y, Demmler M, Massler H, Sanchez E, et al. A simplified broadband large signal non quasi-static table-based FET model. IEEE Trans Microwave Theory Tech March 2000;48(3):395–405.
[51] Filicori F, Santarelli A, Traverso PA, Raffo A, Vannini G, Pagani M. Non-linear RF device modelling in the presence of low-frequency dispersive phenomena. Int J RF Microwave Comput Aided Eng January 2006;16(1):81–94.
[52] Root DE. Nonlinear charge modeling for FET large-signal simulation and its importance for IP3 and ACPR in communication circuits. In: Proc 44th IEEE Midwest Circuits Systems Symp, vol. 2. (Dayton, OH, USA) August 2001. pp. 768–72.
[53] Avolio G, Schreurs D, Raffo A, Angelov I, Crupi G, Vannini G, et al. Waveform-based large-signal identification of transistor models. In: Proc IEEE MTT-S Int Microwave Symp (Montréal, Canada), June 2012. pp. 1–3.
[54] Avolio G, Schreurs D, Raffo A, Crupi G, Angelov I, Vannini G, et al. Identification technique of FET model based on vector nonlinear measurements. Electron Lett November 2011;47(24):1323–4.
[55] Root DE, Fan S. Experimental evaluation of large-signal modeling assumptions based on vector analysis of bias-dependent *S*-parameter data from MESFETs and HEMTs. In: Proc IEEE MTT-S Int Microwave Symp (Albuquerque, NM, USA), June 1992. pp. 255–8.

[56] Follmann R, Kother D, Lauer A, Stahlmann R, Wolff I. Consistent large signal implementation of capacitances driven by two steering voltages for FET modeling. In: Proc 35th Eur Microwave Conf, vol. 2. (Paris, France), October 2005. pp. 1149–52.

[57] Kallfass I, Schumacher H, Brazil TJ. A unified approach to charge-conservative capacitance modelling in HEMTs. IEEE Microwave Wireless Compon Lett December 2006;16(12):678–80.

[58] Aaen P, Plá JA, Wood J. Modeling and characterization of RF and microwave power FETs (Cambridge, UK): Cambridge University Press; 2007.

[59] Rudolph M, Fager C, Root DE. Nonlinear transistor model parameter extraction techniques (Cambridge, UK): Cambridge University Press; 2012.

[60] Raffo A, Santarelli A, Traverso PA, Pagani M, Vannini G, Filicori F. Accurate modelling of electron device I/V characteristics through a simplified large-signal measurement setup. Int J RF Microwave Comput Aided Eng September 2005;15(5):441–52.

[61] Menozzi R. Off-state breakdown of GaAs PHEMT's: review and new data. IEEE Trans Device Mater Reliab March 2004;4(1):54–62.

[62] Dieci D, Sozzi G, Menozzi R, Tediosi E, Lanzieri C, Canali C. Electric-field-related reliability of AlGaAs/GaAs power HFETs: bias dependence and correlation with breakdown. IEEE Trans Electron Dev September 2001;48(9):1929–37.

[63] Borgarino M, Menozzi R, Baeyens Y, Cova P, Fantini F. Hot electron degradation of the DC and RF characteristics of AlGaAs/InGaAs/GaAs PHEMT's. IEEE Trans Electron Dev February 1998;45(2):366–72.

[64] Vadalà V, Bosi G, Raffo A, Vannini G, Avolio G, Schreurs D. Influence of the gate current dynamic behaviour on GaAs HEMT reliability issues. In: Proc Eur Microwave Int Circuits Conf (Amsterdam, The Netherlands), October 2012. pp. 258–61.

[65] Musio A, Vadalà V, Scappaviva F, Raffo A, Di Falco S, Vannini G. A new approach to Class-E power amplifier design. In: Proc Int Nonlin Microwave Millimeter Wave Circuits (Wien, Austria), April 2011. pp. 1–4.

[66] Di Falco S, Raffo A, Vadalà V, Vannini G. Power amplifier design accounting for input large-signal matching. In: Proc. Eur Microwave Int Circuits Conf (Amsterdam, The Netherlands), October 2012. pp. 465–8.

[67] Resca D, Raffo A, Di Falco S, Scappaviva F, Vadalà V, Vannini G. X-band GaN Power Amplifier for Future Generation SAR Systems [to be published].

[68] Raffo A, Avolio G, Schreurs DMM-P, Di Falco S, Vadalà V, Scappaviva F, et al. On the evaluation of the high-frequency load line in active devices. Int J Microwave Wireless Tech January 2011;3(1):19–24.

[69] Avolio G, Schreurs DMM-P, Raffo A, Crupi G, Vannini G, Nauwelaers B. Waveforms-only based nonlinear de-embedding in active devices. IEEE Microwave Wireless Compon Lett April 2012;22(4):215–17.

[70] Vadalà V, Avolio G, Raffo A, Schreurs DMM-P, Vannini G. Nonlinear embedding and de-embedding techniques for large-signal FET measurements. Microwave Opt Technol Lett December 2011;54(12):2835–8.

[71] Raffo A, Vadalà V, Avolio G, Bosi G, Nalli A, Schreurs DMMP, et al. Linear versus nonlinear de-embedding: experimental investigation. In: 81st ARFTG Conf Dig (Seattle, WA, USA), June 2013. pp. 1–5.

[72] Vanaverbeke F, De Raedt W, Schreurs D, Vanden Bossche M. Real-time non-linear de-embedding. In: 77th ARFTG Conf Dig (Baltimore, MD, USA), June 2011. pp. 1–6.

Index

Note: Page numbers with "*f*" denote figures; "*t*" tables.

D

G

H

I

K

L

M

N

O

S

W

X

Y

Z

Printed and bound by CPI Group (UK) Ltd, Croydon, CR0 4YY
08/05/2025
01864900-0005